IMDG CODE

INTERNATIONAL MARITIME DANGEROUS GOODS CODE

Incorporating Amendment 34-08

2008 EDITION • VOLUME 1

INTERNATIONAL
MARITIME
ORGANIZATION

London, 2008

Published in 2008
by the INTERNATIONAL MARITIME ORGANIZATION
4 Albert Embankment, London SE1 7SR

Printed in the United Kingdom by Polestar Wheatons Ltd, Exeter

2   4   6   8   10   9   7   5   3   1

ISBN: 978-92-801-4241-9

IMO PUBLICATION
Sales number: IG200E

Foreword

The International Convention for the Safety of Life at Sea, 1974 (SOLAS), as amended, deals with various aspects of maritime safety and contains in chapter VII the mandatory provisions governing the carriage of dangerous goods in packaged form or in solid form in bulk. The carriage of dangerous goods is prohibited except in accordance with the relevant provisions of chapter VII, which are amplified by the International Maritime Dangerous Goods (IMDG) Code.

Regulation II-2/19 of the SOLAS Convention, as amended, specifies the special requirements for a ship intended to carry dangerous goods, the keel of which was laid or which was at a similar stage of construction on or after 1 July 2002.

The International Convention for the Prevention of Pollution from Ships, 1973, as modified by the Protocol of 1978 relating thereto (MARPOL 73/78), deals with various aspects of prevention of marine pollution, and contains in its Annex III the mandatory provisions for the prevention of pollution by harmful substances carried by sea in packaged form. Regulation 1(2) prohibits the carriage of harmful substances in ships except in accordance with the provisions of Annex III, which are also amplified by the IMDG Code.

In accordance with the Provisions concerning Reports on Incidents Involving Harmful Substances (Protocol I to MARPOL 73/78), incidents involving losses of such substances from ships must be reported by the master or other person having charge of the ship concerned. Each substance defined as harmful to the marine environment is identified as a marine pollutant in column 4 of its entry in the Dangerous Goods List and in the Index of the IMDG Code by the letter **P**.

The IMDG Code that was adopted by resolution A.716(17) and amended by Amendments 27 to 30 was recommended to Governments for adoption or for use as the basis for national regulations in pursuance of their obligations under regulation VII/1.4 of the 1974 SOLAS Convention, as amended, and regulation 1(3) of Annex III of MARPOL 73/78. The IMDG Code, as amended, attained mandatory status from 1 January 2004 under the umbrella of SOLAS, 1974; however, some parts of the Code continue to be recommendatory. Observance of the Code harmonizes the practices and procedures followed in the carriage of dangerous goods by sea and ensures compliance with the mandatory provisions of the SOLAS Convention and of Annex III of MARPOL 73/78.

The Code, which sets out in detail the requirements applicable to each individual substance, material or article, has undergone many changes, both in layout and content, in order to keep pace with the expansion and progress of industry. IMO's Maritime Safety Committee (MSC) is authorized by the Organization's Assembly to adopt amendments to the Code, thus enabling IMO to respond promptly to developments in transport.

The MSC at its eighty-fourth session agreed that, in order to facilitate the multimodal transport of dangerous goods, the provisions of the IMDG Code, 2008, may be applied from 1 January 2009 on a voluntary basis, pending their official entry into force on 1 January 2010 without any transitional period. This is described in resolution MSC.262(84) and the Preamble to this Code. It needs to be emphasized that, in the context of the language of the Code, the words "shall", "should" and "may", when used in the Code, mean that the relevant provisions are "mandatory", "recommendatory" and "optional", respectively.

The IMDG Code is also available as a fully searchable database on CD-ROM (including the items within its Supplement). Intranet and Internet (subscription) versions are also available. For more information, please visit the IMO Publishing Service website at www.imo.org to see a live demonstration of the CD-ROM version and obtain details of how online subscription to the IMDG Code will work. If and when required, the IMO website will also include any files that show errata or corrigenda to this edition of the IMDG Code.

Contents

Volume 1

Contents

Contents

Volume 2

Preamble

1.   Carriage of dangerous goods by sea is regulated in order reasonably to prevent injury to persons or damage to ships and their cargoes. Carriage of marine pollutants is primarily regulated to prevent harm to the marine environment. The objective of the IMDG Code is to enhance the safe carriage of dangerous goods while facilitating the free unrestricted movement of such goods.

2.   Over the years, many maritime countries have taken measures to regulate the transport of dangerous goods by sea. The various regulations, codes and practices, however, differed in their framework and, in particular, in the identification and labelling of such goods. Both the terminology used and the provisions for packaging and stowage varied from country to country and created difficulties for all directly or indirectly concerned with the transport of dangerous goods by sea.

3.   The need for international regulation of the transport of dangerous goods by sea was recognized by the 1929 International Conference on Safety of Life at Sea (SOLAS), which recommended that rules on the subject shall have international effect. The classification of dangerous goods and certain general provisions concerning their transport in ships were adopted by the 1948 SOLAS Conference. This Conference also recommended further study with the object of developing international regulations.

4.   Meanwhile, the Economic and Social Council of the United Nations had appointed an ad hoc Committee of Experts on the Transport of Dangerous Goods (UN Committee of Experts), which had been actively considering the international aspect of the transport of dangerous goods by all modes of transport. This committee completed a report in 1956 dealing with classification, listing and labelling of dangerous goods and with the transport documents required for such goods. This report, with subsequent modifications, offered a general framework to which existing regulations could be harmonized and within which they could be further developed. The primary goal being world-wide uniformity for regulations concerning the transport of dangerous goods by sea as well as other modes of transport.

5.   As a further step towards meeting the need for international rules governing the transport of dangerous goods in ships, the 1960 SOLAS Conference, in addition to laying down a general framework of provisions in chapter VII of the SOLAS Convention, invited IMO (Recommendation 56) to undertake a study with a view to establishing a unified international code for the transport of dangerous goods by sea. This study would be pursued in co-operation with the UN Committee of Experts and should take account of existing maritime practices and procedures. The Conference further recommended that the unified code be prepared by IMO and that it be adopted by the Governments that were Parties to the 1960 Convention.

6.   To implement Recommendation 56, IMO's Maritime Safety Committee (MSC) appointed a working group drawn from those countries having considerable experience in the transport of dangerous goods by sea. Preliminary drafts for each class of substances, materials and articles were subsequently brought under close scrutiny by the working group to take into account throughout the practices and procedures of a number of maritime countries in order to make the Code as widely acceptable as possible. This new International Maritime Dangerous Goods (IMDG) Code was approved by the MSC and recommended to Governments by the Assembly of IMO in 1965.

7.   During another SOLAS Conference held in 1974, chapter VII of the Convention remained essentially unchanged. Since that date, revisions and amendments to chapter VII adopted by the MSC entered into force in 1986, 1992, 1994, 1996, 2001, 2003, 2004 and 2006. Although invoked by a footnote reference in regulation 1 of chapter VII, the IMDG Code itself had only recommendatory status until 31 December 2003.

8.   All the substances, material and articles set out by class in a series of individual schedules and all the supporting recommendations prepared by the UN Committee of Experts were regularly reviewed by the MSC and its subsidiary bodies for inclusion in the IMDG Code with necessary modifications for the sea mode. The questions of excluding goods not in fact transported by sea, of including further goods, or of transferring goods between classes, as necessary, of prescribing suitable packagings, of marking, labelling and placarding, of documentation and of transport in portable tanks have been dealt with in continuous consultation with the UN Committee of Experts.

9.   At the International Conference on Marine Pollution, 1973, the need was recognized to preserve the marine environment. It was further recognized that negligent or accidental release of marine pollutants transported by sea in packaged form should be minimized. Consequently, provisions were established and adopted by the Conference, and are contained in Annex III of the International Convention for the Prevention of Pollution from Ships, 1973, as modified by the Protocol of 1978 relating thereto (MARPOL 73/78). The Marine Environment Protection Committee (MEPC) decided in 1985 that Annex III should be implemented through the IMDG Code. This decision was also endorsed by the MSC in 1985. Amendments agreed by the MEPC and MSC to Annex III to MARPOL 73/78 entered into force in 1994, 1996 and 2002. MEPC adopted resolution MEPC.156(55), a revised text to take into account the GHS criteria.

10   The UN Committee of Experts has continued to meet until the present day and its published "Recommendations on the Transport of Dangerous Goods" ("Orange Book") are updated biennially. In 1996, the MSC agreed that the IMDG Code should be reformatted consistent with the format of the UN Recommendations on the Transport of Dangerous Goods. The consistency in format of the UN Recommendations, the IMDG Code and other dangerous goods transport regulations is intended to enhance user-friendliness, compliance with the regulations, and the safe transport of dangerous goods.

11   The reformatted IMDG Code lays down basic principles. Detailed provisions for individual substances, material and articles and for good practice are included in a "Dangerous Goods List". This list shall be consulted when attempting to locate relevant transport information for any substance, material or article.

12   At its seventy-fifth session in May 2002, the MSC confirmed its earlier decision to make the IMDG Code mandatory with the issue of Amendment 31. This second revision in reformatted style entered into force from 1 January 2004 and became mandatory without any transitional period. However, Governments could apply that Amendment in whole or in part on a voluntary basis from 1 January 2003. Accordingly, the MSC adopted resolution MSC.123(75) with the appropriate amendments to chapters VI and VII of SOLAS 74, as amended, denoting the changed status of the IMDG Code.

12bis   At its seventy-eighth session in May 2004, the MSC adopted Amendment 32 to the mandatory IMDG Code which entered into force from 1 January 2006 without any transitional period. However, in accordance with resolution MSC.157(78), Governments were encouraged to apply this Amendment in whole or in part on a voluntary basis from 1 January 2005.

12tris   At its eighty-first session in May 2006, the MSC adopted Amendment 33 to the mandatory IMDG Code which will enter into force from 1 January 2008 without any transitional period. However, in accordance with resolution MSC.205(81), Governments are encouraged to apply this Amendment in whole or part on a voluntary basis from 1 January 2007.

13   In order to keep the Code up to date from the maritime transport operational aspect, it will be necessary for the MSC to continue to take into account technological developments, as well as changes to chemical classifications and the related consignment provisions that primarily concern the shipper/consignor. The two-year periodicity of amendments to the UN Recommendations on Transport of Dangerous Goods, which it is expected will continue, will thus provide the source of most future updating of the IMDG Code, also on a two-year basis.

14   The MSC will also have regard to future implications for the carriage of dangerous goods by sea, in particular, arising from any acceptance by the UN Conference on Environmental Development (UNCED) of common criteria for the classification of chemicals for all human purposes on the basis of a Global Harmonization System (GHS).

15   Attention is drawn to IMO document FAL.6/Circ.14, a list of existing publications relevant to areas and topics relating to ship/port interface matters.

16   Advice on emergency procedures and for initial management of chemical poisoning and diagnosis that may be used in conjunction with the IMDG Code is published separately in "The EmS Guide: Emergency Response Procedures for Ships Carrying Dangerous Goods" (see MSC/Circ.1025 and MSC/Circ.1025/Add.1) and in the "Medical First Aid Guide for Use in Accidents Involving Dangerous Goods" (see MSC/Circ.857 and DSC 3/15/Add.2), respectively.

17   In addition, referring to Part D of chapter VII of the SOLAS Convention, a ship transporting INF cargo, as defined in regulation VII/14.2, shall comply with the requirements of the International Code for the Safe Carriage of Packaged Irradiated Nuclear Fuel, Plutonium and High-Level Radioactive Wastes on board Ships (INF Code).

1

PART 1

GENERAL PROVISIONS, DEFINITIONS AND TRAINING

Chapter 1.1

General provisions

1.1.0   Introductory note

It should be noted that other international and national modal regulations exist and that those regulations may recognize all or part of the provisions of this Code. In addition, port authorities and other bodies and organizations should recognize the Code and may use it as a basis for their storage and handling bye-laws within loading and discharge areas.

1.1.1   Application and implementation of the Code

1.1.1.1   The provisions contained in this Code are applicable to all ships to which the International Convention for the Safety of Life at Sea, 1974 (SOLAS 74), as amended, applies and which are carrying dangerous goods as defined in regulation 1 of part A of chapter VII of that Convention.

1.1.1.2   The provisions of regulation II-2/19 of that Convention apply to passenger ships and to cargo ships constructed on or after 1 July 2002.

For:

    .1   a passenger ship constructed on or after 1 September 1984 but before 1 July 2002; or

    .2   a cargo ship of 500 gross tons or over constructed on or after 1 September 1984 but before 1 July 2002; or

    .3   a cargo ship of less than 500 gross tons constructed on or after 1 February 1992 but before 1 July 2002,

the requirements of regulation II-2/54 of SOLAS, 1974, as amended by resolutions MSC.1(XLV), MSC.6(48), MSC.13(57), MSC.22(59), MSC.24(60), MSC.27(61), MSC.31(63) and MSC.57(67), apply (see II-2/1.2).

For cargo ships of less than 500 gross tons constructed on or after 1 September 1984 and before 1 February 1992, it is recommended that Contracting Governments extend such application to these cargo ships as far as possible.

1.1.1.3   All ships, irrespective of type and size, carrying substances, material or articles identified in this Code as marine pollutants are subject to the provisions of this Code.

1.1.1.4   In certain parts of this Code, a particular action is prescribed, but the responsibility for carrying out the action is not specifically assigned to any particular person. Such responsibility may vary according to the laws and customs of different countries and the international conventions into which these countries have entered. For the purpose of this Code, it is not necessary to make this assignment, but only to identify the action itself. It remains the prerogative of each Government to assign this responsibility.

1.1.1.5   Although this Code is legally treated as a mandatory instrument under chapter VII of SOLAS 74, as amended, the following provisions of the Code remain recommendatory:

    .1   paragraphs 1.3.1.4 to 1.3.1.7 (Training);

    .2   chapter 1.4 (Security provisions) except 1.4.1.1, which is mandatory;

    .3   section 2.1.0 of chapter 2.1 (class 1 – explosives, Introductory notes);

    .4   section 2.3.3 of chapter 2.3 (Determination of flashpoint);

    .5   columns (15) and (17) of the Dangerous Goods List in chapter 3.2;

    .6   section 5.4.5 of chapter 5.4 (Multimodal Dangerous Goods Form), insofar as the layout of the form is concerned;

    .7   chapter 7.3 (Special provisions in the event of an incident and fire precautions involving dangerous goods only);

    .8   section 7.9.3 (Contact information for the main designated national competent authorities); and

    .9   appendix B.

1

1.1.2 Conventions

1.1.2.1 International Convention for the Safety of Life at Sea, 1974

Part A of chapter VII of the International Convention for the Safety of Life at Sea, 1974 (SOLAS 1974), as amended, deals with the carriage of dangerous goods in packaged form, and is reproduced in full:

CHAPTER VII
Carriage of Dangerous Goods

Part A
Carriage of Dangerous Goods in Packaged Form

Regulation 1
Definitions

For the purpose of this chapter, unless expressly provided otherwise:

1   *IMDG Code* means the International Maritime Dangerous Goods (IMDG) Code adopted by the Maritime Safety Committee of the Organization by resolution MSC.122(75), as may be amended by the Organization, provided that such amendments are adopted, brought into force and take effect in accordance with the provisions of article VIII of the present Convention concerning the amendment procedures applicable to the annex other than chapter I.

2   *Dangerous goods* mean the substances, materials and articles covered by the IMDG Code.

3   *Packaged form* means the form of containment specified in the IMDG Code.

Regulation 2
Application *

1   Unless expressly provided otherwise, this part applies to the carriage of dangerous goods in packaged form in all ships to which the present regulations apply and in cargo ships of less than 500 gross tonnage.

2   The provisions of this part do not apply to ships' stores and equipment.

3   The carriage of dangerous goods in packaged form is prohibited except in accordance with the provisions of this chapter.

4   To supplement the provisions of this part, each Contracting Government shall issue, or cause to be issued, detailed instructions on emergency response and medical first aid relevant to incidents involving dangerous goods in packaged form, taking into account the guidelines developed by the Organization.[†]

Regulation 3
Requirements for the carriage of dangerous goods

The carriage of dangerous goods in packaged form shall comply with the relevant provisions of the IMDG Code.

Regulation 4
Documents

1   In all documents relating to the carriage of dangerous goods in packaged form by sea, the Proper Shipping Name of the goods shall be used (trade names alone shall not be used) and the correct description given in accordance with the classification set out in the IMDG Code.

---

* Refer to:
   .1   part D which contains special requirements for the carriage of INF cargo; and
   .2   regulation II-2/19 which contains special requirements for ships carrying dangerous goods.
[†] Refer to:
   .1   the *Emergency Response Procedures for Ships Carrying Dangerous Goods (EmS Guide)* (MSC/Circ.1025, as amended); and
   .2   the *Medical First Aid Guide for Use in Accidents Involving Dangerous Goods (MFAG)* (MSC/Circ.857),
published by the Organization.

2   The transport documents prepared by the shipper shall include, or be accompanied by, a signed certificate or a declaration that the consignment, as offered for carriage, is properly packaged, marked, labelled or placarded, as appropriate, and in proper condition for carriage.

3   The person(s) responsible for the packing/loading of dangerous goods in a cargo transport unit[*] shall provide a signed container/vehicle packing certificate stating that the cargo in the unit has been properly packed and secured and that all applicable transport requirements have been met. Such a certificate may be combined with the document referred to in paragraph 2.

4   Where there is due cause to suspect that a cargo transport unit in which dangerous goods are packed is not in compliance with the requirements of paragraph 2 or 3, or where a container/vehicle packing certificate is not available, the cargo transport unit shall not be accepted for carriage.

5   Each ship carrying dangerous goods in packaged form shall have a special list or manifest setting forth, in accordance with the classification set out in the IMDG Code, the dangerous goods on board and the location thereof. A detailed stowage plan, which identifies by class and sets out the location of all dangerous goods on board, may be used in place of such a special list or manifest. A copy of one of these documents shall be made available before departure to the person or organization designated by the port State authority.

Regulation 5
Cargo Securing Manual

Cargo, cargo units[†] and cargo transport units, shall be loaded, stowed and secured throughout the voyage in accordance with the Cargo Securing Manual approved by the Administration. The Cargo Securing Manual shall be drawn up to a standard at least equivalent to the guidelines developed by the Organization.[‡]

Regulation 6
Reporting of incidents involving dangerous goods

1   When an incident takes place involving the loss or likely loss overboard of dangerous goods in packaged form into the sea, the master, or other person having charge of the ship, shall report the particulars of such an incident without delay and to the fullest extent possible to the nearest coastal State. The report shall be drawn up based on general principles and guidelines developed by the Organization.[§]

2   In the event of the ship referred to in paragraph 1 being abandoned, or in the event of a report from such a ship being incomplete or unobtainable, the company, as defined in regulation IX/1.2, shall, to the fullest extent possible, assume the obligations placed upon the master by this regulation.

1.1.2.2   **International Convention for the Prevention of Pollution from Ships, 1973/78**

1.1.2.2.1   Annex III of the International Convention for the Prevention of Pollution from Ships, 1973, as modified by the Protocol of 1978 relating thereto (MARPOL 73/78), deals with the prevention of pollution by harmful substances carried by sea in packaged form and is reproduced in full, as revised by the Marine Environment Protection Committee.[¶]

---

[*] Refer to the International Maritime Dangerous Goods (IMDG) Code, adopted by the Organization by resolution MSC.122(75) as amended.

[†] As defined in the Code of Safe Practice for Cargo Stowage and Securing (CSS Code), adopted by the Organization by resolution A.715(17), as amended.

[‡] Refer to Guidelines for the preparation of the Cargo Securing Manual (MSC/Circ.745).

[§] Refer to the General principles for ship reporting systems and ship reporting requirements, including guidelines for reporting incidents involving dangerous goods, harmful substances and/or marine pollutants, adopted by the Organization by resolution A.851(20).

[¶] The revised text of Annex III was adopted by resolution MEPC.156(55) and will enter into force on 1 January 2010, which is the mandatory entry into force date of amendment 34-08 to the IMDG Code.

Annex III
Regulations for the Prevention of Pollution by Harmful Substances Carried by Sea in Packaged Form

Regulation 1
Application

1 Unless expressly provided otherwise, the regulations of this Annex apply to all ships carrying harmful substances in packaged form.

   .1 For the purpose of this Annex, "harmful substances" are those substances which are identified as marine pollutants in the International Maritime Dangerous Goods Code (IMDG Code) or which meet the criteria in the appendix of this Annex.

   .2 For the purposes of this Annex, "packaged form" is defined as the forms of containment specified for harmful substances in the IMDG Code.

2 The carriage of harmful substances is prohibited, except in accordance with the provisions of this Annex.

3 To supplement the provisions of this Annex, the Government of each Party to the Convention shall issue, or cause to be issued, detailed requirements on packing, marking, labelling, documentation, stowage, quantity limitations and exceptions for preventing or minimizing pollution of the marine environment by harmful substances.[*]

4 For the purposes of this Annex, empty packagings which have been used previously for the carriage of harmful substances shall themselves be treated as harmful substances unless adequate precautions have been taken to ensure that they contain no residue that is harmful to the marine environment.

5 The requirements of this Annex do not apply to ship's stores and equipment.

Regulation 2
Packing

Packages shall be adequate to minimize the hazard to the marine environment, having regard to their specific contents.

Regulation 3
Marking and labelling

1 Packages containing a harmful substance shall be durably marked with the correct technical name (trade names alone shall not be used) and, further, shall be durably marked or labelled to indicate that the substance is a marine pollutant. Such identification shall be supplemented where possible by any other means, for example, by use of the relevant United Nations Number.

2 The method of marking the correct technical name and of affixing labels on packages containing a harmful substance shall be such that this information will still be identifiable on packages surviving at least three months' immersion in the sea. In considering suitable marking and labelling, account shall be taken of the durability of the materials used and of the surface of the package.

3 Packages containing small quantities of harmful substances may be exempted from the marking requirements.[†]

Regulation 4[‡]
Documentation

1 In all documents relating to the carriage of harmful substances by sea where such substances are named, the correct technical name of each such substance shall be used (trade names alone shall

---

[*] Refer to the IMDG Code adopted by the Organization by resolution MSC.122(75), as amended.

[†] Refer to the specific exemptions provided for in the IMDG Code adopted by resolution MSC.122(75), as amended.

[‡] Reference to "documents" in this regulation does not preclude the use of electronic data processing (EDP) and electronic data interchange (EDI) transmission techniques as an aid to paper documentation.

not be used) and the substance further identified by the addition of the words "MARINE POLLUTANT".

2   The shipping documents supplied by the shipper shall include, or be accompanied by, a signed certificate or declaration that the shipment offered for carriage is properly packaged and marked, labelled or placarded as appropriate and in proper condition for carriage to minimize the hazard to the marine environment.

3   Each ship carrying harmful substances shall have a special list or manifest setting forth the harmful substances on board and the location thereof. A detailed stowage plan which sets out the location of the harmful substances on board may be used in place of such special list or manifest. Copies of such documents shall also be retained on shore by the owner of the ship or his representative until the harmful substances are unloaded. A copy of one of these documents shall be made available before departure to the person or organization designated by the port State authority.

4   At any stopover, where any loading or unloading operations, even partial, are carried out, a revision of the documents listing the harmful substances taken on board, indicating their location on board or showing a detailed stowage plan, shall be made available before departure to the person or organization designated by the port State authority.

5   When the ship carries a special list or manifest or a detailed stowage plan, required for the carriage of dangerous goods by the International Convention for the Safety of Life at Sea, 1974, as amended, the documents required by this regulation may be combined with those for dangerous goods. Where documents are combined, a clear distinction shall be made between dangerous goods and harmful substances covered by this Annex.

Regulation 5
Stowage

Harmful substances shall be properly stowed and secured so as to minimize the hazards to the marine environment without impairing the safety of the ship and persons on board.

Regulation 6
Quantity limitations

Certain harmful substances may, for sound scientific and technical reasons, need to be prohibited for carriage or be limited as to the quantity which may be carried aboard any one ship. In limiting the quantity, due consideration shall be given to size, construction and equipment of the ship, as well as the packaging and the inherent nature of the substances.

Regulation 7
Exceptions

1   Jettisoning of harmful substances carried in packaged form shall be prohibited, except where necessary for the purpose of securing the safety of the ship or saving life at sea.

2   Subject to the provisions of the present Convention, appropriate measures based on the physical, chemical and biological properties of harmful substances shall be taken to regulate the washing of leakages overboard, provided that compliance with such measures would not impair the safety of the ship and persons on board.

Regulation 8
*Port State control on operational requirements**

1   A ship when in a port or an offshore terminal of another Party is subject to inspection by officers duly authorized by such Party concerning operational requirements under this Annex, where there are clear grounds for believing that the master or crew are not familiar with essential shipboard procedures relating to the prevention of pollution by harmful substances.

2   In the circumstances given in paragraph 1 of this regulation, the Party shall take such steps as will ensure that the ship shall not sail until the situation has been brought to order in accordance with the requirements of this Annex.

---

* Refer to the Procedures for port State control adopted by the Organization by resolution A.787(19) and amended by resolution A.882(21).

3   Procedures relating to the port State control prescribed in article 5 of the present Convention shall apply to this regulation.

4   Nothing in this regulation shall be construed to limit the rights and obligations of a Party carrying out control over operational requirements specifically provided for in the present Convention.

Appendix to Annex III
Criteria for the identification of harmful substances in packaged form

For the purposes of this Annex, substances identified by any one of the following criteria are harmful substances:[*]

Category: Acute 1	
96 hr LC_{50} (for fish)	$\leqslant 1$ mg/ℓ and/or
48 hr EC_{50} (for crustacea)	$\leqslant 1$ mg/ℓ and/or
72 or 96 hr ErC_{50} (for algae or other aquatic plants)	$\leqslant 1$ mg/ℓ

Category: Chronic 1

96 hr LC_{50} (for fish)	$\leqslant 1$ mg/ℓ and/or
48 hr EC_{50} (for crustacea)	$\leqslant 1$ mg/ℓ and/or
72 or 96 hr ErC_{50} (for algae or other aquatic plants)	$\leqslant 1$ mg/ℓ

and the substance is not rapidly degradable and/or the log $K_{ow} \geqslant 4$ (unless the experimentally determined BCF < 500).

Category: Chronic 2

96 hr LC_{50} (for fish)	>1 to $\leqslant 10$ mg/ℓ and/or
48 hr EC_{50} (for crustacea)	>1 to $\leqslant 10$ mg/ℓ and/or
72 or 96 hr ErC_{50} (for algae or other aquatic plants)	>1 to $\leqslant 10$ mg/ℓ

and the substance is not rapidly degradable and/or the log $K_{ow} \geqslant 4$ (unless the experimentally determined BCF < 500), unless the chronic toxicity NOECs are > 1 mg/ℓ.

1.1.3   Dangerous goods forbidden from transport

1.1.3.1   Unless provided otherwise by this Code, the following are forbidden from transport:

Any substance or article which, as presented for transport, is liable to explode, dangerously react, produce a flame or dangerous evolution of heat or dangerous emission of toxic, corrosive or flammable gases or vapours under normal conditions of transport.

In chapter 3.3, special provision 900 lists certain substances, which are forbidden for transport.

---

[*] The criteria are based on those developed by the United Nations Globally Harmonized System of Clasification and Labelling of Chemicals (GHS), as amended. For definitions of acronyms or terms used in this appendix, refer to the relevant paragraphs of the IMDG Code.

Chapter 1.2

Definitions, units of measurement and abbreviations

1.2.1   Definitions

The following is a list of definitions of general applicability that are used throughout this Code. Additional definitions of a highly specific nature are presented in the relevant chapters.

For the purposes of this Code:

Aerosols or *aerosol dispensers* means non-refillable receptacles meeting the provisions of 6.2.4, made of metal, glass or plastics and containing a gas compressed, liquefied or dissolved under pressure, with or without a liquid, paste or powder, and fitted with a release device allowing the contents to be ejected as solid or liquid particles in suspension in a gas, as a foam, paste or powder or in a liquid state or in a gaseous state.

Alternative arrangement means an approval granted by the competent authority for a portable tank or MEGC that has been designed, constructed or tested to technical requirements or testing methods other than those specified in this Code (see, for instance, 6.7.5.11.1).

Animal material means animal carcasses, animal body parts, or animal foodstuffs.

Approval

Multilateral approval, for the transport of class 7 material, means approval by the relevant competent authority of the country of origin of the design or shipment, as applicable, and also, where the consignment is to be transported through or into any other country, approval by the competent authority of that country. The term "through or into" specifically excludes "over", i.e. the approval and notification requirements shall not apply to a country over which radioactive material is carried in an aircraft, provided that there is no scheduled stop in that country.

Unilateral approval, for the transport of class 7 material, means an approval of a design which is required to be given by the competent authority of the country of origin of the design only.

Bags means flexible packagings made of paper, plastic film, textiles, woven material, or other suitable materials.

Barge-carrying ship means a ship specially designed and equipped to transport shipborne barges.

Barge feeder vessel means a vessel specially designed and equipped to transport shipborne barges to or from a barge-carrying ship.

Boxes means packagings with complete rectangular or polygonal faces, made of metal, wood, plywood, reconstituted wood, fibreboard, plastics, or other suitable material. Small holes for purposes such as ease of the handling or opening of the box or to meet classification provisions are permitted as long as they do not compromise the integrity of the packaging during transport.

Bulk containers are containment systems (including any liner or coating) intended for the transport of solid substances which are in direct contact with the containment system. Packagings, intermediate bulk containers (IBCs), large packagings and portable tanks are not included.

Bulk containers:

  – are of a permanent character and accordingly strong enough to be suitable for repeated use;

  – are specially designed to facilitate the transport of goods by one or more means of transport without intermediate reloading;

  – are fitted with devices permitting ready handling; and

  – have a capacity of not less than 1 cubic metre.

Examples of bulk containers are freight containers, offshore bulk containers, skips, bulk bins, swap bodies, trough-shaped containers, roller containers, load compartments of vehicles.

Bundles of cylinders are assemblies of cylinders that are fastened together and which are interconnected by a manifold and transported as a unit. The total water capacity shall not exceed 3000 litres except that bundles intended for the transport of gases of class 2.3 shall be limited to 1000 litres water capacity.

Cargo transport unit means a road freight vehicle, a railway freight wagon, a freight container, a road tank vehicle, a railway tank wagon or a portable tank.

Carrier means any person, organization or Government undertaking the transport of dangerous goods by any means of transport. The term includes both carriers for hire or reward (known as *common* or *contract carriers* in some countries) and carriers on own account (known as *private carriers* in some countries).

Cellular ship means a ship in which containers are loaded under deck into specially designed slots giving a permanent stowage of the container during sea transport. Containers loaded on deck in such a ship are specially stacked and secured on fittings.

Closed cargo transport unit, with the exception of class 1, means a unit which totally encloses the contents by permanent structures. Cargo transport units with fabric sides or tops are not closed cargo transport units; for definition of class 1 cargo transport unit see 7.1.7.1.1.

Closed ro–ro cargo space means a ro–ro cargo space which is neither an open ro–ro cargo space nor a weather deck.

Closure means a device which closes an opening in a receptacle.

Combination packagings means a combination of packagings for transport purposes, consisting of one or more inner packagings secured in an outer packaging in accordance with 4.1.1.5.

Competent authority means any body or authority designated or otherwise recognized as such for any purpose in connection with this Code.

Compliance assurance means a systematic programme of measures applied by a competent authority which is aimed at ensuring that the provisions of this Code are met in practice.

Composite packagings means packagings consisting of an outer packaging and an inner receptacle so constructed that the inner receptacle and the outer packaging form an integral packaging. Once assembled, it remains thereafter an integrated single unit; it is filled, stored, transported and emptied as such.

Confinement system, for the transport of class 7 material, means the assembly of fissile material and packaging components specified by the designer and agreed to by the competent authority as intended to preserve criticality safety.

Consignee means any person, organization or Government which is entitled to take delivery of a consignment.

Consignment means any package or packages, or load of dangerous goods, presented by a consignor for transport.

Consignor means any person, organization or Government which prepares a consignment for transport.

Containment system, for the transport of class 7 material, means the assembly of components of the packaging specified by the designer as intended to retain the radioactive material during transport.

Control temperature means the maximum temperature at which certain substances (such as organic peroxides and self-reactive and related substances) can be safely transported during a prolonged period of time.

Conveyance means:

.1   for transport by road or rail: any vehicle,

.2   for transport by water: any ship, or any cargo space or defined deck area of a ship,

.3   for transport by air: any aircraft.

Crates are outer packagings with incomplete surfaces.

Criticality safety index (CSI) assigned to a package, overpack or freight container containing fissile material, for the transport of class 7 material, means a number which is used to provide control over the accumulation of packages, overpacks or freight containers containing fissile material.

Critical temperature is the temperature above which the substance cannot exist in the liquid state.

Cryogenic receptacles are transportable thermally insulated receptacles for refrigerated liquefied gases, of a water capacity of not more than 1000 litres.

Cylinders are transportable pressure receptacles of a water capacity not exceeding 150 litres.

Defined deck area means the area, of the weather deck of a ship, or of a vehicle deck of a roll-on/roll-off ship, which is allocated for the stowage of dangerous goods.

Design, for the transport of class 7 material, means the description of special form radioactive material, low dispersible radioactive material, package or packaging which enables such an item to be fully identified. The description may include specifications, engineering drawings, reports demonstrating compliance with regulatory requirements, and other relevant documentation.

Drums means flat-ended or convex-ended cylindrical packagings made of metal, fibreboard, plastics, plywood or other suitable materials. This definition also includes packagings of other shapes, such as round taper-necked packagings, or pail-shaped packagings. Wooden barrels and jerricans are not covered by this definition.

Elevated temperature substance means a substance which is transported or offered for transport:

- in the liquid state at a temperature at or above 100°C
- in the liquid state with a flashpoint above 60°C that is intentionally heated to a temperature above its flashpoint; or
- in the solid state at a temperature at or above 240°C.

Emergency temperature means the temperature at which emergency procedures shall be implemented.

Exclusive use, for the transport of class 7 material, means the sole use, by a single consignor, of a conveyance or of a large freight container, in respect of which all initial, intermediate and final loading and unloading is carried out in accordance with the directions of the consignor or consignee.

Filling ratio means the ratio of the mass of gas to the mass of water at 15°C that would fill completely a pressure receptacle fitted ready for use.

Flashpoint means the lowest temperature of a liquid at which its vapour forms an ignitable mixture with air.

Freight container means an article of transport equipment that is of a permanent character and accordingly strong enough to be suitable for repeated use; specially designed to facilitate the transport of goods, by one or more modes of transport, without intermediate reloading; designed to be secured and/or readily handled, having fittings for these purposes, and approved in accordance with the International Convention for Safe Containers (CSC), 1972, as amended. The term "freight container" includes neither vehicle nor packaging. However, a freight container that is carried on a chassis is included.

For freight containers for the transport of radioactive material, a freight container may be used as a packaging. A small freight container is that which has either any overall outer dimension less than 1.5 m, or an internal volume of not more than 3 m^3. Any other freight container is considered to be a large freight container.

GHS means the second revised edition of the *Globally Harmonized System of Classification and Labelling of Chemicals*, published by the United Nations as document ST/SG/AC.10/30/Rev.2.

IMO type 4 tank means a road tank vehicle for the transport of dangerous goods of classes 3 to 9 and includes a semi-trailer with a permanently attached tank or a tank attached to a chassis, with at least four twist locks that take account of ISO standards, (i.e. ISO International Standard 1161:1984).

IMO type 6 tank means a road tank vehicle for the transport of non-refrigerated liquefied gases of class 2 and includes a semi-trailer with a permanently attached tank or a tank attached to a chassis which is fitted with items of service equipment and structural equipment necessary for the transport of gases.

IMO type 8 tank means a road tank vehicle for the transport of refrigerated liquefied gases of class 2 and includes a semi-trailer with a permanently attached thermally insulated tank fitted with items of service equipment and structural equipment necessary for the transport of refrigerated liquefied gases.

Inner packagings means packagings for which an outer packaging is required for transport.

Inner receptacles means receptacles which require an outer packaging in order to perform their containment function.

Inspection body means an independent inspection and testing body approved by the competent authority.

Intermediate bulk containers (IBCs) means rigid or flexible portable packagings, other than specified in chapter 6.1, that:

.1   have a capacity of:

   .1   not more than 3.0 m^3 (3000 litres) for solids and liquids of packing groups II and III;

   .2   not more than 1.5 m^3 for solids of packing group I when packed in flexible, rigid plastics, composite, fibreboard or wooden IBCs;

   .3   not more than 3.0 m^3 for solids of packing group I when packed in metal IBCs;

   .4   not more than 3.0 m^3 for radioactive material of class 7;

.2   are designed for mechanical handling; and

.3   are resistant to the stresses produced in handling and transport, as determined by tests.

Remanufactured IBCs are metal, rigid plastics or composite IBCs that:

.1   are produced as a UN type from a non-UN type; or

.2   are converted from one UN design type to another UN design type.

Remanufactured IBCs are subject to the same provisions of this Code that apply to new IBCs of the same type (see also design type definition in 6.5.6.1.1).

Repaired IBCs are metal, rigid plastics or composite IBCs that, as a result of impact or for any other cause (e.g. corrosion, embrittlement or other evidence of reduced strength as compared to the design type) are restored so as to conform to the design type and to be able to withstand the design type tests. For the purposes of this Code, the replacement of the rigid inner receptacle of a composite IBC with a receptacle conforming to the original manufacturer's specification is considered repair. However, routine maintenance of rigid IBCs (see definition below) is not considered repair. The bodies of rigid plastics IBCs and the inner receptacles of composite IBCs are not repairable. Flexible IBCs are not repairable, unless approved by the competent authority.

Routine maintenance of flexible IBCs is the routine performance on plastics or textile flexible IBCs of operations, such as:

.1 cleaning; or

.2 replacement of non-integral components, such as non-integral liners and closure ties, with components conforming to the original manufacturer's specification;

provided that these operations do not adversely affect the containment function of the flexible IBC or alter the design type.

Note: For rigid IBCs, see "Routine maintenance of rigid IBCs".

Routine maintenance of rigid IBCs is the routine performance on metal, rigid plastics or composite IBCs of operations such as:

.1 cleaning;

.2 removal and reinstallation or replacement of body closures (including associated gaskets), or of service equipment, conforming to the original manufacturer's specifications, provided that the leaktightness of the IBC is verified; or

.3 restoration of structural equipment not directly performing a dangerous goods containment or discharge pressure retention function so as to conform to the design type (e.g. the straightening of legs or lifting attachments) provided that the containment function of the IBC is not affected.

Note: For flexible IBCs, see "Routine maintenance of flexible IBCs".

Intermediate packagings means packagings placed between inner packagings, or articles, and an outer packaging.

Jerricans means metal or plastics packagings of rectangular or polygonal cross-section.

Large packagings means packagings consisting of an outer packaging which contains articles or inner packagings and which:

.1 are designed for mechanical handling; and

.2 exceed 400 kg net mass or 450 ℓ capacity but have a volume of not more than 3 m^3.

Liner means a separate tube or bag inserted into a packaging (including IBCs and large packagings) but not forming an integral part of it, including the closures of its openings.

Liquids are dangerous goods which at 50°C have a vapour pressure of not more than 300 kPa (3 bar), which are not completely gaseous at 20°C and at a pressure of 101.3 kPa, and which have a melting point or initial melting point of 20°C or less at a pressure of 101.3 kPa. A viscous substance for which a specific melting point cannot be determined shall be subjected to the ASTM D 4359-90 test; or to the test for determining fluidity (penetrometer test) prescribed in section 2.3.4 of Annex A of the European Agreement concerning the International Carriage of Dangerous Goods by Road (ADR).*

Long international voyage means an international voyage that is not a short international voyage.

Manual of Tests and Criteria means the United Nations publication entitled "Recommendations on the Transport of Dangerous Goods, Manual of Tests and Criteria" as amended.

Maximum capacity as used in 6.1.4 means the maximum inner volume of receptacles or packagings expressed in litres.

Maximum net mass as used in 6.1.4 means the maximum net mass of contents in a single packaging or maximum combined mass of inner packagings and the contents thereof and is expressed in kilograms.

Maximum normal operating pressure, for the transport of class 7 material, means the maximum pressure above atmospheric pressure at mean sea-level that would develop in the containment system in a period of one year under the conditions of temperature and solar radiation corresponding to environmental conditions in the absence of venting, external cooling by an ancillary system, or operational controls during transport.

---

* United Nations publication ECE/TRANS/185 (Sales No. E.06.VIII.1).

Multiple-element gas containers (MEGCs) are multimodal assemblies of cylinders, tubes and bundles of cylinders which are interconnected by a manifold and which are assembled within a framework. The MEGC includes service equipment and structural equipment necessary for the transport of gases.

Offshore bulk container means a bulk container specially designed for repeated use for the transport of dangerous goods to, from and between offshore facilities. An offshore bulk container is designed and constructed in accordance with MSC/Circ.860 "Guidelines for the approval of containers handled in open seas".

Open cargo transport unit means a unit which is not a closed cargo transport unit.

Open ro–ro cargo space means a ro–ro cargo space either open at both ends, or open at one end and provided with adequate natural ventilation effective over its entire length through permanent openings in the side plating or deckhead to the satisfaction of the Administration.

Outer packaging means the outer protection of a composite or combination packaging together with any absorbent materials, cushioning and any other components necessary to contain and protect inner receptacles or inner packagings.

Overpack means an enclosure used by a single consignor to contain one or more packages and to form one unit for the convenience of handling and stowage during transport. Examples of overpacks are a number of packages either:

.1 placed or stacked on to a load board, such as a pallet, and secured by strapping, shrink-wrapping, stretch-wrapping, or other suitable means; or

.2 placed in a protective outer packaging such as a box or crate.

Overstowed means that a package or container is directly stowed on top of another.

Package means the complete product of the packing operation, consisting of the packaging and its contents prepared for transport.

Packaging means one or more receptacles and any other components or materials necessary for the receptacles to perform their containment and other safety functions.

Pressure drums are welded transportable pressure receptacles of a water capacity exceeding 150 litres and of not more than 1000 litres (e.g. cylindrical receptacles equipped with rolling hoops, spheres on skids).

Pressure receptacles is a collective term that includes cylinders, tubes, pressure drums, closed cryogenic receptacles and bundles of cylinders.

Quality assurance means a systematic programme of controls and inspections applied by any organization or body which is aimed at providing adequate confidence that the standard of safety prescribed in this Code is achieved in practice.

Radiation level, for the transport of class 7 material, means the corresponding dose rate expressed in millisieverts per hour.

Receptacles means containment vessels for receiving and holding substances or articles, including any means of closing.

Reconditioned packagings include:

.1 metal drums that:

    .1 are cleaned to original materials of construction, with all former contents, internal and external corrosion, and external coatings and labels removed;

    .2 are restored to original shape and contour, with chimes (if any) straightened and sealed, and all non-integral gaskets replaced; and

    .3 are inspected after cleaning, but before painting, with rejection of packagings with visible pitting, significant reduction in material thickness, metal fatigue, damaged threads or closures, or other significant defects;

.2 plastic drums and jerricans that:

    .1 are cleaned to original materials of construction, with all former contents, external coatings and labels removed;

    .2 have all non-integral gaskets replaced; and

    .3 are inspected after cleaning, with rejection of packagings with visible damage such as tears, creases or cracks, or damaged threads or closures, or other significant defects.

Recycled plastics material means material recovered from used industrial packagings that has been cleaned and prepared for processing into new packagings. The specific properties of the recycled material used for production of new packagings shall be assured and documented regularly as part of a quality assurance programme recognized by the competent authority. The quality assurance programme shall include a record of proper pre-sorting and verification that each batch of recycled plastics material has the proper melt flow rate, density, and tensile yield strength, consistent with that of the design type manufactured from such recycled material. This necessarily includes knowledge about the packaging material from which the recycled plastics have been derived, as well as awareness of the prior contents of those packagings if those prior contents might reduce the capability of new packagings produced using that material. In addition, the packaging manufacturer's quality assurance programme under 6.1.1.3 shall include performance of the mechanical design type test in 6.1.5 on packagings manufactured from each batch of recycled plastics material. In this testing, stacking performance may be verified by appropriate dynamic compression testing rather than static load testing.

Note: ISO 16103:2005 "Packaging – Transport packages for dangerous goods – Recycled plastics material", provides additional guidance on procedures to be followed in approving the use of recycled plastics material.

Remanufactured IBCs (see Intermediate bulk containers (IBCs)).

Remanufactured packagings include:

.1 metal drums that:

    .1 are produced as a UN type from a non-UN type;

    .2 are converted from one UN type to another UN type; or

    .3 undergo the replacement of integral structural components (such as non-removable heads); or

.2 plastic drums that:

    .1 are converted from one UN type to another UN type (such as 1H1 to 1H2); or

    .2 undergo the replacement of integral structural components.

Remanufactured drums are subject to the same provisions of this Code that apply to a new drum of the same type.

Repaired IBCs (see Intermediate bulk containers (IBCs)).

Re-used packagings means packagings to be refilled which have been examined and found free of defects affecting the ability to withstand the performance tests; the term includes those which are refilled with the same or similar compatible contents and are transported within distribution chains controlled by the consignor of the product.

Road tank vehicle means a vehicle equipped with a tank with a capacity of more than 450 litres, fitted with pressure-relief devices.

Ro–ro cargo space means spaces not normally subdivided in any way and extending to either a substantial length or the entire length of the ship in which goods (packaged or in bulk, in or on rail or road cars, vehicles (including road or rail tankers), trailers, containers, pallets, demountable tanks or in or on similar stowage units or other receptacles) can be loaded and unloaded normally in a horizontal direction.

Ro–ro ship (roll-on/roll-off ship) means a ship which has one or more decks, either closed or open, not normally subdivided in any way and generally running the entire length of the ship, carrying goods which are normally loaded and unloaded in a horizontal direction.

Routine maintenance of IBCs (see Intermediate bulk containers (IBCs)).

Salvage packagings are special packagings into which damaged, defective, leaking or non-conforming dangerous goods packages, or dangerous goods that have spilled or leaked, are placed for purposes of transport for recovery or disposal.

Self-accelerating decomposition temperature (SADT) means the lowest temperature at which self-accelerating decomposition may occur for a substance in the packaging as used in transport. The self-accelerating decomposition temperature (SADT) shall be determined in accordance with the latest version of the United Nations *Manual of Tests and Criteria*.

Settled pressure means the pressure of the contents of a pressure receptacle in thermal and diffusive equilibrium.

Shipborne barge or *barge* means an independent, non-self-propelled vessel, specially designed and equipped to be lifted in a loaded condition and stowed aboard a barge-carrying ship or barge feeder vessel.

Shipment means the specific movement of a consignment from origin to destination.

Shipper, for the purpose of this Code, has the same meaning as *consignor*.

Short international voyage means an international voyage in the course of which a ship is not more than 200 miles from a port or place in which the passengers and crew could be placed in safety. Neither the distance between the last port of call in the country in which the voyage begins and the final port of destination nor the return voyage shall exceed 600 miles. The final port of destination is the last port of call in the scheduled voyage at which the ship commences its return voyage to the country in which the voyage began.

Sift-proof packagings are packagings impermeable to dry contents, including fine solid material produced during transport.

Solid bulk cargo means any material, other than liquid or gas, consisting of a combination of particles, granules or any larger pieces of material, generally uniform in composition, which is loaded directly into the cargo spaces of a ship without any intermediate form of containment (this includes a material loaded in a barge on a barge-carrying ship).

Solids are dangerous goods, other than gases, that do not meet the definition of *liquids* in this chapter.

Special category space means an enclosed space, above or below deck, intended for the transport of motor vehicles with fuel in their tanks for their own propulsion, into and from which such vehicles can be driven and to which passengers have access.

Tank means a portable tank (including a tank-container), a road tank vehicle, a rail tank wagon or a receptacle to contain solids, liquids, or liquefied gases and has a capacity of not less than 450 litres when used for the transport of gases of class 2.

Test pressure means the required pressure applied during a pressure test for qualification or requalification (for portable tanks, see 6.7.2.1).

Transboundary movement of wastes means any shipment of wastes from an area under the national jurisdiction of one country to or through an area under the national jurisdiction of another country, or to or through an area not under the national jurisdiction of any country, provided at least two countries are concerned by the movement.

Transport index (TI) assigned to a package, overpack or freight container, or to unpackaged LSA-I or SCO-I, for the transport of class 7 material, means a number which is used to provide control over radiation exposure.

Tubes are seamless transportable pressure receptacles of a water capacity exceeding 150 litres and of not more than 3000 litres.

Unit load means that a number of packages are either:

.1 placed or stacked on and secured by strapping, shrink-wrapping, or other suitable means to a load board, such as a pallet;

.2 placed in a protective outer enclosure, such as a pallet box;

.3 permanently secured together in a sling.

Vehicle means a road vehicle (including an articulated vehicle, i.e. a tractor and semi-trailer combination) or railroad car or railway wagon. Each trailer shall be considered as a separate vehicle.

Wastes means substances, solutions, mixtures, or articles containing or contaminated with one or more constituents which are subject to the provisions of this Code and for which no direct use is envisaged but which are transported for dumping, incineration, or other methods of disposal.

Water-reactive means a substance which, in contact with water, emits flammable gas.

Weather deck means a deck which is completely exposed to the weather from above and from at least two sides.

Wooden barrels means packagings made of natural wood, of round cross-section, having convex walls, consisting of staves and heads and fitted with hoops.

Working pressure means the settled pressure of a compressed gas at a reference temperature of 15°C in a full pressure receptacle.

1.2.1.1 Clarifying examples for certain defined terms

The following explanations and examples are meant to assist in clarifying the use of some of the packaging terms defined in this chapter.

The definitions in this chapter are consistent with the use of the defined terms throughout the Code. However, some of the defined terms are commonly used in other ways. This is particularly evident in respect of the term ''inner receptacle'' which has often been used to describe the ''inners'' of a combination packaging.

The ''inners'' of ''combination packagings'' are always termed ''inner packagings'', not ''inner receptacles''. A glass bottle is an example of such an ''inner packaging''.

The ''inners'' of ''composite packagings'' are normally termed ''inner receptacles''. For example, the ''inner'' of a 6HA1 composite packaging (plastics material) is such an ''inner receptacle'' since it is normally not designed to perform a containment function without its ''outer packaging'' and is not, therefore, an ''inner packaging''.

1.2.2 Units of measurement

1.2.2.1 The following units of measurement* are applicable in this Code:

Measurement of:	SI unit[†]		Acceptable alternative unit		Relationship between units
Length	m	(metre)	–		–
Area	m^2	(square metre)	–		–
Volume	m^3	(cubic metre)	ℓ[‡]	(litre)	$1\ \ell$ $= 10^{-3}\ m^3$
Time	s	(second)	min	(minute)	1 min = 60 s
			h	(hour)	1 h = 3600 s
			d	(day)	1 d = 86400 s
Mass	kg	(kilogram)	g	(gram)	$1\ g$ $= 10^{-3}\ kg$
			t	(ton)	$1\ t$ $= 10^3\ kg$
Mass density	kg/m^3		kg/ℓ		$1\ kg/\ell$ $= 10^3\ kg/m^3$
Temperature	K	(kelvin)	°C	(degree Celsius)	0°C = 273.15 K
Difference of temperature	K	(kelvin)	°C	(degree Celsius)	1°C = 1 K
Force	N	(newton)	–		$1\ N$ $= 1\ kg{\cdot}m/s^2$
Pressure	Pa	(pascal)	bar	(bar)	1 bar $= 10^5$ Pa
					$1\ Pa$ $= 1\ N/m^2$
Stress	N/m^2		N/mm^2		$1\ N/mm^2 = 1\ MPa$
Work			kWh	(kilowatt hour)	1 kWh = 3.6 MJ
Energy	J	(joule)			1 J = 1 N·m = 1 W·s
Quantity of heat			eV	(electronvolt)	$1\ eV = 0.1602 \times 10^{-18}\ J$
Power	W	(watt)	–		1 W = 1 J/s = 1 N·m/s
Kinematic viscosity	m^2/s		mm^2/s		$1\ mm^2/s = 10^{-6}\ m^2/s$
Dynamic viscosity	Pa·s		mPa·s		$1\ mPa{\cdot}s = 10^{-3}\ Pa{\cdot}s$
Activity	Bq	(becquerel)	–		
Dose equivalent	Sv	(sievert)	–		
Conductivity	S/m	(siemens/metre)	–		

* The following round figures are applicable for the conversion of the units hitherto used into SI units.

† The International System of Units (SI) is the result of decisions taken at the General Conference on Weights and Measures (Address: Pavillon de Breteuil, Parc de St-Cloud, F-92312 Sèvres).

‡ The abbreviation "L" for litre may also be used in place of the abbreviation "ℓ", when a typewriter/word-processor cannot distinguish between figure "1" and letter "ℓ".

Force
1 kg = 9.807 N
1 N = 0.102 kg

Stress
$1 \text{ kg/mm}^2 = 9.807 \text{ N/mm}^2$
$1 \text{ N/mm}^2 = 0.102 \text{ kg/mm}^2$

Pressure
$1 \text{ Pa} = 1 \text{ N/m}^2 = 10^{-5} \text{ bar}$ $= 1.02 \times 10^{-5} \text{ kg/cm}^2$ $= 0.75 \times 10^{-2} \text{ torr}$
$1 \text{ bar} = 10^5 \text{ Pa}$ $= 1.02 \text{ kg/cm}^2$ $= 750 \text{ torr}$
$1 \text{ kg/cm}^2 = 9.807 \times 10^4 \text{ Pa}$ $= 0.9807 \text{ bar}$ $= 736 \text{ torr}$
$1 \text{ torr} = 1.33 \times 10^2 \text{ Pa}$ $= 1.33 \times 10^{-3} \text{ bar}$ $= 1.36 \times 10^{-3} \text{ kg/cm}^2$

Energy, work, quantity of heat
$1 \text{ J} = 1 \text{ N·m}$ $= 0.278 \times 10^{-6} \text{ kWh}$ $= 0.102 \text{ kg·m}$ $= 0.239 \times 10^{-3} \text{ kcal}$
$1 \text{ kWh} = 3.6 \times 10^6 \text{ J}$ $= 367 \times 10^3 \text{ kg·m}$ $= 860 \text{ kcal}$
$1 \text{ kg·m} = 9.807 \text{ J}$ $= 2.72 \times 10^{-6} \text{ kWh}$ $= 2.34 \times 10^{-3} \text{ kcal}$
$1 \text{ kcal} = 4.19 \times 10^3 \text{ J}$ $= 1.16 \times 10^{-3} \text{ kWh}$ $= 427 \text{ kg·m}$

Power
1 W = 0.102 kg·m/s   = 0.86 kcal/h
1 kg·m/s = 9.807 W   = 8.43 kcal/h
1 kcal/h = 1.16 W   = 0.119 kg·m/s

Kinematic viscosity
$1 \text{ m}^2\text{/s} = 10^4 \text{ St (stokes)}$
$1 \text{ St} = 10^{-4} \text{ m}^2\text{/s}$

Dynamic viscosity
$1 \text{ Pa·s} = 1 \text{ N·s/m}^2$ $= 10 \text{ P (poise)}$ $= 0.102 \text{ kg·s/m}^2$
$1 \text{ P} = 0.1 \text{ Pa·s}$ $= 0.1 \text{ N·s/m}^2$ $= 1.02 \times 10^{-2} \text{ kg·s/m}^2$
$1 \text{ kg·s/m}^2 = 9.807 \text{ Pa·s}$ $= 9.807 \text{ N·s/m}^2$ $= 98.07 \text{ P}$

The decimal multiples and sub-multiples of a unit may be formed by prefixes or symbols, having the following meanings, placed before the name or symbol of the unit:

Multiplying factor		Prefix	Symbol
1 000 000 000 000 000 000	$= 10^{18}$ quintillion	exa	E
1 000 000 000 000 000	$= 10^{15}$ quadrillion	peta	P
1 000 000 000 000	$= 10^{12}$ trillion	tera	T
1 000 000 000	$= 10^{9}$ billion	giga	G
1 000 000	$= 10^{6}$ million	mega	M
1 000	$= 10^{3}$ thousand	kilo	k
100	$= 10^{2}$ hundred	hecto	h
10	$= 10^{1}$ ten	deca	da
0.1	$= 10^{-1}$ tenth	deci	d
0.01	$= 10^{-2}$ hundredth	centi	c
0.001	$= 10^{-3}$ thousandth	milli	m
0.000 001	$= 10^{-6}$ millionth	micro	μ
0.000 000 001	$= 10^{-9}$ billionth	nano	n
0.000 000 000 001	$= 10^{-12}$ trillionth	pico	p
0.000 000 000 000 001	$= 10^{-15}$ quadrillionth	femto	f
0.000 000 000 000 000 001	$= 10^{-18}$ quintillionth	atto	a

Note: 10^9 = 1 billion is United Nations usage in English. By analogy, so is 10^{-9} = 1 billionth.

1.2.2.2   [Reserved]

1.2.2.3   Whenever the mass of a package is mentioned, the gross mass is meant unless otherwise stated. The mass of containers or tanks used for the transport of goods is not included in the gross mass.

1.2.2.4   Unless expressly stated otherwise, the sign "%" represents:

.1   in the case of mixtures of solids or of liquids, and also in the case of solutions and of solids wetted by a liquid: a percentage mass based on the total mass of the mixture, the solution or the wetted solid;

.2   in the case of mixtures of compressed gases: when filled by pressure, the proportion of the volume indicated as a percentage of the total volume of the gaseous mixture, or, when filled by mass, the proportion of the mass indicated as a percentage of the total mass of the mixture;

.3   in the case of mixtures of liquefied gases and gases dissolved under pressure: the proportion of the mass indicated as a percentage of the total mass of the mixture.

1.2.2.5 Pressures of all kinds relating to receptacles (such as test pressure, internal pressure, safety-valve opening pressure) are always indicated in gauge pressure (pressure in excess of atmospheric pressure); however, the vapour pressure of substances is always expressed in absolute pressure.

1.2.2.6 **Tables of equivalence**

1.2.2.6.1 *Mass conversion tables*

1.2.2.6.1.1 *Conversion factors*

Multiply	by	to obtain
Grams	0.03527	Ounces
Grams	0.002205	Pounds
Kilograms	35.2736	Ounces
Kilograms	2.2046	Pounds
Ounces	28.3495	Grams
Pounds	16	Ounces
Pounds	453.59	Grams
Pounds	0.45359	Kilograms
Hundredweight	112	Pounds
Hundredweight	50.802	Kilograms

1.2.2.6.1.2 *Pounds to kilograms and vice versa*

When the central value in any row of these mass conversion tables is taken to be in pounds, its equivalent value in kilograms is shown on the left; when the central value is in kilograms, its equivalent in pounds is shown on the right.

kg	← lb	→ kg	lb	kg	← lb	→ kg	lb	kg	← lb	→ kg	lb
0.227	0.5		1.10	22.7	50		110	90.7	200		441
0.454	1		2.20	24.9	55		121	95.3	210		463
0.907	2		4.41	27.2	60		132	99.8	220		485
1.36	3		6.61	29.5	65		143	102	225		496
1.81	4		8.82	31.8	70		154	104	230		507
2.27	5		11.0	34.0	75		165	109	240		529
2.72	6		13.2	36.3	80		176	113	250		551
3.18	7		15.4	38.6	85		187	118	260		573
3.63	8		17.6	40.8	90		198	122	270		595
4.08	9		19.8	43.1	95		209	125	275		606
4.54	10		22.0	45.4	100		220	127	280		617
4.99	11		24.3	47.6	105		231	132	290		639
5.44	12		26.5	49.9	110		243	136	300		661
5.90	13		28.7	52.2	115		254	159	350		772
6.35	14		30.9	54.4	120		265	181	400		882
6.80	15		33.1	56.7	125		276	204	450		992
7.26	16		35.3	59.0	130		287	227	500		1102
7.71	17		37.5	61.2	135		298	247	545		1202
8.16	18		39.7	63.5	140		309	249	550		1213
8.62	19		41.9	65.8	145		320	272	600		1323
9.07	20		44.1	68.0	150		331	318	700		1543
11.3	25		55.1	72.6	160		353	363	800		1764
13.6	30		66.1	77.1	170		375	408	900		1984
15.9	35		77.2	79.4	175		386	454	1000		2205
18.1	40		88.2	81.6	180		397				
20.4	45		99.2	86.2	190		419				

1.2.2.6.2   *Liquid measure conversion tables*

1.2.2.6.2.1   *Conversion factors*

Multiply	by	to obtain
Litres	0.2199	Imperial gallons
Litres	1.759	Imperial pints
Litres	0.2643	US gallons
Litres	2.113	US pints
Gallons	8	Pints
Imperial gallons	4.546	Litres
Imperial gallons Imperial pints	1.20095	US gallons US pints
Imperial pints	0.568	Litres
US gallons	3.7853	Litres
US gallons US pints	0.83268	Imperial gallons Imperial pints
US pints	0.473	Litres

1.2.2.6.2.2   *Imperial pints to litres and vice versa*

When the central value in any row of these liquid measure conversion tables is taken to be in pints, its equivalent value in litres is shown on the left; when the central value is in litres, its equivalent in pints is shown on the right.

ℓ	← → pt ℓ	pt
0.28	0.5	0.88
0.57	1	1.76
0.85	1.5	2.64
1.14	2	3.52
1.42	2.5	4.40
1.70	3	5.28
1.99	3.5	6.16
2.27	4	7.04
2.56	4.5	7.92
2.84	5	8.80
3.12	5.5	9.68
3.41	6	10.56
3.69	6.5	11.44
3.98	7	12.32
4.26	7.5	13.20
4.55	8	14.08

1.2.2.6.2.3 *Imperial gallons to litres and vice versa*

When the central value in any row of these liquid measure conversion tables is taken to be in gallons, its equivalent value in litres is shown on the left; when the central value is in litres, its equivalent in gallons is shown on the right.

ℓ	$\leftarrow$ gal $\rightarrow$ ℓ	gal	ℓ	$\leftarrow$ gal $\rightarrow$ ℓ	gal
2.27	0.5	0.11	159.11	35	7.70
4.55	1	0.22	163.65	36	7.92
9.09	2	0.44	168.20	37	8.14
13.64	3	0.66	172.75	38	8.36
18.18	4	0.88	177.29	39	8.58
22.73	5	1.10	181.84	40	8.80
27.28	6	1.32	186.38	41	9.02
31.82	7	1.54	190.93	42	9.24
36.37	8	1.76	195.48	43	9.46
40.91	9	1.98	200.02	44	9.68
45.46	10	2.20	204.57	45	9.90
50.01	11	2.42	209.11	46	10.12
54.55	12	2.64	213.66	47	10.34
59.10	13	2.86	218.21	48	10.56
63.64	14	3.08	222.75	49	10.78
68.19	15	3.30	227.30	50	11.00
72.74	16	3.52	250.03	55	12.09
77.28	17	3.74	272.76	60	13.20
81.83	18	3.96	295.49	65	14.29
86.37	19	4.18	318.22	70	15.40
90.92	20	4.40	340.95	75	16.49
95.47	21	4.62	363.68	80	17.60
100.01	22	4.84	386.41	85	18.69
104.56	23	5.06	409.14	90	19.80
109.10	24	5.28	431.87	95	20.89
113.65	25	5.50	454.60	100	22.00
118.19	26	5.72	613.71	135	29.69
122.74	27	5.94	681.90	150	32.98
127.29	28	6.16	909.20	200	43.99
131.83	29	6.38	1022.85	225	49.48
136.38	30	6.60	1136.50	250	54.97
140.92	31	6.82	1363.80	300	65.99
145.47	32	7.04	1591.10	350	76.96
150.02	33	7.26	1818.40	400	87.99
154.56	34	7.48	2045.70	450	98.95

1.2.2.6.3   *Temperature conversion tables*
Degrees Fahrenheit to degrees Celsius and vice versa
When the central value in any row of these temperature conversion tables is taken to be in °F, its equivalent value in °C is shown on the left; when the central value is in °C, its equivalent in °F is shown on the right.

General formula: $°F = (°C \times \frac{9}{5}) + 32;$ $°C = (°F - 32) \times \frac{5}{9}$

°C	← °F °C →	°F	°C	← °F °C →	°F	°C	← °F °C →	°F
−73.3	−100	−148	−21.1	−6	21.2	1.1	34	93.2
−67.8	−90	−130	−20.6	−5	23.0	1.7	35	95
−62.2	−80	−112	−20.0	−4	24.8	2.2	36	96.8
−56.7	−70	−94	−19.4	−3	26.6	2.8	37	98.6
−51.1	−60	−76	−18.9	−2	28.4	3.3	38	100.4
−45.6	−50	−58	−18.3	−1	30.2	3.9	39	102.2
−40	−40	−40	−17.8	0	32.0	4.4	40	104
−39.4	−39	−38.2	−17.2	1	33.8	5	41	105.8
−38.9	−38	−36.4	−16.7	2	35.6	5.6	42	107.6
−38.3	−37	−34.6	−16.1	3	37.4	6.1	43	109.4
−37.8	−36	−32.8	−15.6	4	39.2	6.7	44	111.2
−37.2	−35	−31	−15.0	5	41.0	7.2	45	113
−36.7	−34	−29.2	−14.4	6	42.8	7.8	46	114.8
−36.1	−33	−27.4	−13.9	7	44.6	8.3	47	116.6
−35.6	−32	−25.6	−13.3	8	46.4	8.9	48	118.4
−35	−31	−23.8	−12.8	9	48.2	9.4	49	120.2
−34.4	−30	−22	−12.2	10	50.0	10.0	50	122.0
−33.9	−29	−20.2	−11.7	11	51.8	10.6	51	123.8
−33.3	−28	−18.4	−11.1	12	53.6	11.1	52	125.6
−32.8	−27	−16.6	−10.6	13	55.4	11.7	53	127.4
−32.2	−26	−14.8	−10.0	14	57.2	12.2	54	129.2
−31.7	−25	−13	−9.4	15	59.0	12.8	55	131.0
−31.1	−24	−11.2	−8.9	16	60.8	13.3	56	132.8
−30.6	−23	−9.4	−8.3	17	62.6	13.9	57	134.6
−30	−22	−7.6	−7.8	18	64.4	14.4	58	136.4
−29.4	−21	−5.8	−7.2	19	66.2	15.0	59	138.2
−28.9	−20	−4	−6.7	20	68	15.6	60	140.0
−28.3	−19	−2.2	−6.1	21	69.8	16.1	61	141.8
−27.8	−18	−0.4	−5.6	22	71.6	16.7	62	143.6
−27.2	−17	1.4	−5	23	73.4	17.2	63	145.4
−26.7	−16	3.2	−4.4	24	75.2	17.8	64	147.2
−26.1	−15	5	−3.9	25	77	18.3	65	149.0
−25.6	−14	6.8	−3.3	26	78.8	18.9	66	150.8
−25.0	−13	8.6	−2.8	27	80.6	19.4	67	152.6
−24.4	−12	10.4	−2.2	28	82.4	20.0	68	154.4
−23.9	−11	12.2	−1.7	29	84.2	20.6	69	156.2
−23.3	−10	14.0	−1.1	30	86	21.1	70	158.0
−22.8	−9	15.8	−0.6	31	87.8	21.7	71	159.8
−22.2	−8	17.6	0	32	89.6	22.2	72	161.6
−21.7	−7	19.4	0.6	33	91.4	22.8	73	163.4

°C	←°F	→°C	°F	°C	←°F	→°C	°F	°C	←°F	→°C	°F
23.3	74		165.2	37.8	100		212	52.2	126		258.8
23.9	75		167.0	38.3	101		213.8	52.8	127		260.6
24.4	76		168.8	38.9	102		215.6	53.3	128		262.4
25.0	77		170.6	39.4	103		217.4	53.9	129		264.2
25.6	78		172.4	40	104		219.2	54.4	130		266.0
26.1	79		174.2	40.6	105		221	55.0	131		267.8
26.7	80		176.0	41.1	106		222.8	55.6	132		269.6
27.2	81		177.8	41.7	107		224.6	56.1	133		271.4
27.8	82		179.6	42.2	108		226.4	56.7	134		273.2
28.3	83		181.4	42.8	109		228.2	57.2	135		275.0
28.9	84		183.2	43.3	110		230	57.8	136		276.8
29.4	85		185	43.9	111		231.8	58.3	137		278.6
30	86		186.8	44.4	112		233.6	58.9	138		280.4
30.6	87		188.6	45	113		235.4	59.4	139		282.2
31.1	88		190.4	45.6	114		237.2	60.0	140		284.0
31.7	89		192.2	46.1	115		239.0	65.6	150		302.0
32.2	90		194	46.7	116		240.8	71.1	160		320.0
32.8	91		195.8	47.2	117		242.6	76.7	170		338.0
33.3	92		197.6	47.8	118		244.4	82.2	180		356.0
33.9	93		199.4	48.3	119		246.2	87.8	190		374.0
34.4	94		201.2	48.9	120		248.0	93.3	200		392.0
35	95		203	49.4	121		249.8	98.9	210		410.0
35.6	96		204.8	50.0	122		251.6	104.4	220		428.0
36.1	97		206.6	50.6	123		253.4	110.0	230		446.0
36.7	98		208.4	51.1	124		255.2	115.6	240		464.0
37.2	99		210.2	51.7	125		257.0	121.1	250		482.0

1.2.3 List of abbreviations

ASTM	American Society for Testing and Materials (ASTM International, 100 Barr Harbor Drive, P.O. Box C700, West Conshohocken, PA, 19428-2959, United States of America)
BC Code	Code of Safe Practice for Solid Bulk Cargoes
CGA	Compressed Gas Association (CGA, 4221 Walney Road, 5th Floor, Chantilly VA 20151-2923, United States of America)
CSC	International Convention for Safe Containers, 1972, as amended
DSC	IMO Sub-Committee on Dangerous Goods, Solid Cargoes and Containers
ECOSOC	Economic and Social Council (UN)
EmS	*The EmS Guide: Emergency Response Procedures for Ships Carrying Dangerous Goods*
EN (standard)	European standard published by the European Committee for Standardization (CEN) (CEN, 36 rue de Stassart, B-1050 Brussels, Belgium)
FAO	Food and Agriculture Organization (FAO; Viale delle Terme di Caracalla, 00100 Rome, Italy)
HNS Convention	International Convention on Liability and Compensation for Damage in Connection with the Transport of Hazardous and Noxious Substances (IMO)
IAEA	International Atomic Energy Agency (IAEA, P.O. Box 100, A -1400 Vienna, Austria)
ICAO	International Civil Aviation Organization (ICAO, 999 University Street, Montreal, Quebec H3C 5H7, Canada)

IEC	International Electrotechnical Commission (IEC, 3 rue de Varembé, P.O. Box 131, CH-1211 Geneva 20, Switzerland)
ILO	International Labour Organization/Office (ILO, 4 route des Morillons, CH-1211 Geneva 22, Switzerland)
IMGS	*International Medical Guide for Ships*
IMO	International Maritime Organization (IMO, 4 Albert Embankment, London SE1 7SR, United Kingdom)
IMDG Code	International Maritime Dangerous Goods Code
INF Code	International Code for the Safe Carriage of Packaged Irradiated Nuclear Fuel, Plutonium and High-Level Radioactive Wastes on board Ships
ISO (standard)	An international standard published by the International Organization for Standardization (ISO,1 rue de Varembé, CH-1204 Geneva 20, Switzerland)
MARPOL 73/78	International Convention for the Prevention of Pollution from Ships, 1973/78, as amended
MAWP	Maximum allowable working pressure
MEPC	Marine Environment Protection Committee (IMO)
MFAG	*Medical First Aid Guide for Use in Accidents Involving Dangerous Goods*
MSC	Maritime Safety Committee (IMO)
N.O.S.	not otherwise specified
SADT	Self-accelerating decomposition temperature
SOLAS 74	International Convention for the Safety of Life at Sea, 1974, as amended
UNECE	United Nations Economic Commission for Europe (UNECE, Palais des Nations, 8–14 avenue de la Paix, CH-1211 Geneva 10, Switzerland)
UN Number	Four-digit United Nations Number is assigned to dangerous, hazardous and harmful substances, materials and articles most commonly transported
UNEP	United Nations Environment Programme (United Nations Avenue, Gigiri, PO Box 30552, 00100, Nairobi, Kenya)
UNESCO/IOC	UN Educational, Scientific and Cultural Organization/Intergovernmental Oceanographic Commission (UNESCO/IOC, 1 rue Miollis, 75732 Paris Cedex 15, France)
WHO	World Health Organization (Avenue Appia 20, CH-1211 Geneva 27, Switzerland)
WMO	World Meteorological Organization (WMO, 7bis, avenue de la Paix, Case postale No. 2300, CH-1211 Geneva 2, Switzerland)

Chapter 1.3

Training

1.3.0 Introductory note

The successful application of regulations concerning the transport of dangerous goods and the achievement of their objectives are greatly dependent on the appreciation by all persons concerned of the risks involved and on a detailed understanding of the regulations. This can only be achieved by properly planned and maintained initial and retraining programmes for all persons concerned with the transport of dangerous goods. The provisions of paragraphs 1.3.1.4 to 1.3.1.7 remain recommendatory (see 1.1.1.5).

1.3.1 Training of shore-side personnel

1.3.1.1   Shore-based personnel* engaged in the transport of dangerous goods intended to be transported by sea shall receive training in the contents of dangerous goods provisions commensurate with their responsibilities. Training requirements specific to security of dangerous goods in chapter 1.4 shall also be addressed.

Entities engaging shore-based personnel in such activities shall determine which staff will be trained, what levels of training they require and the training methods used to enable them to comply with the provisions of the IMDG Code. This training shall be provided or verified upon employment in a position involving dangerous goods transport. For personnel who have not yet received the required training, the entities shall ensure that those personnel may only perform functions under the direct supervision of a trained person. The training shall be periodically supplemented with refresher training to take account of changes in regulations and practice. The competent authority, or its authorized body, may audit the entity to verify the effectiveness of the system in place, in providing training of staff commensurate with their role and responsibilities in the transport chain.

1.3.1.2   Shore-based personnel such as those who:

- classify dangerous goods and identify Proper Shipping Names of dangerous goods;
- pack dangerous goods;
- mark, label or placard dangerous goods;
- load/unload Cargo Transport Units;
- prepare transport documents for dangerous goods;
- offer dangerous goods for transport;
- accept dangerous goods for transport;
- handle dangerous goods in transport;
- prepare dangerous goods loading/stowage plans;
- load/unload dangerous goods into/from ships;
- carry dangerous goods in transport;
- enforce or survey or inspect for compliance with applicable rules and regulations; or
- are otherwise involved in the transport of dangerous goods as determined by the competent authority

shall receive the following training:

1.3.1.2.1   *General awareness/familiarization training:*

.1   each person shall receive training designed to provide familiarity with the general provisions of dangerous goods transport provisions;

---

* For the training of officers and ratings responsible for cargo handling on ships carrying dangerous and hazardous substances in solid form in bulk, or in packaged form, see the STCW Code, as amended.

.2 such training shall include a description of the classes of dangerous goods; labelling, marking, placarding, packing, stowage, segregation and compatibility provisions; a description of the purpose and content of the dangerous goods transport documents (such as the Multimodal Dangerous Goods Form and the Container/Vehicle Packing Certificate); and a description of available emergency response documents.

1.3.1.2.2 *Function-specific training:* Each person shall receive detailed training concerning specific dangerous goods transport provisions which are applicable to the function that person performs. An indicative list, for guidance purposes only, of some of the functions typically found in dangerous goods transport operations by sea and training requirements is given in paragraph 1.3.1.6.

1.3.1.3 Details of all the training undertaken shall be kept by both the employer and the employee. Training records shall be made available to the competent authority if requested.

1.3.1.4 *Safety training*: Commensurate with the risk of exposure in the event of a release and the functions performed, each person should receive training on:

.1 methods and procedures for accident avoidance, such as proper use of package-handling equipment and appropriate methods of stowage of dangerous goods;

.2 available emergency response information and how to use it;

.3 general dangers presented by the various classes of dangerous goods and how to prevent exposure to those hazards, including, if appropriate, the use of personal protective clothing and equipment; and

.4 immediate procedures to be followed in the event of an unintentional release of dangerous goods, including any emergency response procedures for which the person is responsible and personal protection procedures to be followed.

1.3.1.5 Recommended training needs for shore-side personnel involved in the transport of dangerous goods under the IMDG Code

The following indicative table is for information purposes only as every entity is arranged differently and may have varied roles and responsibilities within that entity.

Function	Specific training requirements	Numbers in this column refer to the list of related codes and publications in 1.3.1.7
1 Classify dangerous goods and identify Proper Shipping Name	Classification requirements, in particular – the structure of the description of substances – the classes of dangerous goods and the principles of their classification – the nature of the dangerous substances and articles transported (their physical, chemical and toxicological properties) – the procedure for classifying solutions and mixtures – identification by Proper Shipping Name – use of Dangerous Goods List	.1, .4, .5 and .12
2 Pack dangerous goods	Classes Packaging requirements – type of packages (IBC, large packaging, tank container and bulk container) – UN marking for approved packagings – segregation requirements – limited quantities and excepted quantities Marking and labelling First aid measures Emergency response procedures Safe handling procedures	.1 and .4
3 Mark, label or placard dangerous goods	Classes Marking, labelling and placarding requirements – primary and subsidiary risk labels – marine pollutants – limited quantities and excepted quantities	.1

Function	Specific training requirements	Numbers in this column refer to the list of related codes and publications in 1.3.1.7
4 Load/unload cargo transport units*	Documentation Classes Marking, labelling and placarding Stowage requirements, where applicable Segregation requirements Cargo securing requirements (as contained in the IMO/ILO/UNECE Guidelines) Emergency response procedures First aid measures CSC requirements Safe handling procedures	.1, .6, .7 and .8
5 Prepare transport documents for dangerous goods	Documentation requirements – transport document – container/vehicle packing certificate – competent authorities' approval – waste transport documentation – special documentation, where appropriate	.1
6 Offer dangerous goods for transport	Thorough knowledge of the IMDG Code Local requirements at loading and discharge ports – port byelaws – national transport regulations	.1 to .10 and .12
7 Accept dangerous goods for transport	Thorough knowledge of the IMDG Code Local requirements at loading, transiting and discharge ports – port byelaws, in particular quantity limitations – national transport regulations	.1 to .12
8 Handle dangerous goods in transport	Classes and their hazards Marking, labelling and placarding Emergency response procedures First aid measures Safe handling procedures such as – use of equipment – appropriate tools – safe working loads CSC requirements, local requirements at loading, transit and discharge ports Port byelaws, in particular, quantity limitation National transport regulations	.1, .2, .3, .6, .7, .8 and .10
9 Prepare dangerous goods loading/stowage plans	Documentation Classes Stowage requirements Segregation requirements Document of compliance Relevant IMDG Code parts, local requirements at loading, transit and discharge ports Port byelaws, in particular, quantity limitations	.1, .10, .11 and .12
10 Load/unload dangerous goods into/from ships	Classes and their hazards Marking, labelling and placarding Emergency response procedures First aid measures Safe handling procedures such as – use of equipment – appropriate tools – safe working loads Cargo securing requirements CSC requirements, local requirements at loading, transit and discharge ports Port byelaws, in particular, quantity limitation National transport regulations	.1, .2, .3, .7, .9, .10 and .12

* Definition as per IMO/ILO/UNECE Guidelines for Packing of Cargo Transport Units.

Function	Specific training requirements	Numbers in this column refer to the list of related codes and publications in 1.3.1.7

11 Carry dangerous goods	Documentation
	Classes
	Marking, labelling and placarding
	Stowage requirements, where applicable
	Segregation requirements
	Local requirements at loading, transit and discharge ports
	  – port byelaws, in particular, quantity limitations
	  – national transport regulations
	Cargo securing requirements (as contained in the IMO/ILO/UNECE Guidelines)
	Emergency response procedures
	First aid measures
	CSC requirements
	Safe handling procedures
		.1, .2, .3, .6, .7, .10, .11 and .12

12 Enforce or survey or inspect for compliance with applicable rules and regulations	Knowledge of IMDG Code and relevant guidelines and safety procedures	.1 to .12
13 Are otherwise involved in the transport of dangerous goods, as determined by the competent authority	As required by the competent authority commensurate with the task assigned	–

1.3.1.6 Indicative table describing sections of the IMDG Code or other relevant instruments that may be appropriate to be considered in any training for the transport of dangerous goods

#	Function	1	2	2.0	3	4	5	6	6*	7.1	7.2	7.3	7.4	7.5	7.6	7.7	7.8	7.9	SOLAS chapter II-2/19	Port Byelaws	National transport regulations	CSC	Guidelines for Packing of Cargo Transport Units	Emergency Response Procedures	First aid measures	Safe handling procedures
1	Classify	X	X		X		X										X	X								
2	Pack in packages	X		X	X	X	X	X				X				X	X	X						X	X	X
3	Mark, label, placard			X	X		X																			
4	Load/unload cargo transport units	X		X	X	X	X			X	X	X		X	X	X	X						X	X	X	X
5	Prepare transport documents	X		X	X		X										X	X						X	X	
6	Offer for transport	X	X		X	X	X	X		X	X	X	X	X	X	X	X	X		X	X	X	X	X	X	
7	Accept for transport	X	X		X	X	X	X		X	X	X	X	X	X	X	X	X	X	X	X	X	X	X	X	
8	Handle in transport	X		X	X		X		X														X	X	X	X
9	Prepare loading/stowage plans	X		X	X	X	X			X	X	X	X	X	X	X	X	X	X				X	X		
10	Load/unload from ships	X	X		X		X										X		X	X			X	X	X	X
11	Carry	X		X	X	X	X			X	X	X	X	X	X	X	X	X	X	X	X	X	X	X	X	X

Remarks:
* Only sections 6.1.2, 6.1.3, 6.5.2, 6.6.3, 6.7.2.20, 6.7.3.16 and 6.7.4.15 apply

1.3.1.7   Related Codes and publications which may be appropriate for function-specific training

.1   International Maritime Dangerous Goods (IMDG) Code, as amended

.2   The EmS Guide: Emergency Response Procedures for Ships Carrying Dangerous Goods (EmS), as amended

.3   Medical First Aid Guide for Use in Accidents Involving Dangerous Goods (MFAG), as amended

.4   United Nations Recommendations on the Transport of Dangerous Goods – Model Regulations, as amended

.5   United Nations Recommendations on the Transport of Dangerous Goods – Manual of Tests and Criteria, as amended

.6   IMO/ILO/UNECE Guidelines for Packing of Cargo Transport Units (CTUs)

.7   Recommendations on the Safe Transport of Dangerous Cargoes and Related Activities in Port Areas

.8   International Convention for Safe Containers (CSC), 1972, as amended

.9   Code of Safe Practice for Cargo Stowage and Securing (CSS Code), as amended

.10   MSC.1/Circ.1265 Recommendations on the safe use of pesticides in ships applicable to the fumigation of cargo transport units

.11   International Convention for the Safety of Life at Sea (SOLAS) 1974, as amended

.12   International Convention for the Prevention of Pollution from Ships 1973 as modified by the Protocol of 1978 (MARPOL 73/78), as amended.

Chapter 1.4
Security provisions

1.4.0 Introductory note

The provisions of this chapter address the security of dangerous goods in transport by sea. National competent authorities may apply additional security provisions, which should be considered when offering or transporting dangerous goods. The provisions of this chapter remain recommendatory except 1.4.1.1 (see 1.1.1.5).

1.4.1 General provisions for companies, ships and port facilities

1.4.1.1 The relevant provisions of chapter XI-2 of SOLAS 74, as amended, and of part A of the International Ship and Port Facility Security (ISPS) Code apply to companies, ships and port facilities engaged in the transport of dangerous goods and to which regulation XI-2 of SOLAS 74, as amended, apply taking into account the guidance given in part B of the ISPS Code.

1.4.1.2 For cargo ships of less than 500 gross tons engaged in the transport of dangerous goods, it is recommended that Contracting Governments to SOLAS 74, as amended, consider security provisions for these cargo ships.

1.4.1.3 Any shore-based company personnel, ship-based personnel and port facility personnel engaged in the transport of dangerous goods should be aware of the security requirements for such goods, in addition to those specified in the ISPS Code, and commensurate with their responsibilities.

1.4.1.4 The training of the company security officer, shore-based company personnel having specific security duties, port facility security officer and port facility personnel having specific duties, engaged in the transport of dangerous goods, should also include elements of security awareness related to those goods.

1.4.1.5 All shipboard personnel and port facility personnel who are not mentioned in 1.4.1.4 and are engaged in the transport of dangerous goods should be familiar with the provisions of the relevant security plans related to those goods, commensurate with their responsibilities.

1.4.2 General provisions for shore-side personnel

1.4.2.1 For the purpose of this subsection, *shore-side personnel* covers individuals mentioned in 1.3.1.2. However, the provisions of 1.4.2 do not apply to:

– the company security officer and appropriate shore-based company personnel mentioned in 13.1 of part A of the ISPS Code,

– the ship security officer and the shipboard personnel mentioned in 13.2 and 13.3 of part A of the ISPS Code,

– the port facility security officer, the appropriate port facility security personnel and the port facility personnel having specific security duties mentioned in 18.1 and 18.2 of part A of the ISPS Code.

For the training of those officers and personnel, refer to the International Ship and Port Facility Security (ISPS) Code.

1.4.2.2 Shore-side personnel engaged in transport by sea of dangerous goods should consider security provisions for the transport of dangerous goods commensurate with their responsibilities.

1.4.2.3 Security training

1.4.2.3.1 The training of shore-side personnel, as specified in chapter 1.3, shall also include elements of security awareness.

1.4.2.3.2 Security awareness training should address the nature of security risks, recognizing security risks, methods to address and reduce risks and actions to be taken in the event of a security breach. It should include awareness of security plans (if appropriate, refer to 1.4.3) commensurate with the responsibilities of individuals and their part in implementing security plans.

1.4.2.3.3   Such training should be provided or verified upon employment in a position involving dangerous goods transport and should be periodically supplemented with retraining.

1.4.2.3.4   Records of all security training undertaken should be kept by the employer and made available to the employee if requested.

1.4.3   Provisions for high consequence dangerous goods

1.4.3.1   For the purposes of this section, high consequence dangerous goods are those which have the potential for misuse in a terrorist incident and which may, as a result, produce serious consequences such as mass casualties or mass destruction. The following is an indicative list of high consequence dangerous goods:

Class 1	Division 1.1 explosives
Class 1	Division 1.2 explosives
Class 1	Division 1.3 compatibility group C explosives
Class 1	Division 1.4. UN Nos. 0104, 0237, 0255, 0267, 0289, 0361, 0365, 0366, 0440, 0441, 0455, 0456 and 0500
Class 1	Division 1.5 explosives
Class 2.1	Flammable gases in quantities greater than 3000 ℓ in a road tank vehicle, a railway tank wagon or a portable tank
Class 2.3	Toxic gases
Class 3	Flammable liquids of packing groups I and II in quantities greater than 3000 ℓ in a road tank vehicle, a railway tank wagon or a portable tank
Class 3	Desensitized liquid explosives
Class 4.1	Desensitized solid explosives
Class 4.2	Goods of packing group I in quantities greater than 3000 kg or 3000 ℓ in a road tank vehicle, a railway tank wagon, a portable tank or a bulk container
Class 4.3	Goods of packing group I in quantities greater than 3000 kg or 3000 ℓ in a road tank vehicle, a railway tank wagon, a portable tank or a bulk container
Class 5.1	Oxidizing liquids of packing group I in quantities greater than 3000 ℓ in a road tank vehicle, a railway tank wagon or a portable tank
Class 5.1	Perchlorates, ammonium nitrate, ammonium nitrate fertilizers and ammonium nitrate emulsions or suspensions or gels in quantities greater than 3000 kg or 3000 ℓ in a road tank vehicle, a railway tank wagon, a portable tank or a bulk container
Class 6.1	Toxic substances of packing group I
Class 6.2	Infectious substances of category A (UN Nos. 2814 and 2900)
Class 7	Radioactive material in quantities greater than $3000A_1$ (special form) or $3000A_2$, as applicable, in Type B(U) or Type B(M) or Type C packages
Class 8	Corrosive substances of packing group I in quantities greater than 3000 kg or 3000 ℓ in a road tank vehicle, a railway tank wagon, a portable tank or a bulk container

1.4.3.2   The provisions of this section do not apply to ships and to port facilities (see the ISPS Code for ship security plan and for port facility security plan).

1.4.3.3   Consignors and others engaged in the transport of high consequence dangerous goods should adopt, implement and comply with a security plan that addresses at least the elements specified in 1.4.3.4.

1.4.3.4   The security plan should comprise at least the following elements:

.1   specific allocation of responsibilities for security to competent and qualified persons with appropriate authority to carry out their responsibilities;

.2   records of dangerous goods or types of dangerous goods transported;

.3   review of current operations and assessment of vulnerabilities, including intermodal transfer, temporary transit storage, handling and distribution, as appropriate;

.4   clear statements of measures, including training, policies (including response to higher threat conditions, new employee/employment verification, etc.), operating practices (e.g. choice/use of routes where known, access to dangerous goods in temporary storage, proximity to vulnerable infrastructure, etc.), equipment and resources that are to be used to reduce security risks;

.5 effective and up-to-date procedures for reporting and dealing with security threats, breaches of security or security-related incidents;

.6 procedures for the evaluation and testing of security plans and procedures for periodic review and update of the plans;

.7 measures to ensure the security of transport information contained in the plan; and

.8 measures to ensure that the distribution of transport information is limited as far as possible. (Such measures shall not preclude provision of transport documentation required by chapter 5.4 of this Code.)

1.4.3.5 For radioactive material, the provisions of this chapter are deemed to be complied with when the provisions of the Convention on Physical Protection of Nuclear Material and of IAEA INFCIRC/225 (Rev.4) are applied.

Chapter 1.5

General provisions concerning class 7

1.5.1   Scope and application

1.5.1.1   The provisions of this Code establish standards of safety which provide an acceptable level of control of the radiation, criticality and thermal hazards to persons, property and the environment that are associated with the transport of radioactive material. These provisions are based on the IAEA Regulations for the Safe Transport of Radioactive Material (2005 Edition), Safety Standards Series No. TS-R-1, IAEA, Vienna (2005). Explanatory material on the 1996 edition of TS-R-1 can be found in "Advisory Material for the IAEA Regulations for the Safe Transport of Radioactive Material*", Safety Standard Series No. TS-G-1.1 (ST-2), IAEA, Vienna (2002).

1.5.1.2   The objective of the provisions of this Code is to protect persons, property and the environment from the effects of radiation during the transport of radioactive material. This protection is achieved by requiring:

.1   Containment of the radioactive contents;

.2   Control of external radiation levels;

.3   Prevention of criticality; and

.4   Prevention of damage caused by heat.

These provisions are satisfied firstly by applying a graded approach to contents limits for packages and conveyances and to performance standards applied to package designs depending upon the hazard of the radioactive contents. Secondly, they are satisfied by imposing requirements on the design and operation of packages and on the maintenance of packagings, including a consideration of the nature of the radioactive contents. Finally, they are satisfied by requiring administrative controls including, where appropriate, approval by competent authorities.

1.5.1.3   The provisions of this Code apply to the transport of radioactive material by sea, including transport which is incidental to the use of the radioactive material. Transport comprises all operations and conditions associated with and involved in the movement of radioactive material; these include the design, manufacture, maintenance and repair of packaging, and the preparation, consigning, loading, transport including in-transit storage, unloading and receipt at the final destination of loads of radioactive material and packages. A graded approach is applied to the performance standards in the provisions of this Code that is characterized by three general severity levels:

.1   Routine conditions of transport (incident-free);

.2   Normal conditions of transport (minor mishaps); and

.3   Accident conditions of transport.

1.5.1.4   The provisions of this Code shall not apply to:

.1   Radioactive material that is an integral part of the means of transport;

.2   Radioactive material moved within an establishment which is subject to appropriate safety regulations in force in the establishment and where the movement does not involve public roads or railways;

.3   Radioactive material implanted or incorporated into a person or live animal for diagnosis or treatment;

.4   Radioactive material in consumer products which have received regulatory approval, following their sale to the end user;

.5   Natural material and ores containing naturally occurring radionuclides which are either in their natural state, or have only been processed for purposes other than for extraction of the radionuclides, and which are not intended to be processed for use of these radionuclides provided the activity concentration of the material does not exceed 10 times the values specified in 2.7.2.2.1.2, or calculated in accordance with 2.7.2.2.2 to 2.7.2.2.6; and

.6   Non-radioactive solid objects with radioactive substances present on any surfaces in quantities not in excess of the limit set out in the definition for "contamination" in 2.7.1.2.

---

* A revised edition containing explanatory material on the 2005 edition of TS-R-1 is likely to be published by the IAEA in 2008.

1.5.1.5   **Specific provisions for the transport of excepted packages**

1.5.1.5.1   Excepted packages which may contain radioactive material in limited quantities, instruments, manufactured articles and empty packagings as specified in 2.7.2.4.1 may be transported under the following conditions:

     .1   The applicable provisions specified in 2.0.3.5, 2.7.2.4.1.2 to 2.7.2.4.1.6 (as applicable), 4.1.9.1.2, 5.2.1.1, 5.2.1.2, 5.2.1.5.1 to 5.2.1.5.3, 5.4.1.4.1.1 and 7.3.4.2;

     .2   The provisions for excepted packages specified in 6.4.4; and

     .3   If the excepted package contains fissile material, one of the fissile exceptions provided by 2.7.2.3.5 shall apply and the provision of 6.4.7.2 shall be met.

1.5.1.5.2   The following provisions shall not apply to excepted packages and the controls for transport of excepted packages: 1.4.2, 1.4.3, 2.7.2.3.3.1.1, 2.7.2.3.3.2, 4.1.9.1.3, 4.1.9.1.4, 4.1.9.1.6, 4.1.9.1.7, 5.1.3.2, 5.2.2.1.12.1, 5.4.1.5.7.1, 5.4.1.5.7.2, 5.4.1.6, 6.4.6.1, 7.1.14.11 to 7.1.14.14, 7.2.9.1, 7.2.9.2, 7.2.1 and 7.3.4.1.

1.5.2   Radiation protection programme

1.5.2.1   The transport of radioactive material shall be subject to a radiation protection programme which shall consist of systematic arrangements aimed at providing adequate consideration of radiation protection measures.

1.5.2.2   Doses to persons shall be below the relevant dose limits. Protection and safety shall be optimized in order that the magnitude of individual doses, the number of persons exposed, and the likelihood of incurring exposure shall be kept as low as reasonably achievable, economic and social factors being taken into account, within the restrictions that the doses to individuals be subject to dose constraints. A structured and systematic approach shall be adopted and shall include consideration of the interfaces between transport and other activities.

1.5.2.3   The nature and extent of the measures to be employed in the programme shall be related to the magnitude and likelihood of radiation exposures. The programme shall incorporate the provisions in 1.5.2.2 and 1.5.2.4. Programme documents shall be available, on request, for inspection by the relevant competent authority.

1.5.2.4   For occupational exposures arising from transport activities, where it is assessed that the effective dose:

     .1   is likely to be between 1 and 6 mSv in a year, a dose assessment programme via workplace monitoring or individual monitoring shall be conducted;

     .2   is likely to exceed 6 mSv in a year, individual monitoring shall be conducted.

When individual monitoring or workplace monitoring is conducted, appropriate records shall be kept.

Note: For occupational exposures arising from transport activities, where it is assessed that the effective dose is most unlikely to exceed 1 mSv in a year, no special work patterns, detailed monitoring, dose assessment programmes or individual record keeping need be required.

1.5.3   Quality assurance

1.5.3.1   Quality assurance programmes based on international, national or other standards acceptable to the competent authority shall be established and implemented for the design, manufacture, testing, documentation, use, maintenance and inspection of all special form radioactive material, low dispersible radioactive material and packages and for transport and in-transit storage operations to ensure compliance with the relevant provisions of this Code. Certification that the design specification has been fully implemented shall be available to the competent authority. The manufacturer, consignor or user shall be prepared to provide facilities for competent authority inspection during manufacture and use and to demonstrate to any cognizant competent authority that:

     .1   the manufacturing methods and materials used are in accordance with the approved design specifications; and

     .2   all packagings are periodically inspected and, as necessary, repaired and maintained in good condition so that they continue to comply with all relevant requirements and specifications, even after repeated use.

Where competent authority approval is required, such approval shall take into account and be contingent upon the adequacy of the quality assurance programme.

1.5.4   Special arrangement

1.5.4.1   *Special arrangement* shall mean those provisions, approved by the competent authority, under which consignments which do not satisfy all the provisions of this Code applicable to radioactive material may be transported.

1.5.4.2   Consignments for which conformity with any provision applicable to class 7 is impracticable shall not be transported except under special arrangement. Provided the competent authority is satisfied that conformity with the class 7 provisions of this Code is impracticable and that the requisite standards of safety established by this Code have been demonstrated through alternative means, the competent authority may approve special arrangement transport operations for single or a planned series of multiple consignments. The overall level of safety in transport shall be at least equivalent to that which would be provided if all the applicable provisions had been met. For international consignments of this type, multilateral approval shall be required.

1.5.5   Radioactive material possessing other dangerous properties

1.5.5.1   In addition to the radioactive and fissile properties, any subsidiary risk of the contents of a package, such as explosiveness, flammability, pyrophoricity, chemical toxicity and corrosiveness, shall also be taken into account in the documentation, packing, labelling, marking, placarding, stowage, segregation and transport, in order to be in compliance with all relevant provisions for dangerous goods. (See also special provision 172 and, for excepted packages, special provision 290.)

1.5.6   Non-compliance

1.5.6.1   In the event of a non-compliance with any limit in the provisions of this Code applicable to radiation level or contamination,

.1   The consignor shall be informed of the non-compliance

(i)   by the carrier if the non-compliance is identified during transport; or

(ii)   by the consignee if the non-compliance is identified at receipt;

.2   The carrier, consignor or consignee, as appropriate, shall:

(i)   take immediate steps to mitigate the consequences of the non-compliance;

(ii)   investigate the non-compliance and its causes, circumstances and consequences;

(iii)   take appropriate action to remedy the causes and circumstances that led to the non-compliance and to prevent a recurrence of similar circumstances that led to the non-compliance; and

(iv)   communicate to the relevant competent authority(ies) on the causes of the non-compliance and on corrective or preventive actions taken or to be taken; and

.3   The communication of the non-compliance to the consignor and relevant competent authority(ies), respectively, shall be made as soon as practicable and it shall be immediate whenever an emergency exposure situation has developed or is developing.

2

PART 2

CLASSIFICATION

2

Chapter 2.0

Introduction

2

Note: For the purposes of this Code, it has been necessary to classify dangerous goods in different classes, to subdivide a number of these classes and to define and describe characteristics and properties of the substances, materials and articles which would fall within each class or division. Moreover, in accordance with the criteria for the selection of marine pollutants for the purposes of Annex III of the International Convention for the Prevention of Pollution from Ships, 1973, as modified by the Protocol of 1978 relating thereto (MARPOL 73/78), a number of dangerous substances in the various classes have also been identified as substances harmful to the marine environment (MARINE POLLUTANTS).

2.0.0 Responsibilities

The classification shall be made by the shipper/consignor or by the appropriate competent authority where specified in this Code.

2.0.1 Classes, divisions, packing groups

2.0.1.1 Definitions

Substances (including mixtures and solutions) and articles subject to the provisions of this Code are assigned to one of the classes 1–9 according to the hazard or the most predominant of the hazards they present. Some of these classes are subdivided into divisions. These classes or divisions are as listed below:

Class 1: Explosives

    Division 1.1: substances and articles which have a mass explosion hazard

    Division 1.2: substances and articles which have a projection hazard but not a mass explosion hazard

    Division 1.3: substances and articles which have a fire hazard and either a minor blast hazard or a minor projection hazard or both, but not a mass explosion hazard

    Division 1.4: substances and articles which present no significant hazard

    Division 1.5: very insensitive substances which have a mass explosion hazard

    Division 1.6: extremely insensitive articles which do not have a mass explosion hazard

Class 2: Gases

    Class 2.1: flammable gases

    Class 2.2: non-flammable, non-toxic gases

    Class 2.3: toxic gases

Class 3: Flammable liquids

Class 4: Flammable solids; substances liable to spontaneous combustion; substances which, in contact with water, emit flammable gases

    Class 4.1: flammable solids, self-reactive substances and solid desensitized explosives

    Class 4.2: substances liable to spontaneous combustion

    Class 4.3: substances which, in contact with water, emit flammable gases

Class 5: Oxidizing substances and organic peroxides

    Class 5.1: oxidizing substances

    Class 5.2: organic peroxides

Class 6: Toxic and infectious substances

    Class 6.1: toxic substances

    Class 6.2: infectious substances

Class 7: Radioactive material

Class 8: Corrosive substances

Class 9: Miscellaneous dangerous substances and articles

The numerical order of the classes and divisions is not that of the degree of danger.

2.0.1.2 **Marine pollutants and wastes**

2.0.1.2.1 Many of the substances assigned to classes 1 to 9 are deemed as being *marine pollutants* (see chapter 2.10).

2.0.1.2.2 Wastes shall be transported under the provisions of the appropriate class, considering their hazards and the criteria in the Code. Wastes not otherwise subject to the Code but covered under the Basel Convention* may be transported under class 9. Alternatively, the classification may be in accordance with 7.8.4.

2.0.1.3 For packing purposes, substances other than those of classes 1, 2, 5.2, 6.2 and 7, and other than self-reactive substances of class 4.1, are assigned to three packing groups in accordance with the degree of danger they present:

Packing group I: substances presenting high danger;

Packing group II: substances presenting medium danger; and

Packing group III: substances presenting low danger.

The packing group to which a substance is assigned is indicated in the Dangerous Goods List in chapter 3.2.

2.0.1.4 Dangerous goods are determined to present one or more of the dangers represented by classes 1 to 9, marine pollutants and, if applicable, the degree of danger (packing group) on the basis of the provisions in chapters 2.1 to 2.10.

2.0.1.5 Dangerous goods presenting a danger of a single class or division are assigned to that class or division and the packing group, if applicable, determined. When an article or substance is specifically listed by name in the Dangerous Goods List in chapter 3.2, its class or division, its subsidiary risk(s) and, when applicable, its packing group are taken from this list.

2.0.1.6 Dangerous goods meeting the defining criteria of more than one hazard class or division and which are not listed by name in the Dangerous Goods List are assigned to a class or division and subsidiary risk(s) on the basis of the precedence of hazard provisions prescribed in 2.0.3.

2.0.1.7 Known marine pollutants are noted in the Dangerous Goods List and are indicated in the Index.

2.0.2 UN Numbers and Proper Shipping Names

2.0.2.1 Dangerous goods are assigned to UN Numbers and Proper Shipping Names according to their hazard classification and their composition.

2.0.2.2 Dangerous goods commonly transported are listed in the Dangerous Goods List in chapter 3.2. Where an article or substance is specifically listed by name, it shall be identified in transport by the Proper Shipping Name in the Dangerous Goods List. For dangerous goods not specifically listed by name, "generic" or "not otherwise specified" entries are provided (see 2.0.2.7) to identify the article or substance in transport.

Each entry in the Dangerous Goods List is assigned a UN Number. This list also contains relevant information for each entry, such as hazard class, subsidiary risk(s) (if any), packing group (where assigned), packing and tank transport provisions, EmS, segregation and stowage, properties and observations, etc.

Entries in the Dangerous Goods List are of the following four types:

.1 single entries for well-defined substances or articles:
e.g. UN 1090   acetone
      UN 1194   ethyl nitrite solution

.2 generic entries for well-defined groups of substances or articles:
e.g. UN 1133   adhesives
      UN 1266   perfumery product
      UN 2757   carbamate pesticide, solid, toxic
      UN 3101   organic peroxide type B, liquid

---

* Basel Convention on the Control of Transboundary Movements of Hazardous Wastes and their Disposal (1989).

.3   specific N.O.S. entries covering a group of substances or articles of a particular chemical or technical nature:
e.g.   UN 1477   nitrates, inorganic, N.O.S.
        UN 1987   alcohols, N.O.S.

.4   general N.O.S. entries covering a group of substances or articles meeting the criteria of one or more classes:
e.g.   UN 1325   flammable solid, organic, N.O.S.
        UN 1993   flammable liquid, N.O.S.

2.0.2.3   All self-reactive substances of class 4.1 are assigned to one of twenty generic entries in accordance with the classification principles described in 2.4.2.3.3.

2.0.2.4   All organic peroxides of class 5.2 are assigned to one of twenty generic entries in accordance with the classification principles described in 2.5.3.3.

2.0.2.5   A mixture or solution containing a single dangerous substance specifically listed by name in the Dangerous Goods List and one or more substances not subject to this Code shall be assigned a UN Number and Proper Shipping Name of the dangerous substance, except when:

.1   the mixture or solution is specifically identified by name in this Code; or

.2   the entry in this Code specifically indicates that it applies only to the pure or technically pure substance; or

.3   the hazard class or division, physical state or packing group of the solution or mixture is different from that of the dangerous substances; or

.4   there is a significant change in the measures to be taken in emergencies.

In those other cases, except the one described in .1, the mixture or solution shall be treated as a dangerous substance not specifically listed by name in the Dangerous Goods List.

2.0.2.6   When the class, physical state or packing group has changed in comparison with the pure substance, the solution or mixture shall be shipped in accordance with the provisions for the changed hazard under an appropriate N.O.S. entry.

2.0.2.7   Substances or articles which are not specifically listed by name in the Dangerous Goods List shall be classified under a "generic" or "not otherwise specified" (N.O.S.) Proper Shipping Name. The substance or article shall be classified according to the class definitions and test criteria in this part, and the article or substance classified under the generic or "N.O.S." Proper Shipping Name in the Dangerous Goods List which most appropriately describes the article or substance. This means that a substance is only to be assigned to an entry of type .3 – as defined in 2.0.2.2 – if it cannot be assigned to an entry of type .2, and to an entry of type .4 if it cannot be assigned to an entry of type .2 or .3.*

2.0.2.8   When considering a solution or mixture in accordance with 2.0.2.5, due account shall be given to whether the dangerous constituent comprising the solution or mixture has been identified as a marine pollutant. If this is the case, the provisions of chapter 2.10 are also applicable.

2.0.2.9   A mixture or solution, containing one or more substances identified by name in this Code or classified under an N.O.S. or generic entry and one or more substances not subject to the provisions of this Code, is not subject to the provisions of this Code if the hazard characteristics of the mixture or solution are such that they do not meet the criteria (including human experience criteria) for any class.

2.0.3   Classification of substances, mixtures and solutions with multiple hazards (precedence of hazard characteristics)

2.0.3.1   The table of precedence of hazard characteristics in 2.0.3.6 shall be used to determine the class of a substance, mixture or solution having more than one hazard when it is not specifically listed by name in this Code. For substances, mixtures or solutions having multiple hazards which are not specifically listed by name, the most stringent packing group of those assigned to the respective hazards of the goods takes precedence over other packing groups, irrespective of the precedence of hazard table in 2.0.3.6.

* See also the generic or N.O.S. Proper Shipping Name in appendix A.

2.0.3.2 The precedence of hazard table indicates which of the hazards shall be regarded as the primary hazard. The class which appears at the intersection of the horizontal line and the vertical column is the primary hazard and the remaining class is the subsidiary hazard. The packing groups for each of the hazards associated with the substance, mixture or solution shall be determined by reference to the appropriate criteria. The most stringent of the groups so indicated shall then become the packing group of the substance, mixture or solution.

2.0.3.3 The Proper Shipping Name (see 3.1.2) of a substance, mixture or solution when classified in accordance with 2.0.3.1 and 2.0.3.2 shall be the most appropriate N.O.S. ("not otherwise specified") entry in this Code for the class shown as the primary hazard.

2.0.3.4 The precedence of hazard characteristics of the following substances, materials and articles have not been dealt with in the precedence of hazard table, as these primary hazards always take precedence:

.1 substances and articles of class 1;

.2 gases of class 2;

.3 liquid desensitized explosives of class 3;

.4 self-reactive substances and solid desensitized explosives of class 4.1;

.5 pyrophoric substances of class 4.2;

.6 substances of class 5.2;

.7 substances of class 6.1 with a packing group I vapour inhalation toxicity;

.8 substances of class 6.2; and

.9 materials of class 7.

2.0.3.5 Apart from excepted radioactive material (where the other hazardous properties take precedence), radioactive material having other hazardous properties shall always be classified in class 7, with the greatest of the additional hazards being identified.

2.0.3.6 **Precedence of hazards**

Class and Packing Group	4.2	4.3	5.1 I	5.1 II	5.1 III	6.1, I Dermal	6.1, I Oral	6.1 II	6.1 III	8, I Liquid	8, I Solid	8, II Liquid	8, II Solid	8, III Liquid	8, III Solid
3 I*		4.3				3	3	3	3	3	–	3	–	3	–
3 II*		4.3				3	3	3	3	8	–	3	–	3	–
3 III*		4.3				6.1	6.1	6.1	3†	8	–	8	–	3	–
4.1 II*	4.2	4.3	5.1	4.1	4.1	6.1	6.1	4.1	4.1	–	8	–	4.1	–	4.1
4.1 III*	4.2	4.3	5.1	4.1	4.1	6.1	6.1	6.1	4.1	–	8	–	8	–	4.1
4.2 II		4.3	5.1	4.2	4.2	6.1	6.1	4.2	4.2	8	8	4.2	4.2	4.2	4.2
4.2 III		4.3	5.1	5.1	4.2	6.1	6.1	6.1	4.2	8	8	8	8	4.2	4.2
4.3 I			5.1	4.3	4.3	6.1	4.3	4.3	4.3	4.3	4.3	4.3	4.3	4.3	4.3
4.3 II			5.1	4.3	4.3	6.1	4.3	4.3	4.3	8	8	4.3	4.3	4.3	4.3
4.3 III			5.1	5.1	4.3	6.1	6.1	6.1	4.3	8	8	8	8	4.3	4.3
5.1 I						5.1	5.1	5.1	5.1	5.1	5.1	5.1	5.1	5.1	5.1
5.1 II						6.1	5.1	5.1	5.1	8	8	5.1	5.1	5.1	5.1
5.1 III						6.1	6.1	6.1	5.1	8	8	8	8	5.1	5.1
6.1 I, Dermal										8	6.1	6.1	6.1	6.1	6.1
6.1 I, Oral										8	6.1	6.1	6.1	6.1	6.1
6.1 II, Inhalation										8	6.1	6.1	6.1	6.1	6.1
6.1 II, Dermal										8	6.1	8	6.1	6.1	6.1
6.1 II, Oral										8	8	8	6.1	6.1	6.1
6.1 III										8	8	8	8	8	8

* Substances of class 4.1 other than self-reactive substances and solid desensitized explosives and substances of class 3 other than liquid desensitized explosives.

† 6.1 for pesticides.

– Denotes an impossible combination.

For hazards not shown in this table, see 2.0.3.

2.0.4   Transport of samples

2.0.4.1   When the hazard class of a substance is uncertain and it is being transported for further testing, a tentative hazard class, Proper Shipping Name and identification number shall be assigned on the basis of the consignor's knowledge of the substances and application of:

.1   the classification criteria of this Code; and

.2   the precedence of hazards given in 2.0.3.

The most severe packing group possible for the Proper Shipping Name chosen shall be used.

Where this provision is used, the Proper Shipping Name shall be supplemented with the word ''SAMPLE'' (such as FLAMMABLE LIQUID, N.O.S., SAMPLE). In certain instances, where a specific Proper Shipping Name is provided for a sample of a substance considered to meet certain classification criteria (such as UN 3167, GAS SAMPLE, NON-PRESSURIZED, FLAMMABLE), that Proper Shipping Name shall be used. When an N.O.S. entry is used to transport the sample, the Proper Shipping Name need not be supplemented with the technical name as required by special provision 274.

2.0.4.2   Samples of the substance shall be transported in accordance with the provisions applicable to the tentative assigned Proper Shipping Name provided:

.1   the substance is not considered to be a substance prohibited for transport by 1.1.3;

.2   the substance is not considered to meet the criteria for class 1 or considered to be an infectious substance or a radioactive material;

.3   the substance is in compliance with 2.4.2.3.2.4.2 or 2.5.3.2.5.1 if it is a self-reactive substance or an organic peroxide, respectively;

.4   the sample is transported in a combination packaging with a net mass per package not exceeding 2.5 kg; and

.5   the sample is not packed together with other goods.

Chapter 2.1

Class 1 – Explosives

2.1.0   **Introductory notes** (these notes are not mandatory)

Note 1:   Class 1 is a restricted class, that is, only those explosive substances and articles that are listed in the Dangerous Goods List in chapter 3.2 may be accepted for transport. However, the competent authorities retain the right by mutual agreement to approve transport of explosive substances and articles for special purposes under special conditions. Therefore entries have been included in the Dangerous Goods List for "Substances, explosive, not otherwise specified" and "Articles, explosive, not otherwise specified". It is intended that these entries should only be used when no other method of operation is possible.

Note 2:   General entries such as "Explosive, blasting, type A" are used to allow for the transport of new substances. In preparing these provisions, military ammunition and explosives have been taken into consideration to the extent that they are likely to be transported by commercial carriers.

Note 3:   A number of substances and articles in class 1 are described in appendix B. These descriptions are given because a term may not be well-known or may be at variance with its usage for regulatory purposes.

Note 4:   Class 1 is unique in that the type of packaging frequently has a decisive effect on the hazard and therefore on the assignment to a particular division. The correct division is determined by use of the procedures provided in this chapter.

2.1.1   Definitions and general provisions

2.1.1.1   Class 1 comprises:

> .1   explosive substances (a substance which is not itself an explosive but which can form an explosive atmosphere of gas, vapour or dust is not included in class 1), except those which are too dangerous to transport or those where the predominant hazard is one appropriate to another class;

> .2   explosive articles, except devices containing explosive substances in such quantity or of such a character that their inadvertent or accidental ignition or initiation during transport shall not cause any effect external to the device either by projection, fire, smoke, heat or loud noise; and

> .3   substances and articles not mentioned under .1 and .2 which are manufactured with a view to producing a practical, explosive or pyrotechnic effect.

2.1.1.2   Transport of explosive substances which are unduly sensitive, or so reactive as to be subject to spontaneous reaction, is prohibited.

2.1.1.3   **Definitions**

For the purposes of this Code, the following definitions apply:

> .1   *Explosive substance* means a solid or liquid substance (or a mixture of substances) which is in itself capable by chemical reaction of producing gas at such a temperature and pressure and at such a speed as to cause damage to the surroundings. Pyrotechnic substances are included even when they do not evolve gases.

> .2   *Pyrotechnic substance* means a substance or a mixture of substances designed to produce an effect by heat, light, sound, gas or smoke or a combination of these as the result of non-detonative self-sustaining exothermic chemical reactions.

> .3   *Explosive article* means an article containing one or more explosive substances.

> .4   *Mass explosion* means one which affects almost the entire load virtually instantaneously.

2.1.1.4   Hazard divisions

The six hazard divisions of class 1 are:

Division 1.1   Substances and articles which have a mass explosion hazard

Division 1.2   Substances and articles which have a projection hazard but not a mass explosion hazard

Division 1.3   Substances and articles which have a fire hazard and either a minor blast hazard or a minor projection hazard or both, but not a mass explosion hazard

This division comprises substances and articles:
.1   which give rise to considerable radiant heat; or
.2   which burn one after another, producing minor blast or projection effects or both.

Division 1.4   Substances and articles which present no significant hazard
This division comprises substances and articles which present only a small hazard in the event of ignition or initiation during transport. The effects are largely confined to the package and no projection of fragments of appreciable size or range is to be expected. An external fire must not cause virtually instantaneous explosion of almost the entire contents of the package.

Note: Substances and articles in this division are in compatibility group S if they are so packaged or designed that any hazardous effects arising from the accidental functioning are confined within the package unless the package has been degraded by fire, in which case all blast or projection effects are limited to the extent that they do not significantly hinder fire fighting or other emergency response efforts in the immediate vicinity of the package.

Division 1.5   Very insensitive substances which have a mass explosion hazard
This division comprises substances which have a mass explosion hazard but are so insensitive that there is very little probability of initiation or of transition from burning to detonation under normal conditions of transport.

Note: The probability of transition from burning to detonation is greater when large quantities are transported in a ship. As a consequence, the stowage provisions for explosive substances in division 1.1 and for those in division 1.5 are identical.

Division 1.6   Extremely insensitive articles which do not have a mass explosion hazard
This division comprises articles which contain only extremely insensitive detonating substances and which demonstrate a negligible probability of accidental initiation or propagation.

Note: The risk from articles of division 1.6 is limited to the explosion of a single article.

2.1.1.5   Any substance or article having or suspected of having explosive characteristics shall first be considered for classification in class 1 in accordance with the procedures in 2.1.3. Goods are not classified in class 1 when:

.1   unless specially authorized, the transport of an explosive substance is prohibited because sensitivity of the substance is excessive;

.2   the substance or article comes within the scope of those explosive substances and articles which are specifically excluded from class 1 by the definition of this class; or

.3   the substance or article has no explosive properties.

2.1.2   Compatibility groups and classification codes

2.1.2.1   Goods of class 1 are considered to be "compatible" if they can be safely stowed or transported together without significantly increasing either the probability of an accident or, for a given quantity, the magnitude of the effects of such an accident. By this criterion, goods listed in this class have been divided into a number of compatibility groups, each denoted by a letter from A to L (excluding I), N and S. These are described in 2.1.2.2 and 2.1.2.3.

2.1.2.2   Compatibility groups and classification codes

Description of substance or articles to be classified	Compatibility group	Classification code
Primary explosive substance	A	1.1A
Article containing a primary explosive substance and not containing two or more effective protective features. Some articles, such as detonators for blasting, detonator assemblies for blasting and primers, cap-type, are included even though they do not contain primary explosives	B	1.1B 1.2B 1.4B
Propellant explosive substance or other deflagrating explosive substance or article containing such explosive substance	C	1.1C 1.2C 1.3C 1.4C
Secondary detonating explosive substance or black powder or article containing a secondary detonating explosive substance, in each case without means of initiation and without a propelling charge, or article containing a primary explosive substance and containing two or more effective protective features	D	1.1D 1.2D 1.4D 1.5D
Article containing a secondary detonating explosive substance, without means of initiation, with a propelling charge (other than one containing a flammable liquid or gel or hypergolic liquids)	E	1.1E 1.2E 1.4E
Article containing a secondary detonating explosive substance with its own means of initiation, with a propelling charge (other than one containing a flammable liquid or gel or hypergolic liquids) or without a propelling charge	F	1.1F 1.2F 1.3F 1.4F
Pyrotechnic substance, or article containing a pyrotechnic substance, or article containing both an explosive substance and an illuminating, incendiary, tear- or smoke-producing substance (other than a water-activated article or one containing white phosphorus, phosphides, a pyrophoric substance, a flammable liquid or gel, or hypergolic liquids)	G	1.1G 1.2G 1.3G 1.4G
Article containing both an explosive substance and white phosphorus	H	1.2H 1.3H
Article containing both an explosive substance and a flammable liquid or gel	J	1.1J 1.2J 1.3J
Article containing both an explosive substance and a toxic chemical agent	K	1.2K 1.3K
Explosive substance or article containing an explosive substance and presenting a special risk (such as due to water-activation or presence of hypergolic liquids, phosphides or a pyrophoric substance) and needing isolation of each type (see 7.2.7.2.1.4, Note 2)	L	1.1L 1.2L 1.3L
Articles containing only extremely insensitive detonating substances	N	1.6N
Substance or article so packaged or designed that any hazardous effects arising from accidental functioning are confined within the package unless the package has been degraded by fire, in which case all blast or projection effects are limited to the extent that they do not significantly hinder or prohibit fire fighting or other emergency response efforts in the immediate vicinity of the package	S	1.4S

2.1.2.3   Scheme of classification of explosives, combination of hazard division with compatibility group

Hazard division	Compatibility group													Σ A–S
	A	B	C	D	E	F	G	H	J	K	L	N	S	
1.1	1.1A	1.1B	1.1C	1.1D	1.1E	1.1F	1.1G		1.1J		1.1L			9
1.2		1.2B	1.2C	1.2D	1.2E	1.2F	1.2G	1.2H	1.2J	1.2K	1.2L			10
1.3			1.3C			1.3F	1.3G	1.3H	1.3J	1.3K	1.3L			7
1.4		1.4B	1.4C	1.4D	1.4E	1.4F	1.4G						1.4S	7
1.5				1.5D										1
1.6												1.6N		1
Σ1.1–1.6	1	3	4	4	3	4	4	2	3	2	3	1	1	35

2.1.2.4   The definitions of compatibility groups in 2.1.2.2 are intended to be mutually exclusive, except for a substance or article which qualifies for compatibility group S. Since the criterion of compatibility group S is an empirical one, assignment to this group is necessarily linked to the tests for assignment to division 1.4.

2.1.3   Classification procedure

2.1.3.1   Any substance or article having or suspected of having explosive characteristics shall be considered for classification in class 1. Substances and articles classified in class 1 shall be assigned to the appropriate division and compatibility group. Goods of class 1 shall be classified in accordance with the latest version of the United Nations *Manual of Tests and Criteria.*

2.1.3.2   Prior to transport, the classification of all explosive substances and articles, together with the compatibility group assignment and the Proper Shipping Name under which the substance or article is to be transported, shall have been approved by the competent authority of the country of manufacture. A new approval would be required for:

.1   a new explosive substance; or

.2   a new combination or mixture of explosive substances which is significantly different from other combinations or mixtures previously manufactured and approved; or

.3   a new design of an explosive article, an article containing a new explosive substance, or an article containing a new combination or mixture of explosive substances; or

.4   an explosive substance or article with a new design or type of packaging, including a new type of inner packaging.

2.1.3.3   Assessment of the hazard division is usually made on the basis of test results. A substance or article shall be assigned to the hazard division which corresponds to the results of the tests to which the substance or article, as offered for transport, has been subjected. Other test results, and data assembled from accidents which have occurred, may also be taken into account.

2.1.3.4   The competent authority may exclude an article or substance from class 1 by virtue of test results and the class 1 definition.

2.1.3.5   Assignment of fireworks to hazard divisions

2.1.3.5.1   Fireworks shall normally be assigned to hazard divisions 1.1, 1.2, 1.3, and 1.4 on the basis of test data derived from Test Series 6 of the United Nations *Manual of Tests and Criteria*. However, since the range of such articles is very extensive and the availability of test facilities may be limited, assignment to hazard divisions may also be made in accordance with the procedure in 2.1.3.5.2.

2.1.3.5.2   Assignment of fireworks to UN Nos. 0333, 0334, 0335 or 0336 may be made on the basis of analogy, without the need for Test Series 6 testing, in accordance with the default fireworks classification table in 2.1.3.5.5. Such assignment shall be made with the agreement of the competent authority. Items not specified in the table shall be classified on the basis of test data derived from Test Series 6 of the United Nations *Manual of Tests and Criteria.*

Note: The addition of other types of fireworks to column 1 of the table in 2.1.3.5.5 shall only be made on the basis of full test data submitted to the UN Sub-Committee of Experts on the Transport of Dangerous Goods for consideration.

2.1.3.5.3   Where fireworks of more than one hazard division are packed in the same package they shall be classified on the basis of the highest hazard division unless test data derived from Test Series 6 of the United Nations *Manual of Tests and Criteria* indicate otherwise.

2.1.3.5.4   The classification shown in the table in 2.1.3.5.5 applies only for articles packed in fibreboard boxes (4G).

2.1.3.5.5 Default fireworks classification table*

Note 1: References to percentages in the table, unless otherwise stated, are to the mass of all pyrotechnic composition (e.g., rocket motors, lifting charge, bursting charge and effect charge).

Note 2: "Flash composition" in this table refers to pyrotechnic compositions in powder form or as pyrotechnic units as presented in the fireworks, that are used to produce an aural effect, or used as a bursting charge or lifting charge, unless the time taken for the pressure rise is demonstrated to be more than 8 ms for 0.5 g of pyrotechnic composition in Test Series 2(c) (i) "Time/pressure test" of the UN *Manual of Tests and Criteria*.

Note 3: Dimensions in mm refers to:

– for spherical and peanut shells, the diameter of the sphere of the shell;

– for cylinder shells, the length of the shell;

– for a shell in mortar, Roman candle, shot tube firework or mine, the inside diameter of the tube comprising or containing the firework;

– for a bag mine or cylinder mine, the inside diameter of the mortar intended to contain the mine.

* This table contains a list of firework classifications that may be used in the absence of Test Series 6, of the United Nations *Manual of Tests and Criteria*, data (see 2.1.3.5.2).

Type	Includes: / Synonym:	Definition	Specification	Classification
Shell, spherical or cylindrical	Spherical display shell: aerial shell, colour shell, dye shell, multi-break shell, multi-effect shell, nautical shell, parachute shell, smoke shell, star shell; report shell: maroon, salute, sound shell, thunderclap, aerial shell kit	Device with or without propellant charge, with delay fuse and bursting charge, pyrotechnic unit(s) or loose pyrotechnic composition and designed to be projected from a mortar	All report shells	1.1G
			Colour shell: ⩾ 180 mm	1.1G
			Colour shell: < 180 mm with > 25% flash composition, as loose powder and/or report effects	1.1G
			Colour shell: < 180 mm with ⩽ 25% flash composition, as loose powder and/or report effects	1.3G
			Colour shell: ⩽ 50 mm, or ⩽ 60 g pyrotechnic composition, with ⩽ 2% flash composition as loose powder and/or report effects	1.4G
	Peanut shell	Device with two or more spherical aerial shells in a common wrapper propelled by the same propellant charge with separate external delay fuses	The most hazardous spherical aerial shell determines the classification	
	Preloaded mortar, shell in mortar	Assembly comprising a spherical or cylindrical shell inside a mortar from which the shell is designed to be projected	All report shells	1.1G
			Colour shell: ⩾ 180 mm	1.1G
			Colour shell: > 25% flash composition as loose powder and/or report effects	1.1G
			Colour shell: > 50 mm and < 180 mm	1.2G
			Colour shell: ⩽ 50 mm, or < 60 g pyrotechnic composition, with ⩽ 25% flash composition as loose powder and/or report effects	1.3G

Type	Includes: / Synonym:	Definition	Specification	Classification
	Shell of shells (spherical) (Reference to percentages for shell of shells are to the gross mass of the fireworks article)	Device without propellant charge, with delay fuse and bursting charge, containing report shells and inert materials and designed to be projected from a mortar	> 120 mm	1.1G
		Device without propellant charge, with delay fuse and bursting charge, containing report shells ≤ 25 g flash composition per report unit, with ≤ 33% flash composition and ≥ 60% inert materials and designed to be projected from a mortar	≤ 120 mm	1.3G
		Device without propellant charge, with delay fuse and bursting charge, containing colour shells and/or pyrotechnic units and designed to be projected from a mortar	> 300 mm	1.1G
		Device without propellant charge, with delay fuse and bursting charge, containing colour shells ≤ 70 mm and/or pyrotechnic units, with ≤ 25% flash composition and ≤ 60% pyrotechnic composition and designed to be projected from a mortar	> 200 mm and ≤ 300 mm	1.3G
		Device with propellant charge, with delay fuse and bursting charge, containing colour shells ≤ 70 mm and/or pyrotechnic units, with ≤ 25% flash composition and ≤ 60% pyrotechnic composition and designed to be projected from a mortar	≤ 200 mm	1.3G
Battery/ combination	Barrage, bombardos, cakes, finale box, flowerbed, hybrid, multiple tubes, shell cakes, banger batteries, flash banger batteries	Assembly including several elements either containing the same type or several types each corresponding to one of the types of fireworks listed in this table, with one or two points of ignition	The most hazardous firework type determines the classification	
Roman candle	Exhibition candle, candle, bombettes	Tube containing a series of pyrotechnic units consisting of alternate pyrotechnic composition, propellant charge, and transmitting fuse	≥ 50 mm inner diameter, containing flash composition, or <50 mm with >25% flash composition	1.1G
			≥ 50 mm inner diameter, containing no flash composition	1.2G
			< 50 mm inner diameter and ≤ 25% flash composition	1.3G
			≤ 30 mm inner diameter, each pyrotechnic unit ≤ 25 g and ≤ 5% flash composition	1.4G

Type	Includes: / Synonym:	Definition	Specification	Classification
Shot tube	Single shot Roman candle, small preloaded mortar	Tube containing a pyrotechnic unit consisting of pyrotechnic composition, propellant charge with or without transmitting fuse	⩽ 30 mm inner diameter and pyrotechnic unit > 25 g, or > 5% and ⩽ 25% flash composition	1.3G
			⩽ 30 mm inner diameter, pyrotechnic unit ⩽ 25 g and ⩽ 5% flash composition	1.4G
Rocket	Avalanche rocket, signal rocket, whistling rocket, bottle rocket, sky rocket, missile type rocket, table rocket	Tube containing pyrotechnic composition and/or pyrotechnic units, equipped with stick(s) or other means for stabilization of flight, and designed to be propelled into the air	Flash composition effects only	1.1G
			Flash composition > 25% of the pyrotechnic composition	1.1G
			> 20 g pyrotechnic composition and flash composition ⩽ 25 %	1.3G
			⩽ 20 g pyrotechnic composition, black powder bursting charge and ⩽ 0.13 g flash composition per report and ⩽ 1 g in total	1.4G
Mine	Pot-a-feu, ground mine, bag mine, cylinder mine	Tube containing propellant charge and pyrotechnic units and designed to be placed on the ground or to be fixed in the ground. The principal effect is ejection of all the pyrotechnic units in a single burst producing a widely dispersed visual and/or aural effect in the air or: Cloth or paper bag or cloth or paper cylinder containing propellant charge and pyrotechnic units, designed to be placed in a mortar and to function as a mine	> 25% flash composition, as loose powder and/or report effects	1.1G
			⩾ 180 mm and ⩽ 25% flash composition, as loose powder and/or report effects	1.1G
			< 180 mm and ⩽ 25% flash composition, as loose powder and/or report effects	1.3G
			⩽ 150 g pyrotechnic composition, containing ⩽ 5% flash composition as loose powder and/or report effects. Each pyrotechnic unit ⩽ 25 g, each report effect < 2 g ; each whistle, if any, ⩽ 3 g	1.4G
Fountain	Volcanos, gerbs, showers, lances, Bengal fire, flitter sparkle, cylindrical fountains, cone fountains, illuminating torch	Non-metallic case containing pressed or consolidated pyrotechnic composition producing sparks and flame	⩾ 1 kg pyrotechnic composition	1.3G
			< 1 kg pyrotechnic composition	1.4G
Sparkler	Handheld sparklers, non-handheld sparklers, wire sparklers	Rigid wire partially coated (along one end) with slow-burning pyrotechnic composition with or without an ignition tip	Perchlorate based sparklers: > 5 g per item or > 10 items per pack	1.3G
			Perchlorate based sparklers: ⩽ 5 g per item and ⩽ 10 items per pack Nitrate based sparklers: ⩽ 30 g per item	1.4G
Bengal stick	Dipped stick	Non-metallic stick partially coated (along one end) with slow-burning pyrotechnic composition and designed to be held in the hand	Perchlorate based items: > 5 g per item or > 10 items per pack	1.3 G
			Perchlorate based items: ⩽ 5 g per item and ⩽ 10 items per pack; nitrate based items: ⩽ 30 g per item	1.4G

Type	Includes: / Synonym:	Definition	Specification	Classification
Low hazard fireworks and novelties	Table bombs, throwdowns, crackling granules, smokes, fog, snakes, glow worm, serpents, snaps, party poppers	Device designed to produce very limited visible and/or audible effect which contains small amounts of pyrotechnic and/or explosive composition	Throwdowns and snaps may contain up to 1.6 mg of silver fulminate; snaps and party poppers may contain up to 16 mg of potassium chlorate/red phosphorus mixture; other articles may contain up to 5 g of pyrotechnic composition, but no flash composition	1.4G
Spinner	Aerial spinner, helicopter, chaser, ground spinner	Non-metallic tube or tubes containing gas- or spark-producing pyrotechnic composition, with or without noise producing composition, with or without aerofoils attached	Pyrotechnic composition per item > 20 g, containing ≤ 3% flash composition as report effects, or whistle composition ≤ 5 g	1.3G
			Pyrotechnic composition per item ≤ 20 g, containing ≤ 3% flash composition as report effects, or whistle composition ≤ 5 g	1.4G
Wheels	Catherine wheels, Saxon	Assembly including drivers containing pyrotechnic composition and provided with a means of attaching it to a support so that it can rotate	≥ 1 kg total pyrotechnic composition, no report effect, each whistle (if any) ≤ 25 g and ≤ 50 g whistle composition per wheel	1.3G
			< 1 kg total pyrotechnic composition, no report effect, each whistle (if any) ≤ 5 g and ≤ 10 g whistle composition per wheel	1.4G
Aerial wheel	Flying Saxon, UFOs, rising crown	Tubes containing propellant charges and sparks-, flame- and/or noise-producing pyrotechnic compositions, the tubes being fixed to a supporting ring	> 200 g total pyrotechnic composition or > 60 g pyrotechnic composition per driver, < 3% flash composition as report effects, each whistle (if any) ≤ 25 g and ≤ 50 g whistle composition per wheel	1.3G
			≤ 200 g total pyrotechnic composition and ≤ 60 g pyrotechnic composition per driver, ≤ 3% flash composition as report effects, each whistle (if any) ≤ 5 g and ≤ 10 g whistle composition per wheel	1.4G
Selection pack	Display selection box, display selection pack, garden selection box, indoor selection box; assortment	A pack of more than one type each corresponding to one of the types of fireworks listed in this table	The most hazardous firework type determines the classification	
Firecracker	Celebration cracker, celebration roll, string cracker	Assembly of tubes (paper or cardboard) linked by a pyrotechnic fuse, each tube intended to produce an aural effect	Each tube ≤ 140 mg of flash composition or ≤ 1 g black powder	1.4G
Banger	Salute, flash banger, lady cracker	Non-metallic tube containing report composition intended to produce an aural effect	> 2 g flash composition per item	1.1G
			≤ 2 g flash composition per item and ≤ 10 g per inner packaging	1.3G
			≤ 1 g flash composition per item and ≤ 10 g per inner packaging or ≤ 10 g black powder per item	1.4G

2

Chapter 2.2

Class 2 – Gases

2.2.0 Introductory notes

Note 1: "Toxic" has the same meaning as "poisonous".

Note 2: Carbonated beverages are not subject to the provisions of this Code.

2.2.1 Definitions and general provisions

2.2.1.1 A gas is a substance which:

.1 at 50°C has a vapour pressure greater than 300 kPa; or

.2 is completely gaseous at 20°C at a standard pressure of 101.3 kPa.

2.2.1.2 The transport condition of a gas is described according to its physical state as:

.1 *Compressed gas:* a gas which when packaged under pressure for transport is entirely gaseous at –50°C; this category includes all gases with a critical temperature less than or equal to –50°C;

.2 *Liquefied gas:* a gas which when packaged under pressure for transport is partially liquid at temperatures above –50 °C. A distinction is made between:

high pressure liquefied gas: a gas with a critical temperature between –50°C and +65°C, and

low pressure liquefied gas: a gas with a critical temperature above +65°C;

.3 *Refrigerated liquefied gas*: a gas which when packaged for transport is made partially liquid because of its low temperature; or

.4 *Dissolved gas:* a gas which when packaged under pressure for transport is dissolved in a liquid phase solvent.

2.2.1.3 The class comprises compressed gases, liquefied gases, dissolved gases, refrigerated liquefied gases, mixtures of one or more gases with one or more vapours of substances of other classes, articles charged with a gas and aerosols.

2.2.1.4 Gases are normally transported under pressure varying from high pressure in the case of compressed gases to low pressure in the case of refrigerated gases.

2.2.1.5 According to their chemical properties or physiological effects, which may vary widely, gases may be: flammable; non-flammable; non-toxic; toxic; supporters of combustion; corrosive; or may possess two or more of these properties simultaneously.

2.2.1.5.1 Some gases are chemically and physiologically inert. Such gases as well as other gases, normally accepted as non-toxic, will nevertheless be suffocating in high concentrations.

2.2.1.5.2 Many gases of this class have narcotic effects which may occur at comparatively low concentrations or may evolve highly toxic gases when involved in a fire.

2.2.1.5.3 All gases which are heavier than air will present a potential danger if allowed to accumulate in the bottom of cargo spaces.

2.2.2 Class subdivisions

Class 2 is subdivided further according to the primary hazard of the gas during transport:

Note: For UN 1950 AEROSOLS, see also the criteria in special provision 63 and for UN 2037 RECEPTACLES, SMALL, CONTAINING GAS (GAS CARTRIDGES) see also special provision 303.

2.2.2.1 **Class 2.1 Flammable gases**

Gases which at 20°C and a standard pressure of 101.3 kPa:

.1 are ignitable when in a mixture of 13% or less by volume with air; or

.2 have a flammable range with air of at least 12 percentage points regardless of the lower flammable limit. Flammability shall be determined by tests or calculation in accordance with methods adopted by the

International Organization for Standardization (see ISO Standard 10156:1996). Where insufficient data are available to use these methods, tests by a comparable method recognized by a national competent authority may be used.

2.2.2.2   **Class 2.2   Non-flammable, non-toxic gases**

Gases which:

.1   are asphyxiant – gases which dilute or replace the oxygen normally in the atmosphere; or

.2   are oxidizing – gases which may, generally by providing oxygen, cause or contribute to the combustion of other material more than air does. The oxidizing ability shall be determined by tests or by calculation in accordance with methods adopted by ISO (see ISO 10156:1996 and ISO 10156-2:2005); or

.3   do not come under the other classes.

2.2.2.3   **Class 2.3   Toxic gases**

Gases which:

.1   are known to be so toxic or corrosive to humans as to pose a hazard to health; or

.2   are presumed to be toxic or corrosive to humans because they have a LC_{50} value (as defined in 2.6.2.1) equal to or less than 5,000 $m\ell/m^3$ (ppm).

Note: Gases meeting the above criteria owing to their corrosivity are to be classified as toxic with a subsidiary corrosive risk.

2.2.2.4   Gases and gas mixtures with hazards associated with more than one division take the following precedence:

.1   class 2.3 takes precedence over all other classes;

.2   class 2.1 takes precedence over class 2.2.

2.2.2.5   Gases of class 2.2 are not subject to the provisions of this Code if they are transported at a pressure of less than 200 kPa at 20°C and are not liquefied or refrigerated liquefied gases.

2.2.3   Mixtures of gases

For the classification of gas mixtures (including vapours of substances from other classes), the following principles shall be used:

.1   Flammability shall be determined by tests or calculation in accordance with methods adopted by the International Organization for Standardization (see ISO Standard 10156:1996). Where insufficient data are available to use these methods, tests by a comparable method recognized by a national competent authority may be used.

.2   The level of toxicity is determined either by tests to measure the LC_{50} value (as defined in 2.6.2.1) or by a calculation method using the following formula:

$$LC_{50} \quad \text{Toxic (mixture)} = \frac{1}{\sum_{i=1}^{n} \frac{f_i}{T_i}}$$

where: f_i =   mole fraction of the ith component substance of the mixture;
          T_i =   toxicity index of the ith component substance of the mixture (the T_i equals the LC_{50} value when available).

When LC_{50} values are unknown, the toxicity index is determined by using the lowest LC_{50} value of substances of similar physiological and chemical effects, or through testing if this is the only practical possibility.

.3   A gas mixture has a subsidiary risk of corrosivity when the mixture is known by human experience to be destructive to the skin, eyes or mucous membranes or when the LC_{50} value of the corrosive components of the mixture is equal to or less than 5,000 $m\ell/m^3$ (ppm) when the LC_{50} is calculated by the formula:

$$LC_{50} \quad \text{Corrosive (mixture)} = \frac{1}{\sum_{i=1}^{n} \frac{f_{ci}}{T_{ci}}}$$

where: f_{ci} =   mole fraction of the ith corrosive component substance of the mixture;
          T_{ci} =   toxicity index of the ith corrosive component substance of the mixture (the T_{ci} equals the LC_{50} value when available).

.4   Oxidizing ability is determined either by tests or by calculation methods adopted by the International Organization for Standardization (see ISO 10156:1996 and ISO 10156-2:2005).

Chapter 2.3

Class 3 – Flammable liquids

2.3.0 Introductory note

The flashpoint of a flammable liquid may be altered by the presence of an impurity. The substances listed in class 3 in the Dangerous Goods List in chapter 3.2 shall generally be regarded as chemically pure. Since commercial products may contain added substances or impurities, flashpoints may vary, and this may have an effect on classification or determination of the packing group for the product. In the event of doubt regarding the classification or packing group of a substance, the flashpoint of the substance shall be determined experimentally.

2.3.1 Definitions and general provisions

2.3.1.1 Class 3 includes the following substances:

.1 flammable liquids (see 2.3.1.2 and 2.3.1.3);

.2 liquid desensitized explosives (see 2.3.1.4).

2.3.1.2 *Flammable liquids* are liquids, or mixtures of liquids, or liquids containing solids in solution or suspension (such as paints, varnishes, lacquers, etc., but not including substances which, on account of their other dangerous characteristics, have been included in other classes) which give off a flammable vapour at or below 60°C closed-cup test (corresponding to 65.6°C open-cup test), normally referred to as the "flashpoint". This also includes:

.1 liquids offered for transport at temperatures at or above their flashpoint; and

.2 substances transported or offered for transport at elevated temperatures in a liquid state, which give off a flammable vapour at temperatures equal to or below the maximum transport temperature.

2.3.1.3 However, the provisions of this Code need not apply to such liquids with a flashpoint of more than 35°C which do not sustain combustion. Liquids are considered to be unable to sustain combustion for the purposes of the Code if:

.1 they have passed the suitable combustibility test (see the Sustained Combustibility Test prescribed in part III, 32.5.2 of the United Nations *Manual of Tests and Criteria*); or

.2 their fire point according to ISO 2592:1973 is greater than 100°C; or

.3 they are water-miscible solutions with a water content of more than 90%, by mass.

2.3.1.4 *Liquid desensitized explosives* are explosive substances which are dissolved or suspended in water or other liquid substances, to form a homogeneous liquid mixture to suppress their explosive properties. Entries in the Dangerous Goods List for liquid desensitized explosives are UN 1204, UN 2059, UN 3064, UN 3343, UN 3357 and UN 3379.

2.3.2 Assignment of packing group

2.3.2.1 The criteria in 2.3.2.6 are used to determine the hazard grouping of a liquid that presents a risk due to flammability.

2.3.2.1.1 For liquids whose only risk is flammability, the packing group for the substance is the hazard grouping shown in 2.3.2.6.

2.3.2.1.2 For a liquid with additional risk(s), the hazard group determined from 2.3.2.6 and the hazard group based on the severity of the additional risk(s) shall be considered, and the classification and packing group determined in accordance with the provisions in chapter 2.0.

2.3.2.2 Viscous substances such as paints, enamels, lacquers, varnishes, adhesives and polishes having a flashpoint of less than 23°C may be placed in packing group III in conformity with the procedures prescribed in part III, chapter 32.3, of the United Nations *Manual of Tests and Criteria* on the basis of:

.1 the viscosity, expressed as the flowtime in seconds;

.2   the closed-cup flashpoint;

.3   a solvent separation test.

2.3.2.3   Viscous flammable liquids such as paints, enamels, varnishes, adhesives and polishes with a flashpoint of less than 23°C are included in packing group III provided that:

.1   less than 3% of the clear solvent layer separates in the solvent separation test;

.2   the mixture or any separated solvent does not meet the criteria for class 6.1 or class 8.

.3   the viscosity and flashpoint are in accordance with the following table:

Flow time t in seconds	Jet diameter in mm	Flashpoint in °C c.c.
$20 < t \leqslant 60$	4	above 17
$60 < t \leqslant 100$	4	above 10
$20 < t \leqslant 32$	6	above 5
$32 < t \leqslant 44$	6	above −1
$44 < t \leqslant 100$	6	above −5
$100 < t$	6	−5 and below

.4   the capacity of the receptacle used does not exceed 30 ℓ.

2.3.2.4   Substances classified as flammable liquids due to their being transported or offered for transport at elevated temperatures are included in packing group III.

2.3.2.5   Viscous substances which:

– have a flashpoint of 23°C or above and less than or equal to 60°C;

– are not toxic, corrosive or environmentally hazardous;

– contain not more than 20% nitrocellulose, provided the nitrocellulose contains not more than 12.6% nitrogen by dry mass; and

– are packed in receptacles not exceeding 30 ℓ capacity

are not subject to the provisions for the marking, labelling and testing of packages in chapters 4.1, 5.2 and 6.1, if:

.1   in the solvent separation test (see part III, 32.5.1 of the United Nations *Manual of Tests and Criteria*) the height of the separated layer of solvent is less than 3% of the total height; and

.2   the flowtime in the viscosity test (see part III, 32.4.3 of the United Nations *Manual of Tests and Criteria*) with a jet diameter of 6 mm is equal to or greater than:

.1   60 s; or

.2   40 s if the viscous substance contains not more than 60% of class 3 substances.

The following statement shall be included in the transport document: "Transport in accordance with 2.3.2.5 of the IMDG Code." (see 5.4.1.5.10).

2.3.2.6   **Hazard grouping based on flammability**

Flammable liquids are grouped for packing purposes according to their flashpoint, their boiling point, and their viscosity. This table shows the relationship between two of these characteristics.

Packing group	Flashpoint in °C closed cup (c.c.)	Initial boiling point in °C
I	–	$\leqslant 35$
II	< 23	> 35
III	$\geqslant 23$ to $\leqslant 60$	> 35

2.3.3   Determination of flashpoint

Note: The provisions of this section are not mandatory.

2.3.3.1   The flashpoint of a flammable liquid is the lowest temperature of the liquid at which its vapour forms an ignitable mixture with air. It gives a measure of the risk of formation of explosive or ignitable mixtures when the liquid escapes from its packing. A flammable liquid cannot be ignited so long as its temperature remains below the flashpoint.

2

Note: Do not confuse the flashpoint with the ignition temperature, which is the temperature to which an explosive vapour–air mixture must be heated to cause actual explosion. There is no relationship between the flashpoint and the ignition temperature.

2.3.3.2   The flashpoint is not an exact physical constant for a given liquid. It depends to some extent on the construction of the test apparatus used and on the testing procedure. Therefore, when providing flashpoint data, specify the name of the test apparatus.

2.3.3.3   Several standard apparatuses are in current use. They all operate on the same principle: a specified quantity of the liquid is introduced into a receptacle at a temperature well below the flashpoint to be expected, then slowly heated; periodically, a small flame is brought near to the surface of the liquid. The flashpoint is the lowest temperature at which a "flash" is observed.

2.3.3.4   The test methods can be divided into two groups, depending on the use in an apparatus of an open receptacle (open-cup methods) or a closed one which is only opened to admit the flame (closed-cup methods). As a rule, the flashpoints found in an open-cup test are a few degrees higher than in a closed-cup test.

2.3.3.5   In general, reproducibility in closed-cup apparatus is better than in open-cup.

2.3.3.5.1   It is therefore recommended that flashpoints, especially in the range around 23°C, shall be determined by means of closed-cup (c.c) methods.

2.3.3.5.2   Flashpoint data in this Code are generally based on closed-cup methods. In countries where it is customary to determine flashpoints by the open-cup method, the temperatures given by that method would need to be reduced to correspond with those in this Code.

2.3.3.6   The following list of documents describe methods used in certain countries to determine the flashpoint of substances in class 3:

France   (Association française de normalisation, AFNOR,
Tour Europe, 92049 Paris La Défense):
French Standard NF M 07-019
French Standards NF M 07-011 / NF T 30-050 / NF T 66-009
French Standard NF M 07-036

Germany   (Deutsches Institut für Normung,
Burggrafenstr. 6, D-10787 Berlin):
Standard DIN 51755 (flashpoints below 65°C)
Standard DIN EN 22719 (flashpoints above 5°C)
Standard DIN 53213 (for varnishes, lacquers and similar viscous liquids
with flashpoints below 65°C)

Netherlands   ASTM D93-96
ASTM D3278-96
ISO 1516
ISO 1523
ISO 3679
ISO 3680

Russian Federation   (State Committee of the Council of Ministers for Standardization,
113813, GSP, Moscow, M-49, Leninsky Prospect, 9):
GOST 12.1.044-84

United Kingdom   (British Standards Institution, Linford Wood, Milton Keynes, MK14 6LE):
British Standard BS EN 22719
British Standard BS 2000 Part 170

United States of America   (American Society for Testing and Materials,
1916 Race Street, Philadelphia, PA 19103):
ASTM D 3828-93, Standard Test Methods for Flash Point by Small Scale Closed Tester
ASTM D 56-93, Standard Test Method for Flash Point by Tag Closed Tester
ASTM D 3278-96, Standard Test Methods for Flash Point of Liquids by Setaflash
    Closed Cup Apparatus
ASTM D 0093-96, Standard Test Methods for Flash Point by Pensky–Martens
    Closed Cup Tester

Chapter 2.4

Class 4 – Flammable solids; substances liable to spontaneous combustion; substances which, in contact with water, emit flammable gases

2.4.0 Introductory note

Since organometallic substances can be classified in classes 4.2 or 4.3 with additional subsidiary risks, depending on their properties, a specific classification flowchart for these substances is given in 2.4.5.

2.4.1 Definition and general provisions

2.4.1.1 In this Code, class 4 deals with substances, other than those classified as explosives, which, under conditions of transport, are readily combustible or may cause or contribute to a fire. Class 4 is subdivided as follows:

Class 4.1 – Flammable solids
Solids which, under conditions encountered in transport, are readily combustible or may cause or contribute to fire through friction; self-reactive substances (solids and liquids) which are liable to undergo a strongly exothermic reaction; solid desensitized explosives which may explode if not diluted sufficiently;

Class 4.2 – Substances liable to spontaneous combustion
Substances (solids and liquids) which are liable to spontaneous heating under normal conditions encountered in transport, or to heating up in contact with air, and being then liable to catch fire;

Class 4.3 – Substances which, in contact with water, emit flammable gases
Substances (solids and liquids) which, by interaction with water, are liable to become spontaneously flammable or to give off flammable gases in dangerous quantities.

2.4.1.2 As referenced in this chapter, test methods and criteria, with advice on application of the tests, are given in the United Nations *Manual of Tests and Criteria* for the classification of following types of substances of class 4:

.1 flammable solids (class 4.1);

.2 self-reactive substances (class 4.1);

.3 pyrophoric solids (class 4.2);

.4 pyrophoric liquids (class 4.2);

.5 self-heating substances (class 4.2); and

.6 substances which, in contact with water, emit flammable gases (class 4.3).

Test methods and criteria for self-reactive substances are given in part II of the United Nations *Manual of Tests and Criteria*, and test methods and criteria for the other types of substances of class 4 are given in the United Nations *Manual of Tests and Criteria*, part III, chapter 33.

2.4.2 Class 4.1 – Flammable solids, self-reactive substances and solid desensitized explosives

2.4.2.1 General

Class 4.1 includes the following types of substances:

.1 flammable solids (see 2.4.2.2);

.2 self-reactive substances (see 2.4.2.3); and

.3 solid desensitized explosives (see 2.4.2.4).

Some substances (such as celluloid) may evolve toxic and flammable gases when heated or if involved in a fire.

2.4.2.2 **Class 4.1 Flammable solids**

2.4.2.2.1 *Definitions and properties*

2.4.2.2.1.1 For the purpose of this Code, *flammable solids* means readily combustible solids and solids which may cause fire through friction.

2.4.2.2.1.2 *Readily combustible solids* means fibres, powdered, granular, or pasty substances which are dangerous if they can be easily ignited by brief contact with an ignition source such as a burning match, and if the flame spreads rapidly. The danger may come not only from the fire but also from toxic combustion products. Metal powders are especially dangerous because of the difficulty of extinguishing a fire, since normal extinguishing agents such as carbon dioxide or water can increase the hazard.

2.4.2.2.2 *Classification of flammable solids*

2.4.2.2.2.1 Powdered, granular or pasty substances shall be classified as readily combustible solids of class 4.1 when the time of burning of one or more of the test runs, performed in accordance with the test method described in the United Nations *Manual of Tests and Criteria*, part III, 33.2.1, is less than 45 s or the rate of burning is more than 2.2 mm/s. Powders of metals or metal alloys shall be classified in class 4.1 when they can be ignited and the reaction spreads over the whole length of the sample in 10 minutes or less.

2.4.2.2.2.2 Solids which may cause fire through friction shall be classified in class 4.1 by analogy with existing entries (such as matches) until definitive criteria are established.

2.4.2.2.3 *Assignment of packing groups*

2.4.2.2.3.1 Packing groups are assigned on the basis of the test methods referred to in 2.4.2.2.2.1. For readily combustible solids (other than metal powders), packing group II shall be assigned if the burning time is less than 45 s and the flame passes the wetted zone. Packing group II shall be assigned to powders of metal or metal alloys if the zone of reaction spreads over the whole length of the sample in five minutes or less.

2.4.2.2.3.2 Packing groups are assigned on the basis of the test methods referred to in 2.4.2.2.2.1. For readily combustible solids (other than metal powders), packing group III shall be assigned if the burning time is less than 45 s and the wetted zone stops the flame propagation for at least four minutes. Packing group III shall be assigned to metal powders if the reaction spreads over the whole length of the sample in more than five minutes but not more than ten minutes.

2.4.2.2.3.3 For solids which may cause fire through friction, the packing group shall be assigned by analogy with existing entries or in accordance with any appropriate special provision.

2.4.2.2.4 Pyrophoric metal powders, if wetted with sufficient water to suppress their pyrophoric properties, may be classified as class 4.1.

2.4.2.3 **Class 4.1 Self-reactive substances**

2.4.2.3.1 *Definitions and properties*

2.4.2.3.1.1 For the purposes of this Code:

Self-reactive substances are thermally unstable substances liable to undergo a strongly exothermic decomposition even without participation of oxygen (air). Substances are not considered to be self-reactive substances of class 4.1, if:

.1 they are explosives according to the criteria of class 1;

.2 they are oxidizing substances according to the classification procedure for class 5.1 (see 2.5.2) except that mixtures of oxidizing substances which contain 5.0% or more of combustible organic substances shall be subjected to the classification procedure defined in Note 3;

.3 they are organic peroxides according to the criteria of class 5.2;

.4 their heat of decomposition is less than 300 J/g; or

.5 their self-accelerating decomposition temperature (SADT) (see 2.4.2.3.4) is greater than 75°C for a 50 kg package.

Note 1: The heat of decomposition may be determined using any internationally recognized method such as differential scanning calorimetry and adiabatic calorimetry.

Note 2: Any substance which shows the properties of a self-reactive substance shall be classified as such, even if this substance gives a positive test result according to 2.4.3.2 for inclusion in class 4.2.

Note 3: Mixtures of oxidizing substances meeting the criteria of class 5.1 which contain 5.0% or more of combustible organic substances, which do not meet the criteria mentioned in .1, .3, .4 or .5 above, shall be subjected to the self-reactive substance classification procedure.

A mixture showing the properties of a self-reactive substance, type B to F, shall be classified as a self-reactive substance of class 4.1.

A mixture showing the properties of a self-reactive substance, type G, according to the principle of 2.4.2.3.3.2.7 shall be considered for classification as a substance of class 5.1 (see 2.5.2).

2.4.2.3.1.2   The decomposition of self-reactive substances can be initiated by heat, contact with catalytic impurities (such as acids, heavy-metal compounds, bases), friction or impact. The rate of decomposition increases with temperature and varies with the substance. Decomposition, particularly if no ignition occurs, may result in the evolution of toxic gases or vapours. For certain self-reactive substances, the temperature shall be controlled. Some self-reactive substances may decompose explosively, particularly if confined. This characteristic may be modified by the addition of diluents or by the use of appropriate packagings. Some self-reactive substances burn vigorously. Self-reactive substances are, for example, some compounds of the types listed below:

.1   aliphatic azo compounds ($-C-N=N-C-$);

.2   organic azides ($-C-N_3$);

.3   diazonium salts ($-CN_2^+ Z^-$);

.4   *N*-nitroso compounds ($-N-N=O$); and

.5   aromatic sulphohydrazides ($-SO_2-NH-NH_2$).

This list is not exhaustive and substances with other reactive groups and some mixtures of substances may have similar properties.

2.4.2.3.2   *Classification of self-reactive substances*

2.4.2.3.2.1   Self-reactive substances are classified into seven types according to the degree of danger they present. The types of self-reactive substance range from type A, which may not be accepted for transport in the packaging in which it is tested, to type G, which is not subject to the provisions for self-reactive substances of class 4.1. The classification of types B to F is directly related to the maximum quantity allowed in one packaging.

2.4.2.3.2.2   Self-reactive substances permitted for transport in packagings are listed in 2.4.2.3.2.3, those permitted for transport in IBCs are listed in packing instruction IBC520 and those permitted for transport in portable tanks are listed in portable tank instruction T23. For each permitted substance listed, the appropriate generic entry of the Dangerous Goods List (UN 3221 to UN 3240) is assigned, and appropriate subsidiary risks and remarks providing relevant transport information are given. The generic entries specify:

.1   self-reactive substance type (B to F);

.2   physical state (liquid or solid); and

.3   temperature control, when required (2.4.2.3.4).

2.4.2.3.2.3   *List of currently assigned self-reactive substances in packagings*

In the column "Packing Method" codes "OP1" to "OP8" refer to packing methods in packing instruction P520. Self-reactive substances to be transported shall fulfill the classification and the control and emergency temperatures (derived from the SADT) as listed. For substances permitted in IBCs, see packing instruction IBC520, and for those permitted in tanks, see portable tank instruction T23.

Note: The classification given in this table is based on the technically pure substance (except where a concentration of less than 100% is specified). For other concentrations, the substances may be classified differently following the procedures in 2.4.2.3.3 and 2.4.2.3.4.

UN generic entry	SELF-REACTIVE SUBSTANCE	Concen- tration (%)	Packing method	Control temper- ature (°C)	Emergency temper- ature (°C)	Remarks
3222	2-DIAZO-1-NAPHTHOL-4-SULPHONYL CHLORIDE	100	OP5			(2)
	2-DIAZO-1-NAPHTHOL-5-SULPHONYL CHLORIDE	100	OP5			(2)
3223	SELF-REACTIVE LIQUID, SAMPLE		OP2			(8)
3224	AZODICARBONAMIDE FORMULATION TYPE C	<100	OP6			(3)
	2,2′-AZODI(ISOBUTYRONITRILE) as a water-based paste	≤50	OP6			
	N,N′-DINITROSO-*N,N*′-DIMETHYL-TEREPHTHALAMIDE, as a paste	72	OP6			
	N,N′-DINITROSOPENTAMETHYLENETETRAMINE	82	OP6			(7)
	SELF-REACTIVE SOLID, SAMPLE		OP2			(8)

UN generic entry	SELF-REACTIVE SUBSTANCE	Concentration (%)	Packing method	Control temperature (°C)	Emergency temperature (°C)	Remarks
3226	AZODICARBONAMIDE FORMULATION TYPE D	<100	OP7			(5)
	1,1′-AZODI(HEXAHYDROBENZONITRILE)	100	OP7			
	BENZENE-1,3-DISULPHONYL HYDRAZIDE as a paste	52	OP7			
	BENZENESULPHONYL HYDRAZIDE	100	OP7			
	4-(BENZYL(ETHYL)AMINO)-3-ETHOXY-BENZENEDIAZONIUM ZINC CHLORIDE	100	OP7			
	3-CHLORO-4-DIETHYLAMINOBENZENE-DIAZONIUM ZINC CHLORIDE	100	OP7			
	2-DIAZO-1-NAPHTHOL-4-SULPHONIC ACID ESTER	100	OP7			
	2-DIAZO-1-NAPHTHOL-5-SULPHONIC ACID ESTER	100	OP7			
	2-DIAZO-1-NAPHTHOLSULPHONIC ACID ESTER MIXTURE TYPE D	<100	OP7			(9)
	2,5-DIETHOXY-4-(4-MORPHOLINYL)-BENZENEDIAZONIUM SULPHATE	100	OP7			
	DIPHENYLOXIDE-4,4′-DISULPHONYL HYDRAZIDE	100	OP7			
	4-DIPROPYLAMINOBENZENEDIAZONIUM ZINC CHLORIDE	100	OP7			
	4-METHYLBENZENESULPHONYLHYDRAZIDE	100	OP7			
	SODIUM 2-DIAZO-1-NAPHTHOL-4-SULPHONATE	100	OP7			
	SODIUM 2-DIAZO-1-NAPHTHOL-5-SULPHONATE	100	OP7			
3228	ACETONE–PYROGALLOL COPOLYMER 2-DIAZO-1-NAPHTHOL-5-SULPHONATE	100	OP8			
	4-(DIMETHYLAMINO)BENZENEDIAZONIUM TRICHLOROZINCATE(–1)	100	OP8			
	2,5-DIBUTOXY-4-(4-MORPHOLINYL)-BENZENEDIAZONIUM TETRACHLOROZINCATE(2:1)	100	OP8			
3232	AZODICARBONAMIDE FORMULATION TYPE B, TEMPERATURE CONTROLLED	<100	OP5			(1) (2)
3233	SELF-REACTIVE LIQUID, SAMPLE, TEMPERATURE CONTROLLED		OP2			(8)
3234	AZODICARBONAMIDE FORMULATION TYPE C, TEMPERATURE CONTROLLED	<100	OP6			(4)
	2,2′-AZODI(ISOBUTYRONITRILE)	100	OP6	+40	+45	
	3-METHYL-4-(PYRROLIDIN-1-YL)BENZENE-DIAZONIUM TETRAFLUOROBORATE	95	OP6	+45	+50	
	SELF-REACTIVE SOLID, SAMPLE, TEMPERATURE CONTROLLED		OP2			(8)
	TETRAMINEPALLADIUM(II) NITRATE	100	OP6	+30	+35	
3235	2,2′-AZODI(ETHYL-2-METHYLPROPIONATE)	100	OP7	+20	+25	
3236	AZODICARBONAMIDE FORMULATION TYPE D, TEMPERATURE CONTROLLED	<100	OP7			(6)
	2,2′-AZODI(2,4-DIMETHYL-4-METHOXY-VALERONITRILE)	100	OP7	–5	+5	
	2,2′-AZODI(2,4-DIMETHYLVALERONITRILE)	100	OP7	+10	+15	
	2,2′-AZODI(2-METHYLBUTYRONITRILE)	100	OP7	+35	+40	
	4-(BENZYL(METHYL)AMINO)-3-ETHOXY-BENZENEDIAZONIUM ZINC CHLORIDE	100	OP7	+40	+45	
	2,5-DIETHOXY-4-MORPHOLINO-BENZENEDIAZONIUM ZINC CHLORIDE	67–100	OP7	+35	+40	
	2,5-DIETHOXY-4-MORPHOLINO-BENZENEDIAZONIUM ZINC CHLORIDE	66	OP7	+40	+45	

UN generic entry	SELF-REACTIVE SUBSTANCE	Concen-tration (%)	Packing method	Control temper-ature (°C)	Emergency temper-ature (°C)	Remarks
3236 (cont.)	2,5-DIETHOXY-4-MORPHOLINOBENZENE-DIAZONIUM TETRAFLUOROBORATE	100	OP7	+30	+35	
	2,5-DIETHOXY-4-(PHENYLSULPHONYL)-BENZENEDIAZONIUM ZINC CHLORIDE	67	OP7	+40	+45	
	2,5-DIMETHOXY-4-(4-METHYLPHENYL-SULPHONYL)BENZENEDIAZONIUM ZINC CHLORIDE	79	OP7	+40	+45	
	4-DIMETHYLAMINO-6-(2-DIMETHYLAMINO-ETHOXY)TOLUENE-2-DIAZONIUM ZINC CHLORIDE	100	OP7	+40	+45	
	2-(N,N-ETHOXYCARBONYLPHENYLAMINO)-3-METHOXY-4-(N-METHYL-N-CYCLOHEXYLAMINO)-BENZENEDIAZONIUM ZINC CHLORIDE	63–92	OP7	+40	+45	
	2-(N,N-ETHOXYCARBONYLPHENYLAMINO)-3-METHOXY-4-(N-METHYL-N-CYCLOHEXYLAMINO)-BENZENEDIAZONIUM ZINC CHLORIDE	62	OP7	+35	+40	
	N-FORMYL-2-(NITROMETHYLENE)-1,3-PERHYDROTHIAZINE	100	OP7	+45	+50	
	2-(2-HYDROXYETHOXY)-1-(PYRROLIDIN-1-YL)-BENZENE-4-DIAZONIUM ZINC CHLORIDE	100	OP7	+45	+50	
	3-(2-HYDROXYETHOXY)-4-(PYRROLIDIN-1-YL)-BENZENEDIAZONIUM ZINC CHLORIDE	100	OP7	+40	+45	
	2-(N,N-METHYLAMINOETHYLCARBONYL)-4-(3,4-DIMETHYLPHENYLSULPHONYL)-BENZENEDIAZONIUM HYDROGEN SULPHATE	96	OP7	+45	+50	
	4-NITROSOPHENOL	100	OP7	+35	+40	
3237	DIETHYLENEGLYCOL BIS(ALLYLCARBONATE) + DI-ISOPROPYL PEROXYDICARBONATE	⩾88 + ⩽12	OP8	–10	0	

Remarks

(1) Azodicarbonamide formulations which fulfil the criteria of 2.4.2.3.3.2.2. The control and emergency temperatures shall be determined by the procedure given in 7.7.2.

(2) "EXPLOSIVE" subsidiary risk label (Model No 1, see 5.2.2.2.2) required.

(3) Azodicarbonamide formulations which fulfil the criteria of 2.4.2.3.3.2.3.

(4) Azodicarbonamide formulations which fulfil the criteria of 2.4.2.3.3.2.3. The control and emergency temperatures shall be determined by the procedure given in 7.7.2.

(5) Azodicarbonamide formulations which fulfil the criteria of 2.4.2.3.3.2.4.

(6) Azodicarbonamide formulations which fulfil the criteria of 2.4.2.3.3.2.4. The control and emergency temperatures shall be determined by the procedure given in 7.7.2.

(7) With a compatible diluent having a boiling point of not less than 150°C.

(8) See 2.4.2.3.2.4.2.

(9) This entry applies to mixtures of esters of 2-diazo-1-naphthol-4-sulphonic acid and 2-diazo-1-naphthol-5-sulphonic acid meeting the criteria of 2.4.2.3.3.2.4.

2.4.2.3.2.4 Classification of self-reactive substances not listed in 2.4.2.3.2.3, packing instruction IBC520 or portable tank instruction T23 and assignment to a generic entry shall be made by the competent authority of the country of origin on the basis of a test report. Principles applying to the classification of such substances are provided in 2.4.2.3.3. The applicable classification procedures, test methods and criteria, and an example of a suitable test report, are given in the United Nations *Manual of Tests and Criteria*, part II. The statement of approval shall contain the classification and the relevant transport conditions.

.1 Activators, such as zinc compounds, may be added to some self-reactive substances to change their reactivity. Depending on both the type and the concentration of the activator, this may result in a decrease in thermal stability and a change in explosive properties. If either of these properties is altered, the new formulation shall be assessed in accordance with this classification procedure.

.2 Samples of self-reactive substances or formulations of self-reactive substances not listed in 2.4.2.3.2.3, for which a complete set of test results is not available and which are to be transported for further testing or evaluation, may be assigned to one of the appropriate entries for self-reactive substances type C provided the following conditions are met:

.1 the available data indicate that the sample would be no more dangerous than self-reactive substances type B;

.2 the sample is packaged in accordance with packing method OP2 (see applicable packing instruction) and the quantity per cargo transport unit is limited to 10 kg; and

.3 the available data indicate that the control temperature, if any, is sufficiently low to prevent any dangerous decomposition and sufficiently high to prevent any dangerous phase separation.

2.4.2.3.3 *Principles for classification of self-reactive substances*

Note: This section refers only to those properties of self-reactive substances which are decisive for their classification. A flow chart, presenting the classification principles in the form of a graphically arranged scheme of questions concerning the decisive properties together with the possible answers, is given in Figure 2.1(a) in chapter 2.4 of the United Nations *Recommendations on the Transport of Dangerous Goods*. These properties shall be determined experimentally. Suitable test methods with pertinent evaluation criteria are given in the United Nations *Manual of Tests and Criteria,* part II.

2.4.2.3.3.1 A self-reactive substance is regarded as possessing explosive properties when, in laboratory testing, the formulation is liable to detonate, to deflagrate rapidly or to show a violent effect when heated under confinement.

2.4.2.3.3.2 The following principles apply to the classification of self-reactive substances not listed in 2.4.2.3.2.3:

.1 Any substance which can detonate or deflagrate rapidly, as packaged for transport, is prohibited from transport under the provisions for self-reactive substances of class 4.1 in that packaging (defined as SELF-REACTIVE SUBSTANCE TYPE A);

.2 Any substance possessing explosive properties and which, as packaged for transport, neither detonates nor deflagrates rapidly, but is liable to undergo a thermal explosion in that package, shall also bear an "EXPLOSIVE" subsidiary risk label (Model No. 1, see 5.2.2.2.2). Such a substance may be packaged in amounts of up to 25 kg unless the maximum quantity has to be limited to a lower amount to preclude detonation or rapid deflagration in the package (defined as SELF-REACTIVE SUBSTANCE TYPE B);

.3 Any substance possessing explosive properties may be transported without an "EXPLOSIVE" subsidiary risk label when the substance as packaged (maximum 50 kg) for transport cannot detonate or deflagrate rapidly or undergo a thermal explosion (defined as SELF-REACTIVE SUBSTANCE TYPE C);

.4 Any substance which, in laboratory testing:

.1 detonates partially, does not deflagrate rapidly and shows no violent effect when heated under confinement; or

.2 does not detonate at all, deflagrates slowly and shows no violent effect when heated under confinement; or

.3 does not detonate or deflagrate at all and shows a medium effect when heated under confinement

may be accepted for transport in packages of not more than 50 kg net mass (defined as SELF-REACTIVE SUBSTANCE TYPE D);

.5 Any substance which, in laboratory testing, neither detonates nor deflagrates at all and shows low or no effect when heated under confinement may be accepted for transport in packages of not more than 400 kg/450 ℓ (defined as SELF-REACTIVE SUBSTANCE TYPE E);

.6 Any substance which, in laboratory testing, neither detonates in the cavitated state nor deflagrates at all and shows only a low or no effect when heated under confinement as well as low or no explosive power may be considered for transport in IBCs (defined as SELF-REACTIVE SUBSTANCE TYPE F); (for additional provisions see 4.1.7.2.2);

.7 Any substance which, in laboratory testing, neither detonates in the cavitated state nor deflagrates at all and shows no effect when heated under confinement nor any explosive power shall be exempted from classification as a self-reactive substance of class 4.1 provided that the formulation is thermally stable (self-accelerating decomposition temperature 60°C to 75°C for a 50 kg package) and any diluent meets the provisions of 2.4.2.3.5 (defined as SELF-REACTIVE SUBSTANCE TYPE G). If the formulation is not thermally stable or a compatible diluent having a boiling point less than 150°C is used for desensitization, the formulation shall be defined as SELF-REACTIVE LIQUID/SOLID TYPE F.

2.4.2.3.4 *Temperature control provisions*

2.4.2.3.4.1 Self-reactive substances are subject to temperature control in transport if their self-accelerating decomposition temperature (SADT) is less than or equal to 55°C. For currently assigned self-reactive substances, the control and emergency temperatures are shown in 2.4.2.3.2.3. Test methods for determining the SADT are given in the United Nations *Manual of Tests and Criteria*, part II, chapter 28. The test selected shall be conducted in a manner which is representative, both in size and material, of the package to be transported. The temperature control provisions are given in chapter 7.7.

2.4.2.3.5   *Desensitization of self-reactive substances*

2.4.2.3.5.1   In order to ensure safety during transport, self-reactive substances may be desensitized through the use of a diluent. If a diluent is used, the self-reactive substance shall be tested with the diluent present in the concentration and form used in transport.

2.4.2.3.5.2   Diluents which may allow a self-reactive substance to concentrate to a dangerous extent in the event of leakage from a package shall not be used.

2.4.2.3.5.3   The diluent shall be compatible with the self-reactive substance. In this regard, compatible diluents are those solids or liquids which have no detrimental influence on the thermal stability and hazard type of the self-reactive substance.

2.4.2.3.5.4   Liquid diluents in liquid formulations requiring temperature control shall have a boiling point of at least 60°C and a flashpoint not less than 5°C. The boiling point of the liquid shall be at least 50°C higher than the control temperature of the self-reactive substance (see 7.7.2).

2.4.2.4   Class 4.1 Solid desensitized explosives

2.4.2.4.1   *Definitions and properties*

2.4.2.4.1.1   Solid desensitized explosives are explosive substances which are wetted with water or alcohols or are diluted with other substances to form a homogeneous solid mixture to suppress their explosive properties. The desensitizing agent shall be distributed uniformly throughout the substance in the state in which it is to be transported. Where transport under conditions of low temperature is anticipated for substances containing or wetted with water, a suitable and compatible solvent, such as alcohol, may have to be added to lower the freezing point of the liquid. Some of these substances, when in a dry state, are classified as explosives. Where reference is made to a substance which is wetted with water, or some other liquid, it shall be permitted for transport as a class 4.1 substance only when in the wetted condition specified. Entries in the Dangerous Goods List in chapter 3.2 for solid desensitized explosives are UN 1310, UN 1320, UN 1321, UN 1322, UN 1336, UN 1337, UN 1344, UN 1347, UN 1348, UN 1349, UN 1354, UN 1355, UN 1356, UN 1357, UN 1517, UN 1571, UN 2555, UN 2556, UN 2557, UN 2852, UN 2907, UN 3317, UN 3319, UN 3344, UN 3364, UN 3365, UN 3366, UN 3367, UN 3368, UN 3369, UN 3370, UN 3376, UN 3380 and UN 3474.

2.4.2.4.2   Substances that:

.1   have been provisionally accepted into class 1 according to Test Series 1 and 2 but exempted from class 1 by Test Series 6;

.2   are not self-reactive substances of class 4.1;

.3   are not substances of class 5

are also assigned to class 4.1. UN 2956, UN 3241, UN 3242 and UN 3251 are such entries.

2.4.3   Class 4.2 – Substances liable to spontaneous combustion

2.4.3.1   Definitions and properties

2.4.3.1.1   Class 4.2 comprises:

.1   *Pyrophoric substances*, which are substances, including mixtures and solutions (liquid or solid), which, even in small quantities, ignite within 5 minutes of coming into contact with air. These substances are the most liable to spontaneous combustion; and

.2   *Self-heating substances*, which are substances, other than pyrophoric substances, which, in contact with air without energy supply, are liable to self-heating. These substances will ignite only when in large amounts (kilograms) and after long periods of time (hours or days).

2.4.3.1.2   Self-heating of substances, leading to spontaneous combustion, is caused by reaction of the substance with oxygen (in the air) and the heat developed not being conducted away rapidly enough to the surroundings. Spontaneous combustion occurs when the rate of heat production exceeds the rate of heat loss and the autoignition temperature is reached.

2.4.3.1.3   Some substances may also give off toxic gases if involved in a fire.

2.4.3.2 **Classification of class 4.2 substances**

2.4.3.2.1 Solids are considered pyrophoric solids which shall be classified in class 4.2 if, in tests performed in accordance with the test method given in the United Nations *Manual of Tests and Criteria*, part III, 33.3.1.4, the sample ignites in one of the tests.

2.4.3.2.2 Liquids are considered pyrophoric liquids which shall be classified in class 4.2 if, in tests performed in accordance with the test method given in the United Nations *Manual of Tests and Criteria*, part III, 33.3.1.5, the liquid ignites in the first part of the test, or if it ignites or chars the filter paper.

2.4.3.2.3 *Self-heating substances*

2.4.3.2.3.1 A substance shall be classified as a self-heating substance of class 4.2 if, in tests performed in accordance with the test method given in the United Nations *Manual of Tests and Criteria*, part III, 33.3.1.6:

.1 a positive result is obtained using a 25 mm cube sample at 140°C;

.2 a positive result is obtained in a test using a 100 mm cube sample at 140°C and a negative result is obtained in a test using a 100 mm cube sample at 120°C <u>and</u> the substance is to be transported in packages with a volume of more than 3 m^3;

.3 a positive result is obtained in a test using a 100 mm cube sample at 140°C and a negative result is obtained in a test using a 100 mm cube sample at 100°C <u>and</u> the substance is to be transported in packages with a volume of more than 450 ℓ;

.4 a positive result is obtained in a test using a 100 mm cube sample at 140°C <u>and</u> a positive result is obtained using a 100 mm cube sample at 100°C.

Note: Self-reactive substances, except for type G, giving also a positive result with this test method shall not be classified in class 4.2 but in class 4.1 (see 2.4.2.3.1.1).

2.4.3.2.3.2 A substance shall not be classified in class 4.2 if:

.1 a negative result is obtained in a test using a 100 mm cube sample at 140°C;

.2 a positive result is obtained in a test using a 100 mm cube sample at 140°C and a negative result is obtained in a test using a 25 mm cube sample at 140°C, a negative result is obtained in a test using a 100 mm cube sample at 120°C <u>and</u> the substance is to be transported in packages with a volume not more than 3 m^3;

.3 a positive result is obtained in a test using a 100 mm cube sample at 140°C and a negative result is obtained in a test using a 25 mm cube sample at 140°C, a negative result is obtained in a test using a 100 mm cube sample at 100°C <u>and</u> the substance is to be transported in packages with a volume not more than 450 ℓ.

2.4.3.3 **Assignment of packing groups**

2.4.3.3.1 Packing group I shall be assigned to all pyrophoric solids and liquids.

2.4.3.3.2 Packing group II shall be assigned to self-heating substances which give a positive result in a test using a 25 mm cube sample at 140°C.

2.4.3.3.3 Packing group III shall be assigned to self-heating substances if:

.1 a positive result is obtained in a test using a 100 mm cube sample at 140°C and a negative result is obtained in a test using a 25 mm cube sample at 140°C <u>and</u> the substance is to be transported in packages with a volume of more than 3 m^3;

.2 a positive result is obtained in a test using a 100 mm cube sample at 140°C and a negative result is obtained in a test using a 25 mm cube sample at 140°C, a positive result is obtained in a test using a 100 mm cube sample at 120°C <u>and</u> the substance is to be transported in packages with a volume of more than 450 ℓ;

.3 a positive result is obtained in a test using a 100 mm cube sample at 140°C and a negative result is obtained in a test using a 25 mm cube sample at 140°C <u>and</u> a positive result is obtained in a test using a 100 mm cube sample at 100°C.

2.4.4 **Class 4.3 – Substances which, in contact with water, emit flammable gases**

2.4.4.1 **Definitions and properties**

2.4.4.1.1 For the purpose of this Code, the substances in this class are either liquids or solids which, by interaction with water, are liable to become spontaneously flammable or to give off flammable gases in dangerous quantities.

2.4.4.1.2 Certain substances, in contact with water, may emit flammable gases that can form explosive mixtures with air. Such mixtures are easily ignited by all ordinary sources of ignition, for example naked lights, sparking handtools or unprotected light bulbs. The resulting blast wave and flames may endanger people and the environment. The test method referred to in 2.4.4.2 is used to determine whether the reaction of a substance with water leads to the development of a dangerous amount of gases which may be flammable. This test method shall not be applied to pyrophoric substances.

2.4.4.2 **Classification of class 4.3 substances**

2.4.4.2.1 Substances which, in contact with water, emit flammable gases shall be classified in class 4.3 if, in tests performed in accordance with the test method given in the United Nations *Manual of Tests and Criteria*, part III, 33.4.1:

.1 spontaneous ignition takes place in any step of the test procedure; or

.2 there is an evolution of a flammable gas at a rate greater than 1 litre per kilogram of the substance per hour.

2.4.4.3 **Assignment of packing groups**

2.4.4.3.1 Packing group I shall be assigned to any substance which reacts vigorously with water at ambient temperatures and demonstrates generally a tendency for the gas produced to ignite spontaneously, or which reacts readily with water at ambient temperatures such that the rate of evolution of flammable gas is equal to or greater than 10 litres per kilogram of substance over any one minute.

2.4.4.3.2 Packing group II shall be assigned to any substance which reacts readily with water at ambient temperatures such that the maximum rate of evolution of flammable gas is equal to or greater than 20 litres per kilogram of substance per hour, and which does not meet the criteria for packing group I.

2.4.4.3.3 Packing group III shall be assigned to any substance which reacts slowly with water at ambient temperatures such that the maximum rate of evolution of flammable gas is equal to or greater than 1 litre per kilogram of substance per hour, and which does not meet the criteria for packing groups I or II.

2.4.5 **Classification of organometallic substances**

Depending on their properties, organometallic substances may be classified in classes 4.2 or 4.3, as appropriate, in accordance with the following flowchart:

2

Flowchart scheme for organometallic substances[1,2]

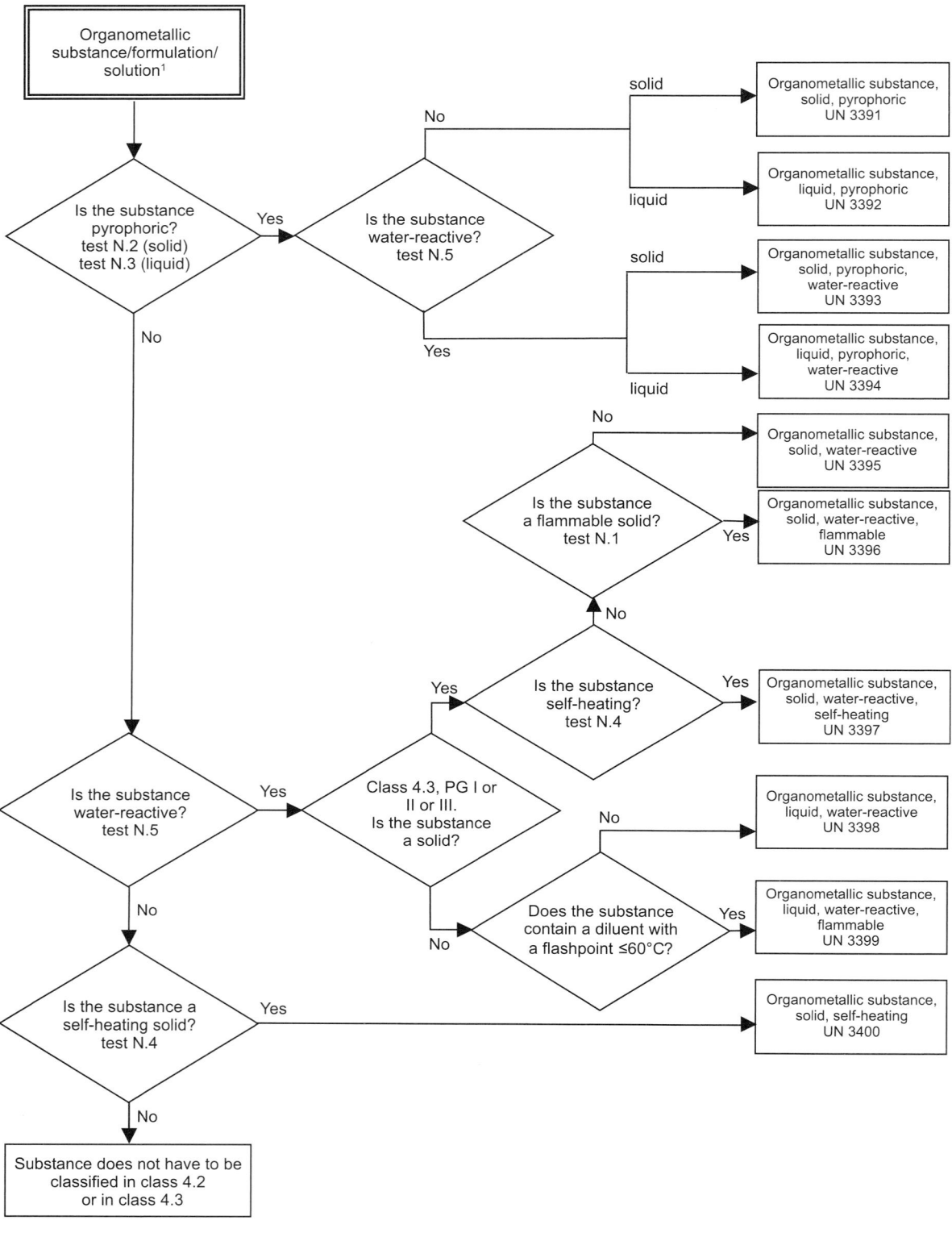

---

[1] If applicable and testing is relevant, taking into account reactivity properties, class 6.1 and class 8 properties shall be considered according to the Precedence of hazards table 2.0.3.6.

[2] Test methods N.1 to N.5 can be found in the United Nations *Manual of Tests and Criteria*, part III, section 33.

Chapter 2.5

Class 5 – Oxidizing substances and organic peroxides

2.5.0 Introductory note

Note: Because of the differing properties exhibited by dangerous goods within classes 5.1 and 5.2, it is impracticable to establish a single criterion for classification in either class. Tests and criteria for assignment to the two classes are addressed in this chapter.

2.5.1 Definitions and general provisions

In this Code, class 5 is divided into two classes as follows:

Class 5.1 – Oxidizing substances
Substances which, while in themselves not necessarily combustible, may, generally by yielding oxygen, cause, or contribute to, the combustion of other material. Such substances may be contained in an article;

Class 5.2 – Organic peroxides
Organic substances which contain the bivalent –O–O– structure and may be considered derivatives of hydrogen peroxide, where one or both of the hydrogen atoms have been replaced by organic radicals. Organic peroxides are thermally unstable substances which may undergo exothermic self-accelerating decomposition. In addition, they may have one or more of the following properties:

- be liable to explosive decomposition;
- burn rapidly;
- be sensitive to impact or friction;
- react dangerously with other substances;
- cause damage to the eyes.

2.5.2 Class 5.1 – Oxidizing substances

Note: For the classification of oxidizing substances to class 5.1, in the event of divergence between test results and known experience, judgement based on known experience shall take precedence over test results.

2.5.2.1 Properties

2.5.2.1.1 Substances of class 5.1 in certain circumstances directly or indirectly evolve oxygen. For this reason, oxidizing substances increase the risk and intensity of fire in combustible material with which they come into contact.

2.5.2.1.2 Mixtures of oxidizing substances with combustible material and even with material such as sugar, flour, edible oils, mineral oils, etc., are dangerous. These mixtures are readily ignited, in some cases by friction or impact. They may burn violently and may lead to explosion.

2.5.2.1.3 There will be a violent reaction between most oxidizing substances and liquid acids, evolving toxic gases. Toxic gases may also be evolved when certain oxidizing substances are involved in a fire.

2.5.2.1.4 The above-mentioned properties are, in general, common to all substances in this class. Additionally, some substances possess specific properties, which shall be taken into account in transport. These properties are shown in the Dangerous Goods List in chapter 3.2.

2.5.2.2 Oxidizing solids

2.5.2.2.1 *Classification of solid substances of class 5.1*

2.5.2.2.1.1 Tests are performed to measure the potential for the solid substance to increase the burning rate or burning intensity of a combustible substance when the two are thoroughly mixed. The procedure is given in the United Nations *Manual of Tests and Criteria*, part III, 34.4.1. Tests are conducted on the substance to be evaluated mixed with dry fibrous cellulose in mixing ratios of 1:1 and 4:1, by mass, of sample to cellulose. The burning characteristics of the mixtures are compared with the standard 3:7 mixture, by mass, of potassium bromate to cellulose. If the burning time is equal to or less than this standard mixture, the burning times shall be compared with those from the packing group I or II reference standards, 3:2 and 2:3 ratios, by mass, of potassium bromate to cellulose respectively.

2.5.2.2.1.2 The classification test results are assessed on the basis of:

.1   the comparison of the mean burning time with those of the reference mixtures; and

.2   whether the mixture of substance and cellulose ignites and burns.

2.5.2.2.1.3 A solid substance is classified in class 5.1 if the 4:1 or 1:1 sample-to-cellulose ratio (by mass) tested exhibits a mean burning time equal to or less than the mean burning time of a 3:7 mixture (by mass) of potassium bromate and cellulose.

2.5.2.2.2 *Assignment of packing groups*

2.5.2.2.2.1 Solid oxidizing substances are assigned to a packing group according to the test procedure in the United Nations *Manual of Tests and Criteria*, part III, 34.4.1, in accordance with the following criteria:

.1   Packing group I: any substance which, in the 4:1 or 1:1 sample-to-cellulose ratio (by mass) tested, exhibits a mean burning time less than the mean burning time of a 3:2 mixture (by mass) of potassium bromate and cellulose;

.2   Packing group II: any substance which, in the 4:1 or 1:1 sample-to-cellulose ratio (by mass) tested, exhibits a mean burning time equal to or less than the mean burning time of a 2:3 mixture (by mass) of potassium bromate and cellulose and the criteria for packing group I are not met;

.3   Packing group III: any substance which, in the 4:1 or 1:1 sample-to-cellulose ratio (by mass) tested, exhibits a mean burning time equal to or less than the mean burning time of a 3:7 mixture (by mass) of potassium bromate and cellulose and the criteria for packing groups I and II are not met;

.4   Not classified as class 5.1: any substance which, in both the 4:1 and 1:1 sample-to-cellulose ratio (by mass) tested, does not ignite and burn, or exhibits mean burning times greater than that of a 3:7 mixture (by mass) of potassium bromate and cellulose.

2.5.2.3 **Oxidizing liquids**

2.5.2.3.1 *Classification of liquid substances of class 5.1*

2.5.2.3.1.1 A test is performed to determine the potential for a liquid substance to increase the burning rate or burning intensity of a combustible substance or for spontaneous ignition to occur when the two are thoroughly mixed. The procedure is given in the United Nations *Manual of Tests and Criteria*, part III, 34.4.2. It measures the pressure rise time during combustion. Whether a liquid is an oxidizing substance of class 5.1 and, if so, whether packing group I, II or III shall be assigned, is decided on the basis of the test result (see also Precedence of hazard characteristics in 2.0.3).

2.5.2.3.1.2 The classification test results are assessed on the basis of:

.1   whether the mixture of substance and cellulose spontaneously ignites;

.2   the comparison of the mean time taken for the pressure to rise from 690 kPa to 2070 kPa gauge with those of the reference substances.

2.5.2.3.1.3 A liquid substance is classified in class 5.1 if the 1:1 mixture, by mass, of substance and cellulose tested exhibits a mean pressure rise time less than or equal to the mean pressure rise time of a 1:1 mixture, by mass, of 65% aqueous nitric acid and cellulose.

2.5.2.3.2 *Assignment of packing groups*

2.5.2.3.2.1 Liquid oxidizing substances are assigned to a packing group according to the test procedure in the United Nations *Manual of Tests and Criteria*, part III, 34.4.2, in accordance with the following criteria:

.1   Packing group I: any substance which, in the 1:1 mixture (by mass) of substance and cellulose tested, spontaneously ignites; or the mean pressure rise time of a 1:1 mixture (by mass) of substance and cellulose is less than that of a 1:1 mixture (by mass) of 50% perchloric acid and cellulose;

.2   Packing group II: any substance which, in the 1:1 mixture (by mass) of substance and cellulose tested, exhibits a mean pressure rise time less than or equal to the mean pressure rise time of a 1:1 mixture (by mass) of 40% aqueous sodium chlorate solution and cellulose; and the criteria for packing group I are not met;

.3   Packing group III: any substance which, in the 1:1 mixture (by mass) of substance and cellulose tested, exhibits a mean pressure rise time less than or equal to the mean pressure rise time of a 1:1 mixture (by mass) of 65% aqueous nitric acid and cellulose; and the criteria for packing groups I and II are not met;

.4   Not classified as class 5.1: any substance which, in the 1:1 mixture (by mass) of substance and cellulose tested, exhibits a pressure rise of less than 2070 kPa gauge; or exhibits a mean pressure rise time greater than the mean pressure rise time of a 1:1 mixture (by mass) of 65% aqueous nitric acid and cellulose.

2.5.3   Class 5.2 – Organic peroxides

2.5.3.1   Properties

2.5.3.1.1   Organic peroxides are liable to exothermic decomposition at normal or elevated temperatures. The decomposition can be initiated by heat, contact with impurities (such as acids, heavy-metal compounds, amines), friction or impact. The rate of decomposition increases with temperature and varies with the organic peroxide formulation. Decomposition may result in the evolution of harmful, or flammable, gases or vapours. For certain organic peroxides the temperature shall be controlled during transport. Some organic peroxides may decompose explosively, particularly if confined. This characteristic may be modified by the addition of diluents or by the use of appropriate packagings. Many organic peroxides burn vigorously.

2.5.3.1.2   Contact of organic peroxides with the eyes is to be avoided. Some organic peroxides will cause serious injury to the cornea, even after brief contact, or will be corrosive to the skin.

2.5.3.2   Classification of organic peroxides

2.5.3.2.1   Any organic peroxide shall be considered for classification in class 5.2, unless the organic peroxide formulation contains:

.1   not more than 1.0% available oxygen from the organic peroxides when containing not more than 1.0% hydrogen peroxide; or

.2   not more than 0.5% available oxygen from the organic peroxides when containing more than 1.0% but not more than 7.0% hydrogen peroxide.

Note: The available oxygen content (%) of an organic peroxide formulation is given by the formula:

$$16 \times \Sigma(n_i \times c_i/m_i)$$

where

n_i =   number of peroxygen groups per molecule of organic peroxide i;
c_i =   concentration (mass %) of organic peroxide i;
m_i =   molecular mass of organic peroxide i.

2.5.3.2.2   Organic peroxides are classified into seven types according to the degree of danger they present. The types of organic peroxide range from type A, which may not be accepted for transport in the packaging in which it is tested, to type G, which is not subject to the provisions for organic peroxides of class 5.2. The classification of types B to F is directly related to the maximum quantity allowed in one packaging.

2.5.3.2.3   Organic peroxides permitted for transport in packagings are listed in 2.5.3.2.4, those permitted for transport in IBCs are listed in packing instruction IBC520 and those permitted for transport in portable tanks are listed in portable tank instruction T23. For each permitted substance listed, the generic entry of the Dangerous Goods List (UN 3101 to UN 3120) is assigned, appropriate subsidiary risks and remarks providing relevant transport information are given. The generic entries specify:

.1   organic peroxide type (B to F);

.2   physical state (liquid or solid); and

.3   temperature control, when required (see 2.5.3.4).

2.5.3.2.3.1   Mixtures of the listed formulations may be classified as the same type of organic peroxide as that of the most dangerous component and be transported under the conditions of transport given for this type. However, as two stable components can form a thermally less stable mixture, the self-accelerating decomposition temperature (SADT) of the mixture shall be determined and, if necessary, temperature control applied as required by 2.5.3.4.

2.5.3.2.4   List of currently assigned organic peroxides in packagings

Note: Packing Method codes "OP1" to "OP8" refer to packing methods in packing instruction P520. Peroxides to be transported shall fulfil the classification and the control and emergency temperatures (derived from the SADT) as listed. For substances permitted in IBCs, see packing instruction IBC520, and for those permitted in tanks, see portable tank instruction T23.

Number (generic entry)	ORGANIC PEROXIDE	Concentration (%)	Diluent type A (%)	Diluent type B (%) (1)	Inert solid (%)	Water (%)	Packing method	Control temperature (°C)	Emergency temperature (°C)	Subsidiary risks and remarks
3101	tert-BUTYL PEROXYACETATE	>52 – 77	≥23				OP5			(3)
	1,1-DI-(tert-BUTYLPEROXY)CYCLOHEXANE	>80 – 100					OP5			(3)
	1,1-DI-(tert-BUTYLPEROXY)-3,3,5-TRIMETHYL-CYCLOHEXANE	>90 – 100					OP5			(3)
	METHYL ETHYL KETONE PEROXIDE(S)	see remark (8)	≥48				OP5			(3) (8) (13)
	2,5-DIMETHYL-2,5-DI-(tert-BUTYLPEROXY)-HEXYNE-3	>86 – 100					OP5			(3)
3102	tert-BUTYL MONOPEROXYMALEATE	>52 – 100					OP5			(3)
	3-CHLOROPEROXYBENZOIC ACID	>57 – 86			≥14		OP1			(3)
	DIBENZOYL PEROXIDE	>51 – 100			≤48		OP2			(3)
	DIBENZOYL PEROXIDE	>77 – 94				≥6	OP4			(3)
	DI-4-CHLOROBENZOYL PEROXIDE	≤77				≥23	OP5			(3)
	DI-2,4-DICHLOROBENZOYL PEROXIDE	≤77				≥23	OP5			(3)
	2,2-DIHYDROPEROXYPROPANE	≤27			≥73		OP5			(3)
	2,5-DIMETHYL-2,5-DI-(BENZOYLPEROXY)HEXANE	>82 – 100					OP5			(3)
	DI-(2-PHENOXYETHYL) PEROXYDICARBONATE	>85 – 100					OP5			(3)
	DISUCCINIC ACID PEROXIDE	>72 – 100					OP4			(3) (17)
3103	tert-AMYL PEROXYBENZOATE	≤100					OP5			
	tert-AMYLPEROXY ISOPROPYL CARBONATE	≤77	≥23				OP5			
	n-BUTYL 4,4-DI-(tert-BUTYLPEROXY)VALERATE	>52 – 100					OP5			
	tert-BUTYL HYDROPEROXIDE	>79 – 90				≥10	OP5			(13)
	tert-BUTYL HYDROPEROXIDE + DI-tert-BUTYL PEROXIDE	<82 + >9				≥7	OP5			(13)
	tert-BUTYL MONOPEROXYMALEATE	≤52	≥48				OP6			
	tert-BUTYL PEROXYACETATE	>32 – 52	≥48				OP6			
	tert-BUTYL PEROXYBENZOATE	>77 – 100					OP5			
	tert-BUTYLPEROXY ISOPROPYLCARBONATE	≤77	≥23				OP5			
	tert-BUTYLPEROXY-2-METHYLBENZOATE	≤100					OP5			
	1,1-DI-(tert-AMYLPEROXY)CYCLOHEXANE	≤82	≥18				OP6			

Number (generic entry)	ORGANIC PEROXIDE	Concentration (%)	Diluent type A (%)	Diluent type B (%) (1)	Inert solid (%)	Water (%)	Packing method	Control temperature (°C)	Emergency temperature (°C)	Subsidiary risks and remarks
3103 (cont.)	2,2-DI-(tert-BUTYLPEROXY)BUTANE	≤52	≥48				OP6			
	1,6-DI-(tert-BUTYLPEROXYCARBONYLOXY)-HEXANE	≤72	≥28				OP5			
	1,1-DI-(tert-BUTYLPEROXY)CYCLOHEXANE	>52 – 80	≥20				OP5			(30)
	1,1-DI-(tert-BUTYLPEROXY)CYCLOHEXANE	≤72		≥28			OP5			
	1,1-DI-(tert-BUTYLPEROXY)-3,3,5-TRIMETHYL-CYCLOHEXANE	>57 – 90	≥10				OP5			
	1,1-DI-(tert-BUTYLPEROXY)-3,3,5-TRIMETHYL-CYCLOHEXANE	≤77		≥23			OP5			(30)
	1,1-DI-(tert-BUTYLPEROXY)-3,3,5-TRIMETHYL-CYCLOHEXANE	≤90		≥10			OP5			
	2,5-DIMETHYL-2,5-DI-(tert-BUTYLPEROXY)-HEXYNE-3	>52 – 86	≥14				OP5			(26)
	ETHYL 3,3-DI-(tert-BUTYLPEROXY)BUTYRATE	>77 – 100					OP5			
	ORGANIC PEROXIDE, LIQUID, SAMPLE						OP2			(11)
	CYCLOHEXANONE PEROXIDE(S)	≤91				≥9	OP6			(13)
3104	DIBENZOYL PEROXIDE	≤77				≥23	OP6			
	2,5-DIMETHYL-2,5-DI(BENZOYLPEROXY)HEXANE	≤82				≥18	OP5			
	2,5-DIMETHYL-2,5-DIHYDROPEROXYHEXANE	≤82				≥18	OP6			
	ORGANIC PEROXIDE, SOLID, SAMPLE					≥8	OP2			(11)
3105	ACETYL ACETONE PEROXIDE	≤42	≥48				OP7			(2)
	tert-AMYL PEROXYACETATE	≤62	≥38				OP7			
	tert-AMYL PEROXY-2-ETHYLHEXYL CARBONATE	≤100					OP7			
	tert-AMYL PEROXY-3,5,5-TRIMETHYLHEXANOATE	≤100					OP7			(3)
	tert-BUTYL HYDROPEROXIDE	≤80	≥20				OP7			(4) (13)
	tert-BUTYL PEROXYBENZOATE	>52 – 77	≥23				OP7			
	tert-BUTYL PEROXYBUTYL FUMARATE	≤52	≥48				OP7			
	tert-BUTYL PEROXYCROTONATE	≤77	≥23				OP7			
	tert-BUTYL PEROXY-2-ETHYLHEXYLCARBONATE	≤100					OP7			
	1-(2-tert-BUTYLPEROXY ISOPROPYL)-3-ISOPROPENYLBENZENE	≤77	≥23				OP7			
	tert-BUTYL PEROXY-3,5,5-TRIMETHYL-HEXANOATE	>32 – 100	≥28				OP7			(5)
	CYCLOHEXANONE PEROXIDE(S)	≤72	≥28				OP7			

2

Number (generic entry)	ORGANIC PEROXIDE	Concentration (%)	Diluent type A (%)	Diluent type B (%)[1]	Inert solid (%)	Water (%)	Packing method	Control temperature (°C)	Emergency temperature (°C)	Subsidiary risks and remarks
3105 (cont.)	2,2-DI-(tert-AMYLPEROXY)BUTANE	≤57	≥43				OP7			
	DI-tert-BUTYL PEROXYAZELATE	≤52	≥48				OP7			
	1,1-DI-(tert-BUTYLPEROXY)CYCLOHEXANE	>42 – 52	≥48				OP7			
	1,1-DI-(tert-BUTYLPEROXY)CYCLOHEXANE + tert-BUTYL PEROXY-2-ETHYLHEXANOATE	≤43 + ≤16	≥41				OP7			
	DI-(tert-BUTYLPEROXY)PHTHALATE	>42 – 52	≥48				OP7			
	2,2-DI-(tert-BUTYLPEROXY)PROPANE	≤52	≥48				OP7			
	2,5-DIMETHYL-2,5-DI-(tert-BUTYLPEROXY)HEXANE	>52 – 100					OP7			
	2,5-DIMETHYL-2,5-DI-(3,5,5-TRIMETHYL-HEXANOYLPEROXY)HEXANE	≤77	≥23				OP7			
	ETHYL 3,3-DI-(tert-AMYLPEROXY)BUTYRATE	≤67	≥33				OP7			
	ETHYL 3,3-DI-(tert-BUTYLPEROXY)BUTYRATE	≤77	≥23				OP7			
	p-MENTHYL HYDROPEROXIDE	>72 – 100					OP7			(13)
	METHYL ETHYL KETONE PEROXIDE(S)	see remark (9)	≥55				OP7			(9)
	METHYL ISOBUTYL KETONE PEROXIDE(S)	≤62	≥19				OP7			(22)
	PEROXYACETIC ACID, TYPE D, stabilized	≤43					OP7			(13) (14) (19)
	PINANYL HYDROPEROXIDE	>56 – 100					OP7			(13)
	1,1,3,3-TETRAMETHYLBUTYL HYDROPEROXIDE	≤100					OP7			
	3,6,9-TRIETHYL-3,6,9-TRIMETHYL-1,4,7-TRIPEROXONANE	≤42	≥58				OP7			(28)
3106	ACETYL ACETONE PEROXIDE	≤32 as a paste					OP7			(20)
	tert-BUTYL PEROXYBENZOATE	≤52			≥48		OP7			
	tert-BUTYL PEROXY-2-ETHYLHEXANOATE + 2,2-DI-(tert-BUTYLPEROXY)BUTANE	≤12 + ≤14	≥14		≥60		OP7			
	tert-BUTYLPEROXY STEARYLCARBONATE	≤100					OP7			
	tert-BUTYL PEROXY-3,5,5-TRIMETHYL-HEXANOATE	≤42			≥58		OP7			
	3-CHLOROPEROXYBENZOIC ACID	≤57			≥3	≥40	OP7			
	3-CHLOROPEROXYBENZOIC ACID	≤77			≥6	≥17	OP7			
	CYCLOHEXANONE PEROXIDE(S)	≤72 as a paste					OP7			
	DIBENZOYL PEROXIDE	≤62			≥28	≥10	OP7			
	DIBENZOYL PEROXIDE	>52 – 62 as a paste					OP7			(5) (20)
	DIBENZOYL PEROXIDE	>35 – 52			≥48		OP7			(20)
	1,1-DI-(tert-BUTYLPEROXY)CYCLOHEXANE	≤42	≥13		≥45		OP7			

2

Number (generic entry)	ORGANIC PEROXIDE	Concentration (%)	Diluent type A (%)	Diluent type B (%) (1)	Inert solid (%)	Water (%)	Packing method	Control temperature (°C)	Emergency temperature (°C)	Subsidiary risks and remarks
3106 (cont.)	DI-(2-tert-BUTYLPEROXYISOPROPYL)BENZENE(S)	>42 – 100			≤57		OP7			(20)
	DI-(tert-BUTYLPEROXY)PHTHALATE	≤52 as a paste					OP7			
	2,2-DI-(tert-BUTYLPEROXY)PROPANE	≤42	≥13		≥45		OP7			(20)
	DI-4-CHLOROBENZOYL PEROXIDE	≤52 as a paste			≥58		OP7			
	2,2-DI-(4,4-DI-(tert-BUTYLPEROXY)CYCLOHEXYL)-PROPANE	≤42					OP7			
	DI-2,4-DICHLOROBENZOYL PEROXIDE	≤52 as a paste with silicon oil					OP7			(24)
	DI-(1-HYDROXYCYCLOHEXYL)PEROXIDE	≤100					OP7			
	DIISOPROPYLBENZENE DIHYDROPEROXIDE	≤82	≥5			≥5	OP7			
	DILAUROYL PEROXIDE	≤100					OP7			
	DI-(4-METHYLBENZOYL) PEROXIDE	≤52 as paste with silicon oil					OP7			
	2,5-DIMETHYL-2,5-DI-(BENZOYLPEROXY)HEXANE	≤82			≥18		OP7			
	2,5-DIMETHYL-2,5-DI-(tert-BUTYLPEROXY)-HEXYNE-3	≤52			≥48		OP7			
	DI-(2-PHENOXYETHYL)PEROXYDICARBONATE	≤85			≥48	≥15	OP7			
	ETHYL 3,3-DI-(tert-BUTYLPEROXY)BUTYRATE	≤52					OP7			
3107	tert-AMYL HYDROPEROXIDE	≤88	≥6			≥6	OP8			
	tert-BUTYL CUMYL PEROXIDE	>42 – 100					OP8			
	tert-BUTYL HYDROPEROXIDE	≤79				>14	OP8			(13) (23)
	CUMYL HYDROPEROXIDE	>90 – 98	≤10				OP8			(13)
	DI-tert-AMYL PEROXIDE	≤100					OP8			
	DIBENZOYL PEROXIDE	>36 – 42	≥18			≤40	OP8			
	DI-tert-BUTYL PEROXIDE	>52 – 100					OP8			(21)
	1,1-DI-(tert-BUTYLPEROXY)CYCLOHEXANE	≤27	≥25				OP8			
	DI-(tert-BUTYLPEROXY)PHTHALATE	≤42	≥58				OP8			
	1,1-DI-(tert-BUTYLPEROXY)-3,3,5-TRIMETHYL-CYCLOHEXANE	≤57	≥43				OP8			
	1,1-DI-(tert-BUTYLPEROXY)-3,3,5-TRIMETHYL-CYCLOHEXANE	≤32	≥26	≥42			OP8			
	2,2-DI-(4,4-DI-(BUTYLPEROXY)CYCLOHEXYL)-PROPANE	≤22		≥78			OP8			
	METHYL ETHYL KETONE PEROXIDE(S)	see remark (10)	≥60				OP8			(10)

2

Number (generic entry)	ORGANIC PEROXIDE	Concentration (%)	Diluent type A (%)	Diluent type B (%) (1)	Inert solid (%)	Water (%)	Packing method	Control temperature (°C)	Emergency temperature (°C)	Subsidiary risks and remarks
3107 (cont.)	3,3,5,7-PENTAMETHYL-1,2,4-TRIOXEPANE	≤100					OP8			(13) (15) (19)
	PEROXYACETIC ACID, TYPE E, stabilized	≤43					OP8			
	POLYETHER POLY-*tert*-BUTYLPEROXY-CARBONATE	≤52		≥48			OP8			
3108	*tert*-BUTYL CUMYL PEROXIDE	≤52			≥48		OP8			
	n-BUTYL 4,4-DI-(*tert*-BUTYLPEROXY)VALERATE	≤52			≥48		OP8			
	tert-BUTYL MONOPEROXYMALEATE	≤52			≥48		OP8			
	tert-BUTYL MONOPEROXYMALEATE	≤52 as a paste			≥58		OP8			
	1-(2-*tert*-BUTYLPEROXYISOPROPYL)-3-ISOPROPENYLBENZENE	≤42					OP8			
	DIBENZOYL PEROXIDE	≤56.5 as a paste				≥15	OP8			
	DIBENZOYL PEROXIDE	≤52 as a paste					OP8			(20)
	2,5-DIMETHYL-2,5-DI-(*tert*-BUTYLPEROXY)HEXANE	≤47 as a paste					OP8			
	2,5-DIMETHYL-2,5-DI-(*tert*-BUTYLPEROXY)HEXANE	≤77			≥23		OP8			
	tert-BUTYL HYDROPEROXIDE	≤72				≥28	OP8			(13)
	tert-BUTYL PEROXYACETATE	≤32		≥68			OP8			
	tert-BUTYL PEROXY-3,5,5-TRIMETHYL-HEXANOATE	≤32		≥68			OP8			
3109	CUMYL HYDROPEROXIDE	≤90	≥10				OP8			(13) (18)
	DIBENZOYL PEROXIDE	≤42 as a stable dispersion in water					OP8			
	DI-*tert*-BUTYL PEROXIDE	≤52		≥48			OP8			(25)
	1,1-DI-(*tert*-BUTYLPEROXY)CYCLOHEXANE	≤42	≥58				OP8			
	1,1-DI-(*tert*-BUTYLPEROXY)CYCLOHEXANE	≤13	≥13	≥74			OP8			
	DILAUROYL PEROXIDE	≤42 as a stable dispersion in water					OP8			
	2,5-DIMETHYL-2,5-DI-(*tert*-BUTYLPEROXY)HEXANE	≤52	≥48				OP8			
	ISOPROPYLCUMYL HYDROPEROXIDE	≤72	≥28				OP8			(13)
	p-MENTHYL HYDROPEROXIDE	≤72	≥28				OP8			(27)
	METHYL ISOPROPYL KETONE PEROXIDE(S)	See remark (31)	≥70				OP8			(31)
	PEROXYACETIC ACID, TYPE F, stabilized	≤43					OP8			
	PINANYL HYDROPEROXIDE	≤56	≥44				OP8			(13) (16) (19)

Number (generic entry)	ORGANIC PEROXIDE	Concentration (%)	Diluent type A (%)	Diluent type B (%) (1)	Inert solid (%)	Water (%)	Packing method	Control temperature (°C)	Emergency temperature (°C)	Subsidiary risks and remarks
3110	DICUMYL PEROXIDE	>52 – 100					OP8			(12)
	1,1-DI-(*tert*-BUTYLPEROXY)-3,3,5-TRIMETHYL-CYCLOHEXANE	≤57			≥43		OP8			
3111	*tert*-BUTYL PEROXYISOBUTYRATE	>52 – 77		≥23			OP5	+15	+20	(3)
	DIISOBUTYRYL PEROXIDE	>32 – 52		≥48			OP5	–20	–10	(3)
	ISOPROPYL *sec*-BUTYL PEROXYDICARBONATE + DI-*sec*-BUTYL PEROXYDICARBONATE + DIISOPROPYL PEROXYDICARBONATE	≤52 + ≤28 + ≤22					OP5	–20	–10	(3)
3112	ACETYL CYCLOHEXANESULPHONYL PEROXIDE	≤82				≥12	OP4	–10	0	(3)
	DICYCLOHEXYL PEROXYDICARBONATE	>91 – 100					OP3	+10	+15	(3)
	DIISOPROPYL PEROXYDICARBONATE	>52 – 100					OP2	–15	–5	(3)
	DI-(2-METHYLBENZOYL) PEROXIDE	≤87				≥13	OP5	+30	+35	(3)
3113	*tert*-AMYL PEROXYPIVALATE	≤77		≥23			OP5	+10	+15	
	tert-BUTYL PEROXYDIETHYLACETATE	≤100					OP5	+20	+25	
	tert-BUTYL PEROXY-2-ETHYLHEXANOATE	>52 – 100					OP6	+20	+25	
	tert-BUTYL PEROXYPIVALATE	>67 – 77	≥23				OP5	0	+10	
	DI-*sec*-BUTYL PEROXYDICARBONATE	>52 – 100					OP4	–20	–10	
	DI-(2-ETHYLHEXYL)PEROXYDICARBONATE	>77 – 100					OP5	–20	–10	
	2,5-DIMETHYL-2,5-DI-(2-ETHYLHEXANOYLPEROXY)-HEXANE	≤100					OP5	+20	+25	
	DI-*n*-PROPYL PEROXYDICARBONATE	≤100					OP3	–25	–15	
	DI-*n*-PROPYL PEROXYDICARBONATE	≤77		≥23			OP5	–20	–10	
	ORGANIC PEROXIDE, LIQUID, SAMPLE, TEMPERATURE CONTROLLED						OP2			(11)
3114	DI-(4-*tert*-BUTYLCYCLOHEXYL)-PEROXYDICARBONATE	≤100					OP6	+30	+35	
	DICYCLOHEXYL PEROXYDICARBONATE	≤91				≥9	OP5	+10	+15	
	DIDECANOYL PEROXIDE	≤100					OP6	+30	+35	
	DI-*n*-OCTANOYL PEROXIDE	≤100					OP5	+10	+15	
	ORGANIC PEROXIDE, SOLID, SAMPLE, TEMPERATURE CONTROLLED						OP2			(11)
3115	ACETYL CYCLOHEXANESULPHONYL PEROXIDE	≤32		≥68			OP7	–10	0	
	tert-AMYL PEROXY-2-ETHYLHEXANOATE	≤100					OP7	+20	+25	
	tert-AMYL PEROXYNEODECANOATE	≤77		≥23			OP7	0	+10	

2

2

Number (generic entry)	ORGANIC PEROXIDE	Concentration (%)	Diluent type A (%)	Diluent type B (%) (1)	Inert solid (%)	Water (%)	Packing method	Control temperature (°C)	Emergency temperature (°C)	Subsidiary risks and remarks
3115 (cont.)	tert-BUTYL PEROXY-2-ETHYLHEXANOATE + 2,2-DI-(tert-BUTYLPEROXY)BUTANE	≤31 + ≤36		≥33			OP7	+35	+40	
	tert-BUTYL PEROXYISOBUTYRATE	≤52		≥48			OP7	+15	+20	
	tert-BUTYL PEROXYNEODECANOATE	>77 – 100					OP7	-5	+5	
	tert-BUTYL PEROXYNEODECANOATE	≤77	≥23				OP7	0	+10	
	tert-BUTYL PEROXYNEOHEPTANOATE	≤77		≥23			OP7	0	+10	
	tert-BUTYL PEROXYPIVALATE	>27 – 67		≥33			OP7	0	+10	
	CUMYL PEROXYNEODECANOATE	≤77		≥23			OP7	-10	0	
	CUMYL PEROXYNEODECANOATE	≤87	≥13				OP7	-10	0	
	CUMYL PEROXYNEOHEPTANOATE	≤77	≥23				OP7	-10	0	
	CUMYL PEROXYPIVALATE	≤77		≥23			OP7	-5	+5	
	DIACETONE ALCOHOL PEROXIDES	≤57		≥26		≥8	OP7	+40	+45	(6)
	DIACETYL PEROXIDE	≤27		≥73			OP7	+20	+25	(7) (13)
	DI-n-BUTYL PEROXYDICARBONATE	>27 – 52		≥48			OP7	-15	-5	
	DI-sec-BUTYL PEROXYDICARBONATE	≤52		≥48			OP7	-15	-5	
	DI-(2-ETHOXYETHYL)PEROXYDICARBONATE	≤52		≥48			OP7	-10	0	
	DI-(2-ETHYLHEXYL)PEROXYDICARBONATE	≤77		≥23			OP7	-15	-5	
	DIISOBUTYRYL PEROXIDE	≤32		≥68			OP7	-20	-10	
	DIISOPROPYL PEROXYDICARBONATE	≤52		≥48			OP7	-20	-10	
	DIISOPROPYL PEROXYDICARBONATE	≤28	≥72				OP7	-15	-5	
	DI-(3-METHOXYBUTYL) PEROXYDICARBONATE	≤52		≥48			OP7	-5	+5	
	DI-(3-METHYLBENZOYL) PEROXIDE + BENZOYL (3-METHYLBENZOYL) PEROXIDE + DIBENZOYL PEROXIDE	≤20 + ≤18 + ≤4		≥58			OP7	+35	+40	
	DI-(2-NEODECANOYLPEROXYISOPROPYL)-BENZENE	≤52	≥48				OP7	-10	0	
	DI-(3,5,5-TRIMETHYLHEXANOYL) PEROXIDE	>38 – 82	≥18				OP7	0	+10	
	1-(2-ETHYLHEXANOYLPEROXY)-1,3-DIMETHYLBUTYL PEROXYPIVALATE	≤52	≥45				OP7	-20	-10	
	tert-HEXYL PEROXYNEODECANOATE	≤71	≥29				OP7	0	+10	
	tert-HEXYL PEROXYPIVALATE	≤72		≥28			OP7	+10	+15	
	3-HYDROXY-1,1-DIMETHYLBUTYL PEROXYNEODECANOATE	≤77	≥23				OP7	-5	+5	

Number (generic entry)	ORGANIC PEROXIDE	Concentration (%)	Diluent type A (%)	Diluent type B (%)(1)	Inert solid (%)	Water (%)	Packing method	Control temperature (°C)	Emergency temperature (°C)	Subsidiary risks and remarks
3115 (cont.)	ISOPROPYL sec-BUTYL PEROXYDICARBONATE + DI-sec-BUTYL PEROXYDICARBONATE + DI-ISOPROPYL PEROXYDICARBONATE	≤32 + ≤15 – 18 + ≤12 – 15	≥38				OP7	–20	–10	
	METHYLCYCLOHEXANONE PEROXIDE(S)	≤67		≥33			OP7	+35	+40	
	1,1,3,3-TETRAMETHYLBUTYL PEROXY-2-ETHYLHEXANOATE	≤100					OP7	+15	+20	
	1,1,3,3-TETRAMETHYLBUTYL PEROXY-NEODECANOATE	≤72		≥28			OP7	–5	+5	
	1,1,3,3-TETRAMETHYLBUTYL PEROXYPIVALATE	≤77	≥23				OP7	0	+10	
3116	DICETYL PEROXYDICARBONATE	≤100					OP7	+30	+35	
	DIMYRISTYL PEROXYDICARBONATE	≤100					OP7	+20	+25	
	DI-n-NONANOYL PEROXIDE	≤100					OP7	0	+10	
	DISUCCINIC ACID PEROXIDE	≤72				≥28	OP7	+10	+15	
3117	tert-BUTYL PEROXY-2-ETHYLHEXANOATE	>32 – 52		≥48			OP8	+30	+35	
	DI-n-BUTYL PEROXYDICARBONATE	≤27		≥73			OP8	–10	0	
	tert-BUTYL PEROXYNEOHEPTANOATE	≤42 as a stable dispersion in water					OP8	0	+10	
	1,1-DIMETHYL-3-HYDROXYBUTYL PEROXY-NEOHEPTANOATE	≤52	≥48				OP8	0	+10	
	DIPROPIONYL PEROXIDE	≤27		≥73			OP8	+15	+20	
	3-HYDROXY-1,1-DIMETHYLBUTYL PEROXY-NEODECANOATE	≤52	≥48				OP8	–5	+5	
3118	tert-BUTYL PEROXY-2-ETHYLHEXANOATE	≤52			≥48		OP8	+20	+25	
	tert-BUTYL PEROXYNEODECANOATE	≤42 as a stable dispersion in water (frozen)					OP8	0	+10	
	DI-n-BUTYL PEROXYDICARBONATE	≤42 as a stable dispersion in water (frozen)					OP8	–15	–5	
3119	DI-2,4-DICHLOROBENZOYL PEROXIDE	≤ 52 as a paste					OP8	+ 20	+ 25	
	PEROXYLAURIC ACID	≤100					OP8	+35	+40	
	tert-AMYL PEROXYNEODECANOATE	≤47	≥53				OP8	0	+10	
	tert-BUTYL PEROXY-2-ETHYLHEXANOATE	≤32		≥68			OP8	+40	+45	
	tert-BUTYL PEROXYNEODECANOATE	≤52 as a stable dispersion in water					OP8	0	+10	
	tert-BUTYL PEROXYNEODECANOATE	≤32	≥68				OP8	0	+10	

2

Number (generic entry)	ORGANIC PEROXIDE	Concentration (%)	Diluent type A (%)	Diluent type B (%) [1]	Inert solid (%)	Water (%)	Packing method	Control temperature (°C)	Emergency temperature (°C)	Subsidiary risks and remarks
3119 (cont.)	*tert*-BUTYL PEROXYPIVALATE	⩽27		⩾73			OP8	+30	+35	
	CUMYL PEROXYNEODECANOATE	⩽52 as a stable dispersion in water					OP8	−10	0	
	DI-(4-*tert*-BUTYLCYCLOHEXYL) PEROXYDICARBONATE	⩽42 as a stable dispersion in water					OP8	+30	+35	
	DICETYL PEROXYDICARBONATE	⩽42 as a stable dispersion in water					OP8	+30	+35	
	DICYCLOHEXYL PEROXYDICARBONATE	⩽42 as a stable dispersion in water					OP8	+15	+20	
	DI-(2-ETHYLHEXYL) PEROXYDICARBONATE	⩽62 as a stable dispersion in water					OP8	−15	−5	
	DIMYRISTYL PEROXYDICARBONATE	⩽42 as a stable dispersion in water					OP8	+20	+25	
	DI-(3,5,5-TRIMETHYLHEXANOYL) PEROXIDE	⩽52 as a stable dispersion in water					OP8	+10	+15	
	DI-(3,5,5-TRIMETHYLHEXANOYL) PEROXIDE	⩽38	⩾62				OP8	+20	+25	
	3-HYDROXY-1,1-DIMETHYLBUTYL PEROXYNEODECANOATE	⩽52 as a stable dispersion in water					OP 8	−5	+ 5	
	1,1,3,3-TETRAMETHYLBUTYL PEROXY-NEODECANOATE	⩽52 as a stable dispersion in water					OP8	−5	+5	
3120	DI-(2-ETHYLHEXYL)PEROXYDICARBONATE	⩽52 as a stable dispersion in water (frozen)					OP8	−15	−5	
Exempt	CYCLOHEXANONE PEROXIDE(S)	⩽32			⩾68					(29)
Exempt	DIBENZOYL PEROXIDE	⩽35			⩾65					(29)
Exempt	DI-(2-*tert*-BUTYLPEROXYISOPROPYL)BENZENE(S)	⩽42			⩾58					(29)
Exempt	DI-4-CHLOROBENZOYL PEROXIDE	⩽ 32			⩾68					(29)
Exempt	DICUMYL PEROXIDE	⩽52			⩾48					(29)

(1) Diluent type B may always be replaced by diluent type A. The boiling point of diluent type B shall be at least 60°C higher than the SADT of the organic peroxide
(2) Available oxygen $\leqslant 4.7\%$
(3) "EXPLOSIVE" subsidiary risk label required. (Model No. 1, see 5.2.2.2.2)
(4) Diluent may be replaced by di-*tert*-butyl peroxide
(5) Available oxygen $\leqslant 9\%$
(6) With $\leqslant 9\%$ hydrogen peroxide; available oxygen $\leqslant 10\%$
(7) Only non-metallic packagings are allowed
(8) Available oxygen $> 10\%$ and $\leqslant 10.7\%$, with or without water
(9) Available oxygen $\leqslant 10\%$, with or without water
(10) Available oxygen $\leqslant 8.2\%$, with or without water
(11) See 2.5.3.2.5.1
(12) Up to 2000 kg per receptacle assigned to ORGANIC PEROXIDE TYPE F on the basis of large-scale trials
(13) "CORROSIVE" subsidiary risk label required (Model No. 8, see 5.2.2.2.2)
(14) Peroxyacetic acid formulations which fulfil the criteria of 2.5.3.3.2.4
(15) Peroxyacetic acid formulations which fulfil the criteria of 2.5.3.3.2.5
(16) Peroxyacetic acid formulations which fulfil the criteria of 2.5.3.3.2.6
(17) Addition of water to this organic peroxide will decrease its thermal stability
(18) No "CORROSIVE" subsidiary risk label required for concentrations below 80%
(19) Mixtures with hydrogen peroxide, water and acid(s)
(20) With diluent type A, with or without water
(21) With $\geqslant 25\%$ diluent type A by mass, and in addition ethylbenzene
(22) With $\geqslant 19\%$ diluent type A by mass, and in addition methyl isobutyl ketone
(23) With $< 6\%$ di-*tert*-butyl peroxide
(24) With $\leqslant 8\%$ 1-isopropylhydroperoxy-4-isopropylhydroxybenzene
(25) Diluent type B with boiling point $> 110°C$
(26) With $< 0.5\%$ hydroperoxides content
(27) For concentrations more than 56%, "CORROSIVE" subsidiary risk label required (Model No. 8, see 5.2.2.2.2)
(28) Available active oxygen $\leqslant 7.6\%$ in diluent type A having a 95% boil-off point in the range 200–260°C
(29) Not subject to the provisions for peroxide, class 5.2
(30) Diluent type B with boiling point $> 130°C$
(31) Active oxygen $\leqslant 6.7\%$

2.5.3.2.5 Classification of organic peroxides not listed in 2.5.3.2.4, packing instruction IBC520 or portable tank instruction T23 and assignment to a generic entry shall be made by the competent authority of the country of origin on the basis of a test report. Principles applying to the classification of such substances are provided in 2.5.3.3. Test methods and criteria and an example of a report are given in the current edition of the United Nations *Manual of Tests and Criteria*, part II. The statement of approval shall contain the classification and the relevant transport conditions (see 5.4.4.1.3).

2.5.3.2.5.1 Samples of new organic peroxides or new formulations of currently assigned organic peroxides for which complete test data are not available and which are to be transported for further testing or evaluation may be assigned to one of the appropriate entries for ORGANIC PEROXIDE TYPE C provided the following conditions are met:

.1 the available data indicate that the sample would be no more dangerous than ORGANIC PEROXIDE TYPE B;

.2 the sample is packaged in accordance with packing method OP2 and the quantity per cargo transport unit is limited to 10 kg; and

.3 the available data indicate that the control temperature, if any, is sufficiently low to prevent any dangerous decomposition and sufficiently high to prevent any dangerous phase separation.

2.5.3.3 **Principles for classification of organic peroxides**

Note: This section refers only to those properties of organic peroxides which are decisive for their classification. A flow chart, presenting the classification principles in the form of a graphically arranged scheme of questions concerning the decisive properties together with the possible answers, is given in Figure 2.2(a) in chapter 2.5 of the United Nations *Recommendations on the Transport of Dangerous Goods*. These properties shall be determined experimentally. Suitable test methods with pertinent evaluation criteria are given in the United Nations *Manual of Tests and Criteria*, part II.

2.5.3.3.1 Any organic peroxide formulation shall be regarded as possessing explosive properties when, in laboratory testing, the formulation is liable to detonate, to deflagrate rapidly or to show a violent effect when heated under confinement.

2.5.3.3.2 The following principles apply to the classification of organic peroxide formulations not listed in 2.5.3.2.4:

.1 Any organic peroxide formulation which can detonate or deflagrate rapidly, as packaged for transport, is prohibited from transport in that packaging under class 5.2 (defined as ORGANIC PEROXIDE TYPE A);

.2 Any organic peroxide formulation possessing explosive properties and which, as packaged for transport, neither detonates nor deflagrates rapidly, but is liable to undergo a thermal explosion in that package, shall bear an "EXPLOSIVE" subsidiary risk label (Model No. 1, see 5.2.2.2.2). Such an organic peroxide may be packaged in amounts of up to 25 kg unless the maximum quantity has to be limited to a lower amount to preclude detonation or rapid deflagration in the package (defined as ORGANIC PEROXIDE TYPE B);

.3 Any organic peroxide formulation possessing explosive properties may be transported without an "EXPLOSIVE" subsidiary risk label when the substance as packaged (maximum 50 kg) for transport cannot detonate or deflagrate rapidly or undergo a thermal explosion (defined as ORGANIC PEROXIDE TYPE C);

.4 Any organic peroxide formulation which, in laboratory testing:

   .1 detonates partially, does not deflagrate rapidly and shows no violent effect when heated under confinement; or

   .2 does not detonate at all, deflagrates slowly and shows no violent effect when heated under confinement; or

   .3 does not detonate or deflagrate at all and shows a medium effect when heated under confinement

is acceptable for transport in packages of not more than 50 kg net mass (defined as ORGANIC PEROXIDE TYPE D);

.5 Any organic peroxide formulation which, in laboratory testing, neither detonates nor deflagrates at all and shows low or no effect when heated under confinement is acceptable for transport in packages of not more than 400 kg/450 ℓ (defined as ORGANIC PEROXIDE TYPE E);

.6 Any organic peroxide formulation which, in laboratory testing, neither detonates in the cavitated state nor deflagrates at all and shows only a low or no effect when heated under confinement as well as low or no explosive power may be considered for transport in IBCs or tanks (defined as ORGANIC PEROXIDE TYPE F); for additional provisions see 4.1.7 and 4.2.1.13;

.7 Any organic peroxide formulation which, in laboratory testing, neither detonates in the cavitated state nor deflagrates at all and shows no effect when heated under confinement nor any explosive power shall be exempted from class 5.2, provided that the formulation is thermally stable (self-accelerating decomposition temperature is 60°C or higher for a 50 kg package) and for liquid formulations diluent type A is used for desensitization (defined as ORGANIC PEROXIDE TYPE G). If the formulation is not thermally stable or a diluent other than type A is used for desensitization, the formulation shall be defined as ORGANIC PEROXIDE TYPE F.

2.5.3.4   Temperature control provisions

2.5.3.4.0   The properties of some organic peroxides require that they be transported under temperature control. Control and emergency temperatures for currently assigned organic peroxides are shown in the list 2.5.3.2.4. The controlled temperature provisions are given in chapter 7.7.

2.5.3.4.1   The following organic peroxides shall be subjected to temperature control during transport:

.1 organic peroxides type B and C with a SADT ≤ 50°C;

.2 organic peroxides type D showing a medium effect when heated under confinement* with a SADT ≤ 50°C or showing a low or no effect when heated under confinement with a SADT ≤ 45°C; and

.3 organic peroxides types E and F with a SADT ≤ 45°C.

2.5.3.4.2   Test methods for determining the SADT are given in the United Nations *Manual of Tests and Criteria*, part II, chapter 28. The test selected shall be conducted in a manner which is representative, both in size and material, of the package to be transported.

2.5.3.4.3   Test methods for determining the flammability are given in the United Nations *Manual of Tests and Criteria*, part III, chapter 32.4. Because organic peroxides may react vigorously when heated, it is recommended to determine their flashpoint using small sample sizes such as described in ISO 3679.

2.5.3.5   Desensitization of organic peroxides

2.5.3.5.1   In order to ensure safety during transport, organic peroxides are in many cases desensitized by organic liquids or solids, inorganic solids or water. Where a percentage of a substance is stipulated, this refers to the percentage by mass, rounded to the nearest whole number. In general, desensitization shall be such that, in case of spillage or fire, the organic peroxide will not concentrate to a dangerous extent.

---

* As determined by test series E as prescribed in the United Nations *Manual of Tests and Criteria*, part II.

2.5.3.5.2 Unless otherwise stated for the individual organic peroxide formulation, the following definitions apply for diluents used for desensitization:

.1 Diluents type A are organic liquids which are compatible with the organic peroxide and which have a boiling point of not less than 150°C. Type A diluents may be used for desensitizing all organic peroxides.

.2 Diluents type B are organic liquids which are compatible with the organic peroxide and which have a boiling point of less than 150°C but not less than 60°C and a flashpoint of not less than 5°C. Type B diluents may be used for desensitization of all organic peroxides provided that the boiling point is at least 60°C higher than the SADT in a 50 kg package.

2.5.3.5.3 Diluents, other than type A or type B, may be added to organic peroxide formulations as listed in 2.5.3.2.4 provided that they are compatible. However, replacement of all or part of a type A or type B diluent by another diluent with differing properties requires that the organic peroxide formulation be re-assessed in accordance with the normal acceptance procedure for class 5.2.

2.5.3.5.4 Water may only be used for the desensitization of organic peroxides which are shown in 2.5.3.2.4 or in the statement of approval according to 2.5.3.2.5 as being with water or as a stable dispersion in water.

2.5.3.5.5 Organic and inorganic solids may be used for desensitization of organic peroxides provided that they are compatible.

2.5.3.5.6 Compatible liquids and solids are those which have no detrimental influence on the thermal stability and hazard type of the organic peroxide formulation.

2

Chapter 2.6

Class 6 – Toxic and infectious substances

2.6.0   Introductory notes

Note 1:   The word "toxic" has the same meaning as "poisonous".

Note 2:   Genetically modified micro-organisms which do not meet the definition of an infectious substance shall be considered for classification in class 9 and assigned to UN 3245.

Note 3:   Toxins from plant, animal or bacterial sources which do not contain any infectious substances, or toxins that are contained in substances which are not infectious substances, shall be considered for classification in class 6.1 and assigned to UN 3172.

2.6.1   Definitions

Class 6 is subdivided into two classes as follows:

Class 6.1 – Toxic substances
These are substances liable either to cause death or serious injury or to harm human health if swallowed or inhaled, or by skin contact.

Class 6.2 – Infectious substances
These are substances known or reasonably expected to contain pathogens. Pathogens are defined as micro-organisms (including bacteria, viruses, rickettsiae, parasites, fungi) and other agents such as prions, which can cause disease in humans or animals.

2.6.2   Class 6.1 – Toxic substances

2.6.2.1   Definitions and properties

2.6.2.1.1   *LD_{50} (median lethal dose) for acute oral toxicity* is the statistically derived single dose of a substance that can be expected to cause death within 14 days in 50 per cent of young adult albino rats when administered by the oral route. The LD_{50} value is expressed in terms of mass of test substance per mass of test animal (mg/kg).

2.6.2.1.2   *LD_{50} for acute dermal toxicity* is that dose of the substance which, administered by continuous contact for 24 hours with the bare skin of the albino rabbit, is most likely to cause death within 14 days in one half of the animals tested. The number of animals tested shall be sufficient to give a statistically significant result and be in conformity with good pharmacological practices. The result is expressed in milligrams per kilogram body mass.

2.6.2.1.3   *LC_{50} for acute toxicity on inhalation* is that concentration of vapour, mist or dust which, administered by continuous inhalation to both male and female young adult albino rats for one hour, is most likely to cause death within 14 days in one half of the animals tested. A solid substance shall be tested if at least 10% (by mass) of its total mass is likely to be dust in the respirable range, such as the aerodynamic diameter of that particle fraction is 10 microns or less. A liquid substance shall be tested if a mist is likely to be generated in a leakage of the transport containment. For both solid and liquid substances, more than 90% (by mass) of a specimen prepared for inhalation toxicity testing shall be in the respirable range as defined above. The result is expressed in milligrams per litre of air for dusts and mists or in millilitres per cubic metre of air (parts per million) for vapours.

2.6.2.1.4   *Properties*

.1   The dangers of poisoning which are inherent in these substances depend upon contact with the human body, that is by inhalation of vapours by unsuspecting persons at some distance from the cargo or the immediate dangers of physical contact with the substance. These have been considered in the context of the probability of accident occurring during transport by sea.

.2   Nearly all toxic substances evolve toxic gases when involved in a fire or when heated to decomposition.

.3   A substance specified as "stabilized" shall not be transported in an unstabilized condition.

2.6.2.2   Assignment of packing groups to toxic substances

2.6.2.2.1   Toxic substances have for packing purposes been apportioned among packing groups according to the degree of their toxic hazards in transport:

.1   Packing group I: substances and preparations presenting a high toxicity risk;

.2   Packing group II: substances and preparations presenting a medium toxicity risk;

.3   Packing group III: substances and preparations presenting a low toxicity risk.

2.6.2.2.2   In making this grouping, account has been taken of human experience in instances of accidental poisoning, and of special properties possessed by any individual substance, such as liquid state, high volatility, any special likelihood of penetration, and special biological effects.

2.6.2.2.3   In the absence of human experience, the grouping has been based on data obtained from animal experiments. Three possible routes of administration have been examined. These routes are exposure through:

– oral ingestion;

– dermal contact; and

– inhalation of dusts, mists or vapours.

2.6.2.2.3.1   For appropriate animal test data for the various routes of exposure, see 2.6.2.1. When a substance exhibits a different order of toxicity by two or more routes of administration, the highest degree of danger indicated by the tests has been used in assigning the packing group.

2.6.2.2.4   The criteria to be applied for grouping a substance according to the toxicity it exhibits by all three routes of administration are presented in the following paragraphs.

2.6.2.2.4.1   The grouping criteria for the oral and dermal routes as well as for inhalation of dusts and mists are shown in the following table:

**Grouping criteria for administration through oral ingestion,
dermal contact and inhalation of dusts and mists**

Packing group	Oral toxicity LD_{50} (mg/kg)	Dermal toxicity LD_{50} (mg/kg)	Inhalation toxicity by dusts and mists LC_{50} (mg/ℓ)
I	$\leqslant 5.0$	$\leqslant 50$	$\leqslant 0.2$
II	> 5.0 and $\leqslant 50$	> 50 and $\leqslant 200$	> 0.2 and $\leqslant 2.0$
III*	> 50 and $\leqslant 300$	> 200 and $\leqslant 1000$	> 2.0 and $\leqslant 4.0$

* Tear gas substances shall be included in packing group II even if their toxicity data correspond to packing group III values.

2.6.2.2.4.2   The criteria for inhalation toxicity of dusts and mists in 2.6.2.2.4.1 are based on LC_{50} data relating to one hour exposures, and where such information is available it shall be used. However, where only LC_{50} data relating to 4-hour exposures to dusts and mists are available, such figures can be multiplied by four and the product substituted in the above criteria, i.e. LC_{50} (4 hours) × 4 is considered the equivalent of LC_{50} (1 hour).

Note: Substances meeting the criteria of class 8 and with an inhalation toxicity of dusts and mists (LC_{50}) leading to packing group I are only accepted for an allocation to class 6.1 if the toxicity through oral ingestion or dermal contact is at least in the range of packing group I or II. Otherwise an allocation to class 8 is made when appropriate (see 2.8.2.2).

2.6.2.2.4.3   Liquids having toxic vapours shall be assigned to the following packing groups, where "*V*" is the saturated vapour concentration in mℓ/m^3 air at 20°C and standard atmospheric pressure:

Packing group I: if $V \geqslant 10\ LC_{50}$ and $LC_{50} \leqslant 1000$ mℓ/m^3.

Packing group II: if $V \geqslant LC_{50}$ and $LC_{50} \leqslant 3000$ mℓ/m^3, and do not meet the criteria for packing group I.

Packing group III: if $V \geqslant \frac{1}{5} LC_{50}$ and $LC_{50} \leqslant 5000$ mℓ/m^3, and do not meet the criteria for packing groups I or II.

Note: Tear gas substances shall be included in packing group II even if their toxicity data correspond to packing group III values.

2.6.2.2.4.4 In figure 2-3 the criteria according to 2.6.2.2.4.3 are expressed in graphical form, as an aid to easy classification. Because of approximations inherent in the use of graphs, substances falling on or near packing group borderlines shall be checked using numerical criteria.

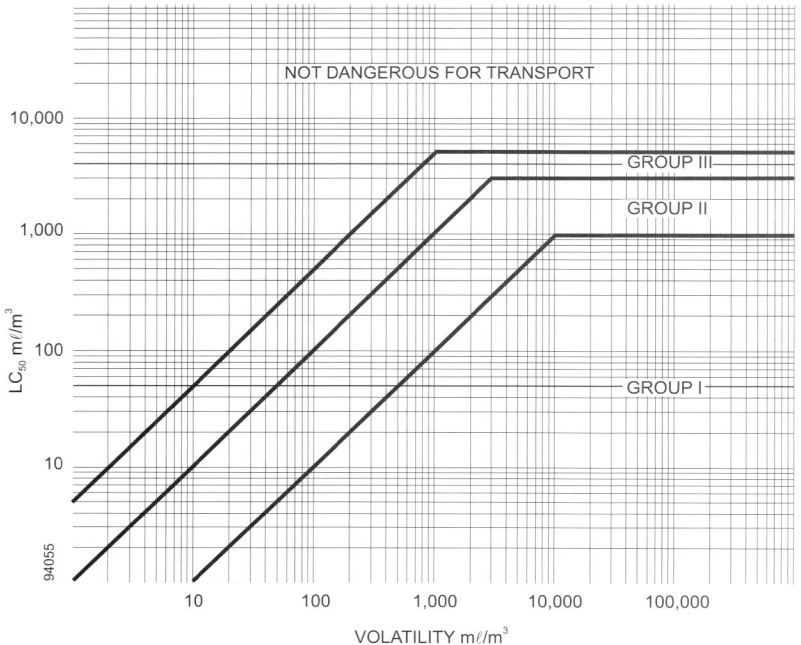

Figure 2-3 – *Inhalation toxicity: packing group borderlines*

2.6.2.2.4.5 The criteria for inhalation toxicity of vapours in 2.6.2.2.4.3 are based on LC_{50} data relating to one hour exposures, and where such information is available it shall be used. However, where only LC_{50} data relating to 4-hour exposures to the vapours are available, such figures can be multiplied by two and the product substituted in the above criteria, i.e. LC_{50} (4 hours) $\times$ 2 is considered the equivalent of LC_{50} (1 hour).

2.6.2.2.4.6 Mixtures of liquids that are toxic by inhalation shall be assigned to packing groups according to 2.6.2.2.4.7 or 2.6.2.2.4.8.

2.6.2.2.4.7 If LC_{50} data are available for each of the toxic substances comprising a mixture, the packing group may be determined as follows:

.1 Estimate the LC_{50} of the mixture using the formula:

$$LC_{50} \text{ (mixture)} = \frac{1}{\sum_{i=1}^{n}\left(\dfrac{f_i}{LC_{50i}}\right)}$$

where: f_i = mole fraction of the ith component substance of the mixture
LC_{50i} = mean lethal concentration of the ith component substance in $m\ell/m^3$.

.2 Estimate the volatility of each component substance comprising the mixture using the formula:

$$V_i = \left(\frac{P_i \times 10^6}{101.3}\right) \text{ } m\ell/m^3$$

where: P_i = the partial pressure of the ith component substance in kPa at 20°C and one atmosphere pressure.

.3 Calculate the ratio of the volatility to the LC_{50} using the formula:

$$R = \sum_{i=1}^{n}\left(\frac{V_i}{LC_{50i}}\right)$$

.4 Using the calculated values of LC_{50} (mixture) and R, the packing group for the mixture is determined:

Packing group I: $R \geqslant 10$ and LC_{50} (mixture) $\leqslant 1000$ $m\ell/m^3$.

Packing group II: $R \geqslant 1$ and LC_{50} (mixture) $\leqslant 3000$ $m\ell/m^3$ and not meeting criteria for packing group I.

Packing group III: $R \geqslant \frac{1}{5}$ and LC_{50} (mixture) $\leqslant 5000$ $m\ell/m^3$ and not meeting criteria for packing groups I or II.

2.6.2.2.4.8 In the absence of LC$_{50}$ data on the toxic constituent substances, the mixture may be assigned a packing group based on the following simplified threshold toxicity tests. When these threshold tests are used, the most restrictive packing group shall be determined and used for transporting the mixture.

.1 A mixture is assigned to packing group I only if it meets both of the following criteria:

– A sample of the liquid mixture is vaporized and diluted with air to create a test atmosphere of 1000 mℓ/m^3 vaporized mixture in air. Ten albino rats (five male and five female) are exposed to the test atmosphere for one hour and observed for 14 days. If five or more of the animals die within the 14-day observation period, the mixture is presumed to have an LC$_{50}$ equal to or less than 1000 mℓ/m^3.

– A sample of the vapour in equilibrium with the liquid mixture at 20°C is diluted with 9 equal volumes of air to form a test atmosphere. Ten albino rats (five male and five female) are exposed to the test atmosphere for one hour and observed for 14-days. If five or more of the animals die within the 14-day observation period, the mixture is presumed to have a volatility equal to or greater than 10 times the mixture LC$_{50}$.

.2 A mixture is assigned to packing group II only if it meets both of the following criteria, and the mixture does not meet the criteria for packing group I:

– A sample of the liquid mixture is vaporized and diluted with air to create a test atmosphere of 3000 mℓ/m^3 vaporized mixture in air. Ten albino rats (five male and five female) are exposed to the test atmosphere for one hour and observed for 14 days. If five or more of the animals die within the 14-day observation period, the mixture is presumed to have an LC$_{50}$ equal to or less than 3000 mℓ/m^3.

– A sample of the vapour in equilibrium with the liquid mixture at 20°C is used to form a test atmosphere. Ten albino rats (five male and five female) are exposed to the test atmosphere for one hour and observed for 14 days. If five or more of the animals die within the 14-day observation period, the mixture is presumed to have a volatility equal to or greater than the mixture LC$_{50}$.

.3 A mixture is assigned to packing group III only if it meets both of the following criteria, and the mixture does not meet the criteria for packing groups I or II:

– A sample of the liquid mixture is vaporized and diluted with air to create a test atmosphere of 5000 mℓ/m^3 vaporized mixture in air. Ten albino rats (five male and five female) are exposed to the test atmosphere for one hour and observed for 14 days. If five or more of the animals die within the 14-day observation period, the mixture is presumed to have an LC$_{50}$ equal to or less than 5000 mℓ/m^3.

– The vapour pressure of the liquid mixture is measured and if the vapour concentration is equal to or greater than 1000 mℓ/m^3, the mixture is presumed to have a volatility equal to or greater than $\frac{1}{5}$ the mixture LC$_{50}$.

2.6.2.3 Methods for determining oral and dermal toxicity of mixtures

2.6.2.3.1 When classifying and assigning the appropriate packing group to mixtures in class 6.1, in accordance with the oral and dermal toxicity criteria in 2.6.2.2, it is necessary to determine the acute LD$_{50}$ of the mixture.

2.6.2.3.2 If a mixture contains only one active substance, and the LD$_{50}$ of that constituent is known, in the absence of reliable acute oral and dermal toxicity data on the actual mixture to be transported, the oral or dermal LD$_{50}$ may be obtained by the following method:

$$\text{LD}_{50} \text{ value of preparation} = \frac{\text{LD}_{50} \text{ value of active substance} \times 100}{\text{percentage of active substance by mass}}$$

2.6.2.3.3 If a mixture contains more than one active constituent, there are three possible approaches that may be used to determine the oral or dermal LD$_{50}$ of the mixture. The preferred method is to obtain reliable acute oral and dermal toxicity data on the actual mixture to be transported. If reliable, accurate data are not available, then either of the following methods may be performed:

.1 Classify the formulation according to the most hazardous constituent of the mixture as if that constituent were present in the same concentration as the total concentration of all active constituents; or

.2 Apply the formula:

$$\frac{C_A}{T_A} + \frac{C_B}{T_B} + \dots \frac{C_Z}{T_Z} = \frac{100}{T_M}$$

where: C = the % concentration of constituent A, B . . . Z in the mixture;
T = the oral LD$_{50}$ value of constituent A, B . . . Z;
T_M = the oral LD$_{50}$ value of the mixture.

2

Note: This formula can also be used for dermal toxicities provided that this information is available on the same species for all constituents. The use of this formula does not take into account any potentiation or protective phenomena.

2.6.2.4 Classification of pesticides

2.6.2.4.1 All active pesticide substances and their preparations for which the LC_{50} and/or LD_{50} values are known and which are classified in class 6.1 shall be classified under appropriate packing groups in accordance with the criteria given in 2.6.2.2. Substances and preparations which are characterized by subsidiary risks shall be classified according to the precedence of hazard table in 2.0.3 with the assignment of appropriate packing groups.

2.6.2.4.2 If the oral or dermal LD_{50} value for a pesticide preparation is not known, but the LD_{50} value of its active substance(s) is known, the LD_{50} value for the preparation may be obtained by applying the procedures in 2.6.2.3.

Note: LD_{50} toxicity data for a number of common pesticides may be obtained from the most current edition of "The WHO Recommended Classification of Pesticides by Hazard and Guidelines to Classification", available from the International Programme on Chemical Safety, World Health Organization (WHO), 1211 Geneva 27, Switzerland. While that publication may be used as a source of LD_{50} data for pesticides, its classification system shall not be used for purposes of transport classification of, or assignment of packing groups to, pesticides, which shall be in accordance with the provisions of this Code.

2.6.2.4.3 The Proper Shipping Name used in the transport of the pesticide shall be selected from those referenced on the basis of the active ingredient, of the physical state of the pesticide and any subsidiary risks which it may exhibit.

2.6.3 Class 6.2 – Infectious substances

2.6.3.1 Definitions

For the purposes of this Code:

2.6.3.1.1 *Infectious substances* are substances which are known or are reasonably expected to contain pathogens. Pathogens are defined as micro-organisms (including bacteria, viruses, rickettsiae, parasites, fungi) and other agents such as prions, which can cause disease in humans or animals.

2.6.3.1.2 *Biological products* are those products derived from living organisms which are manufactured and distributed in accordance with the requirements of appropriate national authorities, which may have special licensing requirements, and are used either for prevention, treatment, or diagnosis of disease in humans or animals, or for development, experimental or investigation purposes related thereto. They include, but are not limited to, finished or unfinished products such as vaccines.

2.6.3.1.3 *Cultures* are the result of a process by which pathogens are intentionally propagated. This definition does not include human or animal patient specimens as defined in 2.6.3.1.4.

2.6.3.1.4 *Patient specimens* are human or animal materials, collected directly from humans or animals, including, but not limited to, excreta, secreta, blood and its components, tissue and tissue fluid swabs, and body parts being transported for purposes such as research, diagnosis, investigational activities, disease treatment and prevention.

2.6.3.1.5 *Genetically modified micro-organisms and organisms* are micro-organisms and organisms in which genetic material has been purposely altered through genetic engineering in a way that does not occur naturally.

2.6.3.1.6 *Medical or clinical wastes* are wastes derived from the medical treatment of animals or humans or from bio-research.

2.6.3.2 Classification of infectious substances

2.6.3.2.1 Infectious substances shall be classified in class 6.2 and assigned to UN 2814, UN 2900, UN 3291 or UN 3373, as appropriate.

2.6.3.2.2 Infectious substances are divided into the following categories:

2.6.3.2.2.1 *Category A:* An infectious substance which is transported in a form that, when exposure to it occurs, is capable of causing permanent disability, life-threatening or fatal disease in otherwise healthy humans or animals. Indicative examples of substances that meet these criteria are given in the table in this paragraph.

> **Note:** An exposure occurs when an infectious substance is released outside the protective packaging, resulting in physical contact with humans or animals.

(a) Infectious substances meeting these criteria which cause disease in humans or in both humans and animals shall be assigned to UN 2814. Infectious substances which cause disease only in animals shall be assigned to UN 2900.

(b) Assignment to UN 2814 or UN 2900 shall be based on the known medical history and symptoms of the source human or animal, endemic local conditions, or professional judgement concerning individual circumstances of the human or animal source.

> **Note 1:** The Proper Shipping Name for UN 2814 is INFECTIOUS SUBSTANCE, AFFECTING HUMANS. The Proper Shipping Name for UN 2900 is INFECTIOUS SUBSTANCE, AFFECTING ANIMALS only.

> **Note 2:** The following table is not exhaustive. Infectious substances, including new or emerging pathogens, which do not appear in the table but which meet the same criteria shall be assigned to Category A. In addition, if there is doubt as to whether or not a substance meets the criteria it shall be included in Category A.

> **Note 3:** In the following table, the micro-organism names written in italics are bacteria, mycoplasmas, rickettsia or fungi.

Indicative examples of infectious substances included in category A in any form unless otherwise indicated (2.6.3.2.2.1 (a))

UN Number and Proper Shipping Name	Micro-organism
UN 2814 Infectious substance, affecting humans	*Bacillus anthracis* (cultures only) *Brucella abortus* (cultures only) *Brucella melitensis* (cultures only) *Brucella suis* (cultures only) *Burkholderia mallei – Pseudomonas mallei –* Glanders (cultures only) *Burkholderia pseudomallei – Pseudomonas pseudomallei* (cultures only) *Chlamydia psittaci –* avian strains (cultures only) *Clostridium botulinum* (cultures only) *Coccidioides immitis* (cultures only) *Coxiella burnetii* (cultures only) Crimean-Congo hemorrhagic fever virus Dengue virus (cultures only) Eastern equine encephalitis virus (cultures only) *Escherichia coli*, verotoxigenic (cultures only) Ebola virus Flexal virus *Francisella tularensis* (cultures only) Guanarito virus Hantaan virus Hantavirus causing hemorragic fever with renal syndrome Hendra virus Hepatitis B virus (cultures only) Herpes B virus (cultures only) Human immunodeficiency virus (cultures only) Highly pathogenic avian influenza virus (cultures only) Japanese Encephalitis virus (cultures only) Junin virus Kyasanur Forest disease virus Lassa virus Machupo virus Marburg virus Monkeypox virus *Mycobacterium tuberculosis* (cultures only)

UN Number and Proper Shipping Name	Micro-organism
UN 2814 Infectious substance, affecting humans *(cont.)*	Nipah virus Omsk hemorrhagic fever virus Poliovirus (cultures only) Rabies virus (cultures only) *Rickettsia prowazekii* (cultures only) *Rickettsia rickettsii* (cultures only) Rift Valley fever virus (cultures only) Russian spring–summer encephalitis virus (cultures only) Sabia virus *Shigella dysenteriae* type 1 (cultures only) Tick-borne encephalitis virus (cultures only) Variola virus Venezuelan equine encephalitis virus (cultures only) West Nile virus (cultures only) Yellow fever virus (cultures only) *Yersinia pestis* (cultures only)
UN 2900 Infectious substance, affecting animals only	African swine fever virus (cultures only) Avian paramyxovirus Type 1 – Velogenic Newcastle disease virus (cultures only) Classical swine fever virus (cultures only) Foot and mouth disease virus (cultures only) Lumpy skin disease virus (cultures only) *Mycoplasma mycoides* – Contagious bovine pleuropneumonia (cultures only) Peste des petits ruminants virus (cultures only) Rinderpest virus (cultures only) Sheep-pox virus (cultures only) Goatpox virus (cultures only) Swine vesicular disease virus (cultures only) Vesicular stomatitis virus (cultures only)

2.6.3.2.2 *Category B:* An infectious substance which does not meet the criteria for inclusion in Category A. Infectious substances in Category B shall be assigned to UN 3373.

Note: The Proper Shipping Name for UN 3373 is "BIOLOGICAL SUBSTANCE, CATEGORY B".

2.6.3.2.3 *Exemptions*

2.6.3.2.3.1 Substances which do not contain infectious substances or substances which are unlikely to cause disease in humans or animals are not subject to the provisions of this Code, unless they meet the criteria for inclusion in another class.

2.6.3.2.3.2 Substances containing micro-organisms which are non-pathogenic to humans or animals are not subject to the provisions of this Code unless they meet the criteria for inclusion in another class.

2.6.3.2.3.3 Substances in a form that any present pathogens have been neutralized or inactivated such that they no longer pose a health risk are not subject to the provisions of this Code unless they meet the criteria for inclusion in another class.

2.6.3.2.3.4 Environmental samples (including food and water samples) which are not considered to pose a significant risk of infection are not subject to the provisions of this Code unless they meet the criteria for inclusion in another class.

2.6.3.2.3.5 Dried blood spots, collected by applying a drop of blood onto absorbent material, or faecal occult blood screening tests and blood or blood components which have been collected for the purposes of transfusion or for the preparation of blood products to be used for transfusion or transplantation and any tissues or organs intended for use in transplants are not subject to this Code.

2.6.3.2.3.6 Human or animal specimens for which there is minimal likelihood that pathogens are present are not subject to the provisions of this Code if the specimen is transported in a packaging which will prevent any leakage and which is marked with the words "Exempt human specimen" or "Exempt animal specimen", as appropriate. The packaging should meet the following conditions:

(a)   The packaging should consist of three components:

    (i)   a leak-proof primary receptacle(s);

    (ii)   a leak-proof secondary packaging; and

    (iii)   an outer packaging of adequate strength for its capacity, mass and intended use, and with at least one surface having minimum dimensions of 100 mm × 100 mm;

(b)   For liquids, absorbent material in sufficient quantity to absorb the entire contents should be placed between the primary receptacle(s) and the secondary packaging so that, during transport, any release or leak of a liquid substance will not reach the outer packaging and will not compromise the integrity of the cushioning material;

(c)   When multiple fragile primary receptacles are placed in a single secondary packaging, they should be either individually wrapped or separated to prevent contact between them

Note 1: An element of professional judgement is required to determine if a substance is exempt under this paragraph. That judgement should be based on the known medical history, symptoms and individual circumstances of the source, human or animal, and endemic local conditions. Examples of specimens which may be transported under this paragraph include the blood or urine tests to monitor cholesterol levels, blood glucose levels, hormone levels, or prostate specific antibodies (PSA); those required to monitor organ function such as heart, liver or kidney function for humans or animals with non-infectious diseases, or therapeutic drug monitoring; those conducted for insurance or employment purposes and are intended to determine the presence of drugs or alcohol; pregnancy test; biopsies to detect cancer; and antibody detection in humans or animals in the absence of any concern for infection (e.g., evaluation of vaccine-induced immunity, diagnosis of autoimmune disease, etc.).

2.6.3.3   Biological products

2.6.3.3.1   For the purposes of this Code, biological products are divided into the following groups:

(a)   those which are manufactured and packaged in accordance with the requirements of appropriate national authorities and transported for the purposes of final packaging or distribution, and use for personal health care by medical professionals or individuals. Substances in this group are not subject to the provisions of this Code.

(b)   those which do not fall under (a) and are known or reasonably believed to contain infectious substances and which meet the criteria for inclusion in Category A or Category B. Substances in this group shall be assigned to UN 2814, UN 2900 or UN 3373, as appropriate.

Note: Some licensed biological products may present a biohazard only in certain parts of the world. Competent authorities may require that such biological products comply with local requirements for infectious substances or may impose other restrictions.

2.6.3.4   Genetically modified micro-organisms and organisms

2.6.3.4.1   Genetically modified micro-organisms not meeting the definition of infectious substance shall be classified in accordance with chapter 2.9.

2.6.3.5   Medical or clinical wastes

2.6.3.5.1   Medical or clinical wastes containing Category A infectious substances shall be assigned to UN 2814 or UN 2900, as appropriate. Medical or clinical wastes containing infectious substances in Category B shall be assigned to UN 3291.

2.6.3.5.2   Medical or clinical wastes which are reasonably believed to have a low probability of containing infectious substances shall be assigned to UN 3291. For the assignment, international, regional or national waste catalogues may be taken into account.

Note: The Proper Shipping Name for UN 3291 is CLINICAL WASTE, UNSPECIFIED, N.O.S. or (BIO) MEDICAL WASTE, N.O.S. or REGULATED MEDICAL WASTE, N.O.S.

2.6.3.5.3   Decontaminated medical or clinical wastes which previously contained infectious substances are not subject to the provisions of this Code unless they meet the criteria for inclusion in another class.

2.6.3.6   **Infected animals**

2.6.3.6.1   Unless an infectious substance cannot be consigned by any other means, live animals shall not be used to consign such a substance. A live animal which has been intentionally infected and is known or suspected to contain an infectious substance shall only be transported under terms and conditions approved by the competent authority.

2.6.3.6.2   Animal material affected by pathogens of Category A. Animal material affected by pathogens of Category B other than those which would be assigned to Category A if they were in cultures shall be assigned to UN 3373, or which would be assigned to Category A in cultures only, shall be assigned to UN 2814 or UN 2900 as appropriate.

Chapter 2.7

Class 7 – Radioactive material

Note: For class 7, the type of packaging may have a decisive effect on classification.

2.7.1   Definitions

2.7.1.1   *Radioactive material* means any material containing radionuclides where both the activity concentration and the total activity in the consignment exceed the values specified in 2.7.2.2.1 to 2.7.2.2.6.

2.7.1.2   Contamination

Contamination means the presence of a radioactive substance on a surface in quantities in excess of 0.4 Bq/cm^2 for beta and gamma emitters and low-toxicity alpha emitters, or 0.04 Bq/cm^2 for all other alpha emitters.

Non-fixed contamination means contamination that can be removed from a surface during routine conditions of transport.

Fixed contamination means contamination other than non-fixed contamination.

2.7.1.3   Definitions of specific terms

A_1 and A_2

> A_1 means the activity value of special form radioactive material which is listed in the table in 2.7.2.2.1 or derived in 2.7.2.2.2 and is used to determine the activity limits for the provisions of this Code.

> A_2 means the activity value of radioactive material, other than special form radioactive material, which is listed in the table in 2.7.2.2.1 or derived in 2.7.2.2.2 and is used to determine the activity limits for the provisions of this Code.

Fissile material means uranium-233, uranium-235, plutonium-239, plutonium-241, or any combination of these radionuclides. Excepted from this definition is:

.1   Natural uranium or depleted uranium which is unirradiated; and

.2   Natural uranium or depleted uranium which has been irradiated in thermal reactors only.

Low dispersible radioactive material means either a solid radioactive material or a solid radioactive material in a sealed capsule, that has limited dispersibility and is not in powder form.

Low specific activity (LSA) material means radioactive material which by its nature has a limited specific activity, or radioactive material for which limits of estimated average specific activity apply. External shielding materials surrounding the LSA material shall not be considered in determining the estimated average specific activity.

Low toxicity alpha emitters are: natural uranium; depleted uranium; natural thorium; uranium-235 or uranium-238; thorium-232; thorium-228 and thorium-230 when contained in ores or physical and chemical concentrates; or alpha emitters with a half-life of less than 10 days.

Specific activity of a radionuclide means the activity per unit mass of that nuclide. The specific activity of a material shall mean the activity per unit mass of the material in which the radionuclides are essentially uniformly distributed.

Special form radioactive material means either:

.1   An indispersible solid radioactive material; or

.2   A sealed capsule containing radioactive material.

Surface contaminated object (SCO) means a solid object which is not itself radioactive but which has radioactive material distributed on its surfaces.

Unirradiated thorium means thorium containing not more than 10^{-7} g of uranium-233 per gram of thorium-232.

Unirradiated uranium means uranium containing not more than 2×10^3 Bq of plutonium per gram of uranium-235, not more than 9×10^6 Bq of fission products per gram of uranium-235 and not more than 5×10^3 g of uranium-236 per gram of uranium-235.

Uranium – natural, depleted, enriched means the following:

Natural uranium means uranium (which may be chemically separated) containing the naturally occurring distribution of uranium isotopes (approximately 99.28% uranium-238, and 0.72% uranium-235 by mass).

Depleted uranium means uranium containing a lesser mass percentage of uranium-235 than in natural uranium.

Enriched uranium means uranium containing a greater mass percentage of uranium-235 than 0.72%.

In all cases, a very small mass percentage of uranium-234 is present.

2.7.2 Classification

2.7.2.1 General provisions

2.7.2.1.1 Radioactive material shall be assigned to one of the UN Numbers specified in table 2.7.2.1.1 depending on the activity level of the radionuclides contained in a package, the fissile or non-fissile properties of these radionuclides, the type of package to be presented for transport, and the nature or form of the contents of the package, or special arrangements governing the transport operation, in accordance with the provisions laid down in 2.7.2.2 to 2.7.2.5.

Table 2.7.2.1.1 – Assignment of UN Numbers

Excepted packages (1.5.1.5)	
UN 2908	RADIOACTIVE MATERIAL, EXCEPTED PACKAGE – EMPTY PACKAGING
UN 2909	RADIOACTIVE MATERIAL, EXCEPTED PACKAGE – ARTICLES MANUFACTURED FROM NATURAL URANIUM or DEPLETED URANIUM or NATURAL THORIUM
UN 2910	RADIOACTIVE MATERIAL, EXCEPTED PACKAGE – LIMITED QUANTITY OF MATERIAL
UN 2911	RADIOACTIVE MATERIAL, EXCEPTED PACKAGE – INSTRUMENTS or ARTICLES
Low specific activity radioactive material (2.7.2.3.1)	
UN 2912	RADIOACTIVE MATERIAL, LOW SPECIFIC ACTIVITY (LSA-I), non-fissile or fissile – excepted
UN 3321	RADIOACTIVE MATERIAL, LOW SPECIFIC ACTIVITY (LSA-II), non-fissile or fissile – excepted
UN 3322	RADIOACTIVE MATERIAL, LOW SPECIFIC ACTIVITY (LSA-III), non-fissile or fissile – excepted
UN 3324	RADIOACTIVE MATERIAL, LOW SPECIFIC ACTIVITY (LSA-II), FISSILE
UN 3325	RADIOACTIVE MATERIAL, LOW SPECIFIC ACTIVITY (LSA-III), FISSILE
Surface contaminated objects (2.7.2.3.2)	
UN 2913	RADIOACTIVE MATERIAL, SURFACE CONTAMINATED OBJECTS (SCO-I or SCO-II), non-fissile or fissile – excepted
UN 3326	RADIOACTIVE MATERIAL, SURFACE CONTAMINATED OBJECTS (SCO-I or SCO-II), FISSILE

Type A packages (2.7.2.4.4)	
UN 2915	RADIOACTIVE MATERIAL, TYPE A PACKAGE, non-special form, non-fissile or fissile – excepted
UN 3327	RADIOACTIVE MATERIAL, TYPE A PACKAGE, FISSILE, non-special form
UN 3332	RADIOACTIVE MATERIAL, TYPE A PACKAGE, SPECIAL FORM, non-fissile or fissile – excepted
UN 3333	RADIOACTIVE MATERIAL, TYPE A PACKAGE, SPECIAL FORM, FISSILE
Type B(U) package (2.7.2.4.6)	
UN 2916	RADIOACTIVE MATERIAL, TYPE B(U) PACKAGE, non-fissile or fissile – excepted
UN 3328	RADIOACTIVE MATERIAL, TYPE B(U) PACKAGE, FISSILE
Type B(M) package (2.7.2.4.6)	
UN 2917	RADIOACTIVE MATERIAL, TYPE B(M) PACKAGE, non-fissile or fissile – excepted
UN 3329	RADIOACTIVE MATERIAL, TYPE B(M) PACKAGE, FISSILE
Type C package (2.7.2.4.6)	
UN 3323	RADIOACTIVE MATERIAL, TYPE C PACKAGE, non-fissile or fissile – excepted
UN 3330	RADIOACTIVE MATERIAL, TYPE C PACKAGE, FISSILE
Special arrangement (2.7.2.5)	
UN 2919	RADIOACTIVE MATERIAL, TRANSPORTED UNDER SPECIAL ARRANGEMENT, non-fissile or fissile – excepted
UN 3331	RADIOACTIVE MATERIAL, TRANSPORTED UNDER SPECIAL ARRANGEMENT, FISSILE
Uranium hexafluoride (2.7.2.4.5)	
UN 2977	RADIOACTIVE MATERIAL, URANIUM HEXAFLUORIDE, FISSILE
UN 2978	RADIOACTIVE MATERIAL, URANIUM HEXAFLUORIDE, non-fissile or fissile – excepted

2.7.2.2   Determination of activity level

2.7.2.2.1   The following basic values for individual radionuclides are given in table 2.7.2.2.1:

.1 A_1 and A_2 in TBq;

.2   Activity concentration for exempt material in Bq/g; and

.3   Activity limits for exempt consignments in Bq.

Table 2.7.2.2.1 – Basic radionuclides values for individual radionuclides

Radionuclide (atomic number)	A_1 (TBq)	A_2 (TBq)	Activity concentration for exempt material (Bq/g)	Activity limit for an exempt consignment (Bq)
Actinium (89)				
Ac-225 *(a)*	8×10^{-1}	6×10^{-3}	1×10^{1}	1×10^{4}
Ac-227 *(a)*	9×10^{-1}	9×10^{-5}	1×10^{-1}	1×10^{3}
Ac-228	6×10^{-1}	5×10^{-1}	1×10^{1}	1×10^{6}
Silver (47)				
Ag-105	2×10^{0}	2×10^{0}	1×10^{2}	1×10^{6}
Ag-108m *(a)*	7×10^{-1}	7×10^{-1}	1×10^{1} *(b)*	1×10^{6} *(b)*
Ag-110m *(a)*	4×10^{-1}	4×10^{-1}	1×10^{1}	1×10^{6}
Ag-111	2×10^{0}	6×10^{-1}	1×10^{3}	1×10^{6}

Radionuclide (atomic number)	A_1 (TBq)	A_2 (TBq)	Activity concentration for exempt material (Bq/g)	Activity limit for an exempt consignment (Bq)
Aluminium (13)				
Al-26	1×10^{-1}	1×10^{-1}	1×10^{1}	1×10^{5}
Americium (95)				
Am-241	1×10^{1}	1×10^{-3}	1×10^{0}	1×10^{4}
Am-242m *(a)*	1×10^{1}	1×10^{-3}	1×10^{0} *(b)*	1×10^{4} *(b)*
Am-243 *(a)*	5×10^{0}	1×10^{-3}	1×10^{0} *(b)*	1×10^{3} *(b)*
Argon (18)				
Ar-37	4×10^{1}	4×10^{1}	1×10^{6}	1×10^{8}
Ar-39	4×10^{1}	2×10^{1}	1×10^{7}	1×10^{4}
Ar-41	3×10^{-1}	3×10^{-1}	1×10^{2}	1×10^{9}
Arsenic (33)				
As-72	3×10^{-1}	3×10^{-1}	1×10^{1}	1×10^{5}
As-73	4×10^{1}	4×10^{1}	1×10^{3}	1×10^{7}
As-74	1×10^{0}	9×10^{-1}	1×10^{1}	1×10^{6}
As-76	3×10^{-1}	3×10^{-1}	1×10^{2}	1×10^{5}
As-77	2×10^{1}	7×10^{-1}	1×10^{3}	1×10^{6}
Astatine (85)				
At-211 *(a)*	2×10^{1}	5×10^{-1}	1×10^{3}	1×10^{7}
Gold (79)				
Au-193	7×10^{0}	2×10^{0}	1×10^{2}	1×10^{7}
Au-194	1×10^{0}	1×10^{0}	1×10^{1}	1×10^{6}
Au-195	1×10^{1}	6×10^{0}	1×10^{2}	1×10^{7}
Au-198	1×10^{0}	6×10^{-1}	1×10^{2}	1×10^{6}
Au-199	1×10^{1}	6×10^{-1}	1×10^{2}	1×10^{6}
Barium (56)				
Ba-131 *(a)*	2×10^{0}	2×10^{0}	1×10^{2}	1×10^{6}
Ba-133	3×10^{0}	3×10^{0}	1×10^{2}	1×10^{6}
Ba-133m	2×10^{1}	6×10^{-1}	1×10^{2}	1×10^{6}
Ba-140 *(a)*	5×10^{-1}	3×10^{-1}	1×10^{1} *(b)*	1×10^{5} *(b)*
Beryllium (4)				
Be–7	2×10^{1}	2×10^{1}	1×10^{3}	1×10^{7}
Be-10	4×10^{1}	6×10^{-1}	1×10^{4}	1×10^{6}
Bismuth (83)				
Bi-205	7×10^{-1}	7×10^{-1}	1×10^{1}	1×10^{6}
Bi-206	3×10^{-1}	3×10^{-1}	1×10^{1}	1×10^{5}
Bi-207	7×10^{-1}	7×10^{-1}	1×10^{1}	1×10^{6}
Bi-210	1×10^{0}	6×10^{-1}	1×10^{3}	1×10^{6}
Bi-210m *(a)*	6×10^{-1}	2×10^{-2}	1×10^{1}	1×10^{5}
Bi-212 *(a)*	7×10^{-1}	6×10^{-1}	1×10^{1} *(b)*	1×10^{5} *(b)*
Berkelium (97)				
Bk-247	8×10^{0}	8×10^{-4}	1×10^{0}	1×10^{4}
Bk-249 *(a)*	4×10^{1}	3×10^{-1}	1×10^{3}	1×10^{6}
Bromine (35)				
Br-76	4×10^{-1}	4×10^{-1}	1×10^{1}	1×10^{5}
Br-77	3×10^{0}	3×10^{0}	1×10^{2}	1×10^{6}
Br-82	4×10^{-1}	4×10^{-1}	1×10^{1}	1×10^{6}

Radionuclide (atomic number)	A_1 (TBq)	A_2 (TBq)	Activity concentration for exempt material (Bq/g)	Activity limit for an exempt consignment (Bq)
Carbon (6)				
C-11	1×10^0	6×10^{-1}	1×10^1	1×10^6
C-14	4×10^1	3×10^0	1×10^4	1×10^7
Calcium (20)				
Ca-41	Unlimited	Unlimited	1×10^5	1×10^7
Ca-45	4×10^1	1×10^0	1×10^4	1×10^7
Ca-47 *(a)*	3×10^0	3×10^{-1}	1×10^1	1×10^6
Cadmium (48)				
Cd-109	3×10^1	2×10^0	1×10^4	1×10^6
Cd-113m	4×10^1	5×10^{-1}	1×10^3	1×10^6
Cd-115 *(a)*	3×10^0	4×10^{-1}	1×10^2	1×10^6
Cd-115m	5×10^{-1}	5×10^{-1}	1×10^3	1×10^6
Cerium (58)				
Ce-139	7×10^0	2×10^0	1×10^2	1×10^6
Ce-141	2×10^1	6×10^{-1}	1×10^2	1×10^7
Ce-143	9×10^{-1}	6×10^{-1}	1×10^2	1×10^6
Ce-144 *(a)*	2×10^{-1}	2×10^{-1}	1×10^2 *(b)*	1×10^5 *(b)*
Californium (98)				
Cf-248	4×10^1	6×10^{-3}	1×10^1	1×10^4
Cf-249	3×10^0	8×10^{-4}	1×10^0	1×10^3
Cf-250	2×10^1	2×10^{-3}	1×10^1	1×10^4
Cf-251	7×10^0	7×10^{-4}	1×10^0	1×10^3
Cf-252	1×10^{-1}	3×10^{-3}	1×10^1	1×10^4
Cf-253 *(a)*	4×10^1	4×10^{-2}	1×10^2	1×10^5
Cf-254	1×10^{-3}	1×10^{-3}	1×10^0	1×10^3
Chlorine (17)				
Cl-36	1×10^1	6×10^{-1}	1×10^4	1×10^6
Cl-38	2×10^{-1}	2×10^{-1}	1×10^1	1×10^5
Curium (96)				
Cm-240	4×10^1	2×10^{-2}	1×10^2	1×10^5
Cm-241	2×10^0	1×10^0	1×10^2	1×10^6
Cm-242	4×10^1	1×10^{-2}	1×10^2	1×10^5
Cm-243	9×10^0	1×10^{-3}	1×10^0	1×10^4
Cm-244	2×10^1	2×10^{-3}	1×10^1	1×10^4
Cm-245	9×10^0	9×10^{-4}	1×10^0	1×10^3
Cm-246	9×10^0	9×10^{-4}	1×10^0	1×10^3
Cm-247 *(a)*	3×10^0	1×10^{-3}	1×10^0	1×10^4
Cm-248	2×10^{-2}	3×10^{-4}	1×10^0	1×10^3
Cobalt (27)				
Co-55	5×10^{-1}	5×10^{-1}	1×10^1	1×10^6
Co-56	3×10^{-1}	3×10^{-1}	1×10^1	1×10^5
Co-57	1×10^1	1×10^1	1×10^2	1×10^6
Co-58	1×10^0	1×10^0	1×10^1	1×10^6
Co-58m	4×10^1	4×10^1	1×10^4	1×10^7
Co-60	4×10^{-1}	4×10^{-1}	1×10^1	1×10^5

Radionuclide (atomic number)	A_1 (TBq)	A_2 (TBq)	Activity concentration for exempt material (Bq/g)	Activity limit for an exempt consignment (Bq)
Chromium (24)				
Cr-51	3×10^1	3×10^1	1×10^3	1×10^7
Caesium (55)				
Cs-129	4×10^0	4×10^0	1×10^2	1×10^5
Cs-131	3×10^1	3×10^1	1×10^3	1×10^6
Cs-132	1×10^0	1×10^0	1×10^1	1×10^5
Cs-134	7×10^{-1}	7×10^{-1}	1×10^1	1×10^4
Cs-134m	4×10^1	6×10^{-1}	1×10^3	1×10^5
Cs-135	4×10^1	1×10^0	1×10^4	1×10^7
Cs-136	5×10^{-1}	5×10^{-1}	1×10^1	1×10^5
Cs-137 *(a)*	2×10^0	6×10^{-1}	1×10^1 *(b)*	1×10^4 *(b)*
Copper (29)				
Cu-64	6×10^0	1×10^0	1×10^2	1×10^6
Cu-67	1×10^1	7×10^{-1}	1×10^2	1×10^6
Dysprosium (66)				
Dy-159	2×10^1	2×10^1	1×10^3	1×10^7
Dy-165	9×10^{-1}	6×10^{-1}	1×10^3	1×10^6
Dy-166 *(a)*	9×10^{-1}	3×10^{-1}	1×10^3	1×10^6
Erbium (68)				
Er-169	4×10^1	1×10^0	1×10^4	1×10^7
Er-171	8×10^{-1}	5×10^{-1}	1×10^2	1×10^6
Europium (63)				
Eu-147	2×10^0	2×10^0	1×10^2	1×10^6
Eu-148	5×10^{-1}	5×10^{-1}	1×10^1	1×10^6
Eu-149	2×10^1	2×10^1	1×10^2	1×10^7
Eu-150 (short-lived)	2×10^0	7×10^{-1}	1×10^3	1×10^6
Eu-150 (long-lived)	7×10^{-1}	7×10^{-1}	1×10^1	1×10^6
Eu-152	1×10^0	1×10^0	1×10^1	1×10^6
Eu-152m	8×10^{-1}	8×10^{-1}	1×10^2	1×10^6
Eu-154	9×10^{-1}	6×10^{-1}	1×10^1	1×10^6
Eu-155	2×10^1	3×10^0	1×10^2	1×10^7
Eu-156	7×10^{-1}	7×10^{-1}	1×10^1	1×10^6
Fluorine (9)				
F-18	1×10^0	6×10^{-1}	1×10^1	1×10^6
Iron (26)				
Fe-52 *(a)*	3×10^{-1}	3×10^{-1}	1×10^1	1×10^6
Fe-55	4×10^1	4×10^1	1×10^4	1×10^6
Fe-59	9×10^{-1}	9×10^{-1}	1×10^1	1×10^6
Fe-60 *(a)*	4×10^1	2×10^{-1}	1×10^2	1×10^5
Gallium (31)				
Ga-67	7×10^0	3×10^0	1×10^2	1×10^6
Ga-68	5×10^{-1}	5×10^{-1}	1×10^1	1×10^5
Ga-72	4×10^{-1}	4×10^{-1}	1×10^1	1×10^5

Radionuclide (atomic number)	A_1 (TBq)	A_2 (TBq)	Activity concentration for exempt material (Bq/g)	Activity limit for an exempt consignment (Bq)
Gadolinium (64)				
Gd-146 *(a)*	5×10^{-1}	5×10^{-1}	1×10^{1}	1×10^{6}
Gd-148	2×10^{1}	2×10^{-3}	1×10^{1}	1×10^{4}
Gd-153	1×10^{1}	9×10^{0}	1×10^{2}	1×10^{7}
Gd-159	3×10^{0}	6×10^{-1}	1×10^{3}	1×10^{6}
Germanium (32)				
Ge-68 *(a)*	5×10^{-1}	5×10^{-1}	1×10^{1}	1×10^{5}
Ge-71	4×10^{1}	4×10^{1}	1×10^{4}	1×10^{8}
Ge-77	3×10^{-1}	3×10^{-1}	1×10^{1}	1×10^{5}
Hafnium (72)				
Hf-172 *(a)*	6×10^{-1}	6×10^{-1}	1×10^{1}	1×10^{6}
Hf-175	3×10^{0}	3×10^{0}	1×10^{2}	1×10^{6}
Hf-181	2×10^{0}	5×10^{-1}	1×10^{1}	1×10^{6}
Hf-182	Unlimited	Unlimited	1×10^{2}	1×10^{6}
Mercury (80)				
Hg-194 *(a)*	1×10^{0}	1×10^{0}	1×10^{1}	1×10^{6}
Hg-195m *(a)*	3×10^{0}	7×10^{-1}	1×10^{2}	1×10^{6}
Hg-197	2×10^{1}	1×10^{1}	1×10^{2}	1×10^{7}
Hg-197m	1×10^{1}	4×10^{-1}	1×10^{2}	1×10^{6}
Hg-203	5×10^{0}	1×10^{0}	1×10^{2}	1×10^{5}
Holmium (67)				
Ho-166	4×10^{-1}	4×10^{-1}	1×10^{3}	1×10^{5}
Ho-166m	6×10^{-1}	5×10^{-1}	1×10^{1}	1×10^{6}
Iodine (53)				
I-123	6×10^{0}	3×10^{0}	1×10^{2}	1×10^{7}
I-124	1×10^{0}	1×10^{0}	1×10^{1}	1×10^{6}
I-125	2×10^{1}	3×10^{0}	1×10^{3}	1×10^{6}
I-126	2×10^{0}	1×10^{0}	1×10^{2}	1×10^{6}
I-129	Unlimited	Unlimited	1×10^{2}	1×10^{5}
I-131	3×10^{0}	7×10^{-1}	1×10^{2}	1×10^{6}
I-132	4×10^{-1}	4×10^{-1}	1×10^{1}	1×10^{5}
I-133	7×10^{-1}	6×10^{-1}	1×10^{1}	1×10^{6}
I-134	3×10^{-1}	3×10^{-1}	1×10^{1}	1×10^{5}
I-135 *(a)*	6×10^{-1}	6×10^{-1}	1×10^{1}	1×10^{6}
Indium (49)				
In-111	3×10^{0}	3×10^{0}	1×10^{2}	1×10^{6}
In-113m	4×10^{0}	2×10^{0}	1×10^{2}	1×10^{6}
In-114m *(a)*	1×10^{1}	5×10^{-1}	1×10^{2}	1×10^{6}
In-115m	7×10^{0}	1×10^{0}	1×10^{2}	1×10^{6}
Iridium (77)				
Ir-189 *(a)*	1×10^{1}	1×10^{1}	1×10^{2}	1×10^{7}
Ir-190	7×10^{-1}	7×10^{-1}	1×10^{1}	1×10^{6}
Ir-192	1×10^{0} *(c)*	6×10^{-1}	1×10^{1}	1×10^{4}
Ir-194	3×10^{-1}	3×10^{-1}	1×10^{2}	1×10^{5}

2

Radionuclide (atomic number)	A_1 (TBq)	A_2 (TBq)	Activity concentration for exempt material (Bq/g)	Activity limit for an exempt consignment (Bq)
Potassium (19)				
K-40	9×10^{-1}	9×10^{-1}	1×10^2	1×10^6
K-42	2×10^{-1}	2×10^{-1}	1×10^2	1×10^6
K-43	7×10^{-1}	6×10^{-1}	1×10^1	1×10^6
Krypton (36)				
Kr-81	4×10^1	4×10^1	1×10^4	1×10^7
Kr-85	1×10^1	1×10^1	1×10^5	1×10^4
Kr-85m	8×10^0	3×10^0	1×10^3	1×10^{10}
Kr-87	2×10^{-1}	2×10^{-1}	1×10^2	1×10^9
Lanthanum (57)				
La-137	3×10^1	6×10^0	1×10^3	1×10^7
La-140	4×10^{-1}	4×10^{-1}	1×10^1	1×10^5
Lutetium (71)				
Lu-172	6×10^{-1}	6×10^{-1}	1×10^1	1×10^6
Lu-173	8×10^0	8×10^0	1×10^2	1×10^7
Lu-174	9×10^0	9×10^0	1×10^2	1×10^7
Lu-174m	2×10^1	1×10^1	1×10^2	1×10^7
Lu-177	3×10^1	7×10^{-1}	1×10^3	1×10^7
Magnesium (12)				
Mg-28 *(a)*	3×10^{-1}	3×10^{-1}	1×10^1	1×10^5
Manganese (25)				
Mn-52	3×10^{-1}	3×10^{-1}	1×10^1	1×10^5
Mn-53	Unlimited	Unlimited	1×10^4	1×10^9
Mn-54	1×10^0	1×10^0	1×10^1	1×10^6
Mn-56	3×10^{-1}	3×10^{-1}	1×10^1	1×10^5
Molybdenum (42)				
Mo-93	4×10^1	2×10^1	1×10^3	1×10^8
Mo-99 *(a)*	1×10^0	6×10^{-1}	1×10^2	1×10^6
Nitrogen (7)				
N-13	9×10^{-1}	6×10^{-1}	1×10^2	1×10^9
Sodium (11)				
Na-22	5×10^{-1}	5×10^{-1}	1×10^1	1×10^6
Na-24	2×10^{-1}	2×10^{-1}	1×10^1	1×10^5
Niobium (41)				
Nb-93m	4×10^1	3×10^1	1×10^4	1×10^7
Nb-94	7×10^{-1}	7×10^{-1}	1×10^1	1×10^6
Nb-95	1×10^0	1×10^0	1×10^1	1×10^6
Nb-97	9×10^{-1}	6×10^{-1}	1×10^1	1×10^6
Neodymium (60)				
Nd-147	6×10^0	6×10^{-1}	1×10^2	1×10^6
Nd-149	6×10^{-1}	5×10^{-1}	1×10^2	1×10^6
Nickel (28)				
Ni-59	Unlimited	Unlimited	1×10^4	1×10^8
Ni-63	4×10^1	3×10^1	1×10^5	1×10^8
Ni-65	4×10^{-1}	4×10^{-1}	1×10^1	1×10^6

Radionuclide (atomic number)	A_1 (TBq)	A_2 (TBq)	Activity concentration for exempt material (Bq/g)	Activity limit for an exempt consignment (Bq)
Neptunium (93)				
Np-235	4×10^1	4×10^1	1×10^3	1×10^7
Np-236 (short-lived)	2×10^1	2×10^0	1×10^3	1×10^7
Np-236 (long-lived)	9×10^0	2×10^{-2}	1×10^2	1×10^5
Np-237	2×10^1	2×10^{-3}	1×10^0 *(b)*	1×10^3 *(b)*
Np-239	7×10^0	4×10^{-1}	1×10^2	1×10^7
Osmium (76)				
Os-185	1×10^0	1×10^0	1×10^1	1×10^6
Os-191	1×10^1	2×10^0	1×10^2	1×10^7
Os-191m	4×10^1	3×10^1	1×10^3	1×10^7
Os-193	2×10^0	6×10^{-1}	1×10^2	1×10^6
Os-194 *(a)*	3×10^{-1}	3×10^{-1}	1×10^2	1×10^5
Phosphorus (15)				
P-32	5×10^{-1}	5×10^{-1}	1×10^3	1×10^5
P-33	4×10^1	1×10^0	1×10^5	1×10^8
Protactinium (91)				
Pa-230 *(a)*	2×10^0	7×10^{-2}	1×10^1	1×10^6
Pa-231	4×10^0	4×10^{-4}	1×10^0	1×10^3
Pa-233	5×10^0	7×10^{-1}	1×10^2	1×10^7
Lead (82)				
Pb-201	1×10^0	1×10^0	1×10^1	1×10^6
Pb-202	4×10^1	2×10^1	1×10^3	1×10^6
Pb-203	4×10^0	3×10^0	1×10^2	1×10^6
Pb-205	Unlimited	Unlimited	1×10^4	1×10^7
Pb-210 *(a)*	1×10^0	5×10^{-2}	1×10^1 *(b)*	1×10^4 *(b)*
Pb-212 *(a)*	7×10^{-1}	2×10^{-1}	1×10^1 *(b)*	1×10^5 *(b)*
Palladium (46)				
Pd-103 *(a)*	4×10^1	4×10^1	1×10^3	1×10^8
Pd-107	Unlimited	Unlimited	1×10^5	1×10^8
Pd-109	2×10^0	5×10^{-1}	1×10^3	1×10^6
Promethium (61)				
Pm-143	3×10^0	3×10^0	1×10^2	1×10^6
Pm-144	7×10^{-1}	7×10^{-1}	1×10^1	1×10^6
Pm-145	3×10^1	1×10^1	1×10^3	1×10^7
Pm-147	4×10^1	2×10^0	1×10^4	1×10^7
Pm-148m *(a)*	8×10^{-1}	7×10^{-1}	1×10^1	1×10^6
Pm-149	2×10^0	6×10^{-1}	1×10^3	1×10^6
Pm-151	2×10^0	6×10^{-1}	1×10^2	1×10^6
Polonium (84)				
Po-210	4×10^1	2×10^{-2}	1×10^1	1×10^4
Praseodymium (59)				
Pr-142	4×10^{-1}	4×10^{-1}	1×10^2	1×10^5
Pr-143	3×10^0	6×10^{-1}	1×10^4	1×10^6

Radionuclide (atomic number)	A_1 (TBq)	A_2 (TBq)	Activity concentration for exempt material (Bq/g)	Activity limit for an exempt consignment (Bq)
Platinum (78)				
Pt-188 *(a)*	1×10^0	8×10^{-1}	1×10^1	1×10^6
Pt-191	4×10^0	3×10^0	1×10^2	1×10^6
Pt-193	4×10^1	4×10^1	1×10^4	1×10^7
Pt-193m	4×10^1	5×10^{-1}	1×10^3	1×10^7
Pt-195m	1×10^1	5×10^{-1}	1×10^2	1×10^6
Pt-197	2×10^1	6×10^{-1}	1×10^3	1×10^6
Pt-197m	1×10^1	6×10^{-1}	1×10^2	1×10^6
Plutonium (94)				
Pu-236	3×10^1	3×10^{-3}	1×10^1	1×10^4
Pu-237	2×10^1	2×10^1	1×10^3	1×10^7
Pu-238	1×10^1	1×10^{-3}	1×10^0	1×10^4
Pu-239	1×10^1	1×10^{-3}	1×10^0	1×10^4
Pu-240	1×10^1	1×10^{-3}	1×10^0	1×10^3
Pu-241 *(a)*	4×10^1	6×10^{-2}	1×10^2	1×10^5
Pu-242	1×10^1	1×10^{-3}	1×10^0	1×10^4
Pu-244 *(a)*	4×10^{-1}	1×10^{-3}	1×10^0	1×10^4
Radium (88)				
Ra-223 *(a)*	4×10^{-1}	7×10^{-3}	1×10^2 *(b)*	1×10^5 *(b)*
Ra-224 *(a)*	4×10^{-1}	2×10^{-2}	1×10^1 *(b)*	1×10^5 *(b)*
Ra-225 *(a)*	2×10^{-1}	4×10^{-3}	1×10^2	1×10^5
Ra-226 *(a)*	2×10^{-1}	3×10^{-3}	1×10^1 *(b)*	1×10^4 *(b)*
Ra-228 *(a)*	6×10^{-1}	2×10^{-2}	1×10^1 *(b)*	1×10^5 *(b)*
Rubidium (37)				
Rb-81	2×10^0	8×10^{-1}	1×10^1	1×10^6
Rb-83 *(a)*	2×10^0	2×10^0	1×10^2	1×10^6
Rb-84	1×10^0	1×10^0	1×10^1	1×10^6
Rb-86	5×10^{-1}	5×10^{-1}	1×10^2	1×10^5
Rb-87	Unlimited	Unlimited	1×10^4	1×10^7
Rb (nat)	Unlimited	Unlimited	1×10^4	1×10^7
Rhenium (75)				
Re-184	1×10^0	1×10^0	1×10^1	1×10^6
Re-184m	3×10^0	1×10^0	1×10^2	1×10^6
Re-186	2×10^0	6×10^{-1}	1×10^3	1×10^6
Re-187	Unlimited	Unlimited	1×10^6	1×10^9
Re-188	4×10^{-1}	4×10^{-1}	1×10^2	1×10^5
Re-189 *(a)*	3×10^0	6×10^{-1}	1×10^2	1×10^6
Re (nat)	Unlimited	Unlimited	1×10^6	1×10^9
Rhodium (45)				
Rh-99	2×10^0	2×10^0	1×10^1	1×10^6
Rh-101	4×10^0	3×10^0	1×10^2	1×10^7
Rh-102	5×10^{-1}	5×10^{-1}	1×10^1	1×10^6
Rh-102m	2×10^0	2×10^0	1×10^2	1×10^6
Rh-103m	4×10^1	4×10^1	1×10^4	1×10^8
Rh-105	1×10^1	8×10^{-1}	1×10^2	1×10^7

Radionuclide (atomic number)	A_1 (TBq)	A_2 (TBq)	Activity concentration for exempt material (Bq/g)	Activity limit for an exempt consignment (Bq)
Radon (86)				
Rn-222 *(a)*	3×10^{-1}	4×10^{-3}	1×10^{1} *(b)*	1×10^{8} *(b)*
Ruthenium (44)				
Ru-97	5×10^{0}	5×10^{0}	1×10^{2}	1×10^{7}
Ru-103 *(a)*	2×10^{0}	2×10^{0}	1×10^{2}	1×10^{6}
Ru-105	1×10^{0}	6×10^{-1}	1×10^{1}	1×10^{6}
Ru-106 *(a)*	2×10^{-1}	2×10^{-1}	1×10^{2} *(b)*	1×10^{5} *(b)*
Sulphur (16)				
S-35	4×10^{1}	3×10^{0}	1×10^{5}	1×10^{8}
Antimony (51)				
Sb-122	4×10^{-1}	4×10^{-1}	1×10^{2}	1×10^{4}
Sb-124	6×10^{-1}	6×10^{-1}	1×10^{1}	1×10^{6}
Sb-125	2×10^{0}	1×10^{0}	1×10^{2}	1×10^{6}
Sb-126	4×10^{-1}	4×10^{-1}	1×10^{1}	1×10^{5}
Scandium (21)				
Sc-44	5×10^{-1}	5×10^{-1}	1×10^{1}	1×10^{5}
Sc-46	5×10^{-1}	5×10^{-1}	1×10^{1}	1×10^{6}
Sc-47	1×10^{1}	7×10^{-1}	1×10^{2}	1×10^{6}
Sc-48	3×10^{-1}	3×10^{-1}	1×10^{1}	1×10^{5}
Selenium (34)				
Se-75	3×10^{0}	3×10^{0}	1×10^{2}	1×10^{6}
Se-79	4×10^{1}	2×10^{0}	1×10^{4}	1×10^{7}
Silicon (14)				
Si-31	6×10^{-1}	6×10^{-1}	1×10^{3}	1×10^{6}
Si-32	4×10^{1}	5×10^{-1}	1×10^{3}	1×10^{6}
Samarium (62)				
Sm-145	1×10^{1}	1×10^{1}	1×10^{2}	1×10^{7}
Sm-147	Unlimited	Unlimited	1×10^{1}	1×10^{4}
Sm-151	4×10^{1}	1×10^{1}	1×10^{4}	1×10^{8}
Sm-153	9×10^{0}	6×10^{-1}	1×10^{2}	1×10^{6}
Tin (50)				
Sn-113 *(a)*	4×10^{0}	2×10^{0}	1×10^{3}	1×10^{7}
Sn-117m	7×10^{0}	4×10^{-1}	1×10^{2}	1×10^{6}
Sn-119m	4×10^{1}	3×10^{1}	1×10^{3}	1×10^{7}
Sn-121m *(a)*	4×10^{1}	9×10^{-1}	1×10^{3}	1×10^{7}
Sn-123	8×10^{-1}	6×10^{-1}	1×10^{3}	1×10^{6}
Sn-125	4×10^{-1}	4×10^{-1}	1×10^{2}	1×10^{5}
Sn-126 *(a)*	6×10^{-1}	4×10^{-1}	1×10^{1}	1×10^{5}

2

Radionuclide (atomic number)	A_1 (TBq)	A_2 (TBq)	Activity concentration for exempt material (Bq/g)	Activity limit for an exempt consignment (Bq)
Strontium (38)				
Sr-82 *(a)*	2×10^{-1}	2×10^{-1}	1×10^{1}	1×10^{5}
Sr-85	2×10^{0}	2×10^{0}	1×10^{2}	1×10^{6}
Sr-85m	5×10^{0}	5×10^{0}	1×10^{2}	1×10^{7}
Sr-87m	3×10^{0}	3×10^{0}	1×10^{2}	1×10^{6}
Sr-89	6×10^{-1}	6×10^{-1}	1×10^{3}	1×10^{6}
Sr-90 *(a)*	3×10^{-1}	3×10^{-1}	1×10^{2} *(b)*	1×10^{4} *(b)*
Sr-91 *(a)*	3×10^{-1}	3×10^{-1}	1×10^{1}	1×10^{5}
Sr-92 *(a)*	1×10^{0}	3×10^{-1}	1×10^{1}	1×10^{6}
Tritium (1)				
T (H-3)	4×10^{1}	4×10^{1}	1×10^{6}	1×10^{9}
Tantalum (73)				
Ta-178 (long-lived)	1×10^{0}	8×10^{-1}	1×10^{1}	1×10^{6}
Ta-179	3×10^{1}	3×10^{1}	1×10^{3}	1×10^{7}
Ta-182	9×10^{-1}	5×10^{-1}	1×10^{1}	1×10^{4}
Terbium (65)				
Tb-157	4×10^{1}	4×10^{1}	1×10^{4}	1×10^{7}
Tb-158	1×10^{0}	1×10^{0}	1×10^{1}	1×10^{6}
Tb-160	1×10^{0}	6×10^{-1}	1×10^{1}	1×10^{6}
Technetium (43)				
Tc-95m *(a)*	2×10^{0}	2×10^{0}	1×10^{1}	1×10^{6}
Tc-96	4×10^{-1}	4×10^{-1}	1×10^{1}	1×10^{6}
Tc-96m *(a)*	4×10^{-1}	4×10^{-1}	1×10^{3}	1×10^{7}
Tc-97	Unlimited	Unlimited	1×10^{3}	1×10^{8}
Tc-97m	4×10^{1}	1×10^{0}	1×10^{3}	1×10^{7}
Tc-98	8×10^{-1}	7×10^{-1}	1×10^{1}	1×10^{6}
Tc-99	4×10^{1}	9×10^{-1}	1×10^{4}	1×10^{7}
Tc-99m	1×10^{1}	4×10^{0}	1×10^{2}	1×10^{7}
Tellurium (52)				
Te-121	2×10^{0}	2×10^{0}	1×10^{1}	1×10^{6}
Te-121m	5×10^{0}	3×10^{0}	1×10^{2}	1×10^{6}
Te-123m	8×10^{0}	1×10^{0}	1×10^{2}	1×10^{7}
Te-125m	2×10^{1}	9×10^{-1}	1×10^{3}	1×10^{7}
Te-127	2×10^{1}	7×10^{-1}	1×10^{3}	1×10^{6}
Te-127m *(a)*	2×10^{1}	5×10^{-1}	1×10^{3}	1×10^{7}
Te-129	7×10^{-1}	6×10^{-1}	1×10^{2}	1×10^{6}
Te-129m *(a)*	8×10^{-1}	4×10^{-1}	1×10^{3}	1×10^{6}
Te-131m *(a)*	7×10^{-1}	5×10^{-1}	1×10^{1}	1×10^{6}
Te-132 *(a)*	5×10^{-1}	4×10^{-1}	1×10^{2}	1×10^{7}

Radionuclide (atomic number)	A_1 (TBq)	A_2 (TBq)	Activity concentration for exempt material (Bq/g)	Activity limit for an exempt consignment (Bq)
Thorium (90)				
Th-227	1×10^1	5×10^{-3}	1×10^1	1×10^4
Th-228 *(a)*	5×10^{-1}	1×10^{-3}	1×10^0 *(b)*	1×10^4 *(b)*
Th-229	5×10^0	5×10^{-4}	1×10^0 *(b)*	1×10^3 *(b)*
Th-230	1×10^1	1×10^{-3}	1×10^0	1×10^4
Th-231	4×10^1	2×10^{-2}	1×10^3	1×10^7
Th-232	Unlimited	Unlimited	1×10^1	1×10^4
Th-234 *(a)*	3×10^{-1}	3×10^{-1}	1×10^3 *(b)*	1×10^5 *(b)*
Th (nat)	Unlimited	Unlimited	1×10^0 *(b)*	1×10^3 *(b)*
Titanium (22)				
Ti-44 *(a)*	5×10^{-1}	4×10^{-1}	1×10^1	1×10^5
Thallium (81)				
Tl-200	9×10^{-1}	9×10^{-1}	1×10^1	1×10^6
Tl-201	1×10^1	4×10^0	1×10^2	1×10^6
Tl-202	2×10^0	2×10^0	1×10^2	1×10^6
Tl-204	1×10^1	7×10^{-1}	1×10^4	1×10^4
Thulium (69)				
Tm-167	7×10^0	8×10^{-1}	1×10^2	1×10^6
Tm-170	3×10^0	6×10^{-1}	1×10^3	1×10^6
Tm-171	4×10^1	4×10^1	1×10^4	1×10^8
Uranium (92)				
U-230 (fast lung absorption) *(a) (d)*	4×10^1	1×10^{-1}	1×10^1 *(b)*	1×10^5 *(b)*
U-230 (medium lung absorption) *(a) (e)*	4×10^1	4×10^{-3}	1×10^1	1×10^4
U-230 (slow lung absorption) *(a) (f)*	3×10^1	3×10^{-3}	1×10^1	1×10^4
U-232 (fast lung absorption) *(d)*	4×10^1	1×10^{-2}	1×10^0 *(b)*	1×10^3 *(b)*
U-232 (medium lung absorption) *(e)*	4×10^1	7×10^{-3}	1×10^1	1×10^4
U-232 (slow lung absorption) *(f)*	1×10^1	1×10^{-3}	1×10^1	1×10^4
U-233 (fast lung absorption) *(d)*	4×10^1	9×10^{-2}	1×10^1	1×10^4
U-233 (medium lung absorption) *(e)*	4×10^1	2×10^{-2}	1×10^2	1×10^5
U-233 (slow lung absorption) *(f)*	4×10^1	6×10^{-3}	1×10^1	1×10^5
U-234 (fast lung absorption) *(d)*	4×10^1	9×10^{-2}	1×10^1	1×10^4
U-234 (medium lung absorption) *(e)*	4×10^1	2×10^{-2}	1×10^2	1×10^5
U-234 (slow lung absorption) *(f)*	4×10^1	6×10^{-3}	1×10^1	1×10^5
U-235 (all lung absorption types) *(a) (d) (e) (f)*	Unlimited	Unlimited	1×10^1 *(b)*	1×10^4 *(b)*
U-236 (fast lung absorption) *(d)*	Unlimited	Unlimited	1×10^1	1×10^4
U-236 (medium lung absorption) *(e)*	4×10^1	2×10^{-2}	1×10^2	1×10^5
U-236 (slow lung absorption) *(f)*	4×10^1	6×10^{-3}	1×10^1	1×10^4
U-238 (all lung absorption types) *(d) (e) (f)*	Unlimited	Unlimited	1×10^1 *(b)*	1×10^4 *(b)*
U (nat)	Unlimited	Unlimited	1×10^0 *(b)*	1×10^3 *(b)*
U (enriched to 20% or less) *(g)*	Unlimited	Unlimited	1×10^0	1×10^3
U (dep)	Unlimited	Unlimited	1×10^0	1×10^3

2

Radionuclide (atomic number)	A_1 (TBq)	A_2 (TBq)	Activity concentration for exempt material (Bq/g)	Activity limit for an exempt consignment (Bq)
Vanadium (23)				
V-48	4×10^{-1}	4×10^{-1}	1×10^1	1×10^5
V-49	4×10^1	4×10^1	1×10^4	1×10^7
Tungsten (74)				
W-178 *(a)*	9×10^0	5×10^0	1×10^1	1×10^6
W-181	3×10^1	3×10^1	1×10^3	1×10^7
W-185	4×10^1	8×10^{-1}	1×10^4	1×10^7
W-187	2×10^0	6×10^{-1}	1×10^2	1×10^6
W-188 *(a)*	4×10^{-1}	3×10^{-1}	1×10^2	1×10^5
Xenon (54)				
Xe-122 *(a)*	4×10^{-1}	4×10^{-1}	1×10^2	1×10^9
Xe-123	2×10^0	7×10^{-1}	1×10^2	1×10^9
Xe-127	4×10^0	2×10^0	1×10^3	1×10^5
Xe-131m	4×10^1	4×10^1	1×10^4	1×10^4
Xe-133	2×10^1	1×10^1	1×10^3	1×10^4
Xe-135	3×10^0	2×10^0	1×10^3	1×10^{10}
Yttrium (39)				
Y-87 *(a)*	1×10^0	1×10^0	1×10^1	1×10^6
Y-88	4×10^{-1}	4×10^{-1}	1×10^1	1×10^6
Y-90	3×10^{-1}	3×10^{-1}	1×10^3	1×10^5
Y-91	6×10^{-1}	6×10^{-1}	1×10^3	1×10^6
Y-91m	2×10^0	2×10^0	1×10^2	1×10^6
Y-92	2×10^{-1}	2×10^{-1}	1×10^2	1×10^5
Y-93	3×10^{-1}	3×10^{-1}	1×10^2	1×10^5
Ytterbium (70)				
Yb-169	4×10^0	1×10^0	1×10^2	1×10^7
Yb-175	3×10^1	9×10^{-1}	1×10^3	1×10^7
Zinc (30)				
Zn-65	2×10^0	2×10^0	1×10^1	1×10^6
Zn-69	3×10^0	6×10^{-1}	1×10^4	1×10^6
Zn-69m *(a)*	3×10^0	6×10^{-1}	1×10^2	1×10^6
Zirconium (40)				
Zr-88	3×10^0	3×10^0	1×10^2	1×10^6
Zr-93	Unlimited	Unlimited	1×10^3 *(b)*	1×10^7 *(b)*
Zr-95 *(a)*	2×10^0	8×10^{-1}	1×10^1	1×10^6
Zr-97 *(a)*	4×10^{-1}	4×10^{-1}	1×10^1 *(b)*	1×10^5 *(b)*

(a) A_1 and/or A_2 values for these parent radionuclides include contributions from daughter radionuclides with half-lives less than 10 days, as listed in the following:

Mg-28	Al-28
Ar-42	K-42
Ca-47	Sc-47
Ti-44	Sc-44
Fe-52	Mn-52m
Fe-60	Co-60m
Zn-69m	Zn-69

Ge-68	Ga-68
Rb-83	Kr-83m
Sr-82	Rb-82
Sr-90	Y-90
Sr-91	Y-91m
Sr-92	Y-92
Y-87	Sr-87m
Zr-95	Nb-95m
Zr-97	Nb-97m, Nb-97
Mo-99	Tc-99m
Tc-95m	Tc-95
Tc-96m	Tc-96
Ru-103	Rh-103m
Ru-106	Rh-106
Pd-103	Rh-103m
Ag-108m	Ag-108
Ag-110m	Ag-110
Cd-115	In-115m
In-114m	In-114
Sn-113	In-113m
Sn-121m	Sn-121
Sn-126	Sb-126m
Te-118	Sb-118
Te-127m	Te-127
Te-129m	Te-129
Te-131m	Te-131
Te-132	I-132
I-135	Xe-135m
Xe-122	I-122
Cs-137	Ba-137m
Ba-131	Cs-131
Ba-140	La-140
Ce-144	Pr-144m, Pr-144
Pm-148m	Pm-148
Gd-146	Eu-146
Dy-166	Ho-166
Hf-172	Lu-172
W-178	Ta-178
W-188	Re-188
Re-189	Os-189m
Os-194	Ir-194
Ir-189	Os-189m
Pt-188	Ir-188
Hg-194	Au-194
Hg-195m	Hg-195
Pb-210	Bi-210
Pb-212	Bi-212, Tl-208, Po-212

Bi-210m	Tl-206
Bi-212	Tl-208, Po-212
At-211	Po-211
Rn-222	Po-218, Pb-214, At-218, Bi-214, Po-214
Ra-223	Rn-219, Po-215, Pb-211, Bi-211, Po-211, Tl-207
Ra-224	Rn-220, Po-216, Pb-212, Bi-212, Tl-208, Po-212
Ra-225	Ac-225, Fr-221, At-217, Bi-213, Tl-209, Po-213, Pb-209
Ra-226	Rn-222, Po-218, Pb-214, At-218, Bi-214, Po-214
Ra-228	Ac-228
Ac-225	Fr-221, At-217, Bi-213, Tl-209, Po-213, Pb-209
Ac-227	Fr-223
Th-228	Ra-224, Rn-220, Po-216, Pb-212, Bi-212, Tl-208, Po-212
Th-234	Pa-234m, Pa-234
Pa-230	Ac-226, Th-226, Fr-222, Ra-222, Rn-218, Po-214
U-230	Th-226, Ra-222, Rn-218, Po-214
U-235	Th-231
Pu-241	U-237
Pu-244	U-240, Np-240m
Am-242m	Am-242, Np-238
Am-243	Np-239
Cm-247	Pu-243
Bk-249	Am-245
Cf-253	Cm-249

(b) Parent nuclides and their progeny included in secular equilibrium are listed in the following:

Sr-90	Y-90
Zr-93	Nb-93m
Zr-97	Nb-97
Ru-106	Rh-106
Ag-108m	Ag-108
Cs-137	Ba-137m
Ce-144	Pr-144
Ba-140	La-140
Bi-212	Tl-208 (0.36), Po-212 (0.64)
Pb-210	Bi-210, Po-210
Pb-212	Bi-212, Tl-208 (0.36), Po-212 (0.64)
Rn-222	Po-218, Pb-214, Bi-214, Po-214
Ra-223	Rn-219, Po-215, Pb-211, Bi-211, Tl-207
Ra-224	Rn-220, Po-216, Pb-212, Bi-212, Tl-208 (0.36), Po-212 (0.64)
Ra-226	Rn-222, Po-218, Pb-214, Bi-214, Po-214, Pb-210, Bi-210, Po-210
Ra-228	Ac-228
Th-228	Ra-224, Rn-220, Po-216, Pb-212, Bi-212, Tl-208 (0.36), Po-212 (0.64)
Th-229	Ra-225, Ac-225, Fr-221, At-217, Bi-213, Po-213, Pb-209
Th (nat)	Ra-228, Ac-228, Th-228, Ra-224, Rn-220, Po-216, Pb-212, Bi-212, Tl-208 (0.36), Po-212 (0.64)
Th-234	Pa-234m
U-230	Th-226, Ra-222, Rn-218, Po-214
U-232	Th-228, Ra-224, Rn-220, Po-216, Pb-212, Bi-212, Tl-208 (0.36), Po-212 (0.64)

U-235	Th-231
U-238	Th-234, Pa-234m
U (nat)	Th-234, Pa-234m, U-234, Th-230, Ra-226, Rn-222, Po-218, Pb-214, Bi-214, Po-214, Pb-210, Bi-210, Po-210
Np-237	Pa-233
Am-242m	Am-242
Am-243	Np-239

(c) The quantity may be determined from a measurement of the rate of decay or a measurement of the radiation level at a prescribed distance from the source.

(d) These values apply only to compounds of uranium that take the chemical form of UF_6, UO_2F_2 and $UO_2(NO_3)_2$ in both normal and accident conditions of transport.

(e) These values apply only to compounds of uranium that take the chemical form of UO_3, UF_4, UCl_4 and hexavalent compounds in both normal and accident conditions of transport.

(f) These values apply to all compounds of uranium other than those specified in (d) and (e) above.

(g) These values apply to unirradiated uranium only.

2.7.2.2.2 For individual radionuclides which are not listed in table 2.7.2.2.1, the determination of the basic radionuclide values referred to in 2.7.2.2.1 shall require multilateral approval. It is permissible to use an A_2 value calculated using a dose coefficient for the appropriate lung absorption type as recommended by the International Commission on Radiological Protection, if the chemical forms of each radionuclide under both normal and accident conditions of transport are taken into consideration. Alternatively, the radionuclide values in table 2.7.2.2.2 may be used without obtaining competent authority approval.

Table 2.7.2.2.2 – Basic radionuclide values for unknown radionuclides or mixtures

Radioactive contents	A_1 (TBq)	A_2 (TBq)	Activity concentration for exempt material (Bq/g)	Activity limit for exempt consignments (Bq)
Only beta or gamma emitting nuclides are known to be present	0.1	0.02	1×10^1	1×10^4
Alpha emitting nuclides but no neutron emitters are known to be present	0.2	9×10^{-5}	1×10^{-1}	1×10^3
Neutron emitting nuclides are known to be present or no relevant data are available	0.001	9×10^{-5}	1×10^{-1}	1×10^3

2.7.2.2.3 In the calculations of A_1 and A_2 for a radionuclide not in table 2.7.2.2.1, a single radioactive decay chain in which the radionuclides are present in their naturally occurring proportions, and in which no daughter nuclide has a half-life either longer than 10 days or longer than that of the parent nuclide, shall be considered as a single radionuclide; and the activity to be taken into account and the A_1 or A_2 value to be applied shall be those corresponding to the parent nuclide of that chain. In the case of radioactive decay chains in which any daughter nuclide has a half-life either longer than 10 days or greater than that of the parent nuclide, the parent and such daughter nuclides shall be considered as mixtures of different nuclides.

2.7.2.2.4 For mixtures of radionuclides, the determination of the basic radionuclide values referred to in 2.7.2.2.1 may be determined as follows:

$$X_m = \frac{1}{\sum_i \dfrac{f(i)}{X(i)}}$$

where: $f(i)$ is the fraction of activity or activity concentration of radionuclide i in the mixture;

$X(i)$ is the appropriate value of A_1 or A_2, or the activity concentration for exempt material or the activity limit for an exempt consignment, as appropriate, for the radionuclide i; and

X_m is the derived value of A_1 or A_2, or the activity concentration for exempt material or the activity limit for an exempt consignment in the case of a mixture.

2.7.2.2.5 When the identity of each radionuclide is known but the individual activities of some of the radionuclides are not known, the radionuclides may be grouped and the lowest radionuclide value, as appropriate, for the radionuclides in each group may be used in applying the formulae in 2.7.2.2.4 and 2.7.2.4.4. Groups may be based on the total alpha activity and the total beta/gamma activity when these are known, using the lowest radionuclide values for the alpha emitters or beta/gamma emitters, respectively.

2.7.2.2.6 For individual radionuclides or for mixtures of radionuclides for which relevant data are not available, the values shown in table 2.7.2.2.2 shall be used.

2.7.2.3 **Determination of other material characteristics**

2.7.2.3.1 *Low specific activity (LSA) material*

2.7.2.3.1.1 [Reserved]

2.7.2.3.1.2 LSA material shall be in one of three groups:

.1 LSA-I

(i) uranium and thorium ores and concentrates of such ores, and other ores containing naturally occurring radionuclides which are intended to be processed for the use of these radionuclides;

(ii) Natural uranium, depleted uranium, natural thorium or their compounds or mixtures, providing they are unirradiated and in solid or liquid form;

(iii) radioactive material for which the A_2 value is unlimited, excluding material classified as fissile according to 2.7.2.3.5; or

(iv) other radioactive material in which the activity is distributed throughout and the estimated average specific activity does not exceed 30 times the values for activity concentration specified in 2.7.2.2.1 to 2.7.2.2.6, excluding material classified as fissile according to 2.7.2.3.5;

.2 LSA-II

(i) water with tritium concentration up to 0.8 TBq/ℓ; or

(ii) other material in which the activity is distributed throughout and the estimated average specific activity does not exceed $10^{-4}A_2$/g for solids and gases, and $10^{-5}A_2$/g for liquids;

.3 LSA-III – Solids (e.g., consolidated wastes, activated materials), excluding powders, in which:

(i) the radioactive material is distributed throughout a solid or a collection of solid objects, or is essentially uniformly distributed in a solid compact binding agent (such as concrete, bitumen, ceramic, etc.);

(ii) the radioactive material is relatively insoluble, or it is intrinsically contained in a relatively insoluble matrix, so that, even under loss of packaging, the loss of radioactive material per package by leaching when placed in water for seven days would not exceed 0.1A_2; and

(iii) the estimated average specific activity of the solid, excluding any shielding material, does not exceed $2 \times 10^{-3}A_2$/g.

2.7.2.3.1.3 LSA-III material shall be a solid of such a nature that, if the entire contents of a package were subjected to the test specified in 2.7.2.3.1.4, the activity in the water would not exceed 0.1A_2.

2.7.2.3.1.4 LSA-III material shall be tested as follows:

A solid material sample representing the entire contents of the package shall be immersed for 7 days in water at ambient temperature. The volume of water to be used in the test shall be sufficient to ensure that at the end of the 7-day test period the free volume of the unabsorbed and unreacted water remaining shall be at least 10% of the volume of the solid test sample itself. The water shall have an initial pH of 6–8 and a maximum conductivity of 1 mS/m at 20°C. The total activity of the free volume of water shall be measured following the 7-day immersion of the test sample.

2.7.2.3.1.5 Demonstration of compliance with the performance standards in 2.7.2.3.1.4 shall be in accordance with 6.4.12.1 and 6.4.12.2.

2.7.2.3.2 *Surface contaminated object (SCO)*

SCO is classified in one of two groups:

.1 SCO-I: A solid object on which:

(i) the non-fixed contamination on the accessible surface averaged over 300 cm^2 (or the area of the surface if less than 300 cm^2) does not exceed 4 Bq/cm^2 for beta and gamma emitters and low-toxicity alpha emitters, or 0.4 Bq/cm^2 for all other alpha emitters;

     (ii)   the fixed contamination on the accessible surface averaged over 300 cm^2 (or the area of the surface if less than 300 cm^2) does not exceed 4×10^4 Bq/cm^2 for beta and gamma emitters and low-toxicity alpha emitters, or 4×10^3 Bq/cm^2 for all other alpha emitters; and

     (iii)   the non-fixed contamination plus the fixed contamination on the inaccessible surface averaged over 300 cm^2 (or the area of the surface if less than 300 cm^2) does not exceed 4×104 Bq/cm^2 for beta and gamma emitters and low-toxicity alpha emitters, or 4×10^3 Bq/cm^2 for all other alpha emitters;

.2   SCO-II: A solid object on which either the fixed or non-fixed contamination on the surface exceeds the applicable limits specified for SCO-I in 2.7.2.3.2.1 above and on which:

     (i)   the non-fixed contamination on the accessible surface averaged over 300 cm^2 (or the area of the surface if less than 300 cm^2) does not exceed 400 Bq/cm^2 for beta and gamma emitters and low-toxicity alpha emitters, or 40 Bq/cm^2 for all other alpha emitters;

     (ii)   the fixed contamination on the accessible surface averaged over 300 cm^2 (or the area of the surface if less than 300 cm^2) does not exceed 8×10^5 Bq/cm^2 for beta and gamma emitters and low-toxicity alpha emitters, or 8×10^4 Bq/cm^2 for all other alpha emitters; and

     (iii)   the non-fixed contamination plus the fixed contamination on the inaccessible surface averaged over 300 cm^2 (or the area of the surface if less than 300 cm^2) does not exceed 8×10^5 Bq/cm^2 for beta and gamma emitters and low-toxicity alpha emitters, or 8×10^4 Bq/cm^2 for all other alpha emitters.

2.7.2.3.3   *Special form radioactive material*

2.7.2.3.3.1   .1   Special form radioactive material shall have at least one dimension not less than 5 mm.

     .2   When a sealed capsule constitutes part of the special form radioactive material, the capsule shall be so manufactured that it can be opened only by destroying it.

     .3   The design for special form radioactive material requires unilateral approval.

2.7.2.3.3.2   Special form radioactive material shall be of such a nature or shall be so designed that, if it is subjected to the tests specified in 2.7.2.3.3.4 to 2.7.2.3.3.8, it shall meet the following requirements:

     .1   It would not break or shatter under the impact, percussion and bending tests 2.7.2.3.3.5.1, 2.7.2.3.3.5.2, 2.7.2.3.3.5.3, or 2.7.2.3.3.6.1 as applicable;

     .2   It would not melt or disperse in the applicable heat test 2.7.2.3.3.5.4 or 2.7.2.3.3.6.2 as applicable; and

     .3   The activity in the water from the leaching tests specified in 2.7.2.3.3.7 and 2.7.2.3.3.8 would not exceed 2 kBq; or alternatively for sealed sources, the leakage rate for the volumetric leakage assessment test specified in ISO 9978:1992 "Radiation protection – Sealed radioactive sources – Leakage test methods" would not exceed the applicable acceptance threshold acceptable to the competent authority.

2.7.2.3.3.3   Demonstration of compliance with the performance standards in 2.7.2.3.3.2 shall be in accordance with 6.4.12.1 and 6.4.12.2.

2.7.2.3.3.4   Specimens that comprise or simulate special form radioactive material shall be subjected to the impact test, the percussion test, the bending test, and the heat test specified in 2.7.2.3.3.5 or alternative tests as authorized in 2.7.2.3.3.6. A different specimen may be used for each of the tests. Following each test, a leaching assessment or volumetric leakage test shall be performed on the specimen by a method no less sensitive than the methods given in 2.7.2.3.3.7 for indispersible solid material or 2.7.2.3.3.8 for encapsulated material.

2.7.2.3.3.5   The relevant test methods are:

     .1   Impact test: The specimen shall drop onto the target from a height of 9 m. The target shall be as defined in 6.4.14;

     .2   Percussion test: The specimen shall be placed on a sheet of lead which is supported by a smooth solid surface and struck by the flat face of a mild steel bar so as to cause an impact equivalent to that resulting from a free drop of 1.4 kg through 1 m. The lower part of the bar shall be 25 mm in diameter with the edges rounded off to a radius of (3.0 ± 0.3) mm. The lead, of hardness number 3.5 to 4.5 on the Vickers scale and not more than 25 mm thick, shall cover an area greater than that covered by the specimen. A fresh surface of lead shall be used for each impact. The bar shall strike the specimen so as to cause maximum damage;

     .3   Bending test: The test shall apply only to long, slender sources with both a minimum length of 10 cm and a length to minimum width ratio of not less than 10. The specimen shall be rigidly clamped in a horizontal position so that one half of its length protrudes from the face of the clamp. The orientation of the specimen shall be such that the specimen will suffer maximum damage when its free end is struck by the flat face of a steel bar. The bar shall strike the specimen so as to cause an impact equivalent to that resulting from a free vertical drop of 1.4 kg through 1 m. The lower part of the bar shall be 25 mm in diameter with the edges rounded off to a radius of (3.0 ± 0.3) mm;

.4 Heat test: The specimen shall be heated in air to a temperature of 800°C and held at that temperature for a period of 10 minutes and shall then be allowed to cool.

2.7.2.3.3.6 Specimens that comprise or simulate radioactive material enclosed in a sealed capsule may be excepted from:

.1 The tests prescribed in 2.7.2.3.3.5.1 and 2.7.2.3.3.5.2 provided the mass of the special form radioactive material:

(i) is less than 200 g and they are alternatively subjected to the class 4 impact test prescribed in ISO 2919:1999 "Radiation protection – Sealed radioactive sources – General requirements and classification"; or

(ii) is less than 500 g and they are alternatively subjected to the class 5 impact test prescribed in ISO 2919:1999 "Radiation protection – Sealed radioactive sources – General requirements and classification"; and

.2 The test prescribed in 2.7.2.3.3.5.4 provided they are alternatively subjected to the class 6 temperature test specified in ISO 2919:1999 "Radiation protection – Sealed radioactive sources – General requirements and classification".

2.7.2.3.3.7 For specimens which comprise or simulate indispersible solid material, a leaching assessment shall be performed as follows:

.1 The specimen shall be immersed for 7 days in water at ambient temperature. The volume of water to be used in the test shall be sufficient to ensure that at the end of the 7-day test period the free volume of the unabsorbed and unreacted water remaining shall be at least 10% of the volume of the solid test sample itself. The water shall have an initial pH of 6–8 and a maximum conductivity of 1 mS/m at 20°C;

.2 The water with specimen shall then be heated to a temperature of (50 ± 5)°C and maintained at this temperature for 4 hours;

.3 The activity of the water shall then be determined;

.4 The specimen shall then be kept for at least 7 days in still air at not less than 30°C and relative humidity not less than 90%;

.5 The specimen shall then be immersed in water of the same specification as in 2.7.2.3.3.7.1 above and the water with the specimen heated to (50 ± 5)°C and maintained at this temperature for 4 hours;

.6 The activity of the water shall then be determined.

2.7.2.3.3.8 For specimens which comprise or simulate radioactive material enclosed in a sealed capsule, either a leaching assessment or a volumetric leakage assessment shall be performed as follows:

.1 The leaching assessment shall consist of the following steps:

(i) the specimen shall be immersed in water at ambient temperature. The water shall have an initial pH of 6–8 with a maximum conductivity of 1 mS/m at 20°C;

(ii) the water and specimen shall be heated to a temperature of (50 ± 5)°C and maintained at this temperature for 4 hours;

(iii) the activity of the water shall then be determined;

(iv) the specimen shall then be kept for at least 7 days in still air at not less than 30°C and relative humidity of not less than 90%;

(v) the process in (i), (ii) and (iii) shall be repeated.

.2 The alternative volumetric leakage assessment shall comprise any of the tests prescribed in ISO 9978:1992 "Radiation protection – Sealed radioactive sources – Leakage test methods" which are acceptable to the competent authority.

2.7.2.3.4 *Low dispersible material*

2.7.2.3.4.1 The design for low dispersible radioactive material shall require multilateral approval. Low dispersible radioactive material shall be such that the total amount of this radioactive material in a package shall meet the following provisions:

.1 The radiation level at 3 m from the unshielded radioactive material does not exceed 10 mSv/h;

.2 If subjected to the tests specified in 6.4.20.3 and 6.4.20.4, the airborne release in gaseous and particulate forms of up to 100 μm aerodynamic equivalent diameter would not exceed 100A_2. A separate specimen may be used for each test; and

.3 If subjected to the test specified in 2.7.2.3.1.4, the activity in the water would not exceed 100A_2. In the application of this test, the damaging effects of the tests specified in 2.7.2.3.4.1.2 above shall be taken into account.

2.7.2.3.4.2 Low dispersible material shall be tested as follows:

A specimen that comprises or simulates low dispersible radioactive material shall be subjected to the enhanced thermal test specified in 6.4.20.3 and the impact test specified in 6.4.20.4. A different specimen may be used for each of the tests. Following each test, the specimen shall be subjected to the leach test specified in 2.7.2.3.1.4. After each test it shall be determined if the applicable provisions of 2.7.2.3.4.1 have been met.

2.7.2.3.4.3 Demonstration of compliance with the performance standards in 2.7.2.3.4.1 and 2.7.2.3.4.2 shall be in accordance with 6.4.12.1 and 6.4.12.2.

2.7.2.3.5 *Fissile material*

Packages containing fissile radionuclides shall be classified under the relevant entry of table 2.7.2.1.1 for fissile material unless one of the conditions .1 to .4 of this paragraph is met. Only one type of exception is allowed per consignment.

.1 A mass limit per consignment such that:

$$\frac{\text{mass of uranium–235 (g)}}{X} + \frac{\text{mass of other fissile material–235 (g)}}{Y} < 1$$

where X and Y are the mass limits defined in table 2.7.2.3.5, provided that the smallest external dimension of each package is not less than 10 cm and that either:

(i) each individual package contains not more than 15 g of fissile material; for unpackaged material, this quantity limitation shall apply to the consignment being carried in or on the conveyance; or

(ii) the fissile material is a homogeneous hydrogenous solution or mixture where the ratio of fissile nuclides to hydrogen is less than 5% by mass; or

(iii) there are not more than 5 g of fissile material in any 10 litre volume of material.

Neither beryllium nor deuterium shall be present in quantities exceeding 1% of the applicable consignment mass limits provided in table 2.7.2.3.5, except for deuterium in natural concentration in hydrogen.

.2 Uranium enriched in uranium-235 to a maximum of 1% by mass, and with a total plutonium and uranium-233 content not exceeding 1% of the mass of uranium-235, provided that the fissile material is distributed essentially homogeneously throughout the material. In addition, if uranium-235 is present in metallic, oxide or carbide forms, it shall not form a lattice arrangement;

.3 Liquid solutions of uranyl nitrate enriched in uranium-235 to a maximum of 2% by mass, with a total plutonium and uranium-233 content not exceeding 0.002% of the mass of uranium, and with a minimum nitrogen to uranium atomic ratio (N/U) of 2;

.4 Packages containing, individually, a total plutonium mass not more than 1 kg, of which not more than 20% by mass may consist of plutonium-239, plutonium-241 or any combination of those radionuclides.

Table 2.7.2.3.5 – Consignment mass limits for exceptions from the requirements for packages containing fissile material

Fissile material	Fissile material mass (g) mixed with substances having an average hydrogen density less than or equal to water	Fissile material mass (g) mixed with substances having an average hydrogen density greater than water
Uranium-235 *(X)*	400	290
Other fissile material *(Y)*	250	180

2.7.2.4 **Classification of packages or unpacked material**

The quantity of radioactive material in a package shall not exceed the relevant limits for the package type as specified below.

2.7.2.4.1 *Classification as excepted package*

2.7.2.4.1.1 Packages may be classified as excepted packages if:

.1 They are empty packagings having contained radioactive material;

.2 They contain instruments or articles in limited quantities;

.3 They contain articles manufactured of natural uranium, depleted uranium or natural thorium; or

.4 They contain radioactive material in limited quantities.

2.7.2.4.1.2 A package containing radioactive material may be classified as an excepted package provided that the radiation level at any point on its external surface does not exceed 5 μSv/h.

Table 2.7.2.4.1.2 – Activity limits for excepted packages

Physical state of contents	Instruments or article		Material package limits[a]
	Item limits[a]	Package limits[a]	
(1)	(2)	(3)	(4)
Solids special form other form	$10^{-2} A_1$ $10^{-2} A_2$	A_1 A_2	$10^{-3} A_1$ $10^{-3} A_2$
Liquids	$10^{-3} A_2$	$10^{-1} A_2$	$10^{-4} A_2$
Gases tritium special form other forms	$2 \times 10^{-2} A_2$ $10^{-3} A_1$ $10^{-3} A_2$	$2 \times 10^{-1} A_2$ $10^{-2} A_1$ $10^{-2} A_2$	$2 \times 10^{-2} A_2$ $10^{-3} A_1$ $10^{-3} A_2$

[a] For mixtures of radionuclides, see 2.7.2.2.4 to 2.7.2.2.6.

2.7.2.4.1.3 Radioactive material which is enclosed in or is included as a component part of an instrument or other manufactured article may be classified under UN 2911, RADIOACTIVE MATERIAL, EXCEPTED PACKAGE – INSTRUMENTS or ARTICLES provided that:

.1 the radiation level at 10 cm from any point on the external surface of any unpackaged instrument or article is not greater than 0.1 mSv/h; and

.2 each instrument or manufactured article bears the marking "RADIOACTIVE" except:

(i) radioluminescent time-pieces or devices;

(ii) consumer products that either have received regulatory approval according to 1.5.1.4.4 or do not individually exceed the activity limit for an exempt consignment in table 2.7.2.2.1 (column 5), provided such products are transported in a package that bears the marking "RADIOACTIVE" on an internal surface in such a manner that warning of the presence of radioactive material is visible on opening the package; and

.3 the active material is completely enclosed by non-active components (a device performing the sole function of containing radioactive material shall not be considered to be an instrument or manufactured article); and

.4 the limits specified in columns 2 and 3 of table 2.7.2.4.1.2 are met for each individual item and each package, respectively.

2.7.2.4.1.4 Radioactive material with an activity not exceeding the limit specified in column 4 of Table 2.7.2.4.1.2, may be classified under UN 2910, RADIOACTIVE MATERIAL, EXCEPTED PACKAGE – LIMITED QUANTITY OF MATERIAL provided that:

.1 the package retains its radioactive contents under routine conditions of transport; and

.2 the package bears the marking "RADIOACTIVE" on an internal surface in such a manner that a warning of the presence of radioactive material is visible on opening the package.

2.7.2.4.1.5 An empty packaging which had previously contained radioactive material with an activity not exceeding the limit specified in column 4 of table 2.7.2.4.1.2 may be classified under UN 2908, RADIOACTIVE MATERIAL, EXCEPTED PACKAGE – EMPTY PACKAGING, provided that:

.1 it is in a well-maintained condition and securely closed;

.2 the outer surface of any uranium or thorium in its structure is covered with an inactive sheath made of metal or some other substantial material;

.3 the level of internal non-fixed contamination, when averaged over any 300 cm^2, does not exceed:

(i) 400 Bq/cm^2 for beta and gamma emitters and low-toxicity alpha emitters; and

(ii) 40 Bq/cm^2 for all other alpha emitters; and

.4 any labels which may have been displayed on it in conformity with 5.2.2.1.12.1 are no longer visible.

2.7.2.4.1.6 Articles manufactured of natural uranium, depleted uranium or natural thorium and articles in which the sole radioactive material is unirradiated natural uranium, unirradiated depleted uranium or unirradiated natural thorium may be classified under UN 2909, RADIOACTIVE MATERIAL, EXCEPTED PACKAGE – ARTICLES MANUFACTURED FROM NATURAL URANIUM or DEPLETED URANIUM or NATURAL THORIUM, provided that the outer surface of the uranium or thorium is enclosed in an inactive sheath made of metal or some other substantial material.

2.7.2.4.2 *Classification as Low specific activity (LSA) material*

Radioactive material may only be classified as LSA material if the conditions of 2.7.2.3.1 and 4.1.9.2 are met.

2.7.2.4.3 *Classification as Surface contaminated object (SCO)*

Radioactive material may be classified as SCO if the conditions of 2.7.2.3.2.1 and 4.1.9.2 are met.

2.7.2.4.4 *Classification as Type A package*

Packages containing radioactive material may be classified as Type A packages provided that the following conditions are met:

Type A packages shall not contain activities greater than the following:

.1   For special form radioactive material – A_1; or

.2   For all other radioactive material – A_2.

For mixtures of radionuclides whose identities and respective activities are known, the following condition shall apply to the radioactive contents of a Type A package:

$$\sum_i \frac{B(i)}{A_1(i)} + \sum_j \frac{C(j)}{A_2(j)} \le 1$$

where: $B(i)$ is the activity of radionuclide i as special form radioactive material;
$A_1(i)$ is the A_1 value for radionuclide i;
$C(j)$ is the activity of radionuclide j> as other than special form radioactive material; and
$A_2(j)$ is the A_2 value for radionuclide j.

2.7.2.4.5 *Classification of uranium hexafluoride*

Uranium hexafluoride shall only be assigned to UN No. 2977, RADIOACTIVE MATERIAL, URANIUM HEXAFLUORIDE, FISSILE, or 2978, RADIOACTIVE MATERIAL, URANIUM HEXAFLUORIDE, non-fissile or fissile – excepted.

2.7.2.4.5.1 Packages containing uranium hexafluoride shall not contain:

.1   a mass of uranium hexafluoride different from that authorized for the package design;

.2   a mass of uranium hexafluoride greater than a value that would lead to an ullage smaller than 5% at the maximum temperature of the package as specified for the plant systems where the package shall be used; or

.3   uranium hexafluoride other than in solid form or at an internal pressure above atmospheric pressure when presented for transport.

2.7.2.4.6 *Classification as Type B(U), Type B(M) or Type C packages*

2.7.2.4.6.1 Packages not otherwise classified in 2.7.2.4 (2.7.2.4.1 to 2.7.2.4.5) shall be classified in accordance with the competent authority approval certificate for the package issued by the country of origin of design.

2.7.2.4.6.2 A package may only be classified as a Type B(U) if it does not contain:

.1   activities greater than those authorized for the package design;

.2   radionuclides different from those authorized for the package design; or

.3   contents in a form, or a physical or chemical state, different from those authorized for the package design

as specified in the certificate of approval.

2.7.2.4.6.3 A package may only be classified as a Type B(M) if it does not contain:

.1   activities greater than those authorized for the package design;

.2   radionuclides different from those authorized for the package design; or

.3   contents in a form, or a physical or chemical state, different from those authorized for the package design, as specified in the certificate of approval.

2.7.2.4.6.4   A package may only be classified as a Type C if it does not contain:

.1   activities greater than those authorized for the package design;

.2   radionuclides different from those authorized for the package design; or

.3   contents in a form, or physical or chemical state, different from those authorized for the package design,

as specified in the certificate of approval.

2.7.2.5   **Special arrangements**

Radioactive material shall be classified as transported under special arrangement when it is intended to be transported in accordance with 1.5.4.

Chapter 2.8

Class 8 – Corrosive substances

2.8.1 Definition and properties

2.8.1.1 Definition

Class 8 substances (corrosive substances) means substances which, by chemical action, will cause severe damage when in contact with living tissue or, in the case of leakage, will materially damage, or even destroy, other goods or the means of transport.

2.8.1.2 Properties

2.8.1.2.1 In cases where particularly severe personal damage is to be expected, a note to that effect is made in the Dangerous Goods List in chapter 3.2 in the wording "causes (severe) burns to skin, eyes and mucous membranes".

2.8.1.2.2 Many substances are sufficiently volatile to evolve vapour irritating to the nose and eyes. If so, this fact is mentioned in the Dangerous Goods List in chapter 3.2 in the wording "vapour irritates mucous membranes".

2.8.1.2.3 A few substances may produce toxic gases when decomposed by very high temperatures. In these cases the statement "when involved in a fire, evolves toxic gases" appears in the Dangerous Goods List in chapter 3.2.

2.8.1.2.4 In addition to direct destructive action in contact with skin or mucous membranes, some substances in this class are toxic or harmful. Poisoning may result if they are swallowed, or if their vapour is inhaled; some of them even may penetrate the skin. Where appropriate, a statement is made to that effect in the Dangerous Goods List in chapter 3.2.

2.8.1.2.5 All substances in this class have a more or less destructive effect on materials such as metals and textiles.

2.8.1.2.5.1 In the Dangerous Goods List, the term "corrosive to most metals" means that any metal likely to be present in a ship, or in its cargo, may be attacked by the substance or its vapour.

2.8.1.2.5.2 The term "corrosive to aluminium, zinc, and tin" implies that iron or steel is not damaged in contact with the substance.

2.8.1.2.5.3 A few substances in this class can corrode glass, earthenware and other siliceous materials. Where appropriate, this is stated in the Dangerous Goods List in chapter 3.2.

2.8.1.2.6 Many substances in this class only become corrosive after having reacted with water, or with moisture in the air. This fact is indicated in the Dangerous Goods List in chapter 3.2 by the words "in the presence of moisture. . .". The reaction of water with many substances is accompanied by the liberation of irritating and corrosive gases. Such gases usually become visible as fumes in the air.

2.8.1.2.7 A few substances in this class generate heat in reaction with water or organic materials, including wood, paper, fibres, some cushioning materials and certain fats and oils. Where appropriate, this is indicated in the Dangerous Goods List in chapter 3.2.

2.8.1.2.8 A substance which is designated as "stabilized" shall not be transported in the unstabilized state.

2.8.2 Assignment of packing groups

2.8.2.1 Substances and preparations of class 8 are divided among the three packing groups according to their degree of hazard in transport as follows:

Packing group I: Very dangerous substances and preparations;

Packing group II: Substances and preparations presenting medium danger;

Packing group III: Substances and preparations presenting minor danger.

The packing group to which a substance has been assigned is given in the Dangerous Goods List in chapter 3.2.

2.8.2.2 Allocation of substances listed in the Dangerous Goods List in chapter 3.2 to the packing groups in class 8 has been on the basis of experience, taking into account such additional factors as inhalation risk (see 2.8.2.3) and reactivity with water (including the formation of dangerous decomposition products). New substances, including mixtures, can be assigned to packing groups on the basis of the length of time of contact necessary to produce full thickness destruction of human skin in accordance with the criteria in 2.8.2.5. Liquids, and solids which may become liquid during transport, which are judged not to cause full thickness destruction of human skin shall still be considered for their potential to cause corrosion in certain metal surfaces in accordance with the criteria in 2.8.2.5.3.2.

2.8.2.3 A substance or preparation meeting the criteria of class 8 and having an inhalation toxicity of dusts and mists (LC_{50}) in the range of packing group I, but toxicity through oral ingestion or dermal contact only in the range of packing group III or less, shall be allocated to class 8 (see Note under 2.6.2.2.4.2).

2.8.2.4 In assigning the packing group to a substance in accordance with 2.8.2.2, account shall be taken of human experience in instances of accidental exposure. In the absence of human experience, the grouping shall be based on data obtained from experiments in accordance with OECD Guideline 404.[*]

2.8.2.5 Packing groups are assigned to corrosive substances in accordance with the following criteria:

.1 Packing group I is assigned to substances that cause full thickness destruction of intact skin tissue within an observation period of up to 60 minutes starting after an exposure time of 3 minutes or less.

.2 Packing group II is assigned to substances that cause full thickness destruction of intact skin tissue within an observation period of up to 14 days starting after an exposure time of more than 3 but not more than 60 minutes.

.3 Packing group III is assigned to substances that:

.1 cause full thickness destruction of intact skin tissue within an observation period of up to 14 days starting after an exposure time of more than 60 minutes but not more than 4 hours; or

.2 are judged not to cause full thickness destruction of intact skin tissue but which exhibit a corrosion rate on either steel or aluminium surfaces exceeding 6.25 mm a year at a test temperature of 55°C when tested on both materials. For the purposes of testing steel, type S235JR+CR (1.0037 resp. St 37-2), S275J2G3+CR (1.0144 resp. St 44-3), ISO 3574:1999, Unified Numbering System (UNS) G10200 or SAE 1020, and for testing aluminium, non-clad, types 7075-T6 or AZ5GU T6 shall be used. An acceptable test is prescribed in the United Nations *Manual of Tests and Criteria*, part III, Section 37.

Note: Where an initial test on either steel or aluminium indicates the substance being tested is corrosive, the follow-up test on the other metal is not required.

---

[*] OECD Guidelines for testing of chemicals No. 404 "Acute Dermal Irritation/Corrosion" 1992.

Chapter 2.9

Miscellaneous dangerous substances and articles (Class 9) and environmentally hazardous substances

Note 1: For the purposes of this Code, the environmentally hazardous substances (aquatic environment) criteria contained in this chapter apply to the classification of marine pollutants (see 2.10).

Note 2: Although the environmentally hazardous substances (aquatic environment) criteria apply to all hazard classes (see 2.10.2.3 and 2.10.2.5), the criteria have been included in this chapter.

2.9.1 Definitions

2.9.1.1 *Class 9 substances and articles (miscellaneous dangerous substances and articles)* are substances and articles which, during transport, present a danger not covered by other classes.

2.9.1.2 *Genetically modified micro-organisms (GMMOs) and genetically modified organisms (GMOs)* are micro-organisms and organisms in which genetic material has been purposely altered through genetic engineering in a way that does not occur naturally.

2.9.2 Assignment to class 9

2.9.2.1 Class 9 includes, *inter alia*:

.1 substances and articles not covered by other classes which experience has shown, or may show, to be of such a dangerous character that the provisions of part A of chapter VII of SOLAS 1974, as amended, shall apply.

.2 substances not subject to the provisions of part A in chapter VII of the aforementioned Convention, but to which the provisions of Annex III of MARPOL 73/78, as amended, apply.

.3 substances that are transported or offered for transport at temperatures equal to, or exceeding, 100°C, in a liquid state, and solids that are transported or offered for transport at temperatures equal to or exceeding 240°C.

.4 GMMOs and GMOs which do not meet the definition of infectious substances (see 2.6.3) but which are capable of altering animals, plants or microbiological substances in a way not normally the result of natural reproduction. They shall be assigned to UN 3245. GMMOs or GMOs are not subject to the provisions of this Code when authorized for use by the competent authorities of the countries of origin, transit and destination.

2.9.3 Environmentally hazardous substances (aquatic environment)

2.9.3.1 General definitions

2.9.3.1.1 *Environmentally hazardous substances* include, *inter alia*, liquid or solid substances pollutant to the aquatic environment and solutions and mixtures of such substances (such as preparations and wastes).

For the purposes of this section,

Substance means chemical elements and their compounds in the natural state or obtained by any impurities deriving from the process used, but excluding any solvent which may be separated without affecting the stability of the substance or changing its composition.

2.9.3.1.2 The aquatic environment may be considered in terms of the aquatic organisms that live in the water, and the aquatic ecosystem of which they are part.* The basis, therefore, of the identification of hazard is the aquatic toxicity of the substance or mixture, although this may be modified by further information on the degradation and bioaccumulation behaviour.

---

* This does not address aquatic pollutants for which there may be a need to consider effects beyond the aquatic environment, such as the impacts on human health, etc.

2.9.3.1.3 While the following classification procedure is intended to apply to all substances and mixtures, it is recognized that in some cases, e.g., metals or poorly soluble inorganic compounds, special guidance will be necessary.*

2.9.3.1.4 The following definitions apply for acronyms or terms used in this section:

BCF	Bioconcentration Factor;
BOD	Biochemical Oxygen Demand;
COD	Chemical Oxygen Demand;
GLP	Good Laboratory Practices;
EC_{50}	the effective concentration of substance that causes 50% of the maximum response;
ErC_{50}	EC_{50} in terms of reduction of growth;
K_{ow}	octanol/water partition coefficient;
LC_{50} (50% lethal concentration)	the concentration of a substance in water which causes the death of 50% (one half) in a group of test animals;
$L(E)C_{50}$	LC_{50} or EC_{50};
NOEC	No Observed Effect Concentration;
OECD Test Guidelines	Test guidelines published by the Organization for Economic Co-operation and Development (OECD).

2.9.3.2   Definitions and data requirements

2.9.3.2.1 The basic elements for classification of environmentally hazardous substances (aquatic environment) are:

– acute aquatic toxicity;

– potential for or actual bioaccumulation;

– degradation (biotic or abiotic) for organic chemicals; and

– chronic aquatic toxicity.

2.9.3.2.2 While data from internationally harmonized test methods are preferred, in practice, data from national methods may also be used where they are considered as equivalent. In general, freshwater and marine species toxicity data can be considered as equivalent data and are preferably to be derived using OECD Test Guidelines or equivalent according to the principles of Good Laboratory Practices (GLP). Where such data are not available, classification shall be based on the best available data.

2.9.3.2.3 *Acute aquatic toxicity* shall normally be determined using a fish 96 hour LC_{50} (OECD Test Guideline 203 or equivalent), a crustacea species 48 hour EC_{50} (OECD Test Guideline 202 or equivalent) and/or an algal species 72 or 96 hour EC_{50} (OECD Test Guideline 201 or equivalent). These species are considered as surrogates for all aquatic organisms. Data on other species such as *Lemna* may also be considered if the test methodology is suitable.

2.9.3.2.4 *Bioaccumulation* means net result of uptake, transformation and elimination of a substance in an organism due to all routes of exposure (i.e. air, water, sediment/soil and food). The potential for bioaccumulation shall normally be determined by using the octanol/water partition coefficient, usually reported as a log K_{ow} determined according to OECD Test Guideline 107 or 117. While this represents a potential to bioaccumulate, an experimentally determined Bioconcentration Factor (BCF) provides a better measure and shall be used in preference when available. A BCF shall be determined according to OECD Test Guideline 305.

2.9.3.2.5 *Environmental degradation* may be biotic or abiotic (e.g. hydrolysis) and the criteria used reflect this fact. Ready biodegradation is most easily defined using the OECD biodegradability tests (OECD Test Guideline 301 (A–F)). A pass level in these tests may be considered as indicative of rapid degradation in most aquatic environments. As these are freshwater tests, use of results from OECD Test Guideline 306, which is more suitable for the marine environment, is also included. Where such data are not available, a BOD (5 days)/COD ratio $\geqslant 0.5$ is considered as indicative of rapid degradation. Abiotic degradation such as hydrolysis, primary degradation, both abiotic and biotic, degradation in non-aquatic media and proven rapid degradation in the environment may all be considered in defining rapid degradability.[†]

---

* This can be found in annex 10 of the Globally Harmonized System of Classification and Labelling of Chemicals (GHS).

[†] Special guidance on data interpretation is provided in chapter 4.1 and annex 9 of the Globally Harmonized System of Classification and Labelling of Chemicals (GHS).

2.9.3.2.5.1 Substances are considered rapidly degradable in the environment if the following criteria are met:

.1 In 28-day ready biodegradation studies, the following levels of degradation are achieved:

(i) tests based on dissolved organic carbon: 70%;

(ii) tests based on oxygen depletion or carbon dioxide generation: 60% of theoretical maxima;

These levels of biodegradation shall be achieved within 10 days of the start of degradation, which point is taken as the time when 10% of the substance has been degraded; or

.2 In those cases where only BOD and COD data are available, when the ratio of BOD_5/COD is $\geqslant 0.5$; or

.3 If other convincing scientific evidence is available to demonstrate that the substance or mixture can be degraded (biotically and/or abiotically) in the aquatic environment to a level above 70% within a 28-day period.

2.9.3.2.6 *Chronic toxicity* data are less available than acute data and the range of testing procedures less standardized. Data generated according to the OECD Test Guidelines 210 (Fish Early Life Stage) or 211 (*Daphnia* Reproduction) and 201 (Algal Growth Inhibition) may be accepted. Other validated and internationally accepted tests may also be used. The "No Observed Effect Concentrations" (NOECs) or other equivalent $L(E)C_x$ shall be used.

2.9.3.3 **Substance classification categories and criteria**

2.9.3.3.1 Substances shall be classified as "environmentally hazardous substances (aquatic environment)" if they satisfy the criteria for Acute 1, Chronic 1 or Chronic 2, according to the following tables:

Acute toxicity

Category: Acute 1	
96 hr LC_{50} (for fish)	$\leqslant 1$ mg/ℓ and/or
48 hr EC_{50} (for crustacea)	$\leqslant 1$ mg/ℓ and/or
72 or 96 hr ErC_{50} (for algae or other aquatic plants)	$\leqslant 1$ mg/ℓ

Chronic toxicity

Category: Chronic 1	
96 hr LC_{50} (for fish)	$\leqslant 1$ mg/ℓ and/or
48 hr EC_{50} (for crustacea)	$\leqslant 1$ mg/ℓ and/or
72 or 96 hr ErC_{50} (for algae or other aquatic plants)	$\leqslant 1$ mg/ℓ
and the substance is not rapidly degradable and/or the log $K_{ow} \geqslant 4$ (unless the experimentally determined BCF < 500)	

Category: Chronic 2	
96 hr LC_{50} (for fish)	>1 to $\leqslant 10$ mg/ℓ and/or
48 hr EC_{50} (for crustacea)	>1 to $\leqslant 10$ mg/ℓ and/or
72 or 96 hr ErC_{50} (for algae or other aquatic plants)	>1 to $\leqslant 10$ mg/ℓ
and the substance is not rapidly degradable and/or the log $K_{ow} \geqslant 4$ (unless the experimentally determined BCF <500), unless the chronic toxicity NOECs are > 1 mg/ℓ	

The classification flowchart below outlines the process to be followed.

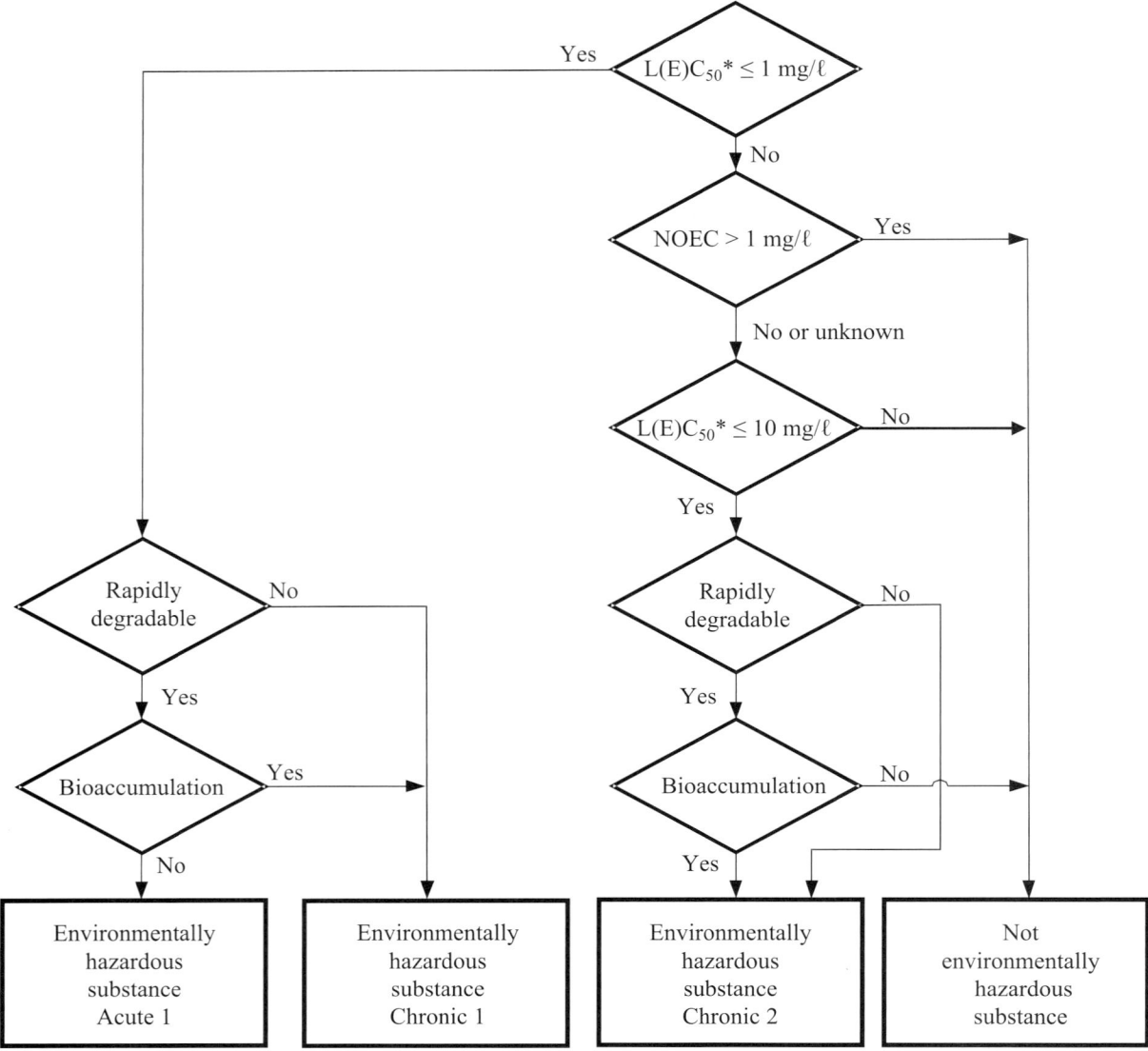

* Lowest value of 96-hour LC_{50}, 48-hour EC_{50} or 72-hour ErC_{50}, as appropriate.

2.9.3.4   Mixtures classification categories and criteria

2.9.3.4.1   The classification system for mixtures covers the classification categories which are used for substances, meaning acute category 1 and chronic categories 1 and 2. In order to make use of all available data for purposes of classifying the aquatic environmental hazards of the mixture, the following assumption is made and is applied, where appropriate:

The "relevant ingredients" of a mixture are those which are present in a concentration of 1% by mass or greater, unless there is a presumption (e.g., in the case of highly toxic ingredients) that an ingredient present at less than 1% can still be relevant for classifying the mixture for aquatic environmental hazards.

2.9.3.4.2   The approach for classification of aquatic environmental hazards is tiered and dependent upon the type of information available for the mixture itself and its ingredients. Elements of the tiered approach include:

.1   classification based on tested mixtures;

.2   classification based on bridging principles;

.3   the use of "summation of classified ingredients" and/or an "additivity formula".

Figure 2.9.1 below outlines the process to be followed.

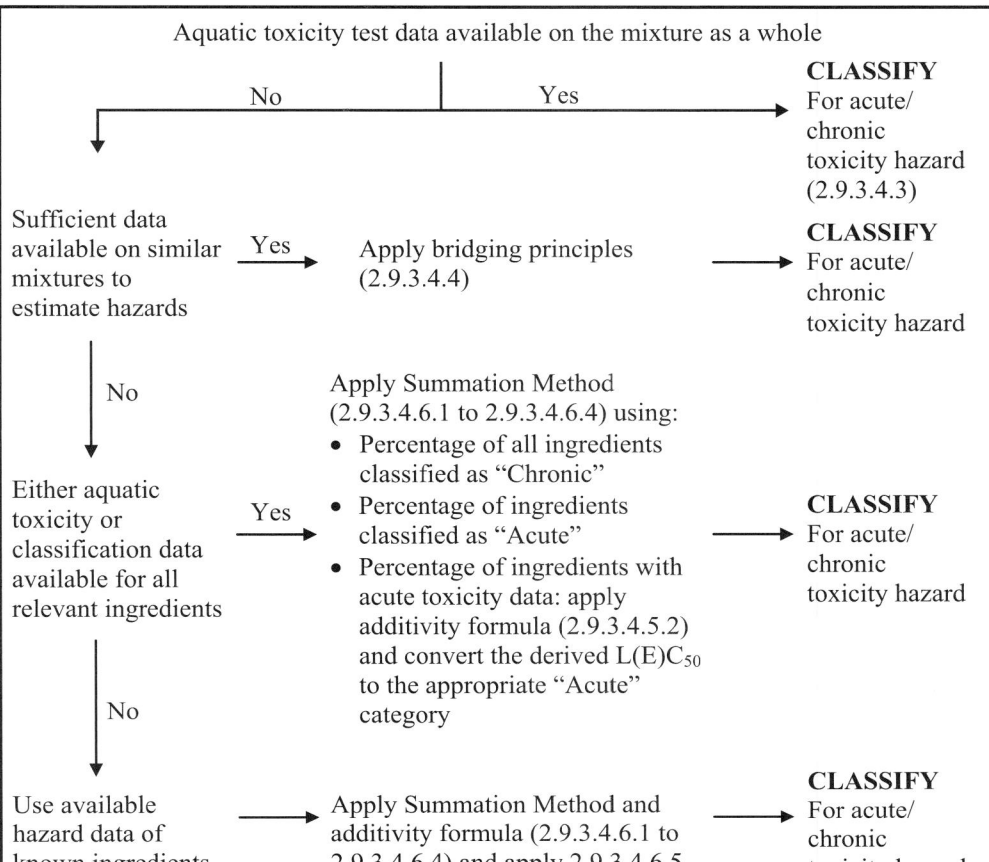

Figure 2.9.1 –Tiered approach to classification of mixtures for acute and chronic aquatic environmental hazards

2.9.3.4.3   *Classification of mixtures when data are available for the complete mixture*

2.9.3.4.3.1   When the mixture as a whole has been tested to determine its aquatic toxicity, it shall be classified according to the criteria that have been agreed for substances, but only for acute toxicity. The classification is based on the data for fish, crustacea and algae/plants. Classification of mixtures by using LC_{50} or EC_{50} data for the mixture as a whole is not possible for chronic categories since both toxicity data and environmental fate data are needed, and there are no degradability and bioaccumulation data for mixtures as a whole. It is not possible to apply the criteria for chronic classification because the data from degradability and bio-accumulation tests of mixtures cannot be interpreted; they are meaningful only for single substances.

2.9.3.4.3.2   When there are acute toxicity test data (LC_{50} or EC_{50}) available for the mixture as a whole, these data as well as information with respect to the classification of ingredients for chronic toxicity shall be used to complete the classification for tested mixtures as follows. When chronic (long-term) toxicity data (NOEC) are also available, these shall be used in addition.

.1 $L(E)C_{50}$ (LC_{50} or EC_{50}) of the tested mixture $\leqslant$ 1 mg/ℓ and NOEC of the tested mixture $\leqslant$ 1.0 mg/ℓ or unknown:

– classify mixture as category Acute 1;

– apply summation of classified ingredients approach (see 2.9.3.4.6.3 and 2.9.3.4.6.4) for chronic classification (Chronic 1, 2, or no need of chronic classification).

.2 $L(E)C_{50}$ of the tested mixture $\leqslant$ 1 mg/ℓ and NOEC of the tested mixture $>$ 1.0 mg/ℓ:

– classify mixture as category Acute 1;

– apply summation of classified ingredients approach (see 2.9.3.4.6.3 and 2.9.3.4.6.4) for classification as category Chronic 1. If the mixture is not classified as category Chronic 1, then there is no need for chronic classification.

.3 L(E)C$_{50}$ of the tested mixture $>$ 1 mg/ℓ, or above the water solubility, and NOEC of the tested mixture $\leqslant$ 1.0 mg/ℓ or unknown:

– no need to classify for acute toxicity;

– apply summation of classified ingredients approach (see 2.9.3.4.6.3 and 2.9.3.4.6.4) for chronic classification or no need for chronic classification.

.4 L(E)C$_{50}$ of the tested mixture $>$ 1 mg/ℓ, or above the water solubility, and NOEC of the tested mixture $>$ 1.0 mg/ℓ:

– No need to classify for acute or chronic toxicity.

2.9.3.4.4   ***Bridging principles***

2.9.3.4.4.1   Where the mixture itself has not been tested to determine its aquatic environmental hazard, but there are sufficient data on the individual ingredients and similar tested mixtures to adequately characterize the hazards of the mixture, these data shall be used in accordance with the following agreed bridging rules. This ensures that the classification process uses the available data to the greatest extent possible in characterizing the hazards of the mixture without the necessity for additional testing in animals.

2.9.3.4.4.2   *Dilution*

2.9.3.4.4.2.1 If a mixture is formed by diluting another classified mixture or a substance with a diluent which has an equivalent or lower aquatic hazard classification than the least toxic original ingredient and which is not expected to affect the aquatic hazards of other ingredients, then the mixture shall be classified as equivalent to the original mixture or substance.

2.9.3.4.4.2.2 If a mixture is formed by diluting another classified mixture or a substance with water or other totally non-toxic material, the toxicity of the mixture shall be calculated from the original mixture or substance.

2.9.3.4.4.3   *Batching*

2.9.3.4.4.3.1 The aquatic hazard classification of one production batch of a complex mixture shall be assumed to be substantially equivalent to that of another production batch of the same commercial product and produced by or under the control of the same manufacturer, unless there is reason to believe there is significant variation such that the aquatic hazard classification of the batch has changed. If the latter occurs, new classification is necessary.

2.9.3.4.4.4   *Concentration of mixtures which are classified with the most severe classification categories (Chronic 1 and Acute 1)*

2.9.3.4.4.4.1 If a mixture is classified as Chronic 1 and/or Acute 1, and ingredients of the mixture which are classified as Chronic 1 and/or Acute 1 are further concentrated, the more concentrated mixture shall be classified with the same classification category as the original mixture without additional testing.

2.9.3.4.4.5   *Interpolation within one toxicity category*

2.9.3.4.4.5.1 If mixtures A and B are in the same classification category and mixture C is made in which the toxicologically active ingredients have concentrations intermediate to those in mixtures A and B, then mixture C shall be in the same category as A and B. Note that the identity of the ingredients is the same in all three mixtures.

2.9.3.4.4.6   *Substantially similar mixtures*

2.9.3.4.4.6.1 Given the following:

.1 Two mixtures:

(i)   A + B

(ii)   C + B

.2 The concentration of ingredient B is the same in both mixtures;

.3 The concentration of ingredient A in mixture (i) equals that of component C in mixture (ii);

.4 Classifications for A and C are available and are the same, i.e. they are in the same hazard category and are not expected to affect the aquatic toxicity of B,

then there shall be no need to test mixture (ii) if mixture (i) is already characterized by testing and both mixtures are classified in the same category.

2.9.3.4.5 *Classification of mixtures when data are available for all components or only for some components of the mixture*

2.9.3.4.5.1 The classification of a mixture shall be based on summation of the classification of its ingredients. The percentage of ingredients classified as "Acute" or "Chronic" will feed straight into the summation method. Details of the summation method are described in 2.9.3.4.6.1 to 2.9.3.4.6.4.1.

2.9.3.4.5.2 Mixtures are often made of a combination of both ingredients that are classified (as Acute 1 and/or Chronic 1, 2) and those for which adequate test data are available. When adequate toxicity data are available for more than one ingredient in the mixture, the combined toxicity of those ingredients shall be calculated using the following additivity formula, and the calculated toxicity shall be used to assign that portion of the mixture an acute toxicity hazard which is then subsequently used in applying the summation method.

$$\frac{\sum C_i}{L(E)C_{50m}} = \sum_n \frac{C_i}{L(E)C_{50i}}$$

where:

C_i = concentration of ingredient i (mass percentage);

$L(E)C_{50i}$ = (mg/ℓ) LC_{50} or EC_{50} for ingredient i;

n = number of ingredients, and i is running from 1 to n; and

$L(E)C_m$ = $L(E)C_{50}$ of the part of the mixture with test data

2.9.3.4.5.3 When applying the additivity formula for part of the mixture, it is preferable to calculate the toxicity of this part of the mixture using for each substance toxicity values that relate to the same species (i.e. fish, daphnia or algae) and then to use the highest toxicity (lowest value) obtained (i.e., use the most sensitive of the three species). However, when toxicity data for each ingredient are not available in the same species, the toxicity value of each ingredient shall be selected in the same manner that toxicity values are selected for the classification of substances, i.e., the higher toxicity (from the most sensitive test organism) is used. The calculated acute toxicity shall then be used to classify this part of the mixture as Acute 1, using the same criteria as described for substances.

2.9.3.4.5.4 If a mixture is classified in more than one way, the method yielding the more conservative result shall be used.

2.9.3.4.6 *Summation method*

2.9.3.4.6.1 *Classification procedure*

2.9.3.4.6.1.1 In general, a more severe classification for mixtures overrides a less severe classification, e.g., a classification with Chronic 1 overrides a classification with Chronic 2. As a consequence, the classification procedure is already completed if the result of the classification is Chronic 1. A more severe classification than Chronic 1 is not possible and it is not necessary therefore to undergo the further classification procedure.

2.9.3.4.6.2 *Classification for the acute category 1*

2.9.3.4.6.2.1 All ingredients classified as Acute 1 shall be considered. If the sum of these ingredients is greater than or equal to 25%, the whole mixture shall be classified as category Acute 1. If the result of the calculation is a classification of the mixture as category Acute 1, the classification process is completed.

2.9.3.4.6.2.2 The classification of mixtures for acute hazards based on this summation of classified ingredients is summarized in table 2.9.1 below.

Table 2.9.1 – Classification of a mixture for acute hazards, based on summation of classified ingredients

Sum of ingredients classified as:	Mixture is classified as:
Acute 1 × M^* ⩾ 25%	Acute 1

*For explanation of the M factor, see 2.9.3.4.6.4.

2.9.3.4.6.3 *Classification for the chronic categories 1, 2*

2.9.3.4.6.3.1 First, all ingredients classified as Chronic 1 are considered. If the sum of these ingredients is greater than or equal to 25%, the mixture shall be classified as category Chronic 1. If the result of the calculation is a classification of the mixture as category Chronic 1, the classification procedure is completed.

2.9.3.4.6.3.2 In cases where the mixture is not classified as Chronic 1, classification of the mixture as Chronic 2 is considered. A mixture shall be classified as Chronic 2 if 10 times the sum of all ingredients classified as Chronic 1 plus the sum of all ingredients classified as Chronic 2 is greater than or equal to 25%. If the result of the calculation is classification of the mixture as Chronic 2, the classification process is completed.

2.9.3.4.6.3.3 The classification of mixtures for chronic hazards, based on this summation of classified ingredients, is summarized in table 2.9.2 below.

Table 2.9.2 – Classification of a mixture for chronic hazards, based on summation of classified ingredients

Sum of ingredients classified as:	Mixture is classified as:
Chronic 1 $\times$ M^* $\geqslant$ 25%	Chronic 1
$(M \times 10 \times$ Chronic 1) + Chronic 2 $\geqslant$ 25%	Chronic 2

*For explanation of the M factor, see 2.9.3.4.6.4.

2.9.3.4.6.4 *Mixtures with highly toxic ingredients*

2.9.3.4.6.4.1 Acute category 1 ingredients with toxicities well below 1 mg/ℓ may influence the toxicity of the mixture and are given increased weight in applying the summation of classification approach. When a mixture contains ingredients classified as acute or chronic category 1, the tiered approach described in 2.9.3.4.6.2 and 2.9.3.4.6.3 shall be applied using a weighted sum by multiplying the concentrations of acute category 1 ingredients by a factor, instead of merely adding up the percentages. This means that the concentration of "Acute 1" in the left column of table 2.9.1 and the concentration of "Chronic 1" in the left column of table 2.9.2 are multiplied by the appropriate multiplying factor. The multiplying factors to be applied to these ingredients are defined using the toxicity value, as summarized in table 2.9.3 below. Therefore, in order to classify a mixture containing Acute 1 and/or Chronic 1 ingredients, the classifier needs to be informed of the value of the M factor in order to apply the summation method. Alternatively, the additivity formula (2.9.3.4.5.2) may be used when toxicity data are available for all highly toxic ingredients in the mixture and there is convincing evidence that all other ingredients, including those for which specific acute toxicity data are not available, are of low or no toxicity and do not significantly contribute to the environmental hazard of the mixture.

Table 2.9.3 – Multiplying factors for highly toxic ingredients of mixtures

L(E)C$_{50}$ value	Multiplying factor (M)
$0.1 < $ L(E)C$_{50}$ $\leqslant 1$	1
$0.01 < $ L(E)C$_{50}$ $\leqslant 0.1$	10
$0.001 < $ L(E)C$_{50}$ $\leqslant 0.01$	100
$0.0001 < $ L(E)C$_{50}$ $\leqslant 0.001$	1000
$0.00001 < $ L(E)C$_{50}$ $\leqslant 0.0001$	10000
(continue in factor 10 intervals)	

2.9.3.4.6.5 *Classification of mixtures with ingredients without any useable information*

2.9.3.4.6.5.1 In the event that no useable information on acute and/or chronic aquatic hazard is available for one or more relevant ingredients, it is concluded that the mixture cannot be attributed (a) definitive hazard category(ies). In this event, the mixture shall be classified based on the known ingredients only with the additional statement that: "x percent of the mixture consists of ingredient(s) of unknown hazards to the aquatic environment."

2.9.3.5 **Substances or mixtures dangerous to the aquatic environment not otherwise classified under the provisions of this Code**

2.9.3.5.1 Substances or mixtures dangerous to the aquatic environment not otherwise classified under this Code shall be designated:

UN 3077 ENVIRONMENTALLY HAZARDOUS SUBSTANCE, SOLID, N.O.S. or

UN 3082 ENVIRONMENTALLY HAZARDOUS SUBSTANCE, LIQUID, N.O.S.

They shall be assigned to Packing Group III.

Chapter 2.10

Marine pollutants

2.10.1 Definition

Marine pollutants means substances which are subject to the provisions of Annex III of MARPOL 73/78, as amended.

2.10.2 General provisions

2.10.2.1 Marine pollutants shall be transported under the provisions of Annex III of MARPOL 73/78, as amended.

2.10.2.2 The Index indicates by the symbol **P** in the column headed **MP** those substances, materials and articles that are identified as marine pollutants.

2.10.2.3 Marine pollutants shall be transported under the appropriate entry according to their properties if they fall within the criteria of any of the classes 1 to 8. If they do not fall within the criteria of any of these classes, they shall be transported under the entry: ENVIRONMENTALLY HAZARDOUS SUBSTANCE, SOLID, N.O.S., UN 3077 or ENVIRONMENTALLY HAZARDOUS SUBSTANCE, LIQUID, N.O.S., UN 3082, as appropriate, unless there is a specific entry in class 9.

2.10.2.4 Column 4 of the Dangerous Goods List also provides information on marine pollutants using the symbol **P**.

2.10.2.5 When a substance, material or article possesses properties that meet the criteria of a marine pollutant but is not identified in this Code, such substance, material or article shall be transported as a marine pollutant in accordance with the Code.

2.10.2.6 With the approval of the competent authority (see 7.9.2), substances, materials or articles that are identified as marine pollutants in this Code but which no longer meet the criteria as a marine pollutant need not be transported in accordance with the provisions of this Code applicable to marine pollutants.

2.10.3 Classification

2.10.3.1 Marine pollutants shall be classified in accordance with chapter 2.9.3.

Part 3 is in Volume 2

PART 4

PACKING AND TANK PROVISIONS

4

Chapter 4.1

Use of packagings, including intermediate bulk containers (IBCs) and large packagings

4.1.0   **Definitions**

Effectively closed: liquid-tight closure.

Hermetically sealed: vapour-tight closure.

Securely closed: so closed that dry contents cannot escape during normal handling; the minimum provisions for any closure.

4.1.1   **General provisions for the packing of dangerous goods in packagings, including IBCs and large packagings**

> **Note**: For the packing of goods of classes 2, 6.2 and 7, the general provisions of this section only apply as indicated in 4.1.8.2 (class 6.2), 4.1.9.1.5 (class 7) and in the applicable packing instructions of 4.1.4 (P201 and LP02 for class 2 and P620, P621, P650, IBC620 and LP621 for class 6.2).

4.1.1.1   Dangerous goods shall be packed in good quality packagings, including IBCs and large packagings, which shall be strong enough to withstand the shocks and loadings normally encountered during transport, including trans-shipment between cargo transport units and between cargo transport units and warehouses as well as any removal from a pallet or overpack for subsequent manual or mechanical handling. Packagings, including IBCs and large packagings, shall be constructed and closed so as to prevent any loss of contents when prepared for transport which may be caused under normal conditions of transport, by vibration, or by changes in temperature, humidity or pressure (resulting from altitude, for example). Packagings, including IBCs and large packagings, shall be closed in accordance with the information provided by the manufacturer. No dangerous residue shall adhere to the outside of packages, IBCs and large packagings during transport. These provisions apply, as appropriate, to new, re-used, reconditioned or remanufactured packagings, and to new, re-used, repaired or remanufactured IBCs, and to new or re-used large packagings.

4.1.1.2   Parts of packagings, including IBCs and large packagings, which are in direct contact with dangerous goods:

     .1   shall not be affected or significantly weakened by those dangerous goods; and

     .2   shall not cause a dangerous effect, such as catalysing a reaction or reacting with the dangerous goods.

     Where necessary, they shall be provided with a suitable inner coating or treatment.

4.1.1.3   Unless otherwise provided elsewhere in this Code, each packaging, including IBCs and large packagings, except inner packagings, shall conform to a design type successfully tested in accordance with the provisions of 6.1.5, 6.3.2, 6.5.4 or 6.6.5, as applicable. However, IBCs manufactured before 1 January 2011 and conforming to a design type which has not passed the vibration test of 6.5.6.13 or which has not passed the drop test criteria of 6.5.6.9.5.4 may still be used.

4.1.1.4   When filling packagings, including IBCs and large packagings, with liquids,* sufficient ullage (outage) shall be left to ensure that neither leakage nor permanent distortion of the packaging occurs as a result of an expansion of the liquid caused by temperatures likely to occur during transport. Unless specific provisions are prescribed, liquids shall not completely fill a packaging at a temperature of 55°C. However, sufficient ullage shall be left in an IBC to ensure that at the mean bulk temperature of 50°C it is not filled to more than 98% of its water capacity.[†]

4.1.1.4.1   For air transport, packagings intended to contain liquids shall also be capable of withstanding a pressure differential without leakage as specified in the international regulations for air transport.

---

* With respect to ullage limits only, the provisions applicable for packagings for solid substances may be used if the viscous substance has an outflow time via a DIN-cup with a 4 mm diameter outlet exceeding 10 minutes at 20°C (corresponding to an outflow time via a Ford cup 4 of more than 690 seconds at 20°C, or to a viscosity of more than 2680 centistokes at 20°C).

[†] For a differing temperature, the maximum degree of filling may be determined as follows:

$$\text{Degree of filling} = \frac{98}{1 + \alpha(50 - t_F)} \text{ per cent of the capacity of the IBC}$$

In this formula α represents the mean coefficient of cubic expansion of the liquid substance between 15°C and 50°C; that is to say, for a maximum rise in the temperature of 35°C, "α" is calculated according to the formula:

$$\alpha = \frac{d_{15} - d_{50}}{35 \times d_{50}}$$

where d_{15} and d_{50} are the relative densities of the liquid at 15°C and 50°C and t_F is the mean temperature of the liquid at the time of filling.

4.1.1.5 Inner packagings shall be packed in an outer packaging in such a way that, under normal conditions of transport, they cannot break, be punctured or leak their contents into the outer packaging. Inner packagings containing liquids shall be packaged with their closures upward and placed within outer packagings consistent with the orientation markings prescribed in 5.2.1.7 of this Code. Inner packagings that are liable to break or be punctured easily, such as those made of glass, porcelain or stoneware or of certain plastics materials, etc., shall be secured in outer packagings with suitable cushioning material. Any leakage of the contents shall not substantially impair the protective properties of the cushioning material or of the outer packaging.

4.1.1.5.1 Where an outer packaging of a combination packaging or a large packaging has been successfully tested with different types of inner packagings, a variety of such different inner packagings may also be assembled in this outer packaging or large packagings. In addition, provided an equivalent level of performance is maintained, the following variations in inner packagings are allowed without further testing of the package:

.1 Inner packagings of equivalent or smaller size may be used provided:

– the inner packagings are of similar design to the tested inner packagings (such as shape – round, rectangular, etc.);

– the material of construction of inner packagings (glass, plastics, metal, etc.) offers resistance to impact and stacking forces equal to or greater than that of the originally tested inner packaging;

– the inner packagings have the same or smaller openings and the closure is of similar design (such as screw cap, friction lid, etc.);

– sufficient additional cushioning material is used to take up void spaces and to prevent significant movement of the inner packagings;

– inner packagings are oriented within the outer packaging in the same manner as in the tested package; and

.2 A lesser number of the tested inner packagings or of the alternative types of inner packagings identified in .1 above may be used, provided sufficient cushioning is added to fill the void space(s) and to prevent significant movement of the inner packagings.

4.1.1.5.2 Cushioning and absorbent material shall be inert and suited to the nature of the contents.

4.1.1.5.3 The nature and the thickness of the outer packagings shall be such that friction during transport does not generate any heating likely to alter dangerously the chemical stability of the contents.

4.1.1.6 Dangerous goods shall not be packed together in the same outer packaging, or in large packagings, with dangerous or other goods if they react dangerously with each other and cause:

.1 combustion and/or evolution of considerable heat;

.2 evolution of flammable, toxic or asphyxiant gases;

.3 the formation of corrosive substances; or

.4 the formation of unstable substances.

4.1.1.7 The closures of packagings containing wetted or diluted substances shall be such that the percentage of liquid (water, solvent or phlegmatizer) does not fall below the prescribed limits during transport.

4.1.1.7.1 Where two or more closure systems are fitted in series on an IBC, that nearest to the substance being transported shall be closed first.

4.1.1.7.2 Unless otherwise specified in the Dangerous Goods List, packages containing substances which:

.1 evolve flammable gases or vapour;

.2 may become explosive if allowed to dry;

.3 evolve toxic gases or vapour;

.4 evolve corrosive gases or vapour; or

.5 may react dangerously with the atmosphere

should be hermetically sealed.

4.1.1.8 Where pressure may develop in a package by the emission of gas from the contents (as a result of temperature increase or other causes), the packaging or IBC may be fitted with a vent provided that the gas emitted will not cause danger on account of its toxicity, its flammability, the quantity released, etc.

A venting device shall be fitted if dangerous overpressure may develop due to normal decomposition of substances. The vent shall be so designed that, when the packaging or IBC is in the attitude in which it is intended to be transported, leakages of liquid and the penetration of foreign substances are prevented under normal conditions of transport.

4.1.1.8.1   Liquids may only be filled into inner packagings which have an appropriate resistance to internal pressure that may be developed under normal conditions of transport.

4.1.1.9   New, remanufactured or re-used packagings, including IBCs and large packagings, or reconditioned packagings and repaired or routinely maintained IBCs shall be capable of passing the tests prescribed in 6.1.5, 6.3.2, 6.5.4 or 6.6.5, as applicable. Before being filled and handed over for transport, every packaging, including IBCs and large packagings, shall be inspected to ensure that it is free from corrosion, contamination or other damage and every IBC shall be inspected with regard to the proper functioning of any service equipment. Any packaging which shows signs of reduced strength as compared with the approved design type shall no longer be used or shall be so reconditioned that it is able to withstand the design type tests. Any IBC which shows signs of reduced strength as compared with the tested design type shall no longer be used or shall be so repaired or routinely maintained that it is able to withstand the design type tests.

4.1.1.10   Liquids shall be filled only into packagings, including IBCs, which have an appropriate resistance to the internal pressure that may develop under normal conditions of transport. As the vapour pressure of low-boiling-point liquids is usually high, the strength of receptacles for these liquids shall be sufficient to withstand, with an ample factor of safety, the internal pressure likely to be generated. Packagings and IBCs marked with the hydraulic test pressure prescribed in 6.1.3.1(d) and 6.5.2.2.1, respectively, shall be filled only with a liquid having a vapour pressure:

    .1   such that the total gauge pressure in the packaging or IBC (i.e. the vapour pressure of the filling substance plus the partial pressure of air or other inert gases, less 100 kPa) at 55°C, determined on the basis of a maximum degree of filling in accordance with 4.1.1.4 and a filling temperature of 15°C, will not exceed two thirds of the marked test pressure; or

    .2   at 50°C, less than four sevenths of the sum of the marked test pressure plus 100 kPa; or

    .3   at 55°C, less than two thirds of the sum of the marked test pressure plus 100 kPa.

IBCs intended for the transport of liquids shall not be used to carry liquids having a vapour pressure of more than 110 kPa (1.1 bar) at 50°C or 130 kPa (1.3 bar) at 55°C.

Examples of required marked test pressures for packagings, including IBCs, calculated as in 4.1.1.10.3

UN No.	Name	Class	Packing group	Vp_{55} (kPa)	$Vp_{55} \times 1.5$ (kPa)	$(Vp_{55} \times 1.5)$ minus 100 (kPa)	Required minimum test pressure (gauge) under 6.1.5.5.4.3 (kPa)	Minimum test pressure (gauge) to be marked on the packaging (kPa)
2056	Tetrahydrofuran	3	II	70	105	5	100	100
2247	n-Decane	3	III	1.4	2.1	−97.9	100	100
1593	Dichloromethane	6.1	III	164	246	146	146	150
1155	Diethyl ether	3	I	199	299	199	199	250

Note 1: For pure liquids, the vapour pressure at 55°C (Vp_{55}) can often be obtained from scientific tables.

Note 2: The table refers to the use of 4.1.1.10.3 only, which means that the marked test pressure shall exceed 1.5 times the vapour pressure at 55°C less 100 kPa. When, for example, the test pressure for n-decane is determined according to 6.1.5.5.4.1, the minimum marked test pressure may be lower.

Note 3: For diethyl ether, the required minimum test pressure under 6.1.5.5.5 is 250 kPa.

4.1.1.11   Empty packagings, including IBCs and large packagings, that have contained a dangerous substance shall be treated in the same manner as is required by this Code for a filled packaging, unless adequate measures have been taken to nullify any hazard.

4.1.1.12   Every packaging, as specified in chapter 6.1, intended to contain liquids shall successfully undergo a suitable leakproofness test, and be capable of meeting the appropriate test level indicated in 6.1.5.4.3:

    .1   before it is first used for transport;

    .2   after remanufacturing or reconditioning of any packaging, before it is re-used for transport.

For this test, the packaging need not have its closures fitted. The inner receptacle of a composite packaging may be tested without the outer packaging, provided the test results are not affected. This test is not necessary for inner packagings of combination packagings or large packagings.

4.1.1.13 Packagings, including IBCs, used for solids which may become liquid at temperatures likely to be encountered during transport shall also be capable of containing the substance in the liquid state.

4.1.1.14 Packagings, including IBCs, used for powdery or granular substances shall be sift-proof or shall be provided with a liner.

4.1.1.15 For plastics drums and jerricans, rigid plastics IBCs and composite IBCs with plastics inner receptacles, unless otherwise approved by the competent authority, the period of use permitted for the transport of dangerous substances shall be five years from the date of manufacture of the receptacles, except where a shorter period of use is prescribed because of the nature of the substance to be transported.

4.1.1.16 Explosives, self-reactive substances and organic peroxides

Unless specific provision to the contrary is made in this Code, the packagings, including IBCs and large packagings, used for goods of class 1, self-reactive substances of class 4.1 and organic peroxides of class 5.2 shall comply with the provisions for the medium danger group (packing group II).

4.1.1.17 Use of salvage packagings

4.1.1.17.1 Damaged, defective, leaking or non-conforming packages, or dangerous goods that have spilled or leaked may be transported in salvage packagings mentioned in 6.1.5.1.11. This does not prevent the use of a bigger size packaging of appropriate type and performance level under the conditions of 4.1.1.17.2.

4.1.1.17.2 Appropriate measures shall be taken to prevent excessive movement of the damaged or leaking packages within a salvage packaging. When the salvage packaging contains liquids, sufficient inert absorbent material shall be added to eliminate the presence of free liquid.

4.1.1.17.3 Salvage packagings shall not be used as packagings for shipment from premises where the substances or materials are produced.

4.1.1.17.4 The use of salvage packagings for other than emergency purposes during transport (land or sea) requires approval by the competent authority.

4.1.1.17.5 In addition to the general provisions of the Code, the following paragraphs apply specifically to salvage packagings: 5.2.1.3, 5.4.1.5.3, 6.1.2.4, 6.1.5.1.11 and 6.1.5.7.

4.1.1.17.6 Appropriate measures shall be taken to ensure there is no dangerous build-up of pressure.

4.1.1.18 During transport, packagings, including IBCs and large packagings, shall be securely fastened to or contained within the cargo transport unit, so that lateral or longitudinal movement or impact is prevented and adequate external support is provided.

4.1.2 Additional general provisions for the use of IBCs

4.1.2.1 When IBCs are used for the transport of liquids with a flashpoint of 60°C (closed cup) or lower, or of powders liable to dust explosion, measures shall be taken to prevent a dangerous electrostatic discharge.

4.1.2.2.1 Every metal, rigid plastics and composite IBC shall be inspected and tested, as relevant, in accordance with 6.5.4.4 or 6.5.4.5:

.1 before it is put into service;

.2 thereafter at intervals not exceeding two and a half and five years, as appropriate; and

.3 after the repair or remanufacture, before it is re-used for transport.

4.1.2.2.2 An IBC shall not be filled and offered for transport after the date of expiry of the last periodic test or inspection. However, an IBC filled prior to the date of expiry of the last periodic test or inspection may be transported for a period not to exceed three months beyond the date of expiry of the last periodic test or inspection. In addition, an IBC may be transported after the date of expiry of the last periodic test or inspection:

.1 after emptying but before cleaning, for purposes of performing the required test or inspection prior to refilling; and

.2 unless otherwise approved by the competent authority, for a period not to exceed six months beyond the date of expiry of the last periodic test or inspection in order to allow the return of dangerous goods or residues for proper disposal or recycling. Reference to this exemption shall be entered in the transport document.

4.1.2.3 IBCs of type 31HZ2 when transporting liquids shall be filled to at least 80% of the volume of the outer casing and shall be transported in closed cargo transport units.

4.1.2.4   Except for routine maintenance of metal, rigid plastics, composite and flexible IBCs performed by the owner of the IBC, whose State and name or authorized symbol is durably marked on the IBC, the party performing routine maintenance shall durably mark the IBC near the manufacturer's UN design type marking to show:

.1   the State in which the routine maintenance was carried out; and

.2   the name or authorized symbol of the party performing the routine maintenance.

4.1.3   General provisions concerning packing instructions

4.1.3.1   Packing instructions applicable to dangerous goods of classes 1 to 9 are specified in 4.1.4. They are subdivided in three sub-sections depending on the type of packagings to which they apply:

sub-section 4.1.4.1 for packagings other than IBCs and large packagings; these packing instructions are designated by an alphanumeric code comprising the letter "P";

sub-section 4.1.4.2 for IBCs; these are designated by an alphanumeric code comprising the letters "IBC";

sub-section 4.1.4.3 for large packagings; these are designated by an alphanumeric code comprising the letters "LP".

Generally, packing instructions specify that the general provisions of 4.1.1, 4.1.2 and/or 4.1.3, as appropriate, are applicable. They may also require compliance with the special provisions of 4.1.5, 4.1.6, 4.1.7, 4.1.8 or 4.1.9 when appropriate. Special packing provisions may also be specified in the packing instruction for individual substances or articles. They are also designated by an alphanumeric code comprising the letters:

"PP"   for packagings other than IBCs and large packagings
"B"   for IBCs
"L"   for large packagings.

Unless otherwise specified, each packaging shall conform to the applicable provisions of part 6. Generally, packing instructions do not provide guidance on compatibility and the user shall not select a packaging without checking that the substance is compatible with the packaging material selected (such as, most fluorides are unsuitable for glass receptacles). Where glass receptacles are permitted in the packing instructions, porcelain, earthenware and stoneware packagings are also allowed.

4.1.3.2   Column 8 of the Dangerous Goods List shows for each article or substance the packing instruction(s) that shall be used. Column 9 indicates the special packing provisions applicable to specific substances or articles.

4.1.3.3   Each packing instruction shows, where applicable, the acceptable single and combination packagings. For combination packagings, the acceptable outer packagings, inner packagings and, when applicable, the maximum quantity permitted in each inner or outer packaging are shown. *Maximum net mass* and *maximum capacity* are as defined in 1.2.1.

4.1.3.4   The following packagings shall not be used when the substances being transported are liable to become liquid during transport:

Packagings
Drums: 1D and 1G
Boxes: 4C1, 4C2, 4D, 4F, 4G and 4H1
Bags: 5L1, 5L2, 5L3, 5H1, 5H2, 5H3, 5H4, 5M1 and 5M2
Composite: 6HC, 6HD2, 6HG1, 6HG2, 6HD1, 6PC, 6PD1, 6PD2, 6PG1, 6PG2 and 6PH1

Large packagings
Flexible plastics: 51H (outer packaging)

IBCs
For substances of packing group I:
  All types of IBCs
For substances of packing groups II and III:
  Wooden: 11C, 11D and 11F
  Fibreboard: 11G
  Flexible: 13H1, 13H2, 13H3, 13H4, 13H5, 13L1, 13L2, 13L3, 13L4, 13M1 and 13M2
  Composite: 11HZ2 and 21HZ2

4.1.3.5   Where the packing instructions in this chapter authorize the use of a particular type of packaging (such as 4G; 1A2), packagings bearing the same packaging identification code followed by the letters "V", "U" or "W" marked in accordance with the provisions of part 6 (such as "4GV", "4GU" or "4GW"; "1A2V", "1A2U" or "1A2W") may also be used under the same conditions and limitations applicable to the use of that type of packaging according to the relevant packing instructions. For example, a combination packaging marked with the packaging code "4GV" may be used whenever a combination packaging marked "4G" is authorized, provided the provisions in the relevant packing instruction regarding types of inner packagings and quantity limitations are respected.

4.1.3.6   Pressure receptacles for liquids and solids

4.1.3.6.1   Unless otherwise indicated in this Code, pressure receptacles conforming to:

(a) the applicable requirements of chapter 6.2; or

(b) the National or International standards on the design, construction, testing, manufacturing and inspection, as applied by the country in which the pressure receptacles are manufactured, provided that the provisions of 4.1.3.6 and 6.2.3.3 are met,

are authorized for the transport of any liquid or solid substance other than explosives, thermally unstable substances, organic peroxides, self-reactive substances, substances where significant pressure may develop by evolution of chemical reaction and radioactive material (unless permitted in 4.1.9).

This sub-section is not applicable to the substances mentioned in 4.1.4.1, packing instruction P200, table 3.

4.1.3.6.2   Every design type of pressure receptacle shall be approved by the competent authority of the country of manufacture or as indicated in chapter 6.2.

4.1.3.6.3   Unless otherwise indicated, pressure receptacles having a minimum test pressure of 0.6 MPa shall be used.

4.1.3.6.4   Unless otherwise indicated, pressures receptacles may be provided with an emergency pressure relief device designed to avoid bursting in case of overfill or fire accidents.

Pressure receptacle valves shall be designed and constructed in such a way that they are inherently able to withstand damage without release of the contents or shall be protected from damage which could cause inadvertent release of the contents of the pressure receptacle, by one of the methods as given in 4.1.6.1.8 (.1) to (.5).

4.1.3.6.5   The level of filling shall not exceed 95% of the capacity of the pressure receptacle at 50°C. Sufficient ullage (outage) shall be left to ensure that the pressure receptacle will not be liquid-full at a temperature of 55°C.

4.1.3.6.6   Unless otherwise indicated, pressure receptacles shall be subjected to a periodic inspection and test every 5 years. The periodic inspection shall include an external examination, an internal examination or alternative method as approved by the competent authority, a pressure test or equivalent effective non-destructive testing with the agreement of the competent authority, including an inspection of all accessories (e.g., tightness of valves, emergency relief valves of fusible elements). Pressure receptacles shall not be filled after they become due for periodic inspection and test but may be transported after the expiry of the time limit. Pressure receptacle repairs shall meet the requirements of 4.1.6.1.11.

4.1.3.6.7   Prior to filling, the filler shall perform an inspection of the pressure receptacle and ensure that the pressure receptacle is authorized for the substances to be transported and that the provisions of this Code have been met. Shut-off valves shall be closed after filling and remain closed during transport. The consignor shall verify that the closures and equipment are not leaking.

4.1.3.6.8   Refillable pressure receptacles shall not be filled with a substance different from that previously contained unless the necessary operations for change of service have been performed.

4.1.3.6.9   Marking of pressure receptacles for liquids and solids according to 4.1.3.6 (not conforming to the requirements of chapter 6.2) shall be in accordance with the requirements of the competent authority of the country of manufacturing.

4.1.3.7   Packagings, including IBCs and large packagings, not specifically authorized in the applicable packing instruction shall not be used for the transport of a substance or article unless specifically approved by the competent authority and provided:

.1 the alternative packaging complies with the general provisions of this chapter;

.2 when the packing instruction indicated in the Dangerous Goods List so specifies, the alternative packaging meets the provisions of part 6;

.3 the competent authority determines that the alternative packaging provides at least the same level of safety as if the substance were packed in accordance with a method specified in the particular packing instruction indicated in the Dangerous Goods List; and

.4   a copy of the competent authority approval accompanies each consignment or the transport document includes an indication that alternative packaging was approved by the competent authority.

Note: The competent authorities granting such approvals shall take action to amend the Code to include the provisions covered by the approval as appropriate.

4.1.3.8   **Unpackaged articles other than class 1 articles**

4.1.3.8.1   Where large and robust articles cannot be packaged in accordance with the requirements of chapter 6.1 or 6.6 and they have to be transported empty, uncleaned and unpackaged, the competent authority may approve such transport. In doing so, the competent authority shall take into account that:

.1   Large and robust articles shall be strong enough to withstand the shocks and loadings normally encountered during transport, including trans-shipment between cargo transport units and between cargo transport units and warehouses, as well as any removal from a pallet for subsequent manual or mechanical handling;

.2   All closures and openings shall be sealed so that there can be no loss of contents which might be caused under normal conditions of transport, by vibration, or by changes in temperature, humidity or pressure (resulting from altitude, for example). No dangerous residue shall adhere to the outside of the large and robust articles;

.3   Parts of large and robust articles, which are in direct contact with dangerous goods:

.3.1   shall not be affected or significantly weakened by those dangerous goods; and

.3.2   shall not cause a dangerous effect, e.g. catalysing a reaction or reacting with the dangerous goods;

.4   Large and robust articles containing liquids shall be stowed and secured to ensure that neither leakage nor permanent distortion of the article occurs during transport;

.5   They shall be fixed in cradles or crates or other handling devices in such a way that they will not become loose during normal conditions of transport.

4.1.3.8.2   Unpackaged articles approved by the competent authority in accordance with the provisions of 4.1.3.8.1 shall be subject to the consignment procedures of part 5. In addition the consignor of such articles shall ensure that a copy of any such approval is transported with the large and robust articles.

Note: A large and robust article may include flexible fuel containment systems, military equipment, machinery or equipment containing dangerous goods above the limited quantity thresholds.

4.1.3.9   Where, in 4.1.3.6 and in the individual packing instructions, cylinders and other pressure receptacles for gases are authorized for the transport of any liquid or solid substance, use is also authorized of cylinders and pressure receptacles of a kind normally used for gases which conform to the requirements of the competent authority of the country in which the cylinder or pressure receptacle is filled. Valves shall be suitably protected. Pressure receptacles with capacities of 1 ℓ or less shall be packed in outer packagings constructed of suitable material of adequate strength and design in relation to the capacity of the packaging and its intended use and secured or cushioned so as to prevent significant movement within the outer packaging during normal conditions of transport.

4

4.1.4 List of packing instructions

4.1.4.1 Packing instructions concerning the use of packagings (except IBCs and large packagings)

P001	PACKING INSTRUCTION (LIQUIDS)			P001
The following packagings are authorized provided the general provisions of 4.1.1 and 4.1.3 are met:				

Combination packagings		Maximum capacity/net mass (see 4.1.3.3)		
Inner packagings	Outer packagings	Packing group I	Packing group II	Packing group III
Glass 10 ℓ Plastics 30 ℓ Metal 40 ℓ	**Drums** steel (1A2) aluminium (1B2) other metal (1N2) plastics (1H2) plywood (1D) fibre (1G)	75 kg 75 kg 75 kg 75 kg 75 kg 75 kg	400 kg 400 kg 400 kg 400 kg 400 kg 400 kg	400 kg 400 kg 400 kg 400 kg 400 kg 400 kg
	Boxes steel (4A) aluminium (4B) natural wood (4C1, 4C2) plywood (4D) reconstituted wood (4F) fibreboard (4G) expanded plastics (4H1) solid plastics (4H2)	75 kg 75 kg 75 kg 75 kg 75 kg 75 kg 40 kg 75 kg	400 kg 400 kg 400 kg 400 kg 400 kg 400 kg 60 kg 400 kg	400 kg 400 kg 400 kg 400 kg 400 kg 400 kg 60 kg 400 kg
	Jerricans steel (3A2) aluminium (3B2) plastics (3H2)	60 kg 60 kg 30 kg	120 kg 120 kg 120 kg	120 kg 120 kg 120 kg
Single packagings				
Drums steel, non-removable head (1A1) steel, removable head (1A2) aluminium, non-removable head (1B1) aluminium, removable head (1B2) other metal, non-removable head (1N1) other metal, removable head (1N2) plastics, non-removable head (1H1) plastics, removable head (1H2)		250 ℓ prohibited 250 ℓ prohibited 250 ℓ prohibited 250 ℓ* prohibited	450 ℓ 250 ℓ 450 ℓ 250 ℓ 450 ℓ 250 ℓ 450 ℓ 250 ℓ	450 ℓ 250 ℓ 450 ℓ 250 ℓ 450 ℓ 250 ℓ 450 ℓ 250 ℓ
Jerricans steel, non-removable head (3A1) steel, removable head (3A2) aluminium, non-removable head (3B1) aluminium, removable head (3B2) plastics, non-removable head (3H1) plastics, removable head (3H2)		60 ℓ prohibited 60 ℓ prohibited 60 ℓ* prohibited	60 ℓ 60 ℓ 60 ℓ 60 ℓ 60 ℓ 60 ℓ	60 ℓ 60 ℓ 60 ℓ 60 ℓ 60 ℓ 60 ℓ
Composite packagings Plastics receptacle in steel or aluminium drum (6HA1, 6HB1)		250 ℓ*	250 ℓ	250 ℓ
Plastics receptacle in fibre, plastics or plywood drum (6HG1, 6HH1, 6HD1)		120 ℓ*	250 ℓ	250 ℓ
Plastics receptacle in steel or aluminium crate or box or plastics receptacle in wood, plywood, fibreboard or solid plastics box (6HA2, 6HB2, 6HC, 6HD2, 6HG2 or 6HH2)		60 ℓ*	60 ℓ	60 ℓ
Glass receptacle in steel, aluminium, fibre, plywood, solid plastics or expanded plastics drum (6PA1, 6PB1, 6PG1, 6PD1, 6PH1 or 6PH2) or in a steel, aluminium, wood or fibreboard box or in a wickerwork hamper (6PA2, 6PB2, 6PC, 6PG2 or 6PD2)		60 ℓ	60 ℓ	60 ℓ
Pressure receptacles, provided that the general provisions of 4.1.3.6 are met				

* Not permitted for class 3, packing group I.

P001	PACKING INSTRUCTION (LIQUIDS) *(continued)*	P001

Special packing provisions:

PP1   For UN Nos. 1133, 1210, 1263 and 1866 and for adhesives, printing inks, printing ink related materials, paints, paint related materials and resin solutions which are assigned to UN 3082, metal or plastics packagings for substances of packing groups II and III in quantities of 5 litres or less per packaging are not required to meet the performance tests in chapter 6.1 when transported:

        (a)   in palletized loads, a pallet box or a unit load device, such as individual packagings placed or stacked and secured by strapping, shrink- or stretch-wrapping or other suitable means to a pallet. For sea transport, the palletized loads, pallet boxes or unit load devices shall be firmly packed and secured in closed cargo transport units; or

        (b)   as an inner packaging of a combination packaging with a maximum net mass of 40 kg.

PP2   For UN 3065, wooden barrels with a maximum capacity of 250 litres and which do not meet the provisions of chapter 6.1 may be used.

PP4   For UN 1774, packagings shall meet the packing group II performance level.

PP5   For UN 1204, packagings shall be so constructed that explosion is not possible by reason of increased internal pressure. Gas cylinders and gas receptacles shall not be used for these substances.

PP10   For UN 1791, for packing group II, the packaging shall be vented.

PP31   For UN Nos. 1131, 1553, 1693, 1694, 1699, 1701, 2478, 2604, 2785, 3148, 3183, 3184, 3185, 3186, 3187, 3188, 3398 (PG II and III), 3399 (PG II and III), 3413 and 3414, packagings shall be hermetically sealed.

PP33   For UN 1308, for packing groups I and II, only combination packagings with a maximum gross mass of 75 kg are allowed.

PP81   For UN 1790 with more than 60% but not more than 85% hydrogen fluoride and UN 2031 with more than 55% nitric acid, the permitted use of plastics drums and jerricans as single packagings shall be two years from their date of manufacture.

4

P002	PACKING INSTRUCTION (SOLIDS)	P002

The following packagings are authorized provided the general provisions of 4.1.1 and 4.1.3 are met.

Combination packagings		Maximum net mass (see 4.1.3.3)		
Inner packagings	Outer packagings	Packing group I	Packing group II	Packing group III

Glass   10 kg
Plastics[1]   30 kg
Metal   40 kg
Paper[1, 2, 3]   50 kg
Fibre[1, 2, 3]   50 kg
	Drums
	   steel (1A2)
	   aluminium (1B2)
	   other metal (1N2)
	   plastics (1H2)
	   plywood (1D)
	   fibre (1G)
		
		125 kg
		125 kg
		125 kg
		125 kg
		125 kg
		125 kg
			
			400 kg
			400 kg
			400 kg
			400 kg
			400 kg
			400 kg
				
				400 kg
				400 kg
				400 kg
				400 kg
				400 kg
				400 kg

	Boxes
	   steel (4A)
	   aluminium (4B)
	   natural wood (4C1)
	   natural wood with
	      sift-proof walls (4C2)
	   plywood (4D)
	   reconstituted wood (4F)
	   fibreboard (4G)
	   expanded plastics (4H1)
	   solid plastics (4H2)
		
		125 kg
		125 kg
		125 kg
		250 kg
		
		125 kg
		125 kg
		75 kg
		40 kg
		125 kg
			
			400 kg
			400 kg
			400 kg
			400 kg
			
			400 kg
			400 kg
			400 kg
			60 kg
			400 kg
				
				400 kg
				400 kg
				400 kg
				400 kg
				
				400 kg
				400 kg
				400 kg
				60 kg
				400 kg

[1] These inner packagings shall be sift-proof.
 	
[2] These inner packagings shall not be used when the substances being transported may become liquid during transport.
 	
[3] Paper and fibre inner packagings shall not be used for substances of packing group I.
	Jerricans
	   steel (3A2)
	   aluminium (3B2)
	   plastics (3H2)
		
		75 kg
		75 kg
		75 kg
			
			120 kg
			120 kg
			120 kg
				
				120 kg
				120 kg
				120 kg

Single packagings				

Drums
   steel (1A1 or 1A2[4])
   aluminium (1B1 or 1B2[4])
   metal, other than steel or aluminium (1N1 or 1N2[4])
   plastics (1H1 or 1H2[4])
   fibre (1G[5])
   plywood (1D[5])
		
		400 kg
		400 kg
		400 kg
		400 kg
		400 kg
		400 kg
			
			400 kg
			400 kg
			400 kg
			400 kg
			400 kg
			400 kg
				
				400 kg
				400 kg
				400 kg
				400 kg
				400 kg
				400 kg

[4] These packagings shall not be used for substances of packing group I that may become liquid during transport (see 4.1.3.4). [5] These packagings shall not be used when the substances being transported may become liquid during transport (see 4.1.3.4).				

P002	PACKING INSTRUCTION (SOLIDS)		P002

The following packagings are authorized provided the general provisions of 4.1.1 and 4.1.3 are met.

	Maximum net mass (see 4.1.3.3)		
Single packagings	Packing group I	Packing group II	Packing group III
Jerricans			
steel (3A1 or 3A2[4])	120 kg	120 kg	120 kg
aluminium (3B1 or 3B2[4])	120 kg	120 kg	120 kg
plastics (3H1 or 3H2[4])	120 kg	120 kg	120 kg
Boxes			
steel (4A)[5]	Not allowed	400 kg	400 kg
aluminium (4B)[5]	Not allowed	400 kg	400 kg
natural wood (4C1)[5]	Not allowed	400 kg	400 kg
natural wood with sift-proof walls (4C2)[5]	Not allowed	400 kg	400 kg
plywood (4D)[5]	Not allowed	400 kg	400 kg
reconstituted wood (4F)[5]	Not allowed	400 kg	400 kg
fibreboard (4G)[5]	Not allowed	400 kg	400 kg
solid plastics (4H2)[5]	Not allowed	400 kg	400 kg
Bags			
bags (5H3, 5H4, 5L3, 5M2)[5]	Not allowed	50 kg	50 kg
Composite packagings			
Plastics receptacle in steel, aluminium, plywood, fibre or plastics drum (6HA1, 6HB1, 6HG1[5], 6HD1[5], or 6HH1)	400 kg	400 kg	400 kg
Plastics receptacle in steel or aluminium crate or box, wooden box, plywood box, fibreboard box or solid plastics box (6HA2, 6HB2, 6HC, 6HD2[5], 6HG2[5] or 6HH2)	75 kg	75 kg	75 kg
Glass receptacle in steel, aluminium, plywood or fibre drum (6PA1, 6PB1, 6PD1[5] or 6PG1[5]) or in steel, aluminium, wood, or fibreboard box or in wickerwork hamper (6PA2, 6PB2, 6PC, 6PG2[5]or 6PD2[5]) or in solid or expanded plastics packaging (6PH2 or 6PH1[5])	75 kg	75 kg	75 kg

[4] These packagings shall not be used for substances of packing group I that may become liquid during transport (see 4.1.3.4).

[5] These packagings shall not be used when the substances being transported may become liquid during transport (see 4.1.3.4).

Pressure receptacles, provided that the general provisions of 4.1.3.6 are met

Special packing provisions:

PP7   For UN 2000, celluloid may be transported unpacked on pallets, wrapped in plastic film and secured by appropriate means, such as steel bands, as a single commodity in closed cargo transport units. Each pallet shall not exceed 1000 kg.

PP8   For UN 2002, packagings shall be so constructed that explosion is not possible by reason of increased internal pressure. Gas cylinders and gas receptacles shall not be used for these substances.

PP9   For UN 3175, UN 3243 and UN 3244, packagings shall conform to a design type that has passed a leakproofness test at the packing group II performance level. For UN 3175 the leakproofness test is not required when the the liquids are fully absorbed in solid material contained in sealed bags.

PP11   For UN 1309, packing group III, and UN 1361 and UN 1362, 5M1 bags are allowed if they are overpacked in plastic bags and are wrapped in shrink or stretch wrap on pallets.

PP12   For UN 1361, UN 2213 and UN 3077, 5H1, 5L1 and 5M1 bags are allowed when transported in closed cargo transport units.

PP13   For articles classified under UN 2870, only combination packagings meeting the packing group I performance level are authorized.

PP14   For UN 2211, UN 2698 and UN 3314, packagings are not required to meet the performance tests in chapter 6.1.

PP15   For UN 1324 and UN 2623, packagings shall meet the packing group III performance level.

PP20   For UN 2217, any sift-proof, tearproof receptacle may be used.

PP30   For UN 2471, paper or fibre inner packagings are not permitted.

PP31   For UN Nos. 1362, 1463, 1565, 1575, 1626, 1680, 1689, 1698, 1868, 1889, 1932, 2471, 2545, 2546, 2881, 3048, 3088, 3170, 3174, 3181, 3182, 3189, 3190, 3205, 3206, 3341, 3342, 3448, 3449 and 3450, packagings shall be hermetically sealed.

PP34   For UN 2969 (as whole beans), 5H1, 5L1 and 5M1 bags are permitted.

PP37   For UN 2590 and UN 2212, 5M1 bags are permitted. All bags of any type shall be transported in closed cargo transport units or be placed in closed rigid overpacks.

PP38   For UN 1309, bags are permitted only in closed cargo transport units or as unit loads.

PP84   For UN 1057, rigid outer packagings meeting the packing group II performance level shall be used. The packagings shall be designed and constructed and arranged to prevent movement, inadvertent ignition of the devices or inadvertent release of flammable gas or liquid.

PP85   For UN 1748, UN 2208 and UN 2880, bags are not allowed.

P003	PACKING INSTRUCTION	P003

Dangerous goods shall be placed in suitable outer packagings. The packagings shall meet the provisions of 4.1.1.1, 4.1.1.2, 4.1.1.4, 4.1.1.8 and 4.1.3 and be so designed that they meet the construction provisions of 6.1.4. Outer packagings constructed of suitable material of adequate strength and design in relation to the packaging capacity and its intended use shall be used. Where this packing instruction is used for the transport of articles or inner packagings of combination packagings, the packaging shall be designed and constructed to prevent inadvertent discharge of articles during normal conditions of transport.

Special packing provisions:

PP16 For UN 2800, batteries shall be protected from short circuit within the packagings.

PP17 For UN Nos. 1950 and 2037, packages shall not exceed 55 kg net mass for fibreboard packagings or 125 kg net mass for other packagings.

PP18 For UN 1845, packagings shall be designed and constructed to permit the release of carbon dioxide gas to prevent a build-up of pressure that could rupture the packagings.

PP19 For UN Nos. 1327, 1364, 1365, 1856 and 3360, transport as bales is authorized.

PP20 For UN Nos. 1363, 1386, 1408 and 2793, any sift-proof, tearproof receptacle may be used.

PP32 UN Nos. 2857 and 3358 may be transported unpackaged, in crates or in appropriate overpacks.

PP87 For UN 1950 waste aerosols transported in accordance with special provision 327, the packagings shall have a means of retaining any free liquid that might escape during transport, e.g. absorbent material. The packaging shall be adequately ventilated to prevent the creation of flammable atmosphere and the build-up of pressure.

P004	PACKING INSTRUCTION	P004

This instruction applies to UN Nos. 3473, 3476, 3477, 3478 and 3479

The following packagings are authorized provided the general provisions of 4.1.1.1, 4.1.1.2, 4.1.1.3, 4.1.1.6 and 4.1.3 are met:

(1) For fuel cell cartridges, packagings conforming to the packing group II performance level; and

(2) For fuel cell cartridges contained in equipment or packed with equipment, strong outer packagings. Large robust equipment (see 4.1.3.8) containing fuel cell cartridges may be transported unpackaged. When fuel cell cartridges are packed with equipment, they shall be packed in inner packagings or placed in the outer packaging with cushioning material or divider(s) so that the fuel cell cartridges are protected against damage that may be caused by the movement or placement of the contents within the outer packaging. Fuel cell cartridges which are installed in equipment shall be protected against short circuit and the entire system shall be protected against inadvertent operation.

P010	PACKING INSTRUCTION	P010

The following packagings are authorized, provided that the general provisions of 4.1.1 and 4.1.3 are met.

Combination packagings		Maximum net mass (see 4.1.3.3)
Inner packagings	**Outer packagings**	

Glass 1 ℓ
Steel 40 ℓ
	Drums
	  steel (1A2)
	  plastics (1H2)
	  plywood (1D)
	  fibre (1G)
		
		400 kg
		400 kg
		400 kg
		400 kg

	Boxes
	  steel (4A)
	  natural wood (4C1, 4C2)
	  plywood (4D)
	  reconstituted wood (4F)
	  fibreboard (4G)
	  expanded plastics (4H1)
	  solid plastics (4H2)
		
		400 kg
		400 kg
		400 kg
		400 kg
		400 kg
		60 kg
		400 kg

Single packagings		**Maximum capacity (see 4.1.3.3)**
Drums steel, non-removable head (1A1)		450 ℓ
Jerricans steel, non-removable head (3A1)		60 ℓ
Composite packagings plastics receptacle in steel drums (6HA1)		250 ℓ

4

P099	PACKING INSTRUCTION	P099

Only packagings which are approved for these goods by the competent authority may be used (see 4.1.3.7). A copy of the competent authority approval shall accompany each consignment or the transport document shall include an indication that the packaging was approved by the competent authority.

P101	PACKING INSTRUCTION	P101

Only packagings which are approved by the competent authority may be used. The State's distinguishing sign for motor vehicles in international traffic of the country for which the authority acts shall be marked on the transport documents as follows:

"Packaging approved by the competent authority of . . ."

P110(a)	PACKING INSTRUCTION	P110(a)

The following packagings are authorized, provided the general packing provisions of 4.1.1, 4.1.3 and special packing provisions of 4.1.5 are met.

Inner packagings	Intermediate packagings	Outer packagings
Bags plastics textile, plastic coated or lined rubber textile, rubberized textile	**Bags** plastics textile, plastic coated or lined rubber textile, rubberized **Receptacles** plastics metal	**Drums** steel, removable head (1A2) plastics, removable head (1H2)

Additional provisions:

1 The intermediate packagings shall be filled with water-saturated material such as an anti-freeze solution or wetted cushioning.

2 Outer packagings shall be filled with water-saturated material such as an anti-freeze solution or wetted cushioning. Outer packagings shall be constructed and sealed to prevent evaporation of the wetting solution, except for UN 0224 when transported dry.

P110(b)	PACKING INSTRUCTION	P110(b)

The following packagings are authorized, provided the general packing provisions of 4.1.1, 4.1.3 and special packing provisions of 4.1.5 are met.

Inner packagings	Intermediate packagings	Outer packagings
Receptacles metal wood rubber, conductive plastics, conductive **Bags** rubber, conductive plastics, conductive	**Dividing partitions** metal wood plastics fibreboard	**Boxes** natural wood, sift-proof wall (4C2) plywood (4D) reconstituted wood (4F)

Special packing provisions:

PP42 For UN Nos. 0074, 0113, 0114, 0129, 0130, 0135 and 0224, the following conditions shall be met:

.1 inner packagings shall not contain more than 50 g of explosive substance (quantity corresponding to dry substance);

.2 compartments between dividing partitions shall not contain more than one inner packaging, firmly fitted; and

.3 the outer packaging may be partitioned into up to 25 compartments.

P111	PACKING INSTRUCTION	P111

The following packagings are authorized, provided the general packing provisions of 4.1.1, 4.1.3 and special packing provisions of 4.1.5 are met.

Inner packagings	Intermediate packagings	Outer packagings

Bags
  paper, waterproofed
  plastics
  textile, rubberized
Sheets
  plastics
  textile, rubberized
	Not necessary	**Boxes**
		  steel (4A)
		  aluminium (4B)
		  natural wood, ordinary (4C1)
		  natural wood, sift-proof (4C2)
		  plywood (4D)
		  reconstituted wood (4F)
		  fibreboard (4G)
		  plastics, expanded (4H1)
		  plastics, solid (4H2)
		
		Drums
		  steel, removable head (1A2)
		  aluminium, removable head (1B2)
		  plywood (1D)
		  fibreboard (1G)
		  plastics, removable head (1H2)

Special packing provisions:

PP43 For UN 0159, inner packagings are not required when metal (1A2 or 1B2) or plastics (1H2) drums are used as outer packagings.

P112(a)	PACKING INSTRUCTION (Solid wetted, 1.1D)	P112(a)

The following packagings are authorized, provided the general packing provisions of 4.1.1, 4.1.3 and special packing provisions of 4.1.5 are met.

Inner packagings	Intermediate packagings	Outer packagings

Bags
  paper, multiwall, water-resistant
  plastics
  textile
  textile, rubberized
  woven plastics
Receptacles
  metal
  plastics
	Bags
	  plastics
	  textile, plastic coated or lined
	Receptacles
	  metal
	  plastics
		Boxes
		  steel (4A)
		  aluminium (4B)
		  natural wood, ordinary (4C1)
		  natural wood, sift-proof (4C2)
		  plywood (4D)
		  reconstituted wood (4F)
		  fibreboard (4G)
		  plastics, expanded (4H1)
		  plastics, solid (4H2)
		
		Drums
		  steel, removable head (1A2)
		  aluminium, removable head (1B2)
		  plywood (1D)
		  fibre (1G)
		  plastics, removable head (1H2)

Additional provision:

Intermediate packagings are not required if leakproof removable head drums are used as the outer packaging.

Special packing provisions:

PP26 For UN Nos. 0004, 0076, 0078, 0154, 0219 and 0394, packagings shall be lead-free.

PP45 For UN 0072 and UN 0226, intermediate packagings are not required.

P112(b)	PACKING INSTRUCTION (Solid dry, other than powder 1.1D)	P112(b)
The following packagings are authorized, provided the general packing provisions of 4.1.1, 4.1.3 and special packing provisions of 4.1.5 are met.		

Inner packagings	Intermediate packagings	Outer packagings
Bags paper, kraft paper, multiwall, water-resistant plastics textile textile, rubberized woven plastics	**Bags** (for UN 0150 only) plastics textile, plastic coated or lined	**Bags** woven plastics, sift-proof (5H2) woven plastics, water-resistant (5H3) plastics, film (5H4) textile, sift-proof (5L2) textile, water-resistant (5L3) paper, multiwall, water-resistant (5M2) **Boxes** steel (4A) aluminium (4B) natural wood, ordinary (4C1) natural wood, sift-proof (4C2) plywood (4D) reconstituted wood (4F) fibreboard (4G) plastics, expanded (4H1) plastics, solid (4H2) **Drums** steel, removable head (1A2) aluminium, removable head (1B2) plywood (1D) fibre (1G) plastics, removable head (1H2)

Special packing provisions:
PP26 For UN Nos. 0004, 0076, 0078, 0154, 0216, 0219 and 0386, packagings shall be lead-free.
PP46 For UN 0209, bags, sift-proof (5H2) are recommended for flake or prilled TNT in the dry state and a maximum net mass of 30 kg.
PP47 For UN 0222, inner packagings are not required when the outer packaging is a bag.

P112(c)	PACKING INSTRUCTION (Solid dry powder 1.1D)	P112(c)

The following packagings are authorized, provided the general packing provisions of 4.1.1, 4.1.3 and special packing provisions of 4.1.5 are met.

Inner packagings	Intermediate packagings	Outer packagings
Bags paper, multiwall, water-resistant plastics woven plastics **Receptacles** fibreboard metal plastics wood	**Bags** paper, multiwall, water-resistant with inner lining plastics **Receptacles** metal plastics	**Boxes** steel (4A) aluminium (4B) natural wood, ordinary (4C1) natural wood, sift-proof (4C2) plywood (4D) reconstituted wood (4F) fibreboard (4G) plastics, solid (4H2) **Drums** steel, removable head (1A2) aluminium, removable head (1B2) plywood (1D) fibre (1G) plastics, removable head (1H2)

Additional provisions:
1   Inner packagings are not required if drums are used as the outer packaging.
2   The packaging shall be sift-proof.

Special packing provisions:
PP26   For UN Nos. 0004, 0076, 0078, 0154, 0216, 0219 and 0386, packagings shall be lead-free.
PP46   For UN 0209, bags, sift-proof (5H2) are recommended for flake or prilled TNT in the dry state and a maximum net mass of 30 kg.
PP48   For UN 0504, metal packagings shall not be used.

P113	PACKING INSTRUCTION	P113

The following packagings are authorized, provided the general packing provisions of 4.1.1, 4.1.3 and special packing provisions of 4.1.5 are met.

Inner packagings	Intermediate packagings	Outer packagings
Bags paper plastics textile, rubberized **Receptacles** fibreboard metal plastics wood	*Not necessary*	**Boxes** steel (4A) aluminium (4B) natural wood, ordinary (4C1) natural wood, sift-proof walls (4C2) plywood (4D) reconstituted wood (4F) fibreboard (4G) plastics, solid (4H2) **Drums** steel, removable head (1A2) aluminium, removable head (1B2) plywood (1D) fibre (1G) plastics, removable head (1H2)

Additional provision:

The packaging shall be sift-proof.

Special packing provisions:

PP49   For UN 0094 and UN 0305, no more than 50 g of substance shall be packed in an inner packaging.

PP50   For UN 0027, inner packagings are not necessary when drums are used as the outer packaging.

PP51   For UN 0028, paper kraft or waxed paper sheets may be used as inner packagings.

P114(a)	PACKING INSTRUCTION (Solid wetted)	P114(a)

The following packagings are authorized, provided the general packing provisions of 4.1.1, 4.1.3 and special packing provisions of 4.1.5 are met.

Inner packagings	Intermediate packagings	Outer packagings
Bags plastics textile woven plastics **Receptacles** metal plastics	**Bags** plastics textile, plastic coated or lined **Receptacles** metal plastics	**Boxes** steel (4A) natural wood, ordinary (4C1) natural wood, sift-proof walls (4C2) plywood (4D) reconstituted wood (4F) fibreboard (4G) plastics, solid (4H2) **Drums** steel, removable head (1A2) aluminium, removable head (1B2) plywood (1D) fibre (1G) plastics, removable head (1H2)

Additional provision:

Intermediate packagings are not required if leakproof removable head drums are used as the outer packaging.

Special packing provisions:

PP26   For UN Nos. 0077, 0132, 0234, 0235 and 0236, packagings shall be lead-free.

PP43   For UN 0342, inner packagings are not required when metal (1A2 or 1B2) or plastics (1H2) drums are used as outer packagings.

P114(b)	PACKING INSTRUCTION (Solid dry)	P114(b)

The following packagings are authorized, provided the general packing provisions of 4.1.1, 4.1.3 and special packing provisions of 4.1.5 are met.

Inner packagings	Intermediate packagings	Outer packagings
Bags paper, kraft plastics textile, sift-proof woven plastics, sift-proof **Receptacles** fibreboard metal paper plastics woven plastics, sift-proof	*Not necessary*	**Boxes** natural wood, ordinary (4C1) natural wood, sift-proof walls (4C2) plywood (4D) reconstituted wood (4F) fibreboard (4G) **Drums** steel, removable head (1A2) aluminium, removable head (1B2) plywood (1D) fibre (1G) plastics, removable head (1H2)

Special packing provisions:

PP26   For UN Nos. 0077, 0132, 0234, 0235 and 0236, packagings shall be lead-free.

PP48   For UN 0508, metal packagings shall not be used.

PP50   For UN Nos. 0160, 0161 and 0508, inner packagings are not necessary when drums are used as the outer packaging.

PP52   For UN 0160 and UN 0161, when metal drums (1A2 or 1B2) are used as the outer packaging, metal packagings shall be so constructed that the risk of explosion, by reason of increase in internal pressure from internal or external causes, is prevented.

P115	PACKING INSTRUCTION	P115

The following packagings are authorized, provided the general packing provisions of 4.1.1, 4.1.3 and special packing provisions of 4.1.5 are met.

Inner packagings	Intermediate packagings	Outer packagings
Receptacles plastics	**Bags** plastics in metal receptacles **Drums** metal	**Boxes** natural wood, ordinary (4C1) natural wood, sift-proof walls (4C2) plywood (4D) reconstituted wood (4F) **Drums** steel, removable head (1A2) aluminium, removable head (1B2) plywood (1D) fibre (1G) plastics, removable head (1H2)

Special packing provisions:

PP45 For UN 0144, intermediate packagings are not required.

PP53 For UN Nos. 0075, 0143, 0495 and 0497, when boxes are used as the outer packaging, inner packagings shall have taped screw-cap closures and be not more than 5 ℓ capacity each. Inner packagings shall be surrounded with non-combustible absorbent cushioning materials. The amount of absorbent cushioning material shall be sufficient to absorb the liquid contents. Metal receptacles shall be cushioned from each other. Net mass of propellant is limited to 30 kg for each package when outer packagings are boxes.

PP54 For UN Nos. 0075, 0143, 0495 and 0497, when drums are used as the outer packaging and when intermediate packagings are drums, they shall be surrounded with non-combustible cushioning material in a quantity sufficient to absorb the liquid contents. A composite packaging consisting of a plastics receptacle in a metal drum may be used instead of the inner and intermediate packagings. The net volume of propellent in each package shall not exceed 120 ℓ.

PP55 For UN 0144, absorbent cushioning material shall be inserted.

PP56 For UN 0144, metal receptacles may be used as inner packagings.

PP57 For UN Nos. 0075, 0143, 0495 and 0497, bags shall be used as intermediate packagings when boxes are used as outer packagings.

PP58 For UN Nos. 0075, 0143, 0495 and 0497, drums shall be used as intermediate packagings when drums are used as outer packagings.

PP59 For UN 0144, fibreboard boxes (4G) may be used as outer packagings.

PP60 For UN 0144, aluminium drums, removable head (1B2) shall nct be used.

4

P116	PACKING INSTRUCTION	P116

The following packagings are authorized, provided the general packing provisions of 4.1.1, 4.1.3 and special packing provisions of 4.1.5 are met.

Inner packagings	Intermediate packagings	Outer packagings

Bags
    paper, water- and oil-resistant
    plastics
    textile, plastic coated or lined
    woven plastics, sift-proof
Receptacles
    fibreboard, water-resistant
    metal
    plastics
    wood, sift-proof
Sheets
    paper, water-resistant
    paper, waxed
    plastics
	Not necessary	**Bags**
		    woven plastics (5H1)
		    paper, multiwall, water-resistant
		      (5M2)
		    plastics, film (5H4)
		    textile, sift-proof (5L2)
		    textile, water-resistant (5L3)
		Boxes
		    steel (4A)
		    aluminium (4B)
		    natural wood, ordinary (4C1)
		    natural wood, sift-proof walls (4C2)
		    plywood (4D)
		    reconstituted wood (4F)
		    fibreboard (4G)
		    plastics, solid (4H2)
		Drums
		    steel, removable head (1A2)
		    aluminium, removable head (1B2)
		    plywood (1D)
		    fibre (1G)
		    plastics, removable head (1H2)
		Jerricans
		    steel, removable head (3A2)
		    plastics, removable head (3H2)

Special packing provisions:

PP61 For UN Nos. 0082, 0241, 0331 and 0332, inner packagings are not required if leakproof removable head drums are used as the outer packaging.

PP62 For UN Nos. 0082, 0241, 0331 and 0332, inner packagings are not required when the explosive is contained in a material impervious to liquid.

PP63 For UN 0081, inner packagings are not required when contained in rigid plastic which is impervious to nitric esters.

PP64 For UN 0331, inner packagings are not required when bags (5H2, 5H3 or 5H4) are used as outer packagings.

PP65 For UN Nos. 0082, 0241, 0331 and 0332, bags (5H2 or 5H3) may be used as outer packagings.

PP66 For UN 0081, bags shall not be used as outer packagings.

P130	PACKING INSTRUCTION	P130

The following packagings are authorized, provided the general packing provisions of 4.1.1, 4.1.3 and special packing provisions of 4.1.5 are met.

Inner packagings	Intermediate packagings	Outer packagings
Not necessary	*Not necessary*	**Boxes** steel (4A) aluminium (4B) natural wood, ordinary (4C1) natural wood, sift-proof walls (4C2) plywood (4D) reconstituted wood (4F) fibreboard (4G) plastics, expanded (4H1) plastics, solid (4H2) **Drums** steel, removable head (1A2) aluminium, removable head (1B2) plywood (1D) fibre (1G) plastics, removable head (1H2)

Special packing provision:

PP67 The following applies to UN Nos. 0006, 0009, 0010, 0015, 0016, 0018, 0019, 0034, 0035, 0038, 0039, 0048, 0056, 0137, 0138, 0168, 0169, 0171, 0181, 0182, 0183, 0186, 0221, 0243, 0244, 0245, 0246, 0254, 0280, 0281, 0286, 0287, 0297, 0299, 0300, 0301, 0303, 0321, 0328, 0329, 0344, 0345, 0346, 0347, 0362, 0363, 0370, 0412, 0424, 0425, 0434, 0435, 0436, 0437, 0438, 0451, 0488, and 0502: large and robust explosives articles, normally intended for military use, without their means of initiation or with their means of initiation containing at least two effective protective features, may be transported unpackaged. When such articles have propelling charges or are self-propelled, their ignition systems shall be protected against stimuli encountered during normal conditions of transport. A negative result in Test Series 4 on an unpackaged article indicates that the article can be considered for transport unpackaged. Such unpackaged articles may be fixed to cradles or contained in crates or other suitable handling devices.

P131	PACKING INSTRUCTION	P131

The following packagings are authorized, provided the general packing provisions of 4.1.1, 4.1.3 and special packing provisions of 4.1.5 are met.

Inner packagings	Intermediate packagings	Outer packagings
Bags paper plastics **Receptacles** fibreboard metal plastics wood **Reels**	*Not necessary*	**Boxes** steel (4A) aluminium (4B) natural wood, ordinary (4C1) natural wood, sift-proof walls (4C2) plywood (4D) reconstituted wood (4F) fibreboard (4G) **Drums** steel, removable head (1A2) aluminium, removable head (1B2) plywood (1D) fibre (1G) plastics, removable head (1H2)

Special packing provision:

PP68 For UN Nos. 0029, 0267 and 0455, bags and reels shall not be used as inner packagings.

P132(a)	PACKING INSTRUCTION	P132(a)
	(Articles consisting of closed metal, plastics or fibreboard casings that contain a detonating explosive, or consisting of plastics-bonded detonating explosives)	

The following packagings are authorized, provided the general packing provisions of 4.1.1, 4.1.3 and special packing provisions of 4.1.5 are met.

Inner packagings	Intermediate packagings	Outer packagings
Not necessary	*Not necessary*	**Boxes** steel (4A) aluminium (4B) natural wood, ordinary (4C1) natural wood, sift-proof walls (4C2) plywood (4D) reconstituted wood (4F) fibreboard (4G) plastics, solid (4H2)

P132(b)	PACKING INSTRUCTION	P132(b)
	(Articles without closed casings)	

The following packagings are authorized, provided the general packing provisions of 4.1.1, 4.1.3 and special packing provisions of 4.1.5 are met.

Inner packagings	Intermediate packagings	Outer packagings
Receptacles fibreboard metal plastics **Sheets** paper plastics	*Not necessary*	**Boxes** steel (4A) aluminium (4B) natural wood, ordinary (4C1) natural wood, sift-proof walls (4C2) plywood (4D) reconstituted wood (4F) fibreboard (4G) plastics, solid (4H2)

P133	PACKING INSTRUCTION	P133

The following packagings are authorized, provided the general packing provisions of 4.1.1, 4.1.3 and special packing provisions of 4.1.5 are met.

Inner packagings	Intermediate packagings	Outer packagings
Receptacles fibreboard metal plastics wood **Trays, fitted with dividing partitions** fibreboard plastics wood	**Receptacles** fibreboard metal plastics wood	**Boxes** steel (4A) aluminium (4B) natural wood, ordinary (4C1) natural wood, sift-proof walls (4C2) plywood (4D) reconstituted wood (4F) fibreboard (4G) plastics, solid (4H2)

Additional provision:

Receptacles are only required as intermediate packagings when the inner packagings are trays.

Special packing provision:

PP69   For UN Nos. 0043, 0212, 0225, 0268 and 0306, trays shall not be used as inner packagings.

P134	PACKING INSTRUCTION	P134
The following packagings are authorized, provided the general packing provisions of 4.1.1, 4.1.3 and special packing provisions of 4.1.5 are met.			

Inner packagings	Intermediate packagings	Outer packagings

Bags
    water-resistant
Receptacles
    fibreboard
    metal
    plastics
    wood
Sheets
    fibreboard, corrugated
Tubes
    fibreboard
	Not necessary	**Boxes**
		    steel (4A)
		    aluminium (4B)
		    natural wood, ordinary (4C1)
		    natural wood, sift-proof walls (4C2)
		    plywood (4D)
		    reconstituted wood (4F)
		    fibreboard (4G)
		    plastics, expanded (4H1)
		    plastics, solid (4H2)
		
		Drums
		    steel, removable head (1A2)
		    aluminium, removable head (1B2)
		    plywood (1D)
		    fibre (1G)
		    plastics, removable head (1H2)

P135	PACKING INSTRUCTION	P135
The following packagings are authorized, provided the general packing provisions of 4.1.1, 4.1.3 and special packing provisions of 4.1.5 are met.			

Inner packagings	Intermediate packagings	Outer packagings

Bags
    paper
    plastics
Receptacles
    fibreboard
    metal
    plastics
    wood
Sheets
    paper
    plastics
	Not necessary	**Boxes**
		    steel (4A)
		    aluminium (4B)
		    natural wood, ordinary (4C1)
		    natural wood, sift-proof walls (4C2)
		    plywood (4D)
		    reconstituted wood (4F)
		    fibreboard (4G)
		    plastics, expanded (4H1)
		    plastics, solid (4H2)
		
		Drums
		    steel, removable head (1A2)
		    aluminium, removable head (1B2)
		    plywood (1D)
		    fibre (1G)
		    plastics, removable head (1H2)

P136	PACKING INSTRUCTION	P136
The following packagings are authorized, provided the general packing provisions of 4.1.1, 4.1.3 and special packing provisions of 4.1.5 are met.			

Inner packagings	Intermediate packagings	Outer packagings

Bags
    plastics
    textile
Boxes
    fibreboard
    plastics
    wood
Dividing partitions in the outer packagings
	Not necessary	**Boxes**
		    steel (4A)
		    aluminium (4B)
		    natural wood, ordinary (4C1)
		    natural wood, sift-proof walls (4C2)
		    plywood (4D)
		    reconstituted wood (4F)
		    fibreboard (4G)
		    plastics, solid (4H2)
		
		Drums
		    steel, removable head (1A2)
		    aluminium, removable head (1B2)
		    plywood (1D)
		    fibre (1G)
		    plastics, removable head (1H2)

P137	PACKING INSTRUCTION	P137
The following packagings are authorized, provided the general packing provisions of 4.1.1, 4.1.3 and special packing provisions of 4.1.5 are met.		

Inner packagings	Intermediate packagings	Outer packagings

Bags
    plastics
Boxes
    fibreboard
Tubes
    fibreboard
    metal
    plastics
Dividing partitions in the outer packagings
	Not necessary	**Boxes**
		    steel (4A)
		    aluminium (4B)
		    natural wood, ordinary (4C1)
		    natural wood, sift-proof walls (4C2)
		    plywood (4D)
		    reconstituted wood (4F)
		    fibreboard (4G)
		
		Drums
		    steel, removable head (1A2)
		    aluminium, removable head (1B2)
		    plywood (1D)
		    fibre (1G)
		    plastics, removable head (1H2)

Special packing provision:

PP70 For UN Nos. 0059, 0439, 0440 and 0441, when the shaped charges are packed singly, the conical cavity shall face downwards and the package shall be marked "THIS SIDE UP". When the shaped charges are packed in pairs, the conical cavities shall face inwards to minimize the jetting effect in the event of accidental initiation.

P138	PACKING INSTRUCTION	P138
The following packagings are authorized, provided the general packing provisions of 4.1.1, 4.1.3 and special packing provisions of 4.1.5 are met.		

Inner packagings	Intermediate packagings	Outer packagings

Bags
    plastics
	Not necessary	**Boxes**
		    steel (4A)
		    aluminium (4B)
		    natural wood, ordinary (4C1)
		    natural wood, sift-proof walls (4C2)
		    plywood (4D)
		    reconstituted wood (4F)
		    fibreboard (4G)
		    plastics, solid (4H2)
		
		Drums
		    steel, removable head (1A2)
		    aluminium, removable head (1B2)
		    plywood (1D)
		    fibre (1G)
		    plastics, removable head (1H2)

Additional provision:

If the ends of the articles are sealed, inner packagings are not necessary.

P139	PACKING INSTRUCTION	P139

The following packagings are authorized, provided the general packing provisions of 4.1.1, 4.1.3 and special packing provisions of 4.1.5 are met.

Inner packagings	Intermediate packagings	Outer packagings
Bags plastics **Receptacles** fibreboard metal plastics wood **Reels** **Sheets** paper plastics	*Not necessary*	**Boxes** steel (4A) aluminium (4B) natural wood, ordinary (4C1) natural wood, sift-proof walls (4C2) plywood (4D) reconstituted wood (4F) fibreboard (4G) plastics, solid (4H2) **Drums** steel, removable head (1A2) aluminium, removable head (1B2) plywood (1D) fibre (1G) plastics, removable head (1H2)

Special packing provisions:

PP71 For UN Nos. 0065, 0102, 0104, 0289 and 0290, the ends of the detonating cord shall be sealed; for example, by a plug firmly fixed so that the explosive cannot escape. The ends of flexible detonating cord shall be fastened securely.

PP72 For UN 0065 and UN 0289, inner packagings are not required when they are in coils.

P140	PACKING INSTRUCTION	P140

The following packagings are authorized, provided the general packing provisions of 4.1.1, 4.1.3 and special packing provisions of 4.1.5 are met.

Inner packagings	Intermediate packagings	Outer packagings
Bags plastics **Reels** **Sheets** paper, kraft plastics	*Not necessary*	**Boxes** steel (4A) aluminium (4B) natural wood, ordinary (4C1) natural wood, sift-proof walls (4C2) plywood (4D) reconstituted wood (4F) fibreboard (4G) plastics, solid (4H2) **Drums** steel, removable head (1A2) aluminium, removable head (1B2) plywood (1D) fibre (1G) plastics, removable head (1H2)

Special packing provisions:

PP73 For UN 0105, no inner packagings are required if the ends are sealed.

PP74 For UN 0101, the packaging shall be sift-proof except when the fuse is covered by a paper tube and both ends of the tube are covered with removable caps.

PP75 For UN 0101, steel or aluminium boxes or drums shall not be used.

P141	PACKING INSTRUCTION	P141

The following packagings are authorized, provided the general packing provisions of 4.1.1, 4.1.3 and special packing provisions of 4.1.5 are met.

Inner packagings	Intermediate packagings	Outer packagings
Receptacles fibreboard metal plastics wood **Trays, fitted with dividing partitions** plastics wood **Dividing partitions in the outer packagings**	*Not necessary*	**Boxes** steel (4A) aluminium (4B) natural wood, ordinary (4C1) natural wood, sift-proof walls (4C2) plywood (4D) reconstituted wood (4F) fibreboard (4G) plastics, solid (4H2) **Drums** steel, removable head (1A2) aluminium, removable head (1B2) plywood (1D) fibre (1G) plastics, removable head (1H2)

P142	PACKING INSTRUCTION	P142

The following packagings are authorized, provided the general packing provisions of 4.1.1, 4.1.3 and special packing provisions of 4.1.5 are met.

Inner packagings	Intermediate packagings	Outer packagings
Bags paper plastics **Receptacles** fibreboard metal plastics wood **Sheets** paper **Trays, fitted with dividing partitions** plastics	*Not necessary*	**Boxes** steel (4A) aluminium (4B) natural wood, ordinary (4C1) natural wood, sift-proof walls (4C2) plywood (4D) reconstituted wood (4F) fibreboard (4G) plastics, solid (4H2) **Drums** steel, removable head (1A2) aluminium, removable head (1B2) plywood (1D) fibre (1G) plastics, removable head (1H2)

P143	PACKING INSTRUCTION	P143

The following packagings are authorized, provided the general packing provisions of 4.1.1, 4.1.3 and special packing provisions of 4.1.5 are met.

Inner packagings	Intermediate packagings	Outer packagings
Bags paper, kraft plastics textile textile, rubberized **Receptacles** fibreboard metal plastics **Trays, fitted with dividing partitions** plastics wood	*Not necessary*	**Boxes** steel (4A) aluminium (4B) natural wood, ordinary (4C1) natural wood, sift-proof walls (4C2) plywood (4D) reconstituted wood (4F) fibreboard (4G) plastics, solid (4H2) **Drums** steel, removable head (1A2) aluminium, removable head (1B2) plywood (1D) fibre (1G) plastics, removable head (1H2)

Additional provision:
Instead of the above inner and outer packagings, composite packagings (6HH2) (plastics receptacle with outer solid box) may be used.

Special packing provision:
PP76 For UN Nos. 0271, 0272, 0415 and 0491, when metal packagings are used, metal packagings shall be so constructed that the risk of explosion, by reason of increase in internal pressure from internal or external causes, is prevented.

P144	PACKING INSTRUCTION	P144

The following packagings are authorized, provided the general packing provisions of 4.1.1, 4.1.3 and special packing provisions of 4.1.5 are met.

Inner packagings	Intermediate packagings	Outer packagings
Receptacles fibreboard metal plastics **Dividing partitions in the outer packagings**	*Not necessary*	**Boxes** steel (4A) aluminium (4B) natural wood, ordinary with metal liner (4C1) plywood (4D) with metal liner reconstituted wood with metal liner (4F) plastics, expanded (4H1) plastics, solid (4H2) **Drums** steel, removable head (1A2) aluminium, removable head (1B2) plastics, removable head (1H2)

Special packing provision:
PP 77 For UN 0248 and UN 0249, packagings shall be protected against the ingress of water. When water-activated contrivances are transported unpackaged, they shall be provided with at least two independent protective features which prevent the ingress of water.

P200	PACKING INSTRUCTION	P200

For pressure receptacles, the general packing provisions of 4.1.6.1 shall be met. In addition, for MEGCs, the general requirements of 4.2.4 shall be met.

Cylinders, tubes, pressure drums, bundles of cylinders constructed as specified in 6.2 and MEGCs constructed as specified in 6.7.5 are authorized for the transport of a specific substance when specified in the following tables. For some substances, the special packing provisions may prohibit a particular type of cylinder, tube, pressure drum or bundle of cylinders.

(1) Pressure receptacles containing toxic substances with an LC_{50} less than or equal to 200 mℓ/m^3 (ppm) as specified in the table shall not be equipped with any pressure relief device. Pressure relief devices shall be fitted on pressure receptacles used for the transport of UN 1013 carbon dioxide and UN 1070 nitrous oxide. Other pressure receptacles shall be fitted with a pressure relief device if specified by the competent authority of the country of use. The type of pressure relief device, the set-to-discharge pressure and relief capacity of pressure relief devices, if required, shall be specified by the competent authority of the country of use.

(2) The following three tables cover compressed gases (table 1), liquefied and dissolved gases (table 2) and substances not in class 2 (table 3). They provide:

(a) the UN Number, Proper Shipping Name and description, and classification of the substance;

(b) the LC_{50} for toxic substances;

(c) the types of pressure receptacles authorized for the substance, shown by the letter "X";

(d) the maximum test period for periodic inspection of the pressure receptacles;
Note: For pressure receptacles which make use of composite materials, the periodic inspection frequencies shall be as determined by the competent authority which approved the receptacles.

(e) the minimum test pressure of the pressure receptacles;

(f) the maximum working pressure of the pressure receptacles for compressed gases (where no value is given, the working pressure shall not exceed two thirds of the test pressure) or the maximum filling ratio(s) dependent on the test pressure(s) for liquefied and dissolved gases;

(g) special packing provisions that are specific to a substance.

(3) In no case shall pressure receptacles be filled in excess of the limit permitted in the following requirements.

(a) For compressed gases, the working pressure shall be not more than two thirds of the test pressure of the pressure receptacles. Restrictions to this upper limit on working pressure are imposed by special packing provision "o" in (4) below. In no case shall the internal pressure at 65°C exceed the test pressure.

(b) For high pressure liquefied gases, the filling ratio shall be such that the settled pressure at 65°C does not exceed the test pressure of the pressure receptacles.

The use of test pressures and filling ratios other than those in the table is permitted, except where (4), special packing provision "o" applies, provided that:

(i) the criterion of (4), special packing provision "r" is met when applicable; or

(ii) the above criterion is met in all other cases.

For high pressure liquefied gases and gas mixtures for which relevant data are not available, the maximum filling ratio (*FR*) shall be determined as follows:
$$FR = 8.5 \times 10^{-4} \times d_g \times P_h$$
where   *FR* = maximum filling ratio
d_g = gas density (at 15°C, 1 bar) (in g/ℓ)
P_h = minimum test pressure (in bar)

If the density of the gas is unknown, the maximum filling ratio shall be determined as follows:
$$FR = \frac{P_h \times MM \times 10^{-3}}{R \times 338}$$
where   *FR* = maximum filling ratio
P_h = minimum test pressure (in bar)
MM = molecular mass (in g/mol)
R = 8.31451 × 10^{-2} bar·ℓ/mol·K (gas constant)

For gas mixtures, the average molecular mass is to be taken, taking into account the volumetric concentrations of the various components.

(c) For low pressure liquefied gases, the maximum mass of contents per litre of water capacity (filling factor) shall equal 0.95 times the density of the liquid phase at 50°C; in addition, the liquid phase shall not fill the pressure receptacle at any temperature up to 60°C. The test pressure of the pressure receptacle shall be at least equal to the vapour pressure (absolute) of the liquid at 65°C, minus 100 kPa (1 bar).

For low pressure liquefied gases and gas mixtures for which relevant data are not available, the maximum filling ratio shall be determined as follows:
$$FR = (0.0032 \times BP - 0.24) \times d_l$$
where   *FR* = maximum filling ratio
BP = boiling point (in kelvin)
d_l = density of the liquid at boiling point (in kg/ℓ)

P200	PACKING INSTRUCTION *(continued)*	P200

(d) For UN 1001, acetylene, dissolved, and UN 3374 acetylene, solvent free, see (4), special packing provision "p".

(4) Keys for the column "Special packing provisions":

Material compatibility (for gases see ISO 11114-1:1997 and ISO 11114-2:2000)
- a: Aluminium alloy pressure receptacles are not authorized.
- b: Copper valves shall not be used.
- c: Metal parts in contact with the contents shall not contain more than 65% copper.
- d: When steel pressure receptacles are used, only those bearing the "H" mark shall be authorized.

Requirements for toxic substances with an LC_{50} less than or equal to 200 mℓ/m^3 (ppm)
- k: Valve outlets shall be fitted with gas-tight plugs or caps.
   Each cylinder within a bundle shall be fitted with an individual valve that shall be closed during transport. After filling, the manifold shall be evacuated, purged and plugged.
   Bundles containing UN 1045 fluorine, compressed, may be constructed with isolation valves on groups of cylinders not exceeding 150 litres total water capacity instead of isolation valves on every cylinder.
   Cylinders and individual cylinders in a bundle shall have a test pressure greater than or equal to 200 bar and a minimum wall thickness of 3.5 mm for aluminium alloy or 2 mm for steel. Individual cylinders not complying with this requirement shall be transported in a rigid outer packaging that will adequately protect the cylinder and its fittings and meeting the packing group I performance level. Pressure drums shall have a minimum wall thickness as specified by the competent authority.
   Pressure receptacles shall not be fitted with a pressure relief device.
   Cylinders and individual cylinders in a bundle shall be limited to a maximum water capacity of 85 ℓ.
   Each valve shall have a taper-threaded connection directly to the pressure receptacle and be capable of withstanding the test pressure of the pressure receptacle.
   Each valve shall either be of the packless type with non-perforated diaphragm, or be of a type which prevents leakage through or past the packing.
   Each pressure receptacle shall be tested for leakage after filling.

Gas specific provisions
- l: UN 1040 ethylene oxide may also be packed in hermetically sealed glass or metal inner packagings suitably cushioned in fibreboard, wooden or metal boxes meeting the packing group I performance level. The maximum quantity permitted in any glass inner packaging is 30 g, and the maximum quantity permitted in any metal inner packaging is 200 g. After filling, each inner packaging shall be determined to be leaktight by placing the inner packaging in a hot water bath at a temperature, and for a period of time, sufficient to ensure that an internal pressure equal to the vapour pressure of ethylene oxide at 55°C is achieved. The maximum net mass in any outer packaging shall not exceed 2.5 kg.
- m: Pressure receptacles shall be filled to a working pressure not exceeding 5 bar.
- n: Cylinders and individual cylinders in a bundle shall contain not more than 5 kg of the gas. When bundles containing UN 1045 fluorine, compressed are divided into groups of cylinders in accordance with special packing provision "k" each group shall contain not more than 5 kg of the gas.
- o: In no case shall the working pressure or filling ratio shown in the table be exceeded.
- p: For UN 1001 acetylene, dissolved and UN 3374 acetylene, solvent free: cylinders shall be filled with a homogeneous monolithic porous material; the working pressure and the quantity of acetylene shall not exceed the values prescribed in the approval or in ISO 3807-1:2000 or ISO 3807-2:2000, as applicable.
   For UN 1001 acetylene, dissolved: cylinders shall contain a quantity of acetone or suitable solvent as specified in the approval (see ISO 3807-1:2000 or ISO 3807-2:2000, as applicable); cylinders fitted with pressure relief devices or manifolded together shall be transported vertically.
   The test pressure of 52 bar applies only to cylinders conforming to ISO 3807-2:2000.
- q: The valves of pressure receptacles for pyrophoric gases or flammable mixtures of gases containing more than 1% of pyrophoric compounds shall be fitted with gas-tight plugs or caps. When these pressure receptacles are manifolded in a bundle, each of the pressure receptacles shall be fitted with an individual valve that shall be closed during transport, and the manifold outlet valve shall be fitted with a gas-tight plug or cap.
- r: The filling ratio of this gas shall be limited such that, if complete decomposition occurs, the pressure does not exceed two thirds of the test pressure of the pressure receptacle.
- s: Aluminium alloy pressure receptacles shall be:
   – equipped only with brass or stainless steel valves; and
   – cleaned in accordance with ISO 11621:1997 and not contaminated with oil.
- t: (i) The wall thickness of pressure receptacles shall be not less than 3 mm.
   (ii) Prior to transport, it shall be ensured that the pressure has not risen due to potential hydrogen generation.

P200	PACKING INSTRUCTION *(continued)*	P200

Periodic inspection

u: The interval between periodic tests may be extended to 10 years for aluminium alloy pressure receptacles when the alloy of the pressure receptacle has been subjected to stress corrosion testing as specified in ISO 7866:1999.

v: The interval between periodic inspections for steel cylinders may be extended to 15 years if approved by the competent authority of the country of use.

Requirements for N.O.S. descriptions and for mixtures

z: The construction materials of the pressure receptacles and their accessories shall be compatible with the contents and shall not react to form harmful or dangerous compounds therewith.

The test pressure and filling ratio shall be calculated in accordance with the relevant requirements of (3).

Toxic substances with an LC_{50} less than or equal to 200 mℓ/m^3 shall not be transported in tubes, pressure drums or MEGCs and shall meet the requirements of special packing provision "k". However, UN 1975 nitric oxide and dinitrogen tetroxide mixtures may be transported in pressure drums.

For pressure receptacles containing pyrophoric gases or flammable mixtures of gases containing more than 1% pyrophoric compounds, the requirements of special packing provision "q" shall be met.

The necessary steps shall be taken to prevent dangerous reactions (i.e. polymerization or decomposition) during transport. If necessary, stabilization or addition of an inhibitor shall be required.

Mixtures containing UN 1911 diborane shall be filled to a pressure such that, if complete decomposition of the diborane occurs, two thirds of the test pressure of the pressure receptacle shall not be exceeded. However, UN 1975 nitric oxide and dinitrogen tetroxide mixtures may be transported in pressure drums. Mixtures containing UN 2192 germane, other than mixtures of up to 35% germane in hydrogen or nitrogen or up to 28% germane in helium or argon, shall be filled to a pressure such that, if complete decomposition of the germane occurs, two thirds of the test pressure of the pressure receptacle shall not be exceeded.

P200	PACKING INSTRUCTION *(continued)*											P200
Table 1: COMPRESSED GASES												
UN No.	Proper Shipping Name	Class	Subsidiary risk	LC_{50}, mℓ/m^3	Cylinders	Tubes	Pressure drums	Bundles of cylinders	MEGCs	Test period, years	Test pressure, bar*	Maximum working pressure, bar*	Special packing provisions
---	---	---	---	---	---	---	---	---	---	---	---	---	---
1002	AIR, COMPRESSED	2.2			X	X	X	X	X	10			
1006	ARGON, COMPRESSED	2.2			X	X	X	X	X	10			
1016	CARBON MONOXIDE, COMPRESSED	2.3	2.1	3760	X	X	X	X	X	5			u
1023	COAL GAS, COMPRESSED	2.3	2.1		X	X	X	X	X	5			
1045	FLUORINE, COMPRESSED	2.3	5.1, 8	185	X			X		5	200	30	a, k, n, o
1046	HELIUM, COMPRESSED	2.2			X	X	X	X	X	10			
1049	HYDROGEN, COMPRESSED	2.1			X	X	X	X	X	10			d
1056	KRYPTON, COMPRESSED	2.2			X	X	X	X	X	10			
1065	NEON, COMPRESSED	2.2			X	X	X	X	X	10			
1066	NITROGEN, COMPRESSED	2.2			X	X	X	X	X	10			
1071	OIL GAS, COMPRESSED	2.3	2.1		X	X	X	X	X	5			
1072	OXYGEN, COMPRESSED	2.2	5.1		X	X	X	X		10			s
1612	HEXAETHYL TETRAPHOSPHATE AND COMPRESSED GAS MIXTURE	2.3			X	X	X	X		5			z
1660	NITRIC OXIDE, COMPRESSED	2.3	5.1, 8	115	X			X		5	225	33	k, o
1953	COMPRESSED GAS, TOXIC, FLAMMABLE, N.O.S.	2.3	2.1	⩽5000	X	X	X	X	X	5			z
1954	COMPRESSED GAS, FLAMMABLE, N.O.S	2.1			X	X	X	X	X	10			z
1955	COMPRESSED GAS, TOXIC, N.O.S.	2.3		⩽5000	X	X	X	X	X	5			z
1956	COMPRESSED GAS, N.O.S.	2.2			X	X	X	X	X	10			z
1957	DEUTERIUM, COMPRESSED	2.1			X	X	X	X	X	10			d
1964	HYDROCARBON GAS MIXTURE, COMPRESSED, N.O.S.	2.1			X	X	X	X	X	10			z
1971	METHANE, COMPRESSED or NATURAL GAS, COMPRESSED with high methane content	2.1			X	X	X	X	X	10			
2034	HYDROGEN AND METHANE MIXTURE, COMPRESSED	2.1			X	X	X	X	X	10			d
2190	OXYGEN DIFLUORIDE, COMPRESSED	2.3	5.1, 8	2.6	X			X		5	200	30	a, k, n, o
3156	COMPRESSED GAS, OXIDIZING, N.O.S.	2.2	5.1		X	X	X	X	X	10			z
3303	COMPRESSED GAS, TOXIC, OXIDIZING, N.O.S.	2.3	5.1	⩽5000	X	X	X	X	X	5			z
3304	COMPRESSED GAS, TOXIC, CORROSIVE, N.O.S.	2.3	8	⩽5000	X	X	X	X	X	5			z
3305	COMPRESSED GAS, TOXIC, FLAMMABLE, CORROSIVE, N.O.S.	2.3	2.1, 8	⩽5000	X	X	X	X	X	5			z
3306	COMPRESSED GAS, TOXIC, OXIDIZING, CORROSIVE, N.O.S.	2.3	5.1, 8	⩽5000	X	X	X	X	X	5			z

4

* Where the entries are blank, the maximum working pressure shall not exceed two thirds of the test pressure.

P200	PACKING INSTRUCTION *(continued)*										P200

	Table 2: LIQUEFIED GASES AND DISSOLVED GASES										
UN No.	Proper Shipping Name	Class	Subsidiary risk	LC$_{50}$, ml/m^3	Cylinders	Tubes	Pressure drums	Bundles of cylinders	MEGCs	Test period, years	Test pressure, bar*	Filling ratio	Special packing provisions
1001	ACETYLENE, DISSOLVED	2.1			X			X		10	60 52		c, p
1005	AMMONIA, ANHYDROUS	2.3	8	4000	X	X	X	X	X	5	29	0.54	b
1008	BORON TRIFLUORIDE	2.3	8	387	X	X	X	X	X	5	225 300	0.715 0.86	
1009	BROMOTRIFLUOROMETHANE (REFRIGERANT GAS R 13B1)	2.2			X	X	X	X	X	10	42 120 250	1.13 1.44 1.60	
1010	BUTADIENES, STABILIZED (1,2-butadiene), or	2.1			X	X	X	X	X	10	10	0.59	
1010	BUTADIENES, STABILIZED (1,3-butadiene), or	2.1			X	X	X	X	X	10	10	0.55	
1010	BUTADIENES AND HYDROCARBON MIXTURE, STABILIZED with more than 40% butadienes	2.1			X	X	X	X	X	10			v, z
1011	BUTANE	2.1			X	X	X	X	X	10	10	0.52	v
1012	BUTYLENE (butylenes mixture) or	2.1			X	X	X	X	X	10	10	0.50	z
1012	BUTYLENE (1-butylene) or	2.1			X	X	X	X	X	10	10	0.53	
1012	BUTYLENE (*cis*-2-butylene) or	2.1			X	X	X	X	X	10	10	0.55	
1012	BUTYLENE (*trans*-2-butylene)	2.1			X	X	X	X	X	10	10	0.54	
1013	CARBON DIOXIDE	2.2			X	X	X	X	X	10	190 250	0.68 0.76	
1017	CHLORINE	2.3	5.1, 8	293	X	X	X	X	X	5	22	1.25	a
1018	CHLORODIFLUOROMETHANE (REFRIGERANT GAS R 22)	2.2			X	X	X	X	X	10	27	1.03	
1020	CHLOROPENTAFLUOROETHANE (REFRIGERANT GAS R 115)	2.2			X	X	X	X	X	10	25	1.05	
1021	1-CHLORO-1,2,2,2-TETRAFLUOROETHANE (REFRIGERANT GAS R 124)	2.2			X	X	X	X	X	10	11	1.20	
1022	CHLOROTRIFLUOROMETHANE (REFRIGERANT GAS R 13)	2.2			X	X	X	X	X	10	100 120 190 250	0.83 0.90 1.04 1.11	
1026	CYANOGEN	2.3	2.1	350	X	X	X	X	X	5	100	0.70	u
1027	CYCLOPROPANE	2.1			X	X	X	X	X	10	18	0.55	
1028	DICHLORODIFLUOROMETHANE (REFRIGERANT GAS R 12)	2.2			X	X	X	X	X	10	16	1.15	
1029	DICHLOROFLUOROMETHANE (REFRIGERANT GAS R 21)	2.2			X	X	X	X	X	10	10	1.23	
1030	1,1-DIFLUOROETHANE (REFRIGERANT GAS R 152a)	2.1			X	X	X	X	X	10	16	0.79	
1032	DIMETHYLAMINE, ANHYDROUS	2.1			X	X	X	X	X	10	10	0.59	b
1033	DIMETHYL ETHER	2.1			X	X	X	X	X	10	18	0.58	
1035	ETHANE	2.1			X	X	X	X	X	10	95 120 300	0.25 0.30 0.40	
1036	ETHYLAMINE	2.1			X	X	X	X	X	10	10	0.61	b
1037	ETHYL CHLORIDE	2.1			X	X	X	X	X	10	10	0.80	a
1039	ETHYL METHYL ETHER	2.1			X	X	X	X	X	10	10	0.64	

* Where the entries are blank, the maximum working pressure shall not exceed two thirds of the test pressure.

P200	PACKING INSTRUCTION (continued)												P200
	Table 2: LIQUEFIED GASES AND DISSOLVED GASES (continued)												
UN No.	Proper Shipping Name	Class	Subsidiary risk	LC$_{50}$, ml/m^3	Cylinders	Tubes	Pressure drums	Bundles of cylinders	MEGCs	Test period, years	Test pressure, bar*	Filling ratio	Special packing provisions
1040	ETHYLENE OXIDE or ETHYLENE OXIDE WITH NITROGEN up to a total pressure of 1 MPa (10 bar) at 50°C	2.3	2.1	2900	X	X	X	X	X	5	15	0.78	l
1041	ETHYLENE OXIDE AND CARBON DIOXIDE MIXTURE with more than 9% ethylene oxide but not more than 87%	2.1			X	X	X	X	X	10	190 250	0.66 0.75	
1043	FERTILIZER AMMONIATING SOLUTION with free ammonia	2.2			X		X	X		5			b, z
1048	HYDROGEN BROMIDE, ANHYDROUS	2.3	8	2860	X	X	X	X	X	5	60	1.51	a, d
1050	HYDROGEN CHLORIDE, ANHYDROUS	2.3	8	2810	X	X	X	X	X	5	100 120 150 200	0.30 0.56 0.67 0.74	a, d a, d a, d a, d
1053	HYDROGEN SULPHIDE	2.3	2.1	712	X	X	X	X	X	5	48	0.67	d, u
1055	ISOBUTYLENE	2.1			X	X	X	X	X	10	10	0.52	
1058	LIQUEFIED GASES, non-flammable, charged with nitrogen, carbon dioxide or air	2.2			X	X	X	X	X	10	Test pressure = 1.5 × working pressure		
1060	METHYLACETYLENE AND PROPADIENE MIXTURE, STABILIZED or	2.1			X	X	X	X	X	10			c, z
1060	METHYLACETYLENE AND PROPADIENE MIXTURE, STABILIZED (Propadiene with 1% to 4% methylacetylene)	2.1			X	X	X	X	X	10	22	0.52	c
1061	METHYLAMINE, ANHYDROUS	2.1			X	X	X	X	X	10	13	0.58	b
1062	METHYL BROMIDE with not more than 2% chloropicrin	2.3		850	X	X	X	X	X	5	10	1.51	a
1063	METHYL CHLORIDE (REFRIGERANT GAS R 40)	2.1			X	X	X	X	X	10	17	0.81	a
1064	METHYL MERCAPTAN	2.3	2.1	1350	X	X	X	X	X	5	10	0.78	d, u
1067	DINITROGEN TETROXIDE (NITROGEN DIOXIDE)	2.3	5.1, 8	115	X		X	X		5	10	1.30	k
1069	NITROSYL CHLORIDE	2.3	8	35	X			X		5	13	1.10	k
1070	NITROUS OXIDE	2.2	5.1		X	X	X	X	X	10	180 225 250	0.68 0.74 0.75	
1075	PETROLEUM GASES, LIQUEFIED	2.1			X	X	X	X	X	10			v, z
1076	PHOSGENE	2.3	8	5	X		X	X		5	20	1.23	k
1077	PROPYLENE	2.1			X	X	X	X	X	10	27	0.43	
1078	REFRIGERANT GAS, N.O.S.	2.2			X	X	X	X	X	10			z
1079	SULPHUR DIOXIDE	2.3	8	2520	X	X	X	X	X	5	12	1.23	
1080	SULPHUR HEXAFLUORIDE	2.2			X	X	X	X	X	10	70 140 160	1.06 1.34 1.38	
1081	TETRAFLUOROETHYLENE, STABILIZED	2.1			X	X	X	X	X	10	200		m, o
1082	TRIFLUOROCHLOROETHYLENE, STABILIZED	2.3	2.1	2000	X	X	X	X	X	5	19	1.13	u
1083	TRIMETHYLAMINE, ANHYDROUS	2.1			X	X	X	X	X	10	10	0.56	b
1085	VINYL BROMIDE, STABILIZED	2.1			X	X	X	X	X	10	10	1.37	a

* Where the entries are blank, the maximum working pressure shall not exceed two thirds of the test pressure.

P200	PACKING INSTRUCTION *(continued)*											P200
Table 2: LIQUEFIED GASES AND DISSOLVED GASES *(continued)*												
UN No.	Proper Shipping Name	Class	Subsidiary risk	LC_{50}, ml/m³	Cylinders	Tubes	Pressure drums	Bundles of cylinders	MEGCs	Test period, years	Test pressure, bar*	Filling ratio	Special packing provisions
1086	VINYL CHLORIDE, STABILIZED	2.1			X	X	X	X	X	10	12	0.81	a
1087	VINYL METHYL ETHER, STABILIZED	2.1			X	X	X	X	X	10	10	0.67	
1581	CHLOROPICRIN AND METHYL BROMIDE MIXTURE with more than 2% chloropicrin	2.3		850	X	X	X	X	X	5	10	1.51	a
1582	CHLOROPICRIN AND METHYL CHLORIDE MIXTURE	2.3			X	X	X	X	X	5	17	0.81	a
1589	CYANOGEN CHLORIDE, STABILIZED	2.3	8	80	X			X		5	20	1.03	k
1741	BORON TRICHLORIDE	2.3	8	2541	X	X	X	X	X	5	10	1.19	
1749	CHLORINE TRIFLUORIDE	2.3	5.1, 8	299	X	X	X	X	X	5	30	1.40	a
1858	HEXAFLUOROPROPYLENE (REFRIGERANT GAS R 1216)	2.2			X	X	X	X	X	10	22	1.11	
1859	SILICON TETRAFLUORIDE	2.3	8	450	X	X	X	X	X	5	200 300	0.74 1.10	
1860	VINYL FLUORIDE, STABILIZED	2.1			X	X	X	X	X	10	250	0.64	a
1911	DIBORANE	2.3	2.1	80	X			X		5	250	0.07	d, k, o
1912	METHYL CHLORIDE AND METHYLENE CHLORIDE MIXTURE	2.1			X	X	X	X	X	10	17	0.81	a
1952	ETHYLENE OXIDE AND CARBON DIOXIDE MIXTURE with not more than 9% ethylene oxide	2.2			X	X	X	X	X	10	190 250	0.66 0.75	
1958	1,2-DICHLORO-1,1,2,2-TETRAFLUOROETHANE (REFRIGERANT GAS R 114)	2.2			X	X	X	X	X	10	10	1.30	
1959	1,1-DIFLUOROETHYLENE (REFRIGERANT GAS R 1132a)	2.1			X	X	X	X	X	10	250	0.77	
1962	ETHYLENE	2.1			X	X	X	X	X	10	225 300	0.34 0.38	
1965	HYDROCARBON GAS MIXTURE, LIQUEFIED, N.O.S.	2.1			X	X	X	X	X	10			v, z
1967	INSECTICIDE GAS, TOXIC, N.O.S.	2.3			X	X	X	X	X	5			z
1968	INSECTICIDE GAS, N.O.S.	2.2			X	X	X	X	X	10			z
1969	ISOBUTANE	2.1			X	X	X	X	X	10	10	0.49	v
1973	CHLORODIFLUOROMETHANE AND CHLOROPENTAFLUOROETHANE MIXTURE with fixed boiling point, with approximately 49% chlorodifluoromethane (REFRIGERANT GAS R 502)	2.2			X	X	X	X	X	10	31	1.01	
1974	CHLORODIFLUOROBROMOMETHANE (REFRIGERANT GAS R 12B1)	2.2			X	X	X	X		10	10	1.61	
1975	NITRIC OXIDE AND DINITROGEN TETROXIDE MIXTURE (NITRIC OXIDE AND NITROGEN DIOXIDE MIXTURE)	2.3	5.1, 8	115	X		X	X		5			k, z
1976	OCTAFLUOROCYCLOBUTANE (REFRIGERANT GAS RC 318)	2.2			X	X	X	X	X	10	11	1.32	
1978	PROPANE	2.1			X	X	X	X	X	10	23	0.43	v
1982	TETRAFLUOROMETHANE (REFRIGERANT GAS R 14)	2.2			X	X	X	X	X	10	200 300	0.71 0.90	
1983	1-CHLORO-2,2,2-TRIFLUOROETHANE (REFRIGERANT GAS R 133a)	2.2			X	X	X	X	X	10	10	1.18	

* Where the entries are blank, the maximum working pressure shall not exceed two thirds of the test pressure.

| P200 | PACKING INSTRUCTION (continued) | | | | | | | | | | P200 |

| Table 2: LIQUEFIED GASES AND DISSOLVED GASES (continued) | | | | | | | | | | | |

UN No.	Proper Shipping Name	Class	Subsidiary risk	LC_{50}, mℓ/m^3	Cylinders	Tubes	Pressure drums	Bundles of cylinders	MEGCs	Test period, years	Test pressure, bar*	Filling ratio	Special packing provisions
1984	TRIFLUOROMETHANE (REFRIGERANT GAS R 23)	2.2			X	X	X	X	X	10	190 250	0.88 0.96	
2035	1,1,1-TRIFLUOROETHANE (REFRIGERANT GAS R 143a)	2.1			X	X	X	X	X	10	35	0.73	
2036	XENON	2.2			X	X	X	X	X	10	130	1.28	
2044	2,2-DIMETHYLPROPANE	2.1			X	X	X	X	X	10	10	0.53	
2073	AMMONIA SOLUTION, relative density less than 0.880 at 15°C in water, with more than 35% but not more than 40% ammonia with more than 40% but not more than 50% ammonia	2.2			X X	X X	X X	X X	X X	5 5	10 12	0.80 0.77	b b
2188	ARSINE	2.3	2.1	20	X			X		5	42	1.10	d, k
2189	DICHLOROSILANE	2.3	2.1, 8	314	X	X	X	X	X	5	200	1.08	
2191	SULPHURYL FLUORIDE	2.3		3020	X	X	X	X	X	5	50	1.10	u
2192	GERMANE	2.3	2.1	620	X	X	X	X	X	5	250	0.064	d, q, r
2193	HEXAFLUOROETHANE (REFRIGERANT GAS R 116)	2.2			X	X	X	X	X	10	200	1.13	
2194	SELENIUM HEXAFLUORIDE	2.3	8	50	X			X		5	36	1.46	k
2195	TELLURIUM HEXAFLUORIDE	2.3	8	25	X			X		5	20	1.00	k
2196	TUNGSTEN HEXAFLUORIDE	2.3	8	160	X			X		5	10	3.08	a, k
2197	HYDROGEN IODIDE, ANHYDROUS	2.3	8	2860	X	X	X	X	X	5	23	2.25	a, d
2198	PHOSPHORUS PENTAFLUORIDE	2.3	8	190	X			X		5	200 300	0.90 1.25	k k
2199	PHOSPHINE	2.3	2.1	20	X			X		5	225 250	0.30 0.45	d, k, q d, k, q
2200	PROPADIENE, STABILIZED	2.1			X	X	X	X	X	10	22	0.50	
2202	HYDROGEN SELENIDE, ANHYDROUS	2.3	2.1	2	X			X		5	31	1.60	k
2203	SILANE	2.1			X	X	X	X	X	10	225 250	0.32 0.36	q q
2204	CARBONYL SULPHIDE	2.3	2.1	1700	X	X	X	X	X	5	30	0.87	u
2417	CARBONYL FLUORIDE	2.3	8	360	X	X	X	X	X	5	200 300	0.47 0.70	
2418	SULPHUR TETRAFLUORIDE	2.3	8	40	X			X		5	30	0.91	k
2419	BROMOTRIFLUOROETHYLENE	2.1			X	X	X	X	X	10	10	1.19	
2420	HEXAFLUOROACETONE	2.3	8	470	X	X	X	X	X	5	22	1.08	
2421	NITROGEN TRIOXIDE	2.3	5.1, 8	57	X			X		5			k
2422	OCTAFLUOROBUT-2-ENE (REFRIGERANT GAS R 1318)	2.2			X	X	X	X	X	10	12	1.34	
2424	OCTAFLUOROPROPANE (REFRIGERANT GAS R 218)	2.2			X	X	X	X	X	10	25	1.04	
2451	NITROGEN TRIFLUORIDE	2.2	5.1		X	X	X	X	X	10	200	0.50	
2452	ETHYLACETYLENE, STABILIZED	2.1			X	X	X	X	X	10	10	0.57	c
2453	ETHYL FLUORIDE (REFRIGERANT GAS R 161)	2.1			X	X	X	X	X	10	30	0.57	
2454	METHYL FLUORIDE (REFRIGERANT GAS R 41)	2.1			X	X	X	X	X	10	300	0.63	
2455	METHYL NITRITE	2.2			(see special provision 900)								

* Where the entries are blank, the maximum working pressure shall not exceed two thirds of the test pressure.

P200	PACKING INSTRUCTION *(continued)*											P200

	Table 2: LIQUEFIED GASES AND DISSOLVED GASES *(continued)*											
UN No.	Proper Shipping Name	Class	Subsidiary risk	LC_{50}, ml/m³	Cylinders	Tubes	Pressure drums	Bundles of cylinders	MEGCs	Test period, years	Test pressure, bar*	Filling ratio	Special packing provisions
2517	1-CHLORO-1,1-DIFLUOROETHANE (REFRIGERANT GAS R 142b)	2.1			X	X	X	X	X	10	10	0.99	
2534	METHYLCHLOROSILANE	2.3	2.1, 8	600	X	X	X	X	X	5			z
2548	CHLORINE PENTAFLUORIDE	2.3	5.1, 8	122	X			X		5	13	1.49	a, k
2599	CHLOROTRIFLUOROMETHANE AND TRIFLUOROMETHANE AZEOTROPIC MIXTURE with approximately 60% chlorotrifluoromethane (REFRIGERANT GAS R 503)	2.2			X	X	X	X	X	10	31 / 42 / 100	0.12 / 0.17 / 0.64	
2601	CYCLOBUTANE	2.1			X	X	X	X	X	10	10	0.63	
2602	DICHLORODIFLUOROMETHANE AND DIFLUOROETHANE AZEOTROPIC MIXTURE with approximately 74% dichlorodifluoromethane (REFRIGERANT GAS R 500)	2.2			X	X	X	X	X	10	22	1.01	
2676	STIBINE	2.3	2.1	20	X			X		5	200	0.49	k, r
2901	BROMINE CHLORIDE	2.3	5.1, 8	290	X	X	X	X	X	5	10	1.50	a
3057	TRIFLUOROACETYL CHLORIDE	2.3	8	10	X		X	X		5	17	1.17	k
3070	ETHYLENE OXIDE AND DICHLORODIFLUORO-METHANE MIXTURE with not more than 12.5% ethylene oxide	2.2			X	X	X	X	X	10	18	1.09	
3083	PERCHLORYL FLUORIDE	2.3	5.1	770	X	X	X	X	X	5	33	1.21	u
3153	PERFLUORO(METHYL VINYL ETHER)	2.1			X	X	X	X	X	10	20	0.75	
3154	PERFLUORO(ETHYL VINYL ETHER)	2.1			X	X	X	X	X	10	10	0.98	
3157	LIQUEFIED GAS, OXIDIZING, N.O.S.	2.2	5.1		X	X	X	X	X	10			z
3159	1,1,1,2-TETRAFLUOROETHANE (REFRIGERANT GAS R 134a)	2.2			X	X	X	X	X	10	18	1.05	
3160	LIQUEFIED GAS, TOXIC, FLAMMABLE, N.O.S.	2.3	2.1	≤5000	X	X	X	X	X	5			z
3161	LIQUEFIED GAS, FLAMMABLE, N.O.S.	2.1			X	X	X	X	X	10			z
3162	LIQUEFIED GAS, TOXIC, N.O.S.	2.3		≤5000	X	X	X	X	X	5			z
3163	LIQUEFIED GAS, N.O.S.	2.2			X	X	X	X	X	10			z
3220	PENTAFLUOROETHANE (REFRIGERANT GAS R 125)	2.2			X	X	X	X	X	10	49 / 35	0.95 / 0.87	
3252	DIFLUOROMETHANE (REFRIGERANT GAS R 32)	2.1			X	X	X	X	X	10	48	0.78	
3296	HEPTAFLUOROPROPANE (REFRIGERANT GAS R 227)	2.2			X	X	X	X	X	10	13	1.21	
3297	ETHYLENE OXIDE AND CHLOROTETRA-FLUOROETHANE MIXTURE with not more than 8.8% ethylene oxide	2.2			X	X	X	X	X	10	10	1.16	
3298	ETHYLENE OXIDE AND PENTAFLUOROETHANE MIXTURE with not more than 7.9% ethylene oxide	2.2			X	X	X	X	X	10	26	1.02	
3299	ETHYLENE OXIDE AND TETRAFLUOROETHANE MIXTURE with not more than 5.6% ethylene oxide	2.2			X	X	X	X	X	10	17	1.03	
3300	ETHYLENE OXIDE AND CARBON DIOXIDE MIXTURE with more than 87% ethylene oxide	2.3	2.1	More than 2900	X	X	X	X	X	5	28	0.73	
3307	LIQUEFIED GAS, TOXIC, OXIDIZING, N.O.S.	2.3	5.1	≤5000	X	X	X	X	X	5			z
3308	LIQUEFIED GAS, TOXIC, CORROSIVE, N.O.S.	2.3	8	≤5000	X	X	X	X	X	5			z
3309	LIQUEFIED GAS, TOXIC, FLAMMABLE, CORROSIVE, N.O.S.	2.3	2.1, 8	≤5000	X	X	X	X	X	5			z

* Where the entries are blank, the maximum working pressure shall not exceed two thirds of the test pressure.

IMDG CODE *(Amdt. 34-08)*

P200	PACKING INSTRUCTION *(continued)*												P200

Table 2: LIQUEFIED GASES AND DISSOLVED GASES *(continued)*

UN No.	Proper Shipping Name	Class	Subsidiary risk	LC_{50}, mℓ/m^3	Cylinders	Tubes	Pressure drums	Bundles of cylinders	MEGCs	Test period, years	Test pressure, bar*	Filling ratio	Special packing provisions
3310	LIQUEFIED GAS, TOXIC, OXIDIZING, CORROSIVE, N.O.S.	2.3	5.1, 8	⩽5000	X	X	X	X	X	5			z
3318	AMMONIA SOLUTION, relative density less than 0.880 at 15°C in water, with more than 50% ammonia	2.3	8		X	X	X	X		5			b
3337	REFRIGERANT GAS R 404A	2.2			X	X	X	X	X	10	36	0.82	
3338	REFRIGERANT GAS R 407A	2.2			X	X	X	X	X	10	32	0.94	
3339	REFRIGERANT GAS R 407B	2.2			X	X	X	X	X	10	33	0.93	
3340	REFRIGERANT GAS R 407C	2.2			X	X	X	X	X	10	30	0.95	
3354	INSECTICIDE GAS, FLAMMABLE, N.O.S.	2.1			X	X	X	X	X	10			z
3355	INSECTICIDE GAS, TOXIC, FLAMMABLE, N.O.S.	2.3	2.1		X	X	X	X	X	5			z
3374	ACETYLENE, SOLVENT FREE	2.1			X			X		5	60 52		c, p

Table 3: SUBSTANCES NOT IN CLASS 2

UN No.	Proper Shipping Name	Class	Subsidiary risk	LC_{50}, mℓ/m^3	Cylinders	Tubes	Pressure drums	Bundles of cylinders	MEGCs	Test period, years	Test pressure, bar*	Filling ratio	Special packing provisions
1051	HYDROGEN CYANIDE, STABILIZED containing less than 3% water	6.1	3	40	X			X		5	100	0.55	k
1052	HYDROGEN FLUORIDE, ANHYDROUS	8	6.1	966	X		X	X		5	10	0.84	t
1745	BROMINE PENTAFLUORIDE	5.1	6.1, 8	25	X		X	X		5	10	†	k
1746	BROMINE TRIFLUORIDE	5.1	6.1, 8	50	X		X	X		5	10	†	k
2495	IODINE PENTAFLUORIDE	5.1	6.1, 8	120	X		X	X		5	10	†	k
2983	ETHYLENE OXIDE AND PROPYLENE OXIDE MIXTURE with not more than 30% ethylene oxide	3	6.1		X		X	X		5	10		z

* Where the entries are blank, the maximum working pressure shall not exceed two thirds of the test pressure.
† A minimum ullage of 8% by volume is required.

P201	PACKING INSTRUCTION	P201

This instruction applies to UN 3167, UN 3168 and UN 3169.

The following packagings are authorized:

(1) Compressed gas cylinders and gas receptacles conforming to the construction, testing and filling provisions approved by the competent authority.

(2) In addition, the following packagings are authorized provided that the general provisions of 4.1.1 and 4.1.3 are met:

.1 for non-toxic gases, combination packagings with hermetically sealed inner packagings of glass or metal with a maximum capacity of 5 ℓ per package which meet the packing group III performance level;

.2 for toxic gases, combination packagings with hermetically sealed inner packagings of glass or metal with a maximum capacity of 1 ℓ per package which meet the packing group III performance level.

P202	PACKING INSTRUCTION	P202

(Reserved)

P203	PACKING INSTRUCTION	P203

This instruction applies to class 2 refrigerated liquefied gases in closed cryogenic receptacles. Refrigerated liquefied gases in open cryogenic receptacles shall conform to the construction, testing and filling requirements approved by the competent authority.

For closed cryogenic receptacles, the general provisions of 4.1.6.1 shall be met.

Closed cryogenic receptacles constructed as specified in chapter 6.2 are authorized for the transport of refrigerated liquefied gases.

The closed cryogenic receptacles shall be so insulated that they do not become coated with frost.

(1) Test pressure

 Refrigerated liquids shall be filled in closed cryogenic receptacles with the following minimum test pressures:

 (a) For closed cryogenic receptacles with vacuum insulation, the test pressure shall not be less than 1.3 times the sum of the maximum internal pressure of the filled receptacle, including during filling and discharge, plus 100 kPa (1 bar);

 (b) For other closed cryogenic receptacles, the test pressure shall be not less than 1.3 times the maximum internal pressure of the filled receptacle, taking into account the pressure developed during filling and discharge.

(2) Degree of filling

 For non-flammable, non-toxic refrigerated liquefied gases, the volume of liquid phase at the filling temperature and at a pressure of 100 kPa (1 bar) shall not exceed 98% of the water capacity of the pressure receptacle.

 For flammable refrigerated liquefied gases, the degree of filling shall remain below the level at which the volume of the liquid phase would reach 98% of the water capacity at that temperature, if the contents were raised to the temperature at which the vapour pressure equalled the opening pressure of the relief valve.

(3) Pressure relief devices

 Closed cryogenic receptacles shall be fitted with at least one pressure relief device.

(4) Compatibility

 Materials used to ensure the leakproofness of the joints or for the maintenance of the closures shall be compatible with the contents. In the case of receptacles intended for the transport of oxidizing gases (i.e., with a subsidiary risk 5.1), these materials shall not react with these gases in a dangerous manner.

P300	PACKING INSTRUCTION	P300

This instruction applies to UN 3064.

The following packagings are authorized, provided that the general provisions of 4.1.1 and 4.1.3 are met:
 Combination packagings consisting of inner metal cans of not more than 1 ℓ capacity each and outer wooden boxes (4C1, 4C2, 4D or 4F) containing not more than 5 ℓ of solution.

Additional provisions:

1 Metal cans shall be completely surrounded with absorbent cushioning material.

2 Wooden boxes shall be completely lined with suitable material impervious to water and nitroglycerin.

P301	PACKING INSTRUCTION	P301

This instruction applies to UN 3165.

The following packagings are authorized, provided that the general provisions of 4.1.1 and 4.1.3 are met:

(1) Aluminium pressure vessel made from tubing and having welded heads
 Primary containment of the fuel within this vessel shall consist of a welded aluminium bladder having a maximum internal volume of 46 ℓ. The outer vessel shall have a minimum design gauge pressure of 1,275 kPa and a minimum burst gauge pressure of 2,755 kPa. Each vessel shall be leak-checked during manufacture and before shipment and shall be found leakproof. The complete inner unit shall be securely packed in non-combustible cushioning material, such as vermiculite, in a strong outer tightly closed metal packaging which will adequately protect all fittings. Maximum quantity of fuel per unit and package is 42 ℓ.

(2) Aluminium pressure vessel
 Primary containment of the fuel within this vessel shall consist of a welded vapourtight fuel compartment with an elastomeric bladder having a maximum internal volume of 46 ℓ. The pressure vessel shall have a minimum design gauge pressure of 2,680 kPa and a minimum burst pressure of 5,170 kPa. Each vessel shall be leak-checked during manufacture and before shipment and shall be securely packed in non-combustible cushioning material such as vermiculite, in a strong outer tightly closed metal packaging which will adequately protect all fittings. Maximum quantity of fuel per unit and package is 42 ℓ.

P302	PACKING INSTRUCTION	P302

This instruction applies to UN 3269.

The following packagings are authorized, provided that the general provisions of 4.1.1 and 4.1.3 are met:

Combination packagings which meet the packing group II or III performance level according to the criteria for class 3, applied to the base material.

The base material and the activator (organic peroxide) shall be each separately packed in inner packagings. The components may be placed in the same outer packaging provided they will not interact dangerously in the event of a leakage.

The activator shall have a maximum quantity of 125 mℓ per inner packaging if liquid, and 500 g per inner packaging if solid.

P400	PACKING INSTRUCTION	P400

The following packagings are authorized, provided that the general provisions of 4.1.1 and 4.1.3 are met:

(1) Pressure receptacles, provided that the general provisions of 4.1.3.6 are met. They shall be made of steel and shall be subjected to an initial test and periodic tests every 10 years at a pressure of not less than 1 MPa (10 bar, gauge pressure). During carriage, the liquid shall be under a layer of inert gas with a gauge pressure of not less than 20 kPa (0.2 bar)

(2) Boxes (4A, 4B, 4C1, 4C2, 4D, 4F or 4G), drums (1A2, 1B2, 1N2, 1D or 1G) or jerricans (3A2 or 3B2) enclosing hermetically sealed metal cans with inner packagings of glass or metal, with a capacity of not more than 1 ℓ each, having threaded closures with gaskets. Inner packagings shall be cushioned on all sides with dry, absorbent, non-combustible material in a quantity sufficient to absorb the entire contents. Inner packagings shall not be filled to more than 90% of their capacity. Outer packagings shall have a maximum net mass of 125 kg.

(3) Steel, aluminium or metal drums (1A2, 1B2 or 1N2), jerricans (3A2 or 3B2) or boxes (4A or 4B) with a maximum net mass of 150 kg each with hermetically sealed inner metal cans of not more than 4 ℓ capacity each, with threaded closures fitted with gaskets. Inner packagings shall be cushioned on all sides with dry, absorbent, non-combustible material in a quantity sufficient to absorb the entire contents. Each layer of inner packagings shall be separated by a dividing partition in addition to cushioning material. Inner packagings shall not be filled to more than 90% of their capacity.

Special packing provisions:

PP31 For UN 2870, packagings shall be hermetically sealed.

PP86 For UN 3392 and UN 3394, air shall be eliminated from the vapour space by nitrogen or other means.

P401	PACKING INSTRUCTION	P401

The following packagings are authorized, provided that the general provisions of 4.1.1 and 4.1.3 are met:

(1) Pressure receptacles, provided that the general provisions of 4.1.3.6 are met. They shall be made of steel and subjected to an initial test and periodic tests every 10 years at a pressure of not less than 0.6 MPa (6 bar, gauge pressure). During carriage, the liquid shall be under a layer of inert gas with a gauge pressure of not less than 20 kPa (0.2 bar).

	Inner packaging	Outer packaging
(2) Combination packagings with inner packagings of glass, metal or plastics which have threaded closures, surrounded in inert cushioning and absorbent material in a quantity sufficient to absorb the entire contents.	1 ℓ	30 kg maximum net mass

Special packing provision:

PP31 For UN Nos. 1183, 1242, 1295, 2965 and 2988, packagings shall be hermetically sealed.

P402	PACKING INSTRUCTION	P402

The following packagings are authorized, provided that the general provisions of 4.1.1 and 4.1.3 are met:

(1) Pressure receptacles, provided that the general provisions of 4.1.3.6 are met. They shall be made of steel and subjected to an initial test and periodic tests every 10 years at a pressure of not less than 0.6 MPa (6 bar, gauge pressure). During carriage, the liquid shall be under a layer of inert gas with a gauge pressure of not less than 20 kPa (0.2 bar).

	Inner packaging	Outer packaging maximum net mass
(2) Combination packagings with inner packagings of glass, metal or plastics which have threaded closures, surrounded in inert cushioning and absorbent material in a quantity sufficient to absorb the entire contents.	10 kg (glass) 15 kg (metal or plastics)	125 kg 125 kg

(3) Steel drums (1A1) with a maximum capacity of 250 ℓ.

(4) Composite packagings consisting of plastics receptacle in a steel or aluminium drum (6HA1 or 6HB1) with a maximum capacity of 250 ℓ.

Special packing provision:

PP31 For UN Nos.1389, 1391, 1392, 1420, 1421, 1422, 3184 (PG II), 3185 (PG II), 3187 (PG II), 3188 (PG II), 3398 (PG I) and 3399 (PG I), packagings shall be hermetically sealed.

4

P403	PACKING INSTRUCTION	P403

The following packagings are authorized, provided that the general provisions of 4.1.1 and 4.1.3 are met:

Combination packagings		Maximum net mass
Inner packagings	Outer packagings	

Glass   2 kg
Plastic   15 kg
Metal   20 kg
 	
Inner packagings shall be hermetically sealed (e.g., by taping or by threaded closures).
	Drums
	   steel (1A2)
	   aluminium (1B2)
	   other metal (1N2)
	   plastics (1H2)
	   plywood (1D)
	   fibre (1G)
		400 kg
		400 kg
		400 kg
		400 kg
		400 kg
		400 kg

	Boxes
	   steel (4A)
	   aluminium (4B)
	   natural wood (4C1)
	   natural wood with sift-proof walls
	      (4C2)
	   plywood (4D)
	   reconstituted wood (4F)
	   fibreboard (4G)
	   expanded plastics (4H1)
	   solid plastics (4H2)
		400 kg
		400 kg
		250 kg
		250 kg
		
		250 kg
		125 kg
		125 kg
		60 kg
		250 kg

	Jerricans
	   steel (3A2)
	   aluminium (3B2)
	   plastics (3H2)
		120 kg
		120 kg
		120 kg

Single packagings		

Drums
   steel (1A1, 1A2)
   aluminium (1B1, 1B2)
   metal other than steel or aluminium (1N1, 1N2)
   plastics (1H1, 1H2)
		250 kg
		250 kg
		250 kg
		250 kg

Jerricans
   steel (3A1, 3A2)
   aluminium (3B1, 3B2)
   plastics (3H1, 3H2)
		120 kg
		120 kg
		120 kg

Composite packagings		
Plastics receptacle in steel or aluminium drum (6HA1 or 6HB1)		250 kg
Plastics receptacle in fibre, plastics or plywood drum (6HG1, 6HH1 or 6HD1)		75 kg
Plastics receptacle in steel, aluminium, wood, plywood, fibreboard or solid plastics box (6HA2, 6HB2, 6HC, 6HD2, 6HG2 or 6HH2)		75 kg

Pressure receptacles, provided that the general provisions of 4.1.3.6 are met

Special packing provisions:

PP31   For UN Nos. 1360, 1397, 1402 (PG I), 1404, 1407, 1409, 1410, 1413, 1414, 1415, 1418 (PG I), 1419, 1423, 1426, 1427, 1428, 1432, 1433, 1714, 1870, 2010, 2011, 2012, 2013, 2257, 2463, 2806, 2813 (PG I), 3208, 3209, 3401, 3402, 3403 and 3404, packagings shall be hermetically sealed, except for solid fused material.

PP83   For UN 2813, waterproof bags containing not more than 20 g of substance for the purposes of heat formation may be packaged for transport. Each waterproof bag shall be sealed in a plastics bag and placed within an intermediate packaging. No outer packaging shall contain more than 400 g of substance. Water or liquid which may react with the water-reactive substance shall not be included in the packaging.

P404	PACKING INSTRUCTION	P404

This instruction applies to pyrophoric solids: UN Nos. 1383, 1854, 1855, 2008, 2441, 2545, 2546, 2846, 2881, 3200, 3391 and 3393.

The following packagings are authorized, provided that the general provisions of 4.1.1 and 4.1.3 are met:
(1) Combination packagings
   Outer packagings: (1A2, 1B2, 1N2, 1H2, 1D, 4A, 4B, 4C1, 4C2, 4D, 4F or 4H2)
   Inner packagings: Metal packagings with a maximum net mass of 15 kg each. Inner packagings shall be hermetically sealed and have threaded closures.
(2) Metal packagings: (1A1, 1A2, 1B1, 1N1, 1N2, 3A1, 3A2, 3B1 and 3B2)
   Maximum gross mass: 150 kg
(3) Composite packagings: Plastics receptacles in a steel or aluminium drum (6HA1 or 6HB1)
   Maximum gross mass: 150 kg

Pressure receptacles may be used provided that the general provisions of 4.1.3.6 are met

Special packing provisions:

PP31 For UN Nos. 1383, 1854, 1855, 2008, 2441, 2545, 2546, 2846, 2881 and 3200, packagings shall be hermetically sealed.

PP86 For UN 3391 and UN 3393, air shall be eliminated from the vapour space by nitrogen or other means.

P405	PACKING INSTRUCTION	P405

This instruction applies to UN 1381.

The following packagings are authorized, provided that the general provisions of 4.1.1 and 4.1.3 are met:
(1) For UN 1381, wet phosphorus:
   .1 Combination packagings
      Outer packagings: (4A, 4B, 4C1, 4C2, 4D or 4F); maximum net mass: 75 kg
      Inner packagings:
      (i) hermetically sealed metal cans, with a maximum net mass of 15 kg; or
      (ii) glass inner packagings cushioned on all sides with dry, absorbent, non-combustible material in a quantity sufficient to absorb the entire contents with a maximum net mass of 2 kg; or
   .2 Drums (1A1, 1A2, 1B1, 1B2, 1N1 or 1N2); maximum net mass: 400 kg
      Jerricans (3A1 or 3B1); maximum net mass: 120 kg.
These packagings shall be capable of passing the leakproofness test specified in 6.1.5.4 at the packing group II performance level.
(2) For UN 1381, dry phosphorus:
   .1 When fused, drums (1A2, 1B2 or 1N2) with a maximum net mass of 400 kg; or
   .2 In projectiles or hard-cased articles when transported without class 1 components, as specified by the competent authority.

Special packing provision:
PP31 For UN 1381, packagings shall be hermetically sealed.

P406	PACKING INSTRUCTION	P406

The following packagings are authorized, provided that the general provisions of 4.1.1 and 4.1.3 are met:

(1) Combination packagings
Outer packagings: (4C1, 4C2, 4D, 4F, 4G, 4H1, 4H2, 1G, 1D, 1H2 or 3H2)
Inner packagings shall be water-resistant.

(2) Plastics, plywood or fibreboard drums (1H2, 1D or 1G) or boxes (4A, 4B, 4C1, 4D, 4F, 4C2, 4G and 4H2) with a water-resistant inner bag, plastics film lining or water-resistant coating.

(3) Metal drums (1A1, 1A2, 1B1, 1B2, 1N1 or 1N2), plastics drums (1H1 or 1H2), metal jerricans (3A1, 3A2, 3B1 or 3B2), plastics jerricans (3H1 or 3H2), plastics receptacle in steel or aluminium drums (6HA1 or 6HB1), plastics receptacle in fibre, plastics or plywood drums (6HG1, 6HH1 or 6HD1), plastics receptacle in steel, aluminium, wood, plywood, fibreboard or solid plastics boxes (6HA2, 6HB2, 6HC, 6HD2, 6HG2 or 6HH2).

Additional provisions:

1 Packagings shall be designed and constructed to prevent the loss of water or alcohol content or the content of the phlegmatizer.

2 Packagings shall be so constructed and closed as to avoid an explosive overpressure or pressure build-up of more than 300 kPa (3 bar).

3 The type of packaging and maximum permitted quantity per packaging are limited by the provisions of 2.1.3.4.

Special packing provisions:

PP24 UN Nos. 2852, 3364, 3365, 3366, 3367, 3368 and 3369 shall not be transported in quantities of more than 500 g per package

PP25 UN 1347 shall not be transported in quantities of more than 15 kg per package.

PP26 For UN Nos. 1310, 1320, 1321, 1322, 1344, 1347, 1348, 1349, 1517, 2907, 3317, 3344 and 3376, packagings shall be lead-free.

PP31 For UN Nos. 1310, 1320, 1321, 1322, 1336, 1337, 1344, 1347, 1348, 1349, 1354, 1355, 1356, 1357, 1517, 1571, 2555, 2556, 2557, 2852, 3317, 3364, 3365, 3366, 3367, 3368, 3369, 3370 and 3376, packagings shall be hermetically sealed.

PP48 For UN 3474, metal packagings shall not be used.

PP78 UN 3370 shall not be transported in quantities of more than 11.5 kg per package.

PP80 For UN 2907 and UN 3344, packagings shall meet the packing group II performance level. Packagings meeting the test criteria of packing group I shall not be used.

4

P407	PACKING INSTRUCTION	P407

This instruction applies to UN Nos. 1331, 1944, 1945 and 2254.

The following packagings are authorized, provided that the general provisions of 4.1.1 and 4.1.3 are met:
Combination packagings comprising securely closed inner packagings to prevent accidental ignition under normal conditions of transport. The maximum gross mass of the package shall not exceed 45 kg except for fibreboard boxes, which shall not exceed 30 kg.

Additional provision:

Matches shall be tightly packed.

Special packing provision:

PP27 UN 1331, Strike-anywhere matches, shall not be packed in the same outer packaging with any other dangerous goods other than safety matches or wax Vesta matches, which shall be packed in separate inner packagings. Inner packagings shall not contain more than 700 strike-anywhere matches.

P408	PACKING INSTRUCTION	P408

This instruction applies to UN 3292.

The following packagings are authorized, provided that the general provisions of 4.1.1 and 4.1.3 are met:

(1) For cells:
Outer packagings with sufficient cushioning material to prevent contact between cells and the internal surfaces of the outer packaging and to ensure that no dangerous movement of the cells within the outer packaging occurs in transport. Packagings shall conform to the packing group II performance level.

(2) For batteries:
Batteries may be transported unpacked or in protective enclosures (such as in fully enclosed or wooden slatted crates). The terminals shall not support the mass of other batteries or materials packed with the batteries.

Additional provision:

Batteries shall be protected against short circuit and shall be isolated in such a manner as to prevent short circuits.

P409	PACKING INSTRUCTION	P409
This instruction applies to UN Nos. 2956, 3242 and 3251.		

The following packagings are authorized, provided that the general provisions of 4.1.1 and 4.1.3 are met:

(1) Fibre drum (1G) which may be fitted with a liner or coating; maximum net mass: 50 kg.

(2) Combination packagings: Fibreboard box (4G) with a single inner plastic bag; maximum net mass: 50 kg.

(3) Combination packagings: Fibreboard box (4G) or fibre drum (1G) with inner plastic packagings each containing a maximum of 5 kg; maximum net mass: 25 kg.

P410	PACKING INSTRUCTION		P410
The following packagings are authorized, provided that the general provisions of 4.1.1 and 4.1.3 are met.			

Combination packagings		Maximum net mass	
Inner packagings	Outer packagings	Packing group II	Packing group III

Glass   10 kg
Plastics[1]   30 kg
Metal   40 kg
Paper[1, 2]   10 kg
Fibre[1, 2]   10 kg
	Drums
	  steel (1A2)
	  aluminium (1B2)
	  other metal (1N2)
	  plastics (1H2)
	  plywood (1D)
	  fibre (1G)[1]
		
		400 kg
		400 kg
		400 kg
		400 kg
		400 kg
		400 kg
			
			400 kg
			400 kg
			400 kg
			400 kg
			400 kg
			400 kg

	Boxes
	  steel (4A)
	  aluminium (4B)
	  natural wood (4C1)
	  natural wood with sift-proof
	    walls (4C2)
	  plywood (4D)
	  reconstituted wood (4F)
	  fibreboard (4G)[1]
	  expanded plastics (4H1)
	  solid plastics (4H2)
		
		400 kg
		400 kg
		400 kg
		400 kg
		
		400 kg
		400 kg
		400 kg
		60 kg
		400 kg
			
			400 kg
			400 kg
			400 kg
			400 kg
			
			400 kg
			400 kg
			400 kg
			60 kg
			400 kg

[1] Packagings shall be sift-proof.
 	
[2] These inner packagings shall not be used when the substances being transported may become liquid during transport.
	Jerricans
	  steel (3A2)
	  aluminium (3B2)
	  plastics (3H2)
		
		120 kg
		120 kg
		120 kg
			
			120 kg
			120 kg
			120 kg

Single packagings			

Drums
  steel (1A1 or 1A2)
  aluminium (1B1 or 1B2)
  metal other than steel or aluminium (1N1 or 1N2)
  plastics (1H1 or 1H2)
		
		400 kg
		400 kg
		400 kg
		400 kg
			
			400 kg
			400 kg
			400 kg
			400 kg

Jerricans
  steel (3A1 or 3A2)
  aluminium (3B1 or 3B2)
  plastics (3H1 or 3H2)
		
		120 kg
		120 kg
		120 kg
			
			120 kg
			120 kg
			120 kg

Boxes
  steel (4A)[3]
  aluminium (4B)[3]
  natural wood (4C1)[3]
  natural wood with sift-proof walls (4C2)[3]
  plywood (4D)[3]
  reconstituted wood (4F)[3]
  fibreboard (4G)[3]
  solid plastics (4H2)[3]
		
		400 kg
		400 kg
		400 kg
		400 kg
		400 kg
		400 kg
		400 kg
		400 kg
			
			400 kg
			400 kg
			400 kg
			400 kg
			400 kg
			400 kg
			400 kg
			400 kg

Bags Bags (5H3, 5H4, 5L3, 5M2)[3, 4]		 50 kg	 50 kg
Composite packagings Plastics receptacle in steel, aluminium, plywood, fibre or plastics drum (6HA1, 6HB1, 6HG1, 6HD1 or 6HH1) Plastics receptacle in steel or aluminium crate or box, wooden box, plywood box, fibreboard box or solid plastics box (6HA2, 6HB2, 6HC, 6HD2, 6HG2 or 6HH2) Glass receptacle in steel, aluminium, plywood or fibre drum (6PA1, 6PB1, 6PD1 or 6PG1) or in steel, aluminium, wooden, wickerwork hamper or fibreboard box (6PA2, 6PB2, 6PC, 6PD2 or 6PG2) or in solid or expanded plastics packaging (6PH1 or 6PH2) [3] These packagings shall not be used when the substances being transported may become liquid during transport. [4] These packagings shall only be used for packing group II substances when transported in a closed cargo transport unit.		400 kg 75 kg 75 kg	400 kg 75 kg 75 kg
Pressure receptacles, provided that the general provisions of 4.1.3.6 are met			

P410	PACKING INSTRUCTION *(continued)*	P410

Special packing provisions:

PP31 For UN Nos. 1326, 1339, 1340, 1341, 1343, 1352, 1358, 1373, 1374, 1378, 1379, 1382, 1384, 1385, 1390, 1393, 1394, 1400, 1401, 1405, 1417, 1431, 1437, 1871, 1923, 1929, 2004, 2008, 2318, 2545, 2546, 2624, 2805, 2813, 2830, 2835, 2844, 2881, 2940, 3078, 3088, 3170 (PG II), 3182, 3189, 3190, 3205, 3206, 3208 and 3209, packagings shall be hermetically sealed.

PP39 For UN 1378, for metal packagings a venting device is required.

PP40 For the following UN Nos., falling in PG II, bags are not allowed: 1326, 1340, 1352, 1358, 1374, 1378, 1382, 1390, 1393, 1394, 1396, 1400, 1401, 1402, 1405, 1409, 1417, 1418, 1436, 1437, 1871, 2624, 2805, 2813, 2830, 2835, 3078, 3131, 3132, 3134, 3170, 3182, 3208 and 3209.

PP83 For UN 2813, waterproof bags containing not more than 20 g of substance for the purposes of heat formation may be packaged for transport. Each waterproof bag shall be sealed in a plastics bag and placed within an intermediate packaging. No outer packaging shall contain more than 400 g of substance. Water or liquid which may react with the water-reactive substance shall not be included in the packaging.

P411	PACKING INSTRUCTION	P411

This instruction applies to UN 3270.

The following packagings are authorized, provided that the general provisions of 4.1.1 and 4.1.3 are met:

(1) Fibreboard box with a maximum gross mass of 30 kg.

(2) Other packagings, provided that explosion is not possible by reason of increased internal pressure. Maximum net mass shall not exceed 30 kg.

P500	PACKING INSTRUCTION	P500

This instruction applies to UN 3356.

The following packagings are authorized, provided that the general provisions of 4.1.1 and 4.1.3 are met:

Packagings shall conform to the packing group II performance level.

The generator(s) shall be transported in a package which meets the following provisions when one generator in the package is actuated:

(a) other generators in the package will not be actuated;

(b) packaging material will not ignite; and

(c) the outside surface temperature of the completed package shall not exceed 100°C.

P501	PACKING INSTRUCTION	P501

This instruction applies to UN 2015.

The following packagings are authorized, provided that the general provisions of 4.1.1 and 4.1.3 are met:

Combination packagings	Inner packagings maximum capacity	Outer packagings maximum net mass
(1) Boxes (4A, 4B, 4C1, 4C2, 4D, 4H2) or drums (1A2, 1B2, 1N2, 1H2, 1D) or jerricans (3A2, 3B2, 3H2) with glass, plastics or metal inner packagings	5 ℓ	125 kg
(2) Fibreboard box (4G) or fibre drum (1G), with plastics or metal inner packagings each in a plastics bag	2 ℓ	50 kg

Single packagings	Maximum capacity
Drums steel (1A1) aluminium (1B1) metal other than steel or aluminium (1N1) plastics (1H1)	250 ℓ 250 ℓ 250 ℓ 250 ℓ
Jerricans steel (3A1) aluminium (3B1) plastics (3H1)	60 ℓ 60 ℓ 60 ℓ
Composite packagings	
Plastics receptacle in steel or aluminium drum (6HA1, 6HB1)	250 ℓ
Plastics receptacle in fibre, plastics or plywood drum (6HG1, 6HH1, 6HD1)	250 ℓ
Plastics receptacle in steel or aluminium crate or box or plastics receptacle in wood, plywood, fibreboard or solid plastics box (6HA2, 6HB2, 6HC, 6HD2, 6HG2 or 6HH2)	60 ℓ
Glass receptacle in steel, aluminium, fibre, plywood, solid plastics or expanded plastics drum (6PA1, 6PB1, 6PG1, 6PD1, 6PH1 or 6PH2) or in a steel, aluminium, wood, fibreboard or plywood box (6PA2, 6PB2, 6PC, 6PG2 or 6PD2)	60 ℓ

Additional provisions:

1   Packagings shall have a minimum ullage of 10%.

2   Packagings shall be vented.

P502	PACKING INSTRUCTION	P502

The following packagings are authorized, provided that the general provisions of 4.1.1 and 4.1.3 are met:

Combination packagings		Maximum net mass
Inner packagings	**Outer packagings**	
Glass 5 ℓ Metal 5 ℓ Plastic 5 ℓ	**Drums** steel (1A2) aluminium (1B2) other metal (1N2) plastics (1H2) plywood (1D) fibre (1G)	125 kg 125 kg 125 kg 125 kg 125 kg 125 kg
	Boxes steel (4A) aluminium (4B) natural wood (4C1) natural wood with sift-proof walls (4C2) plywood (4D) reconstituted wood (4F) fibreboard (4G) expanded plastics (4H1) solid plastics (4H2)	125 kg 125 kg 125 kg 125 kg 125 kg 125 kg 125 kg 60 kg 125 kg

Single packagings	Maximum capacity
Drums steel (1A1) aluminium (1B1) plastics (1H1)	250 ℓ 250 ℓ 250 ℓ
Jerricans steel (3A1) aluminium (3B1) plastics (3H1)	60 ℓ 60 ℓ 60 ℓ
Composite packagings	
Plastics receptacle in steel or aluminium drum (6HA1, 6HB1)	250 ℓ
Plastics receptacle in fibre, plastics or plywood drum (6HG1, 6HH1, 6HD1)	250 ℓ
Plastics receptacle in steel or aluminium crate or box or plastics receptacle in wood, plywood, fibreboard or solid plastics box (6HA2, 6HB2, 6HC, 6HD2, 6HG2 or 6HH2)	60 ℓ
Glass receptacle in steel, aluminium, fibre, plywood, solid plastics or expanded plastics drum (6PA1, 6PB1, 6PG1, 6PD1, 6PH1 or 6PH2) or in a steel, aluminium, wood, fibreboard or plywood box (6PA2, 6PB2, 6PC, 6PG2 or 6PD2)	60 ℓ

Special packing provision:

PP28 For UN 1873, only glass inner packagings or receptacles are authorized for combination and composite packagings.

P503	PACKING INSTRUCTION	P503

The following packagings are authorized, provided that the general provisions of 4.1.1 and 4.1.3 are met:

Combination packagings		Maximum net mass
Inner packagings	Outer packagings	

Glass   5 kg
Metal   5 kg
Plastic   5 kg
	Drums
	    steel (1A2)
	    aluminium (1B2)
	    other metal (1N2)
	    plastics (1H2)
	    plywood (1D)
	    fibre (1G)
		125 kg
		125 kg
		125 kg
		125 kg
		125 kg
		125 kg

	Boxes
	    steel (4A)
	    aluminium (4B)
	    natural wood (4C1)
	    natural wood with sift-proof walls
	        (4C2)
	    plywood (4D)
	    reconstituted wood (4F)
	    fibreboard (4G)
	    expanded plastics (4H1)
	    solid plastics (4H2)
		125 kg
		125 kg
		125 kg
		125 kg
		
		125 kg
		125 kg
		40 kg
		60 kg
		125 kg

Single packagings

Metal drums (1A2, 1B2 or 1N2) with a maximum net mass of 250 kg.

Fibreboard (1G) or plywood drums (1D) fitted with inner liners with a maximum net mass of 200 kg.

P504	PACKING INSTRUCTION	P504

The following packagings are authorized, provided that the general provisions of 4.1.1 and 4.1.3 are met:

Combination packagings	Maximum net mass
(1) Outer packagings: (1A2, 1B2, 1N2, 1H2, 1D, 1G, 4A, 4B, 4C1, 4C2, 4D, 4F, 4G, 4H2) Inner packagings: Glass receptacles with a maximum capacity of 5 ℓ	75 kg
(2) Outer packagings: Plastics receptacles with a maximum capacity of 30 ℓ in 1A2, 1B2, 1N2, 1H2, 1D, 1G, 4A, 4B, 4C1, 4C2, 4D, 4F, 4G, 4H2	75 kg
(3) Metal receptacles with a maximum capacity of 40 ℓ in 1G, 4F or 4G	125 kg
(4) Metal receptacles with a maximum capacity of 40 ℓ in 1A2, 1B2, 1N2, 1H2, 1D, 4A, 4B, 4C1, 4C2, 4D, 4H2 outer packagings	225 kg

Single packagings	Maximum capacity

Drums
    steel, non-removable head (1A1)
    aluminium, non-removable head (1B1)
    other metal, non-removable head (1N1)
    plastics, non-removable head (1H1)
	250 ℓ
	250 ℓ
	250 ℓ
	250 ℓ

Jerricans
    steel, non-removable head (3A1)
    aluminium, non-removable head (3B1)
    plastics, non-removable head (3H1)
	60 ℓ
	60 ℓ
	60 ℓ

Composite packagings	
Plastics receptacle in steel or aluminium drum (6HA1, 6HB1)	250 ℓ
Plastics receptacle in fibre, plastics or plywood drum (6HG1, 6HH1, 6HD1)	120 ℓ
Plastics receptacle in steel or aluminium crate or box or plastics receptacle in wood, plywood, fibreboard or solid plastics box (6HA2, 6HB2, 6HC, 6HD2, 6HG2 or 6HH2)	60 ℓ

Glass receptacle in steel, aluminium, fibre, plywood, solid plastics or expanded plastics drum
    (6PA1, 6PB1, 6PG1, 6PD1, 6PH1 or 6PH2) or in a steel, aluminium, wood or fibreboard box
    or in a wickerwork hamper (6PA2, 6PB2, 6PC, 6PG2 or 6PD2)
	60 ℓ

Special packing provisions:

PP10 For UN 2014 and UN 3149, the packaging shall be vented.

PP31 For UN 2626, packagings shall be hermetically sealed.

P520	PACKING INSTRUCTION	P520

This instruction applies to organic peroxides of class 5.2 and self-reactive substances of class 4.1.

The packagings listed below are authorized provided the general provisions of 4.1.1 and 4.1.3 and special provisions of 4.1.7 are met.
The packing methods are designated OP1 to OP8. The packing methods appropriate for the individual currently assigned organic peroxides and self-reactive substances are listed in 2.4.2.3.2.3 and 2.5.3.2.4. The quantities specified for each packing method are the maximum quantities authorized per package. The following packagings are authorized:

(1) Combination packagings with outer packagings comprising boxes (4A, 4B, 4C1, 4C2, 4D, 4F, 4G, 4H1 and 4H2), drums (1A2, 1B2, 1G, 1H2 and 1D) or jerricans (3A2, 3B2 and 3H2);

(2) Single packagings consisting of drums (1A1, 1A2, 1B1, 1B2, 1G, 1H1, 1H2 and 1D) and jerricans (3A1, 3A2, 3B1, 3B2, 3H1 and 3H2);

(3) Composite packagings with plastics inner receptacles (6HA1, 6HA2, 6HB1, 6HB2, 6HC, 6HD1, 6HD2, 6HG1, 6HG2, 6HH1 and 6HH2).

Maximum quantity per packaging/package[1] for packing methods OP1 to OP8

Packing method Maximum quantity	OP1	OP2[1]	OP3	OP4[1]	OP5	OP6	OP7	OP8
Maximum mass (kg) for solids and for combination packagings (liquid and solid)	0.5	0.5/10	5	5/25	25	50	50	400[2]
Maximum contents in litres for liquids[3]	0.5	–	5	–	30	60	60	225[4]

[1] If two values are given, the first applies to the maximum net mass per inner packaging and the second to the maximum net mass of the complete package.

[2] 60 kg for jerricans/200 kg for boxes and, for solids, 400 kg in combination packagings with outer packagings comprising boxes (4C1, 4C2, 4D, 4F, 4G, 4H1 and 4H2) and with inner packagings of plastics or fibre with a maximum net mass of 25 kg.

[3] Viscous liquids shall be treated as solids when they do not meet the criteria provided in the definition for liquids presented in 1.2.1.

[4] 60 ℓ for jerricans.

Additional provisions:

1 Metal packagings, including inner packagings of combination packagings and outer packagings of combination or composite packagings, may only be used for packing methods OP7 and OP8.

2 In combination packagings, glass receptacles may only be used as inner packagings with a maximum content of 0.5 kg for solids or 0.5 ℓ for liquids.

3 In combination packagings, cushioning materials shall not be readily combustible.

4 The packaging of an organic peroxide or self-reactive substance required to bear an EXPLOSIVE subsidiary risk label (Model No. 1, see 5.2.2.2.2) shall also comply with the provisions given in 4.1.5.10 and 4.1.5.11.

Special packing provisions:

PP21 For certain self-reactive substances of types B or C, UN Nos. 3221, 3222, 3223, 3224, 3231, 3232, 3233 and 3234, a smaller packaging than that allowed by packing methods OP5 or OP6 respectively shall be used (see 4.1.6 and 2.4.2.3.2.3).

PP22 UN 3241, 2-bromo-2-nitropropane-1,3-diol, shall be packed in accordance with packing method OP6.

P600	PACKING INSTRUCTION	P600

This instruction applies to UN Nos. 1700, 2016 and 2017.

The following packagings are authorized, provided the general provisions of 4.1.1 and 4.1.3 are met:
Outer packagings: (1A2, 1B2, 1N2, 1H2, 1D, 1G, 4A, 4B, 4C1, 4C2, 4D, 4F, 4G, 4H2) meeting the packing group II performance level. The articles shall be individually packaged and separated from each other using partitions, dividers, inner packagings or cushioning material to prevent inadvertent discharge during normal conditions of transport.
Maximum net mass: 75 kg

P601	PACKING INSTRUCTION	P601

The following packagings are authorized provided the general provisions of 4.1.1 and 4.1.3 are met and the packagings are hermetically sealed:

(1) Combination packagings with a maximum gross mass of 15 kg, consisting of:

- one or more glass inner packaging(s) with a maximum quantity of 1 litre each and filled to not more than 90% of their capacity; the closure(s) of which shall be physically held in place by any means capable of preventing back-off or loosening by impact or vibration during transport, individually placed in

- metal receptacles together with cushioning and absorbent material sufficient to absorb the entire contents of the glass inner packaging(s), further packed in

- 1A2, 1B2, 1N2, 1H2, 1D, 1G, 4A, 4B, 4C1, 4C2, 4D, 4F, 4G or 4H2 outer packagings.

(2) Combination packagings consisting of metal inner packagings not exceeding 5 ℓ in capacity individually packed with absorbent material sufficient to absorb the contents and inert cushioning material in 1A2, 1B2, 1N2, 1H2, 1D, 1G, 4A, 4B, 4C1, 4C2, 4D, 4F, 4G or 4H2 outer packagings with a maximum gross mass of 75 kg. Inner packagings shall not be filled to more than 90% of their capacity. The closure of each inner packaging shall be physically held in place by any means capable of preventing back-off or loosening of the closure by impact or vibration during transport.

(3) Packagings consisting of:
Outer packagings: Steel or plastics drums, removable head (1A2 or 1H2), tested in accordance with the test provisions in 6.1.5 at a mass corresponding to the mass of the assembled package either as a packaging intended to contain inner packagings, or as a single packaging intended to contain solids or liquids, and marked accordingly.
Inner packagings: Drums and composite packagings (1A1, 1B1, 1N1, 1H1 or 6HA1), meeting the provisions of chapter 6.1 for single packagings, subject to the following conditions:

.1 the hydraulic pressure test shall be conducted at a pressure of at least 3 bar (gauge pressure);

.2 the design and production leakproofness tests shall be conducted at a test pressure of 0.30 bar;

.3 they shall be isolated from the outer drum by the use of inert shock-mitigating cushioning material which surrounds the inner packaging on all sides;

.4 their capacity shall not exceed 125 ℓ;

.5 closures shall be of a screw-cap type that are:

(i) physically held in place by any means capable of preventing back-off or loosening of the closure by impact or vibration during transport; and

(ii) provided with a cap seal.

.6 The outer and inner packagings shall be subjected periodically to a leakproofness test according to .2 at intervals of not more than two and a half years; and

.7 The outer and inner packagings shall bear in clearly legible and durable characters:

(i) the date (month, year) of the initial testing and the latest periodic test;

(ii) the name or authorized symbol of the party performing the tests and inspections.

(4) Pressure receptacles, provided that the general provisions of 4.1.3.6 are met. They shall be subjected to an initial test and periodic tests every 10 years at a pressure of not less than 1 MPa (10 bar) (gauge pressure). Pressure receptacles may not be equipped with any pressure relief device. Each pressure receptacle containing a toxic by inhalation liquid with an LC_{50} less than or equal to 200 mℓ/m^3 (ppm) shall be closed with a plug or valve conforming to the following:

(a) Each plug or valve shall have a taper-threaded connection directly to the pressure receptacle and be capable of withstanding the test pressure of the pressure receptacle without damage or leakage;

(b) Each valve shall be of the packless type with non-perforated diaphragm, except that, for corrosive materials, a valve may be of the packed type with an assembly made gas-tight by means of a seal cap with gasket joint attached to the valve body or the pressure receptacle to prevent loss of material through or past the packing;

(c) Each valve outlet shall be sealed by a threaded cap or threaded solid plug and inert gasket material;

(d) The materials of construction for the pressure receptacle, valves, plugs, outlet caps, luting and gaskets shall be compatible with each other and with the lading.

Each pressure receptacle with a wall thickness at any point of less than 2.0 mm and each pressure receptacle that does not have fitted valve protection shall be transported in an outer packaging. Pressure receptacles shall not be manifolded or interconnected.

P602	PACKING INSTRUCTION	P602

The following packagings are authorized, provided the general provisions of 4.1.1 and 4.1.3 are met and the packagings are hermetically sealed:

(1) Combination packagings with a maximum gross mass of 15 kg, consisting of:

– one or more glass inner packaging(s) with a maximum quantity of 1 litre each and filled to not more than 90% of their capacity, the closure(s) of which shall be physically held in place by any means capable of preventing back-off or loosening by impact or vibration during transport, individually placed in

– metal receptacles together with cushioning and absorbent material sufficient to absorb the entire contents of the glass inner packaging(s), further packed in

– 1A2, 1B2, 1N2, 1H2, 1D, 1G, 4A, 4B, 4C1, 4C2, 4D, 4F, 4G or 4H2 outer packagings.

(2) Combination packagings consisting of metal inner packagings individually packed with absorbent material sufficient to absorb the contents and inert cushioning material in 1A2, 1B2, 1N2, 1H2, 1D, 1G, 4A, 4B, 4C1, 4C2, 4D, 4F, 4G or 4H2 outer packagings with a maximum gross mass of 75 kg. Inner packagings shall not be filled to more than 90% of their capacity. The closure of each inner packaging shall be physically held in place by any means capable of preventing back-off or loosening of the closure by impact or vibration during transport. Inner packagings shall not exceed 5 ℓ in capacity.

(3) Drums and composite packagings (1A1, 1B1, 1N1, 1H1, 6HA1 or 6HH1), subject to the following conditions:

.1 the hydraulic pressure test shall be conducted at a pressure of at least 3 bar (gauge pressure);

.2 the design and production leakproofness tests shall be conducted at a test pressure of 0.30 bar; and

.3 closures shall be of a screw-cap type that are:

(i) physically held in place by any means capable of preventing back-off or loosening of the closure by impact or vibration during transport; and

(ii) provided with a cap seal.

(4) Pressure receptacles, provided that the general provisions of 4.1.3.6 are met. They shall be subjected to an initial test and periodic tests every 10 years at a pressure of not less than 1 MPa (10 bar) (gauge pressure). Pressure receptacles may not be equipped with any pressure relief device. Each pressure receptacle containing a toxic by inhalation liquid with an LC_{50} less than or equal to 200 mℓ/m^3 (ppm) shall be closed with a plug or valve conforming to the following:

(a) Each plug or valve shall have a taper-threaded connection directly to the pressure receptacle and be capable of withstanding the test pressure of the pressure receptacle without damage or leakage;

(b) Each valve shall be of the packless type with non-perforated diaphragm, except that, for corrosive materials, a valve may be of the packed type with an assembly made gas-tight by means of a seal cap with gasket joint attached to the valve body or the pressure receptacle to prevent loss of material through or past the packing;

(c) Each valve outlet shall be sealed by a threaded cap or threaded solid plug and inert gasket material;

(d) The materials of construction for the pressure receptacle, valves, plugs, outlet caps, luting and gaskets shall be compatible with each other and with the lading.

Each pressure receptacle with a wall thickness at any point of less than 2.0 mm and each pressure receptacle that does not have fitted valve protection shall be transported in an outer packaging. Pressure receptacles shall not be manifolded or interconnected.

P620	PACKING INSTRUCTION	P620

This instruction applies to UN 2814 and UN 2900.

The following packagings are authorized, provided the special packing provisions of 4.1.8 are met:

Packagings meeting the provisions of chapter 6.3 and approved accordingly consisting of:

.1 Inner packagings comprising:

(i) leakproof primary receptacle(s);

(ii) a leakproof secondary packaging;

(iii) other than for solid infectious substances, an absorbent material in sufficient quantity to absorb the entire contents placed between the primary receptacle(s) and the secondary packaging; if multiple primary receptacles are placed in a single secondary packaging, they shall be either individually wrapped or separated so as to prevent contact between them;

.2 A rigid outer packaging. The smallest external dimension shall be not less than 100 mm.

Additional provisions:

1 Inner packagings containing infectious substances shall not be consolidated with inner packagings containing unrelated types of goods. Complete packages may be overpacked in accordance with the provisions of 1.2.1 and 5.1.2: such an overpack may contain dry ice.

2 Other than for exceptional consignments, such as whole organs which require special packaging, the following additional provisions shall apply:

(a) *Substances consigned at ambient temperatures or at a higher temperature.* Primary receptacles shall be of glass, metal or plastics. Positive means of ensuring a leakproof seal shall be provided, e.g. a heat seal, a skirted stopper or a metal crimp seal. If screw caps are used, they shall be secured by positive means, e.g., tape, paraffin sealing tape or a manufactured locking closure;

(b) *Substances consigned refrigerated or frozen.* Ice, dry ice or other refrigerant shall be placed around the secondary packaging(s) or alternatively in an overpack with one or more complete packages marked in accordance with 6.3.3. Interior supports shall be provided to secure secondary packaging(s) or packages in position after the ice or dry ice has dissipated. If ice is used, the outer packaging or overpack shall be leakproof. If dry ice is used, the outer packaging or overpack shall permit the release of carbon dioxide gas. The primary receptacle and the secondary packaging shall maintain their integrity at the temperature of the refrigerant used;

(c) *Substances consigned in liquid nitrogen.* Plastics primary receptacles capable of withstanding very low temperature shall be used. The secondary packaging shall also be capable of withstanding very low temperatures, and in most cases will need to be fitted over the primary receptacle individually. Provisions for the consignment of liquid nitrogen shall also be fulfilled. The primary receptacle and the secondary packaging shall maintain their integrity at the temperature of the liquid nitrogen.

(d) Lyophilized substances may also be transported in primary receptacles that are flame-sealed glass ampoules or rubber-stoppered glass vials fitted with metal seals.

3 Whatever the intended temperature of the consignment, the primary receptacle or the secondary packaging shall be capable of withstanding, without leakage, an internal pressure producing a pressure differential of not less than 95 kPa and temperatures in the range –40°C to +55°C.

4 Alternative packagings for the transport of animal material may be authorized by the competent authority in accordance with the provisions of 4.1.3.7.

P621	PACKING INSTRUCTION	P621

This instruction applies to UN 3291.

The following packagings are authorized, provided that the general provisions of 4.1.1 and 4.1.3 are met:

(1) Rigid, leakproof packagings meeting the provisions of chapter 6.1 for solids, at the packing group II performance level, provided there is sufficient absorbent material to absorb the entire amount of liquid present and the packaging is capable of retaining liquids.

(2) For packages containing larger quantities of liquid, rigid packagings meeting the provisions of chapter 6.1 at the packing group II performance level for liquids.

Additional provision:

Packagings intended to contain sharp objects such as broken glass and needles shall be resistant to puncture and retain liquids under the performance test conditions in chapter 6.1.

P650	PACKING INSTRUCTION	P650

This instruction applies to UN 3373.

(1) The packaging shall be of good quality, strong enough to withstand the shocks and loadings normally encountered during transport, including transhipment between cargo transport units and between cargo transport units and warehouses as well as any removal from a pallet or overpack for subsequent manual or mechanical handling. Packagings shall be constructed and closed to prevent any loss of contents that might be caused under normal conditions of transport by vibration or by changes in temperature, humidity or pressure.

(2) The packaging shall consist of at least three components:

(a) a primary receptacle;

(b) a secondary packaging; and

(c) an outer packaging.

of which either the secondary or the outer packaging shall be rigid.

(3) Primary receptacles shall be packed in secondary packagings in such a way that, under normal conditions of transport, they cannot break, be punctured or leak their contents into the secondary packaging. Secondary packagings shall be secured in outer packagings with suitable cushioning material. Any leakage of the contents shall not compromise the integrity of the cushioning material or of the outer packaging.

(4) For transport, the mark illustrated below shall be displayed on the external surface of the outer packaging on a background of a contrasting colour and shall be clearly visible and legible. The mark shall be in the form of a square set at an angle of 45° (diamond-shaped) with each side having a length of at least 50 mm, the width of the line shall be at least 2 mm and the letters and numbers shall be at least 6 mm high. The proper shipping name "BIOLOGICAL SUBSTANCE, CATEGORY B" in letters at least 6 mm high shall be marked on the outer packaging adjacent to the diamond-shaped mark.

(5) At least one surface of the outer packaging shall have a minimum dimension of 100 mm × 100 mm.

(6) The completed package shall be capable of successfully passing the drop test in 6.3.5.3 as specified in 6.3.5.2 of this Code at a height of 1.2 m. Following the appropriate drop sequence, there shall be no leakage from the primary receptacle(s) which shall remain protected by absorbent material, when required, in the secondary packaging.

(7) For liquid substances

(a) The primary receptacle(s) shall be leakproof;

(b) The secondary packaging shall be leakproof;

(c) If multiple fragile primary receptacles are placed in a single secondary packaging, they shall either be individually wrapped or separated to prevent contact between them;

(d) Absorbent material shall be placed between the primary receptacle(s) and the secondary packaging. The absorbent material shall be in a quantity sufficient to absorb the entire contents of the primary receptacle(s) so that any release of the liquid substance will not compromise the integrity of the cushioning material or of the outer packaging.

(e) The primary receptacle or the secondary packaging shall be capable of withstanding, without leakage, an internal pressure of 95 kPa (0.95 bar).

(8) For solid substances

(a) The primary receptacle(s) shall be siftproof;

(b) The secondary packaging shall be siftproof;

(c) If multiple fragile primary receptacles are placed in a single secondary packaging, they shall either be individually wrapped or separated to prevent contact between them.

(d) If there is any doubt as to whether or not residual liquid may be present in the primary receptacle during transport then a packaging suitable for liquids, including absorbent materials, shall be used.

P650	PACKING INSTRUCTION *(continued)*	P650

(9) Refrigerated or frozen specimens: Ice, dry ice and liquid nitrogen

    (a) When dry ice or liquid nitrogen is used to keep specimens cold, all applicable provisions of the Code shall be met. When used, ice or dry ice shall be placed outside the secondary packagings or in the outer packaging or an overpack. Interior supports shall be provided to secure the secondary packagings in the original position after the ice or dry ice has dissipated. If ice is used, the outside packaging or overpack shall be leakproof. If carbon dioxide, solid (dry ice) is used, the packaging shall be designed and constructed to permit the release of carbon dioxide gas to prevent a build-up of pressure that could rupture the packagings and the package (the outer packaging or the overpack) shall be marked ''Carbon dioxide, solid'' or ''Dry ice''.

    (b) The primary receptacle and the secondary packaging shall maintain their integrity at the temperature of the refrigerant used as well as the temperatures and the pressures which could result if refrigeration were lost.

(10) When packages are placed in an overpack, the package markings required by this packing instruction shall either be clearly visible or be reproduced on the outside of the overpack.

(11) Infectious substances assigned to UN 3373 which are packed and marked in accordance with this packing instruction are not subject to any other provisions of this Code.

(12) Clear instructions on filling and closing such packages shall be provided by packaging manufacturers and subsequent distributors to the consignor or to the person who prepares the package (e.g., patient) to enable the package to be correctly prepared for transport.

(13) Other dangerous goods shall not be packed in the same packaging as class 6.2 infectious substances unless they are necessary for maintaining the viability, stabilizing or preventing degradation or neutralizing the hazards of the infectious substances. A quantity of 30 ml or less of dangerous goods included in classes 3, 8 or 9 may be packed in each primary receptacle containing infectious substances. When these small quantities of dangerous goods are packed with infectious substances in accordance with this packing instruction, no other provisions of the Code need be met.

Additional provision:

Alternative packagings for the transport of animal material may be authorized by the competent authority in accordance with the provisions of 4.1.3.7.

4

P800	PACKING INSTRUCTION	P800

This instruction applies to UN 2803 and UN 2809.

The following packagings are authorized, provided the general provisions of 4.1.1 and 4.1.3 are met:

(1) Pressure receptacles, provided that the general provisions of 4.1.3.6 are met.

(2) Steel flasks or bottles with threaded closures with a capacity not exceeding 3.0 ℓ; or

(3) Combination packagings which conform to the following provisions:

    (a) Inner packagings shall comprise glass, metal or rigid plastics intended to contain liquids with a maximum net mass of 15 kg each.

    (b) The inner packagings shall be packed with sufficient cushioning material to prevent breakage.

    (c) Either the inner packagings or the outer packagings shall have inner liners or bags of strong leakproof and puncture-resistant material impervious to the contents and completely surrounding the contents to prevent it from escaping from the package irrespective of its position or orientation.

    (d) The following outer packagings and maximum net masses are authorized:

Outer packaging	Maximum net mass
Drums	
steel (1A2)	400 kg
other metal (1N2)	400 kg
plastics (1H2)	400 kg
plywood (1D)	400 kg
fibre (1G)	400 kg
Boxes	
steel (4A)	400 kg
natural wood (4C1)	250 kg
natural wood with sift-proof walls (4C2)	250 kg
plywood (4D)	250 kg
reconstituted wood (4F)	125 kg
fibreboard (4G)	125 kg
expanded plastics (4H1)	60 kg
solid plastics (4H2)	125 kg

Special packing provision:

PP41   For UN 2803, when it is necessary to transport gallium at low temperatures in order to maintain it in a completely solid state, the above packagings may be overpacked in a strong, water-resistant outer packaging which contains dry ice or other means of refrigeration. If a refrigerant is used, all of the above materials used in the packaging of gallium shall be chemically and physically resistant to the refrigerant and shall have impact resistance at the low temperatures of the refrigerant employed. If dry ice is used, the outer packaging shall permit the release of carbon dioxide gas.

P801	PACKING INSTRUCTION	P801

This instruction applies to new and used batteries assigned to UN Nos. 2794, 2795 or 3028.

The following packagings are authorized, provided the general provisions of 4.1.1, except 4.1.1.3, and 4.1.3 are met, except that packagings need not conform to the provisions of part 6:

(1) Rigid outer packagings;

(2) Wooden slatted crates;

(3) Pallets.

Used storage batteries may also be transported loose in stainless steel or plastics battery boxes capable of containing any free liquid.

Additional provisions:

1   Batteries shall be protected against short circuits.

2   Batteries stacked shall be adequately secured in tiers separated by a layer of non-conductive material.

3   Battery terminals shall not support the mass of other superimposed elements.

4   Batteries shall be packaged or secured to prevent inadvertent movement.

5   For UN 2794 and UN 2795, batteries shall be capable of passing a tilt test at an angle of 45° with no spillage of liquid.

P802	PACKING INSTRUCTION	P802

The following packagings are authorized, provided the general provisions of 4.1.1 and 4.1.3 are met:

(1) Combination packagings
     Outer packagings: 1A2, 1B2, 1N2, 1H2, 1D, 4A, 4B, 4C1, 4C2, 4D, 4F, or 4H2; maximum net mass: 75 kg.
     Inner packagings: glass or plastics; maximum capacity: 10 ℓ.

(2) Combination packagings
     Outer packagings: 1A2, 1B2, 1N2, 1H2, 1D, 1G, 4A, 4B, 4C1, 4C2, 4D, 4F, 4G or 4H2;
     maximum net mass: 125 kg.
     Inner packagings: metal; maximum capacity: 40 ℓ

(3) Composite packagings: Glass receptacle in steel, aluminium, plywood or solid plastics drum (6PA1, 6PB1, 6PD1 or 6PH2) or in a steel, aluminium, wood or plywood box (6PA2, 6PB2, 6PC or 6PD2); maximum capacity: 60 ℓ.

(4) Steel drums (1A1) with a maximum capacity of 250 ℓ.

(5) Pressure receptacles may be used provided that the general provisions of 4.1.3.6 are met.

Special packing provisions:

PP79   For UN 1790 with more than 60% but not more than 85% hydrofluoric acid, see P001.

PP81   For UN 1790 with not more than 85% hydrogen fluoride and UN 2031 with more than 55% nitric acid, the permitted use of plastics drums and jerricans as single packagings shall be two years from their date of manufacture.

P803	PACKING INSTRUCTION	P803

This instruction applies to UN 2028.

The following packagings are authorized, provided the general provisions of 4.1.1 and 4.1.3 are met:

(1) Drums (1A2, 1B2, 1N2, 1H2, 1D, 1G);

(2) Boxes (4A, 4B, 4C1, 4C2, 4D, 4F, 4G, 4H2);

Maximum net mass: 75 kg.

The articles shall be individually packaged and separated from each other, using partitions, dividers, inner packagings or cushioning material to prevent inadvertent discharge during normal conditions of transport.

P804	PACKING INSTRUCTION	P804

This instruction applies to UN 1744.

The following packagings are authorized provided the general provisions of 4.1.1 and 4.1.3 are met and the packagings are hermetically sealed:

(1) Combination packagings with a maximum gross mass of 25 kg, consisting of:

– one or more glass inner packaging(s) with a maximum capacity of 1.3 litres each and filled to not more than 90% of their capacity, the closure(s) of which shall be physically held in place by any means capable of preventing back-off or loosening by impact or vibration during transport, together with cushioning and absorbent material sufficient to absorb the entire contents of the glass inner packaging(s), further packed in:

– 1A2, 1B2, 1N2, 1H2, 1D, 1G, 4A, 4B, 4C1, 4C2, 4D, 4F, 4G or 4H2 outer packagings.

(2) Combination packagings consisting of metal or polyvinylidene fluoride (PVDF) inner packagings, not exceeding 5 litres in capacity individually packed with absorbent material sufficient to absorb the contents and inert cushioning material in 1A2, 1B2, 1N2, 1H2, 1D, 1G, 4A, 4B, 4C1, 4C2, 4D, 4F, 4G or 4H2 outer packagings with a maximum gross mass of 75 kg. Inner packagings shall not be filled to more than 90% of their capacity. The closure of each inner packaging shall be physically held in place by any means capable of preventing back-off or loosening of the closure by impact or vibration during transport.

(3) Packagings consisting of:

Outer packagings:
Steel or plastic drums, removable head (1A2 or 1H2) tested in accordance with the test requirements in 6.1.5 at a mass corresponding to the mass of the assembled package either as a packaging intended to contain inner packagings, or as a single packaging intended to contain solids or liquids, and marked accordingly;

Inner packagings:
Drums and composite packagings (1A1, 1B1, 1N1, 1H1 or 6HA1) meeting the requirements of chapter 6.1 for single packagings, subject to the following conditions:

(a) The hydraulic pressure test shall be conducted at a pressure of at least 300 kPa (3 bar) (gauge pressure);

(b) The design and production leakproofness tests shall be conducted at a test pressure of 30 kPa (0.3 bar);

(c) They shall be isolated from the outer drum by the use of inert shock-mitigating cushioning material which surrounds the inner packaging on all sides;

(d) Their capacity shall not exceed 125 litres;

(e) Closures shall be of a screw type that are:

(i) Physically held in place by any means capable of preventing back-off or loosening of the closure by impact or vibration during transport;

(ii) Provided with a cap seal;

(f) The outer and inner packagings shall be subjected periodically to an internal inspection and leakproofness test according to (b) at intervals of not more than two and a half years; and

(g) The outer and inner packagings shall bear in clearly legible and durable characters:

(i) the date (month, year) of the initial test and the latest periodic test and inspection of the inner packaging; and

(ii) the name or authorized symbol of the expert performing the tests and inspections.

(4) Pressure receptacles, provided that the general provisions of 4.1.3.6 are met.

(a) They shall be subjected to an initial test and periodic tests every 10 years at a pressure of not less than 1 MPa (10 bar) (gauge pressure);

(b) They shall be subjected periodically to an internal inspection and leakproofness test at intervals of not more than two and a half years;

(c) They may not be equipped with any pressure relief device;

(d) Each pressure receptacle shall be closed with a plug or valve(s) fitted with a secondary closure device; and

(e) The materials of construction for the pressure receptacle, valves, plugs, outlet caps, luting and gaskets shall be compatible with each other and with the contents.

4

P900	PACKING INSTRUCTION	P900

This instruction applies to UN 2216.

The following packagings are authorized, provided the general provisions of 4.1.1 and 4.1.3 are met:

(1) Packagings according to P002; or

(2) Bags (5H1, 5H2, 5H3, 5H4, 5L1, 5L2, 5L3, 5M1 or 5M2) with a maximum net mass of 50 kg.

Fish meal may also be transported unpackaged when it is packed in closed cargo transport units and the free air space has been restricted to a minimum.

P901	PACKING INSTRUCTION	P901

This instruction applies to UN 3316.

The following packagings are authorized, provided the general provisions of 4.1.1 and 4.1.3 are met:
Packagings conforming to the performance level consistent with the packing group assigned to the kit as a whole (see 3.3.1, special provision 251).

Maximum quantity of dangerous goods per outer packaging: 10 kg.

Additional provision:

Dangerous goods in kits shall be packed in inner packagings which shall not exceed either 250 mℓ or 250 g and shall be protected from other materials in the kit.

P902	PACKING INSTRUCTION	P902

This instruction applies to UN 3268.

The following packagings are authorized, provided the general provisions of 4.1.1 and 4.1.3 are met:
Packagings conforming to the packing group III performance level. The packagings shall be designed and constructed to prevent movement of the articles and inadvertent operation during normal conditions of transport.

The articles may also be transported unpackaged in dedicated handling devices, vehicles, containers or wagons when moved from where they are manufactured to an assembly plant.

Additional provision:

Any pressure vessel shall be in accordance with the requirements of the competent authority for the substance(s) contained in the pressure vessel(s).

P903	PACKING INSTRUCTION	P903

This instruction applies to UN Nos. 3090, 3091, 3480 and 3481.

The following packagings are authorized, provided the general provisions of 4.1.1 and 4.1.3 are met:
Packagings conforming to the packing group II performance level.

In addition, batteries with a strong, impact-resistant outer casing of a gross mass of 12 kg or more, and assemblies of such batteries, may be packed in strong outer packagings, in protective enclosures (e.g., in fully enclosed or wooden slatted crates) unpackaged or on pallets. Batteries shall be secured to prevent inadvertent movement, and the terminals shall not support the weight of other superimposed elements.

When cells and batteries are packed with equipment, they shall be packed in inner fibreboard packagings that meet the provisions for packing group II. When cells and batteries included in class 9 are contained in equipment, the equipment shall be packed in strong outer packagings in such a manner as to prevent accidental operation during transport.

Additional provision:

Batteries shall be protected against short circuit.

P904	PACKING INSTRUCTION	P904

This instruction applies to UN 3245.

The following packagings are authorized, provided the general provisions of 4.1.1 and 4.1.3 are met:

(1) Packagings according to P001 or P002 conforming to the packing group III performance level.

(2) Outer packagings, which need not conform to the packaging test provisions of part 6, but conforming to the following:

    (a) An inner packaging comprising:

        (i) a watertight primary receptacle(s);

        (ii) a watertight secondary packaging which is leakproof;

        (iii) absorbent material placed between the primary receptacle(s) and the secondary packaging. The absorbent material shall be in a quantity sufficient to absorb the entire contents of the primary receptacle(s) so that any release of the liquid substance will not compromise the integrity of the cushioning material or of the outer packaging;

        (iv) if multiple fragile primary receptacles are placed in a single secondary packaging they shall be individually wrapped or separated to prevent contact between them.

    (b) An outer packaging shall be strong enough for its capacity, mass and intended use and with a smallest external dimension of at least 100 mm.

Additional provision:

Dry ice and liquid nitrogen

When carbon dioxide, solid (dry ice) is used as a refrigerant, the packaging shall be designed and constructed to permit the release of the gaseous carbon dioxide to prevent the build-up of pressure that could rupture the packaging.

Substances consigned in liquid nitrogen or dry ice shall be packed in primary receptacles that are capable of withstanding very low temperatures. The secondary packaging shall also be capable of withstanding very low temperatures and, in most cases, will need to be fitted over the primary receptacle individually.

4

P905	PACKING INSTRUCTION	P905

This instruction applies to UN 2990 and UN 3072.

Any suitable packaging is authorized, provided the general provisions of 4.1.1 and 4.1.3 are met, except that packagings need not conform to the provisions of part 6.

When the life-saving appliances are constructed to incorporate or are contained in rigid outer weatherproof casings (such as for lifeboats), they may be transported unpackaged.

Additional provisions:

1   All dangerous substances and articles contained as equipment within the appliances shall be secured to prevent inadvertent movement and in addition:

    (a) signal devices of class 1 shall be packed in plastics or fibreboard inner packagings;

    (b) gases (class 2.2) shall be contained in cylinders as specified by the competent authority, which may be connected to the appliance;

    (c) electric storage batteries (class 8) and lithium batteries (class 9) shall be disconnected or electrically isolated and secured to prevent any spillage of liquid; and

    (d) small quantities of other dangerous substances (for example in classes 3, 4.1 and 5.2) shall be packed in strong inner packagings.

2   Preparation for transport and packaging shall include provisions to prevent any accidental inflation of the appliance.

P906	PACKING INSTRUCTION	P906

This instruction applies to UN Nos. 2315, 3151, 3152 and 3432.

The following packagings are authorized, provided the general provisions of 4.1.1 and 4.1.3 are met:

(1) For liquids and solids containing or contaminated with PCBs or polyhalogenated biphenyls or terphenyls: Packagings in accordance with P001 or P002, as appropriate.

(2) For transformers and condensers and other devices: Leakproof containment system which is capable of containing, in addition to the devices, at least 1.25 times the volume of the liquid PCBs, polyhalogenated biphenyls or terphenyls present in them. There shall be sufficient absorbent material in the packagings to absorb at least 1.1 times the volume of liquid which is contained in the devices. In general, transformers and condensers shall be transported in leakproof metal packagings which are capable of holding, in addition to the transformers and condensers, at least 1.25 times the volume of the liquid present in them.

Notwithstanding the above, liquids and solids not packaged in accordance with P001 and P002 and unpackaged transformers and condensers may be transported in cargo transport units fitted with a leakproof metal tray to a height of at least 800 mm, containing sufficient inert absorbent material to absorb at least 1.1 times the volume of any free liquid.

Additional provision:
Adequate provisions shall be taken to seal the transformers and condensers to prevent leakage during normal conditions of transport.

P907	PACKING INSTRUCTION	P907

If the machinery or apparatus is constructed and designed so that the receptacles containing the dangerous goods are afforded adequate protection, an outer packaging is not required. Dangerous goods in machinery or apparatus shall otherwise be packed in outer packagings constructed of suitable material of adequate strength and design in relation to the packaging capacity and its intended use, and meeting the applicable requirements of 4.1.1.1.

Receptacles containing dangerous goods shall conform to the general provisions in 4.1.1, except that 4.1.1.3, 4.1.1.4, 4.1.1.12 and 4.1.1.14 do not apply. For class 2.2 gases, the inner cylinder or receptacle, its contents and filling density shall be to the satisfaction of the competent authority of the country in which the cylinder or receptacle is filled.

In addition, the manner in which receptacles are contained within the machinery or apparatus shall be such that, under normal conditions of transport, damage to receptacles containing the dangerous goods is unlikely; and in the event of damage to the receptacles containing solid or liquid dangerous goods, no leakage of the dangerous goods from the machinery or apparatus is possible (a leakproof liner may be used to satisfy this requirement). Receptacles containing dangerous goods shall be so installed, secured or cushioned as to prevent their breakage or leakage and so as to control their movement within the machinery or apparatus during normal conditions of transport. Cushioning material shall not react dangerously with the content of the receptacles. Any leakage of the contents shall not substantially impair the protective properties of the cushioning material.

4.1.4.2   **Packing instructions concerning the use of IBCs**

IBC01	PACKING INSTRUCTION	IBC01

The following IBCs are authorized, provided the general provisions of 4.1.1, 4.1.2 and 4.1.3 are met:

Metal (31A, 31B and 31N).

IBC02	PACKING INSTRUCTION	IBC02

The following IBCs are authorized, provided the general provisions of 4.1.1, 4.1.2 and 4.1.3 are met:

(1) Metal (31A, 31B and 31N);

(2) Rigid plastics (31H1 and 31H2);

(3) Composite (31HZ1).

Special packing provisions:
B5 For UN Nos. 1791, 2014, 2984 and 3149, IBCs shall be provided with a device to allow venting during transport. The inlet to the venting device shall be sited in the vapour space of the IBC under maximum filling conditions during transport.

B8 The pure form of this substance shall not be transported in IBCs since it is known to have a vapour pressure of more than 110 kPa at 50°C or 130 kPa at 55°C.

B15 For UN 2031 with more than 55% nitric acid, the permitted use of rigid plastics IBCs and of composite IBCs with a rigid plastics inner receptacle shall be two years from their date of manufacture.

B20 For UN Nos. 1716, 1717, 1736, 1737, 1738, 1742, 1743, 1755, 1764, 1768, 1776, 1778, 1782, 1789, 1790, 1796, 1826, 1830, 1832, 2031, 2308, 2353, 2513, 2584, 2796 and 2817 coming under PG II, IBCs shall be fitted with two shut-off devices.

IBC03	PACKING INSTRUCTION	IBC03

The following IBCs are authorized, provided the general provisions of 4.1.1, 4.1.2 and 4.1.3 are met:

(1) Metal (31A, 31B and 31N);

(2) Rigid plastics (31H1 and 31H2);

(3) Composite (31HZ1 and 31HA2, 31HB2, 31HN2, 31HD2 and 31HH2).

Special packing provisions:

B8   The pure form of this substance shall not be transported in IBCs since it is known to have a vapour pressure of more than 110 kPa at 50°C or 130 kPa at 55°C.

B11 Notwithstanding the provisions of 4.1.1.10, UN 2672 ammonia solution in concentrations not exceeding 25% may be transported in rigid or composite plastics IBCs (31H1, 31H2 and 31HZ1).

IBC04	PACKING INSTRUCTION	IBC04

The following IBCs are authorized, provided the general provisions of 4.1.1, 4.1.2 and 4.1.3 are met:

Metal (11A, 11B, 11N, 21A, 21B, 21N, 31A, 31B and 31N).

Special packing provision:

B1   For packing group I substances, IBCs shall be carried in closed cargo transport units or in freight containers/vehicles, which shall have rigid sides or fences at least to the height of the IBC.

IBC05	PACKING INSTRUCTION	IBC05

The following IBCs are authorized, provided the general provisions of 4.1.1, 4.1.2 and 4.1.3 are met:

(1) Metal (11A, 11B, 11N, 21A, 21B, 21N, 31A, 31B and 31N);

(2) Rigid plastics (11H1, 11H2, 21H1, 21H2, 31H1 and 31H2);

(3) Composite (11HZ1, 21HZ1 and 31HZ1).

Special packing provisions:

B1   For packing group I substances, IBCs shall be carried in closed cargo transport units or in freight containers/vehicles, which shall have rigid sides or fences at least to the height of the IBC.

B2   For solid substances in IBCs other than metal or rigid plastics IBCs, the IBCs shall be carried in closed cargo transport units or in freight containers/vehicles, which shall have rigid sides or fences at least to the height of the IBC.

IBC06	PACKING INSTRUCTION	IBC06

The following IBCs are authorized, provided the general provisions of 4.1.1, 4.1.2 and 4.1.3 are met:

(1) Metal (11A, 11B, 11N, 21A, 21B, 21N, 31A, 31B and 31N);

(2) Rigid plastics (11H1, 11H2, 21H1, 21H2, 31H1 and 31H2);

(3) Composite (11HZ1, 11HZ2, 21HZ1, 21HZ2, 31HZ1, and 31HZ2).

Additional provision:

Composite IBCs 11HZ2 and 21HZ2 shall not be used when the substances being transported may become liquid during transport.

Special packing provisions:

B1   For packing group I substances, IBCs shall be carried in closed cargo transport units or in freight containers/vehicles, which shall have rigid sides or fences at least to the height of the IBC.

B2   For solid substances in IBCs other than metal or rigid plastics IBCs, the IBCs shall be carried in closed cargo transport units or in freight containers/vehicles, which shall have rigid sides or fences at least to the height of the IBC.

B12 For UN 2907, IBCs shall meet the packing group II performance level. IBCs meeting the test criteria of packing group I shall not be used.

4

IBC07	PACKING INSTRUCTION	IBC07

The following IBCs are authorized, provided the general provisions of 4.1.1, 4.1.2 and 4.1.3 are met:
(1) Metal (11A, 11B, 11N, 21A, 21B, 21N, 31A, 31B and 31N);
(2) Rigid plastics (11H1, 11H2, 21H1, 21H2, 31H1 and 31H2);
(3) Composite (11HZ1, 11HZ2, 21HZ1, 21HZ2, 31HZ1 and 31HZ2);
(4) Wooden (11C, 11D and 11F).

Additional provision:
Liners of wooden IBCs shall be sift-proof.

Special packing provisions:
B1 For packing group I substances, IBCs shall be carried in closed cargo transport units or in freight containers/vehicles, which shall have rigid sides or fences at least to the height of the IBC.
B2 For solid substances in IBCs other than metal or rigid plastics IBCs, the IBCs shall be carried in closed cargo transport units or in freight containers/vehicles, which shall have rigid sides or fences at least to the height of the IBC.
B4 Flexible, fibreboard or wooden IBCs shall be sift-proof and water-resistant or shall be fitted with a sift-proof and water-resistant liner.

IBC08	PACKING INSTRUCTION	IBC08

The following IBCs are authorized, provided the general provisions of 4.1.1, 4.1.2 and 4.1.3 are met:
(1) Metal (11A, 11B, 11N, 21A, 21B, 21N, 31A, 31B and 31N);
(2) Rigid plastics (11H1, 11H2, 21H1, 21H2, 31H1 and 31H2);
(3) Composite (11HZ1, 11HZ2, 21HZ1, 21HZ2, 31HZ1 and 31HZ2);
(4) Fibreboard (11G);
(5) Wooden (11C, 11D and 11F);
(6) Flexible (13H1, 13H2, 13H3, 13H4, 13H5, 13L1, 13L2, 13L3, 13L4, 13M1 or 13M2).

Special packing provisions:
B2 For substances, UN 1374 and UN 2590 in IBCs other than metal or rigid plastics IBCs, the IBCs shall be carried in closed cargo transport units or in freight containers/vehicles, which shall have rigid sides or fences at least to the height of the IBC.
B3 Flexible IBCs shall be sift-proof and water-resistant or shall be fitted with a sift-proof and water-resistant liner.
B4 Flexible, fibreboard or wooden IBCs shall be sift-proof and water-resistant or shall be fitted with a sift-proof and water-resistant liner.
B6 For UN Nos. 1327, 1363, 1364, 1365, 1386, 1408, 1841, 2211, 2217, 2793 and 3314, IBCs are not required to meet the IBC testing provisions of chapter 6.5.

IBC99	PACKING INSTRUCTION	IBC99

Only IBCs which are approved for these goods by the competent authority may be used (see 4.1.3.7). A copy of the competent authority approval shall accompany each consignment or the transport document shall include an indication that the packaging was approved by the competent authority.

IBC100	PACKING INSTRUCTION	IBC100

This instruction applies to UN Nos. 0082, 0241, 0331 and 0332.

The following IBCs are authorized, provided the general provisions of 4.1.1, 4.1.2 and 4.1.3 and special provisions of 4.1.5 are met:

(1) Metal (11A, 11B, 11N, 21A, 21B, 21N, 31A, 31B and 31N);
(2) Flexible (13H2, 13H3, 13H4, 13L2, 13L3, 13L4 and 13M2);
(3) Rigid plastics (11H1, 11H2, 21H1, 21H2, 31H1 and 31H2);
(4) Composite (11HZ1, 11HZ2, 21HZ1, 21HZ2, 31HZ1 and 31HZ2).

Additional provisions:
1   IBCs shall only be used for free-flowing substances.
2   Flexible IBCs shall only be used for solids.

Special packing provisions:
B9   For UN 0082, this packing instruction may only be used when the substances are mixtures of ammonium nitrate or other inorganic nitrates with other combustible substances which are not explosive ingredients. Such explosives shall not contain nitroglycerin, similar liquid organic nitrates, or chlorates. Metal IBCs are not authorized.
B10 For UN 0241, this packing instruction may only be used for substances which consist of water as an essential ingredient and high proportions of ammonium nitrate or other oxidizing substances, some or all of which are in solution. The other constituents may include hydrocarbons or aluminium powder, but shall not include nitro-derivatives such as trinitrotoluene. Metal IBCs are not authorized.

IBC520	PACKING INSTRUCTION	IBC520
This instruction applies to organic peroxides and self-reactive substances of type F.		
The IBCs listed below are authorized for the formulations listed, provided the general provisions of 4.1.1, 4.1.2 and 4.1.3 and special provisions of 4.1.7.2 are met. For formulations not listed below, only IBCs which are approved by the competent authority may be used (see 4.1.7.2.2).		

UN No.	Organic peroxide	Type of IBC	Maximum quantity (litres)	Control temperature	Emergency temperature
3109	**ORGANIC PEROXIDE TYPE F, LIQUID** *tert*-Butyl hydroperoxide, not more than 72% with water	31A	1250		
	tert-Butyl peroxyacetate, not more than 32% in diluent type A	31HA1	1000		
	tert-Butyl peroxybenzoate, not more than 32% in diluent type A	31A	1250		
	tert-Butyl peroxy-3,5,5-trimethylhexanoate, not more than 37% in diluent type A	31A 31HA1	1250 1000		
	Cumyl hydroperoxide, not more than 90% in diluent type A	31HA1	1250		
	Dibenzoyl peroxide, not more than 42% as a stable dispersion	31H1	1000		
	Di-*tert*-butyl peroxide, not more than 52% in diluent type A	31A 31HA1	1250 1000		
	1,1-Di-(*tert*-butylperoxy)cyclohexane, not more than 37% in diluent type A	31A	1250		
	1,1-Di-(*tert*-butylperoxy)cyclohexane, not more than 42% in diluent type A	31H1	1000		
	Dilauroyl peroxide, not more than 42%, stable dispersion, in water	31HA1	1000		
	Isopropyl cumyl hydroperoxide, not more than 72% in diluent type A	31HA1	1250		
	p-Menthyl hydroperoxide, not more than 72% in diluent type A	31HA1	1250		
	Peroxyacetic acid, stabilized, not more than 17%	31H1 31HA1 31A	1500		
3110	**ORGANIC PEROXIDE TYPE F, SOLID** Dicumyl peroxide	31A 31H1 31HA1	2000		
3119	**ORGANIC PEROXIDE TYPE F, LIQUID, TEMPERATURE CONTROLLED** *tert*-Amyl peroxypivalate, not more than 32% in diluent type A	31A	1250	+10°C	+15°C
	tert-Butyl peroxy-2-ethylhexanoate, not more than 32% in diluent type B	31HA1 31A	1000 1250	+30°C +30°C	+35°C +35°C
	tert-Butyl peroxyneodecanoate, not more than 32% in diluent type A	31A	1250	0°C	+10°C
	tert-Butyl peroxyneodecanoate, not more than 42%, stable dispersion, in water	31A	1250	–5°C	+5°C
	tert-Butyl peroxyneodecanoate, not more than 52%, stable dispersion, in water	31A	1250	–5°C	+5°C
	tert-Butyl peroxypivalate, not more than 27% in diluent type B	31HA1 31A	1000 1250	+10°C +10°C	+15°C +15°C
	Di-(2-neodecanoylperoxyisopropyl)benzene, not more than 42%, stable dispersion, in water	31A	1250	–15°C	–5°C
	3-Hydroxy-1,1-dimethylbutyl peroxyneodecanoate, not more than 52%, stable dispersion, in water	31A	1250	–15°C	–5°C
	Cumyl peroxyneodecanoate, not more than 52%, stable dispersion, in water	31A	1250	–15°C	–5°C
	Di-(4-*tert*-butylcyclohexyl) peroxydicarbonate, not more than 42%, stable dispersion, in water	31HA1	1000	+30°C	+35°C
	Dicetyl peroxydicarbonate, not more than 42%, stable dispersion, in water	31HA1	1000	+30°C	+35°C
	Dicyclohexyl peroxydicarbonate, not more than 42% as a stable dispersion, in water	31A	1250	+10°C	+15°C

4

IBC520	PACKING INSTRUCTION *(continued)*				IBC520
UN No.	Organic peroxide	Type of IBC	Maximum quantity (litres)	Control temper-ature	Emergency temper-ature
3119 *(cont)*	**ORGANIC PEROXIDE TYPE F, LIQUID, TEMPERATURE CONTROLLED** Di-(2-ethylhexyl) peroxydicarbonate, not more than 62%, stable dispersion, in water	31A	1250	−20°C	−10°C
	Dimyristyl peroxydicarbonate, not more than 42%, stable dispersion, in water	31HA1	1000	+15°C	+20°C
	Di-(3,5,5-trimethylhexanoyl) peroxide, not more than 38% in diluent type A	31HA1 31A	1000 1250	+10°C +10°C	+15°C +15°C
	Di-(3,5,5-trimethylhexanoyl) peroxide, not more than 52%, stable dispersion, in water	31A	1250	+10°C	+15°C
	1,1,3,3-Tetramethylbutyl peroxyneodecanoate, not more than 52%, stable dispersion, in water	31A	1250	−5°C	+5°C
3120	**ORGANIC PEROXIDE, TYPE F, SOLID, TEMPERATURE CONTROLLED**				

Additional provisions:

1   IBCs shall be provided with a device to allow venting during transport. The inlet to the pressure relief device shall be sited in the vapour space of the IBC under maximum filling conditions during transport.

2   To prevent explosive rupture of metal IBCs or composite IBCs with complete metal casing, the emergency relief devices shall be designed to vent all the decomposition products and vapours evolved during self-accelerating decomposition or during a period of not less than one hour of fire-engulfment as calculated by the formula in 4.2.1.13.8. The control and emergency temperatures specified in this packing instruction are based on a non-insulated IBC. When consigning an organic peroxide in an IBC in accordance with this instruction, it is the responsibility of the consignor to ensure that:

(a)   the pressure and emergency relief devices installed on the IBC are designed to take appropriate account of the self-accelerating decomposition of the organic peroxide and of fire engulfment; and

(b)   when applicable, the control and emergency temperatures indicated are appropriate, taking into account the design (such as insulation) of the IBC to be used.

IBC620	PACKING INSTRUCTION	IBC620
This instruction applies to UN 3291.		
The following IBCs are authorized, provided that the general provisions of 4.1.1, 4.1.2 and 4.1.3 are met: Rigid, leakproof IBCs conforming to the packing group II performance level.		

Additional provisions:

1   There shall be sufficient absorbent material to absorb the entire amount of liquid present in the IBC.

2   IBCs shall be capable of retaining liquids.

3   IBCs intended to contain sharp objects such as broken glass and needles shall be resistant to puncture.

4.1.4.3   Packing instructions concerning the use of large packagings

LP01	PACKING INSTRUCTION (LIQUIDS)			LP01
The following large packagings are authorized, provided the general provisions of 4.1.1 and 4.1.3 are met:				
Inner packagings	Large outer packagings	Packing group I	Packing group II	Packing group III
Glass 10 ℓ Plastics 30 ℓ Metal 40 ℓ	Steel (50A) Aluminium (50B) Other metal (50N) Rigid plastics (50H) Natural wood (50C) Plywood (50D) Reconstituted wood (50F) Rigid fibreboard (50G)	Not allowed	Not allowed	3 m³

LP02	PACKING INSTRUCTION (SOLIDS)			LP02
The following large packagings are authorized, provided the general provisions of 4.1.1 and 4.1.3 are met:				

Inner packagings	Large outer packagings	Packing group I	Packing group II	Packing group III
Glass 10 kg Plastics[2] 50 kg Metal 50 kg Paper[1, 2] 50 kg Fibre[1, 2] 50 kg	Steel (50A) Aluminium (50B) Other metal (50N) Rigid plastics (50H) Natural wood (50C) Plywood (50D) Reconstituted wood (50F) Rigid fibreboard (50G) Flexible plastics (51H)[3]	Not allowed	Not allowed	3 m^3

[1]   These packagings shall not be used when the substances being transported may become liquid during transport.
[2]   Packagings shall be sift-proof.
[3]   To be used with flexible inner packagings only.

Special packing provision:

L2   For UN 1950 aerosols, the large packaging shall meet the packing group III perfcrmance level. Large packagings for waste aerosols transported in accordance with special provision 327 shall have in addition a means of retaining any free liquid that might escape during transport e.g., absorbent material.

LP99	PACKING INSTRUCTION	LP99
Only packagings which are approved for these goods by the competent authority may be used (see 4.1.3.7). A copy of the competent authority approval shall accompany each consignment or the transpcrt document shall include an indication that the packaging was approved by the competent authority.		

LP101	PACKING INSTRUCTION		LP101
The following packagings are authorized, provided the general provisions of 4.1.1 and 4.1.3 and special provisions of 4.1.5 are met:			

Inner packagings	Intermediate packagings	Large packagings
Not necessary	*Not necessary*	Steel (50A) Aluminium (50B) Other metal (50N) Rigid plastics (50H) Natural wood (50C) Plywood (50D) Reconstituted wood (50F) Rigid fibreboard (50G)

Special packing provision:

L1   For UN Nos. 0006, 0009, 0010, 0015, 0016, 0018, 0019, 0034, 0035, 0038, 0039, 0048, 0056, 0137, 0138, 0168, 0169, 0171, 0181, 0182, 0183, 0186, 0221, 0243, 0244, 0245, 0246, 0254, 0280, 0281, 0286, 0287, 0297, 0299, 0300, 0301, 0303, 0321, 0328, 0329, 0344, 0345, 0346, 0347, 0362, 0363, 0370, 0412, 0424, 0425, 0434, 0435, 0436, 0437, 0438, 0451, 0488 and 0502: Large and robust explosives articles, normally intended for military use, without their means of initiation or with their means of initiation containing at least two effective protective features, may be transported unpackaged. When such articles have propelling charges or are self-propelled, their ignition systems shall be protected against stimuli encountered during normal conditions of transport. A negative result in Test Series 4 on an unpackaged article indicates that the article can be considered for transport unpackaged. Such unpackaged articles may be fixed to cradles or contained in crates or other suitable handling devices.

LP102	PACKING INSTRUCTION	LP102

The following packagings are authorized, provided the general provisions of 4.1.1 and 4.1.3 and special provisions of 4.1.5 are met:

Inner packagings	Intermediate packagings	Outer packagings

Bags
    water-resistant
Receptacles
    fibreboard
    metal
    plastics
    wood
Sheets
    fibreboard, corrugated
Tubes
    fibreboard
	Not necessary	Steel (50A)
		Aluminium (50B)
		Other metal (50N)
		Rigid plastics (50H)
		Natural wood (50C)
		Plywood (50D)
		Reconstituted wood (50F)
		Rigid fibreboard (50G)

LP621	PACKING INSTRUCTION	LP621

This instruction applies to UN 3291.

The following large packagings are authorized, provided the general provisions of 4.1.1 and 4.1.3 are met:

(1)   For clinical waste placed in inner packagings: Rigid, leakproof large packagings conforming to the provisions of chapter 6.6 for solids, at the packing group II performance level, provided there is sufficient absorbent material to absorb the entire amount of liquid present and the large packaging is capable of retaining liquids.

(2)   For packages containing larger quantities of liquid: Large rigid packagings conforming to the provisions of chapter 6.6, at the packing group II performance level, for liquids.

Additional provision:

Large packagings intended to contain sharp objects such as broken glass and needles shall be resistant to puncture and retain liquids under the performance test conditions in chapter 6.6.

LP902	PACKING INSTRUCTION	LP902

This instruction applies to UN 3268.

The following packagings are authorized, provided the general provisions of 4.1.1 and 4.1.3 are met:

Packagings conforming to the packing group III performance level. The packagings shall be designed and constructed to prevent movement of the articles and inadvertent operation during normal conditions of transport.

The articles may also be transported unpackaged in dedicated handling devices, vehicles, containers or wagons when moved from where they are manufactured to an assembly plant.

Additional provision:

Any pressure vessel shall be in accordance with the requirements of the competent authority for the substance(s) contained in the pressure vessel(s).

4.1.5   Special packing provisions for goods of class 1

4.1.5.1   The general provisions of 4.1.1 shall be met.

4.1.5.2   All packagings for class 1 goods shall be so designed and constructed that:

.1   they will protect the explosives, prevent them escaping and cause no increase in the risk of unintended ignition or initiation when subjected to normal conditions of transport, including foreseeable changes in temperature, humidity and pressure;

.2   the complete package can be handled safely in normal conditions of transport; and

.3   the packages will withstand any loading imposed on them by foreseeable stacking to which they will be subject during transport so that they do not add to the risk presented by the explosives, the containment function of the packagings is not harmed, and they are not distorted in a way or to an extent which will reduce their strength or cause instability of a stack.

4.1.5.3   All explosive substances and articles, as prepared for transport, shall have been classified in accordance with the procedures detailed in 2.1.3.

4.1.5.4   Class 1 goods shall be packed in accordance with the appropriate packing instruction shown in columns 8 and 9 of the Dangerous Goods List, as detailed in 4.1.4.

4.1.5.5   Packagings, including IBCs and large packagings, shall conform to the provisions of chapter 6.1, 6.5 or 6.6, respectively, and shall meet the test provisions of 6.1.5, 6.5.6 or 6.6.5, respectively, for packing group II, subject to 4.1.1.13, 6.1.2.4 and 6.5.1.4.4. Packagings other than metal packagings meeting the test criteria of packing group I may be used. To avoid unnecessary confinement, metal packagings of packing group I shall not be used.

4.1.5.6   The closure device of packagings containing liquid explosives shall ensure a double protection against leakage.

4.1.5.7   The closure device of metal drums shall include a suitable gasket; if a closure device includes a screw-thread, the ingress of explosive substances into the screw-thread shall be prevented.

4.1.5.8   Packagings for water-soluble substances shall be water-resistant. Packagings for desensitized or phlegmatized substances shall be closed to prevent changes in concentration during transport.

4.1.5.9   When the packaging includes a double envelope filled with water which may freeze during transport, a sufficient quantity of an anti-freeze agent shall be added to the water to prevent freezing. Anti-freeze that could create a fire hazard because of its inherent flammability shall not be used.

4.1.5.10   Nails, staples and other closure devices made of metal without protective covering shall not penetrate to the inside of the outer packaging unless the inner packaging adequately protects the explosives against contact with the metal.

4.1.5.11   Inner packagings, fittings and cushioning materials and the placing of explosive substances or articles in packages shall be accomplished in a manner which prevents the explosive substances or articles from becoming loose in the outer packaging under normal conditions of transport. Metallic components of articles shall be prevented from making contact with metal packagings. Articles containing explosive substances not enclosed in an outer casing shall be separated from each other in order to prevent friction and impact. Padding, trays, partitioning in the inner or outer packaging, mouldings or receptacles may be used for this purpose.

4.1.5.12   Packagings shall be made of materials compatible with, and impermeable to, the explosives contained in the package, so that neither interaction between the explosives and the packaging materials nor leakage causes the explosive to become unsafe to transport, or the hazard division or compatibility group to change.

4.1.5.13   The ingress of explosive substances into the recesses of seamed metal packagings shall be prevented.

4.1.5.14   Plastics packagings shall not be liable to generate or accumulate sufficient static electricity so that a discharge could cause the packaged explosive substances or articles to initiate, ignite or function.

4.1.5.15   Large and robust explosives articles, normally intended for military use, without their means of initiation or with their means of initiation containing at least two effective protective features may be transported unpackaged. When such articles have propelling charges or are self-propelled, their ignition systems shall be protected against stimuli encountered during normal conditions of transport. A negative result in Test Series 4 on an unpackaged article indicates that the article can be considered for transport unpackaged. Such unpackaged articles may be fixed to cradles or contained in crates or other suitable handling, storage or launching devices in such a way that they will not become loose during normal conditions of transport. Where such large explosive articles are, as part of their operational safety and suitability tests, subjected to test regimes that meet the provisions of this Code and such tests have been successfully undertaken, the competent authority may approve such articles to be transported under this Code.

4.1.5.16   Explosive substances shall not be packed in inner or outer packagings where the differences in internal and external pressures, due to thermal or other effects, could cause an explosion or rupture of the package.

4.1.5.17   Whenever loose explosive substances or the explosive substance of an uncased or partly cased article may come into contact with the inner surface of metal packagings (1A2, 1B2, 4A, 4B and metal receptacles), the metal packaging shall be provided with an inner liner or coating (see 4.1.1.2).

4.1.5.18   Packing instruction P101 may be used for any explosive provided the package has been approved by a competent authority regardless of whether the packaging complies with the packing instruction assignment in the Dangerous Goods List.

4.1.5.19   Government-owned military dangerous goods, packaged prior to 1 January 1990 in accordance with the provisions of the IMDG Code in effect at that time, may be transported provided the packagings maintain their integrity and the goods are declared as government-owned goods packaged prior to 1 January 1990.

4.1.6   Special packing provisions for goods of class 2

4.1.6.1   General provisions

4.1.6.1.1   This section provides general requirements applicable to the use of pressure receptacles for the transport of class 2 gases and other dangerous goods in pressure receptacles (e.g. UN 1051 Hydrogen cyanide, stabilized). Pressure receptacles shall be constructed and closed so as to prevent any loss of contents which might be caused under normal conditions of transport, including by vibration, or by changes in temperature, humidity or pressure (resulting from change in altitude, for example).

4.1.6.1.2   Parts of pressure receptacles which are in direct contact with dangerous goods shall not be affected or weakened by those dangerous goods and shall not cause a dangerous effect (e.g. catalysing a reaction or reacting with the dangerous goods). The provisions of ISO 11114-1:1997 and ISO 11114-2:2000 shall be met as applicable.

4.1.6.1.3   Pressure receptacles, including their closures, shall be selected to contain a gas or a mixture of gases according to the requirements of 6.2.1.2 and the requirements of the specific packing instructions of 4.1.4.1. This section also applies to pressure receptacles which are elements of MEGCs.

4.1.6.1.4   Refillable pressure receptacles shall not be filled with a gas or gas mixture different from that previously contained unless the necessary operations for change of gas service have been performed. The change of service for compressed and liquefied gases shall be in accordance with ISO 11621:1997, as applicable. In addition, a pressure receptacle that previously contained a class 8 corrosive substance or a substance of another class with a corrosive subsidiary risk shall not be authorized for the transport of a class 2 substance unless the necessary inspection and testing as specified in 6.2.1.6 have been performed.

4.1.6.1.5   Prior to filling, the filler shall perform an inspection of the pressure receptacle and ensure that the pressure receptacle is authorized for the gas to be transported and that the provisions of this Code have been met. Shut-off valves shall be closed after filling and remain closed during transport. The consignor shall verify that the closures and equipment are not leaking.

4.1.6.1.6   Pressure receptacles shall be filled according to the working pressures, filling ratios and provisions specified in the appropriate packing instruction for the specific substance being filled. Reactive gases and gas mixtures shall be filled to a pressure such that if complete decomposition of the gas occurs, the working pressure of the pressure receptacle shall not be exceeded. Bundles of cylinders shall not be filled in excess of the lowest working pressure of any given cylinder in the bundle.

4.1.6.1.7   Pressure receptacles, including their closures, shall conform to the design, construction, inspection and testing requirements detailed in chapter 6.2. When outer packagings are prescribed, the pressure receptacles shall be firmly secured therein. Unless otherwise specified in the detailed packing instructions, one or more inner packagings may be enclosed in an outer packaging.

4.1.6.1.8   Valves shall be designed and constructed in such a way that they are inherently able to withstand damage without release of the contents or shall be protected from damage which could cause inadvertent release of the contents of the pressure receptacle, by one of the following methods:

.1   Valves are placed inside the neck of the pressure receptacle and protected by a threaded plug or cap;

.2   Valves are protected by caps. Caps shall possess vent-holes of sufficient cross-sectional area to evacuate the gas if leakage occurs at the valves;

.3   Valves are protected by shrouds or guards;

.4   Pressure receptacles are transported in frames (e.g. bundles); or

.5   Pressure receptacles are transported in an outer packaging. The packaging as prepared for transport shall be capable of meeting the drop test specified in 6.1.5.3 at the packing group I performance level.

For pressure receptacles with valves as described in .2 and .3, the requirements of ISO 11117:1998 shall be met; for valves with inherent protection, the provisions of annex B of ISO 10297:1999 shall be met.

4.1.6.1.9   Non-refillable pressure receptacles shall:

.1   be transported in an outer packaging, such as a box, or crate, or in shrink-wrapped trays or stretch-wrapped trays;

.2   be of a water capacity less than or equal to 1.25 ℓ when filled with flammable or toxic gas;

.3   not be used for toxic gases with an LC_{50} less than or equal to 200 mℓ/m^3; and

.4   not be repaired after being put into service.

4.1.6.1.10   Refillable pressure receptacles, other than cryogenic receptacles, shall be periodically inspected in accordance with 6.2.1.6 and packing instruction P200. Pressure receptacles shall not be filled after they become due for periodic inspection but may be transported after the expiry of the time limit.

4.1.6.1.11   Repairs shall be consistent with the manufacture and testing requirements of the applicable design and construction standards and are only permitted as indicated in the relevant periodic inspection standards specified in 6.2.2.4. Pressure receptacles, other than the jacket of closed cryogenic receptacles, shall not be subjected to repairs of any of the following:

.1   weld cracks or other weld defects;

.2   cracks in walls;

.3   leaks or defects in the material of the wall, head or bottom.

4.1.6.1.12   Pressure receptacles shall not be offered for filling:

.1   when damaged to such an extent that the integrity of the pressure receptacle or its service equipment may be affected;

.2   unless the pressure receptacle and its service equipment has been examined and found to be in good working order; or

.3   unless the required certification, retest, and filling markings are legible.

4.1.6.1.13   Filled pressure receptacles shall not be offered for transport:

.1   when leaking;

.2   when damaged to such an extent that the integrity of the pressure receptacle or its service equipment may be affected;

.3   unless the pressure receptacle and its service equipment has been examined and found to be in good working order; or

.4   unless the required certification, retest, and filling markings are legible.

4.1.6.1.14   Where in packing instruction P200 cylinders and other pressure receptacles for gases conforming to the requirements of this sub-section and chapter 6.2 are authorized, use is also authorized of cylinders and pressure receptacles which conform to the requirements of the competent authority of the country in which the cylinder or pressure receptacle is filled. Valves shall be suitably protected. Pressure receptacles with capacities of 1 ℓ or less shall be packed in outer packagings constructed of suitable material of adequate strength and design in relation to the capacity of the packaging and its intended use and secured or cushioned so as to prevent significant movement within the outer packaging during normal conditions of transport.

4.1.7   Special packing provisions for organic peroxides (class 5.2) and self-reactive substances of class 4.1

4.1.7.0   General

4.1.7.0.1   For organic peroxides, all receptacles shall be "effectively closed". Where significant internal pressure may develop in a package by the evolution of gas, a vent may be fitted, provided the gas emitted will not cause danger, otherwise the degree of filling shall be limited. Any venting device shall be so constructed that liquid will not escape when the package is in an upright position and it shall be able to prevent ingress of impurities. The outer packaging, if any, shall be so designed as not to interfere with the operation of the venting device.

4.1.7.1 **Use of packagings**

4.1.7.1.1 Packagings for organic peroxides and self-reactive substances shall meet the provisions of chapter 6.1 or of chapter 6.6 at the packing group II performance level. To avoid unnecessary confinement, metal packaging meeting the test criteria of packing group I shall not be used.

4.1.7.1.2 The packing methods for organic peroxides and self-reactive substances are listed in packing instruction P520 and are designated OP1 to OP8. The quantities specified for each packing method are the maximum quantities authorized per package.

4.1.7.1.3 The packing methods appropriate for the individual currently assigned self-reactive substances and organic peroxides are listed in 2.4.2.3.2.3 and 2.5.3.2.4.

4.1.7.1.4 For new organic peroxides, new self-reactive substances or new formulations of currently assigned organic peroxides or self-reactive substances, the following procedure shall be used to assign the appropriate packing method:

    .1 ORGANIC PEROXIDE TYPE B or SELF-REACTIVE SUBSTANCE TYPE B:
    Packing method OP5 shall be assigned, provided that the organic peroxide (or self-reactive substance) satisfies the criteria of 2.4.2.3.2.3 (resp. 2.4.2.3.3.2.2) in a packaging authorized by the packing method. If the organic peroxide (or self-reactive substance) can only satisfy these criteria in a smaller packaging than those authorized by packing method OP5 (viz. one of the packagings listed for OP1 to OP4), then the corresponding packing method with the lower OP number is assigned;

    .2 ORGANIC PEROXIDE TYPE C or SELF-REACTIVE SUBSTANCE TYPE C:
    Packing method OP6 shall be assigned, provided that the organic peroxide (or self-reactive substance) satisfies the criteria of 2.5.3.3.2.3 (resp. 2.4.2.3.3.2.3) in packaging authorized by the packing method. If the organic peroxide (or self-reactive substance) can only satisfy these criteria in a smaller packaging than those authorized by packing method OP6, then the corresponding packing method with the lower OP number is assigned;

    .3 ORGANIC PEROXIDE TYPE D or SELF-REACTIVE SUBSTANCE TYPE D:
    Packing method OP7 shall be assigned to this type of organic peroxide or self-reactive substance;

    .4 ORGANIC PEROXIDE TYPE E or SELF-REACTIVE SUBSTANCE TYPE E:
    Packing method OP8 shall be assigned to this type of organic peroxide or self-reactive substance;

    .5 ORGANIC PEROXIDE TYPE F or SELF-REACTIVE SUBSTANCE TYPE F:
    Packing method OP8 shall be assigned to this type of organic peroxide or self-reactive substance.

4.1.7.2 **Use of intermediate bulk containers**

4.1.7.2.1 The currently assigned organic peroxides specifically listed in packing instruction IBC520 may be transported in IBCs in accordance with this packing instruction.

4.1.7.2.2 Other organic peroxides and self-reactive substances of type F may be transported in IBCs under conditions established by the competent authority of the country of origin when, on the basis of the appropriate tests, that competent authority is satisfied that such transport may be safely conducted. The tests undertaken shall include those necessary:

    .1 to prove that the organic peroxide (or self-reactive substance) complies with the principles for classification;

    .2 to prove the compatibility of all materials normally in contact with the substance during the transport;

    .3 to determine, when applicable, the control and emergency temperatures associated with the transport of the product in the IBC concerned as derived from the SADT;

    .4 to design, when applicable, pressure and emergency relief devices; and

    .5 to determine if any special provisions are necessary for safe transport of the substance.

4.1.7.2.3 For self-reactive substances, temperature control is required according to 2.4.2.3.4. For organic peroxides, temperature control is required according to 2.5.3.4.1. Temperature control provisions are given in chapter 7.7.

4.1.7.2.4 Emergencies to be taken into account are self-accelerating decomposition and fire engulfment. To prevent explosive rupture of metal or composite IBCs with a complete metal casing, the emergency relief devices shall be designed to vent all the decomposition products and vapours evolved during self-accelerating decomposition or during a period of not less than one hour of complete fire engulfment calculated by the equations given in 4.2.1.13.8.

4.1.8 **Special packing provisions for infectious substances of category A (class 6.2, UN 2814 and UN 2900)**

4.1.8.1 Consignors of infectious substances shall ensure that packages are prepared in such a manner that they arrive at their destination in good condition and present no hazard to persons or animals during transport.

4.1.8.2 The definitions in 1.2.1 and the general packing provisions of 4.1.1.1 to 4.1.1.14, except 4.1.1.10 to 4.1.1.12, apply to infectious substances packages. However, liquids shall only be filled into packagings which have an appropriate resistance to the internal pressure that may develop under normal conditions of transport.

4.1.8.3 An itemized list of contents shall be enclosed between the secondary packaging and the outer packaging. When the infectious substances to be transported are unknown, but suspected of meeting the criteria for inclusion in category A, the words "suspected category A infectious substance" shall be shown, in parentheses, following the Proper Shipping Name on the document inside the outer packaging.

4.1.8.4 Before an empty packaging is returned to the consignor, or sent elsewhere, it shall be disinfected or sterilized to nullify any hazard and any label or marking indicating that it had contained an infectious substance shall be removed or obliterated.

4.1.8.5 Provided an equivalent level of performance is maintained, the following variations in the primary receptacles placed within an intermediate packaging are allowed without further testing of the completed package:

.1 Primary receptacles of equivalent or smaller size as compared to the tested primary receptacles may be used provided:

(a) the primary receptacles are of similar design to the tested primary receptacle (such as shape: round, rectangular, etc.);

(b) the material of construction of the primary receptacle (glass, plastics, metal, etc.) offers resistance to impact and stacking forces equal to or greater than that of the originally tested primary receptacle;

(c) the primary receptacles have the same or smaller openings and the closure is of similar design (such as screw cap, friction lid, etc.);

(d) sufficient additional cushioning material is used to take up void spaces and to prevent significant movement of the primary receptacles; and

(e) primary receptacles are oriented within the intermediate packaging in the same manner as in the tested package.

.2 A lesser number of the tested primary receptacles, or of the alternative types of primary receptacles identified in .1 above, may be used provided sufficient cushioning is added to fill the void space(s) and to prevent significant movement of the primary receptacles.

4.1.9 **Special packing provisions for class 7**

4.1.9.1 **General**

4.1.9.1.1 Radioactive material, packagings and packages shall meet the provisions of chapter 6.4. The quantity of radioactive material in a package shall not exceed the limits specified in 2.7.2.2, 2.7.2.4.1, 2.7.2.4.4, 2.7.2.4.5, 2.7.2.4.6 and 4.1.9.3.

The types of packages for radioactive materials covered by the provisions of this Code are:

.1 Excepted package (see 1.5.1.5);

.2 Industrial package Type 1 (Type IP-1 package);

.3 Industrial package Type 2 (Type IP-2 package);

.4 Industrial package Type 3 (Type IP-3 package);

.5 Type A package;

.6 Type B(U) package;

.7 Type B(M) package;

.8 Type C package.

Packages containing fissile material or uranium hexafluoride are subject to additional requirements.

4.1.9.1.2 The non-fixed contamination on the external surfaces of any package shall be kept as low as practicable and, under routine conditions of transport, shall not exceed the following limits:

(a) 4 Bq/cm^2 for beta and gamma emitters and low-toxicity alpha emitters, and

(b) 0.4 Bq/cm^2 for all other alpha emitters.

These limits are applicable when averaged over any area of 300 cm^2 of any part of the surface.

4.1.9.1.3 A package shall not contain any items other than those that are necessary for the use of the radioactive material. The interaction between these items and the package under the conditions of transport applicable to the design shall not reduce the safety of the package.

4.1.9.1.4 Except as provided in 7.1.14.13, the level of non-fixed contamination on the external and internal surfaces of overpacks, cargo transport units, tanks, IBCs and conveyances shall not exceed the limits specified in 4.1.9.1.2.

4.1.9.1.5 Radioactive material with a subsidiary risk shall be transported in packagings, IBCs or tanks fully complying with the provisions of the relevant chapters of part 6 as appropriate, as well as applicable provisions of chapters 4.1 or 4.2 for that subsidiary risk.

4.1.9.1.6 Before the first shipment of any package, the following provisions shall be fulfilled:

.1 If the design pressure of the containment system exceeds 35 kPa (gauge), it shall be ensured that the containment system of each package conforms to the approved design requirements relating to the capability of that system to maintain its integrity under that pressure;

.2 For each Type B(U), Type B(M) and Type C package and for each package containing fissile material, it shall be ensured that the effectiveness of its shielding and containment and, where necessary, the heat transfer characteristics and the effectiveness of the confinement system are within the limits applicable to or specified for the approved design;

.3 For packages containing fissile material, where, in order to comply with the requirements of 6.4.11.1, neutron poisons are specifically included as components of the package, checks shall be performed to confirm the presence and distribution of those neutron poisons.

4.1.9.1.7 Before each shipment of any package, the following provisions shall be fulfilled:

.1 For any package, it shall be ensured that all provisions specified in the relevant provisions of this Code have been satisfied;

.2 It shall be ensured that lifting attachments which do not meet the requirements of 6.4.2.2 have been removed or otherwise rendered incapable of being used for lifting the package, in accordance with 6.4.2.3;

.3 For each package requiring competent authority approval, it shall be ensured that all the requirements specified in the approval certificates have been satisfied;

.4 Each Type B(U), Type B(M) and Type C package shall be held until equilibrium conditions have been approached closely enough to demonstrate compliance with the requirements for temperature and pressure unless an exemption from these requirements has received unilateral approval;

.5 For each Type B(U), Type B(M) and Type C package, it shall be ensured by inspection and/or appropriate tests that all closures, valves, and other openings of the containment system through which the radioactive contents might escape are properly closed and, where appropriate, sealed in the manner for which the demonstrations of compliance with the requirements of 6.4.8.8 and 6.4.10.3 were made;

.6 For each special form radioactive material, it shall be ensured that all the provisions specified in the approval certificate and the relevant provisions of these Regulations have been satisfied;

.7 For packages containing fissile material, the measurement specified in 6.4.11.4(b) and the tests to demonstrate closure of each package as specified in 6.4.11.7 shall be performed where applicable;

.8 For each low dispersible radioactive material, it shall be ensured that all the requirements specified in the approval certificate and the relevant provisions of these Regulations have been satisfied.

4.1.9.1.8 The consignor shall also have a copy of any instructions with regard to the proper closing of the package and any preparation for shipment before making any shipment under the terms of the certificates.

4.1.9.1.9 Except for consignments under exclusive use, the transport index of any package or overpack shall not exceed 10, nor shall the criticality safety index of any package or overpack exceed 50.

4.1.9.1.10 Except for packages or overpacks transported under exclusive use by rail or by road under the conditions specified in 7.1.14.7.1, or under exclusive use and special arrangement by ship under the conditions specified in 7.1.14.9, the maximum radiation level at any point on any external surface of a package or overpack shall not exceed 2 mSv/h.

4.1.9.1.11 The maximum radiation level at any point on any external surface of a package or overpack under exclusive use shall not exceed 10 mSv/h.

4.1.9.1.12 Pyrophoric radioactive material shall be packaged in Type A, Type B(U), Type B(M) or Type C packages and shall also be suitably inerted.

4.1.9.2 **Provisions and controls for transport of LSA material and SCO**

4.1.9.2.1 The quantity of LSA material or SCO in a single Type IP-1 package, Type IP-2 package, Type IP-3 package, or object or collection of objects, whichever is appropriate, shall be so restricted that the external radiation level at 3 m from the unshielded material or object or collection of objects does not exceed 10 mSv/h.

4.1.9.2.2 For LSA material and SCO which is or contains fissile material, the applicable provisions of 6.4.11.1, 7.2.9.4 and 7.2.9.5 shall be met.

4.1.9.2.3 LSA material and SCO in groups LSA-I and SCO-I may be transported unpackaged under the following conditions:

.1 all unpackaged material other than ores containing only naturally occurring radionuclides shall be transported in such a manner that, under routine conditions of transport, there will be no escape of the radioactive contents from the conveyance nor will there be any loss of shielding;

.2 each conveyance shall be under exclusive use, except when only transporting SCO-I on which the contamination on the accessible and the inaccessible surfaces is not greater than ten times the applicable level specified in 2.7.2.3.2; and

.3 for SCO-I where it is suspected that non-fixed contamination exists on inaccessible surfaces in excess of the values specified in 2.7.2.3.2.1(i), measures shall be taken to ensure that the radioactive material is not released into the conveyance.

4.1.9.2.4 LSA material and SCO, except as otherwise specified in 4.1.9.2.3, shall be packaged in accordance with table 4.1.9.2.4.

Table 4.1.9.2.4 – Industrial package provisions for LSA material and SCO

Radioactive contents	Industrial package type	
	Exclusive use	Not under exclusive use

LSA-I
  Solid[a]
  Liquid
	Type IP-1
	Type IP-1
		Type IP-1
		Type IP-2

LSA-II
  Solid
  Liquid and gas
	Type IP-2
	Type IP-2
		Type IP-2
		Type IP-3

LSA-III	Type IP-2	Type IP-3
SCO-I[a]	Type IP-1	Type IP-1
SCO-II	Type IP-2	Type IP-2

[a] Under the conditions specified in 4.1.9.2.3, LSA-I material and SCO-I may be transported unpackaged.

4.1.9.3 **Packages containing fissile material**

Unless not classified as fissile in accordance with 2.7.2.3.5, packages containing fissile material shall not contain:

.1 A mass of fissile material different from that authorized for the package design;

.2 Any radionuclide or fissile material different from those authorized for the package design; or

.3 Contents in a form or physical or chemical state, or in a spatial arrangement, different from those authorized for the package design,

as specified in their certificates of approval where appropriate.

Chapter 4.2

Use of portable tanks and multiple-element gas containers (MEGCs)

The provisions of this chapter also apply to road tank vehicles to the extent indicated in chapter 6.8.

4.2.0 Transitional provisions

4.2.0.1 The provisions for the use and construction of portable tanks in this chapter and chapter 6.7 are based on the United Nations Recommendations on the transport of dangerous goods. IMO type portable tanks and road tank vehicles certified and approved prior to 1 January 2003 in accordance with the provisions of the IMDG Code in force on 1 July 1999 (amendment 29) may continue to be used provided that they are found to meet the applicable periodic inspections and test provisions. They shall meet the provisions set out in columns (13) and (14) of chapter 3.2. Detailed explanation and construction provisions may be found in DSC/Circ.12 (Guidance on the continued use of existing IMO type portable tanks and road tank vehicles for the transport of dangerous goods).

Note: For ease of reference, the following descriptions of existing IMO type tanks are included:
IMO type 1 tank means a portable tank for the transport of substances of classes 3 to 9 fitted with pressure-relief devices, having a maximum allowable working pressure of 1.75 bar and above.
IMO type 2 tank means a portable tank fitted with pressure-relief devices, having a maximum allowable working pressure equal to or above 1.0 bar but below 1.75 bar, intended for the transport of certain dangerous liquids of low hazard and certain solids.
IMO type 4 tank means a road tank vehicle for the transport of dangerous goods of classes 3 to 9 and includes a semi-trailer with a permanently attached tank or a tank attached to a chassis, with at least four twist locks which comply with ISO standards, (e.g. ISO International Standard 1161:1984).
IMO type 5 tank means a portable tank fitted with pressure-relief devices which is used for non-refrigerated gases of class 2.
IMO type 6 tank means a road tank vehicle for the transport of non-refrigerated liquefied gases of class 2 and includes a semi-trailer with a permanently attached tank or a tank attached to a chassis which is fitted with items of service equipment and structural equipment necessary for the transport of gases.
IMO type 7 tank means a thermally insulated portable tank fitted with items of service and structural equipment necessary for the transport of refrigerated liquefied gases. The portable tank shall be capable of being transported, loaded and discharged without the need of removal of its structural equipment, and shall be capable of being lifted when full. It shall not be permanently secured on board the ship.
IMO type 8 tank means a road tank vehicle for the transport of refrigerated liquefied gases of class 2 and includes a semi-trailer with a permanently attached thermally insulated tank fitted with items of service equipment and structural equipment necessary for the transport of refrigerated liquefied gases.

Note: IMO type 4, 6 and 8 road tank vehicles may be constructed after 1 January 2003 in accordance with the provisions of chapter 6.8.

4.2.0.2 UN portable tanks and MEGCs constructed according to a design approval certificate which has been issued before 1 January 2008 may continue to be used provided that they are found to meet the applicable periodic inspection and test provisions.

4.2.1 General provisions for the use of portable tanks for the transport of substances of class 1 and classes 3 to 9

4.2.1.1 This section provides general provisions applicable to the use of portable tanks for the transport of substances of classes 1, 3, 4, 5, 6, 7, 8 and 9. In addition to these general provisions, portable tanks shall conform to the design, construction, inspection and testing provisions detailed in 6.7.2. Substances shall be transported in portable tanks conforming to the applicable portable tank instruction and the portable tank special provisions assigned to each substance in the Dangerous Goods List.

4.2.1.2 During transport, portable tanks shall be adequately protected against damage to the shell and service equipment resulting from lateral and longitudinal impact and overturning. If the shell and service equipment are so constructed as to withstand impact or overturning, it need not be protected in this way. Examples of such protection are given in 6.7.2.17.5.

4.2.1.3 Certain substances are chemically unstable. They are accepted for transport only when the necessary steps have been taken to prevent their dangerous decomposition, transformation or polymerization during transport. To this end, care shall in particular be taken to ensure that shells do not contain any substances liable to promote these reactions.

4.2.1.4 The temperature of the outer surface of the shell, excluding openings and their closures, or of the thermal insulation shall not exceed 70°C during transport. When necessary, the shell shall be thermally insulated.

4.2.1.5 Empty portable tanks not cleaned and not gas-free shall comply with the same provisions as portable tanks filled with the previous substance.

4.2.1.6 Substances shall not be transported in adjoining compartments of shells when they may react dangerously with each other and cause:

.1 combustion and/or evolution of considerable heat;

.2 evolution of flammable, toxic or asphyxiant gases;

.3 the formation of corrosive substances;

.4 the formation of unstable substances;

.5 dangerous rise in pressure.

4.2.1.7 The design approval certificate, the test report and the certificate showing the results of the initial inspection and test for each portable tank issued by the competent authority or its authorized body shall be retained by the authority or body and the owner. Owners shall be able to provide this documentation upon the request of any competent authority.

4.2.1.8 Unless the name of the substance(s) being transported appears on the metal plate described in 6.7.2.20.2, a copy of the certificate specified in 6.7.2.18.1 shall be made available upon the request of a competent authority or its authorized body and readily provided by the consignor, consignee or agent, as appropriate.

4

4.2.1.9 **Degree of filling**

4.2.1.9.1 Prior to filling, the shipper shall ensure that the appropriate portable tank is used and that the portable tank is not loaded with substances which, in contact with the materials of the shell, gaskets, service equipment and any protective linings, are likely to react dangerously with them to form dangerous products or appreciably weaken these materials. The shipper may need to consult the manufacturer of the substance in conjunction with the competent authority for guidance on the compatibility of the substance with the portable tank materials.

4.2.1.9.1.1 Portable tanks shall not be filled in excess of the maximum degree of filling specified in 4.2.1.9.2 to 4.2.1.9.6. The applicability of 4.2.1.9.2, 4.2.1.9.3 or 4.2.1.9.5.1 to individual substances is specified in the applicable portable tank instructions or special provisions in 4.2.5.2.6 or 4.2.5.3 and columns 12, 13 and 14 of the Dangerous Goods List.

4.2.1.9.2 The maximum degree of filling (in %) for general use is determined by the formula:

$$\text{Degree of filling} = \frac{97}{1 + \alpha(t_r - t_f)}$$

4.2.1.9.3 The maximum degree of filling (in %) for liquids of class 6.1 and class 8, in packing groups I and II, and liquids with an absolute vapour pressure of more than 175 kPa (1.75 bar) at 65°C, or for liquids identified as marine pollutants is determined by the formula:

$$\text{Degree of filling} = \frac{95}{1 + \alpha(t_r - t_f)}$$

4.2.1.9.4 In these formulae, α is the mean coefficient of cubical expansion of the liquid between the mean temperature of the liquid during filling (t_f) and the maximum mean bulk temperature during transport (t_r) (both in °C). For liquids transported under ambient conditions, α could be calculated by the formula:

$$\alpha = \frac{d_{15} - d_{50}}{35\,d_{50}}$$

in which d_{15} and d_{50} are the densities of the liquid at 15°C and 50°C, respectively.

4.2.1.9.4.1   The maximum mean bulk temperature (t_r) shall be taken as 50°C except that, for journeys under temperate or extreme climatic conditions, the competent authorities concerned may agree to a lower or require a higher temperature, as appropriate.

4.2.1.9.5   The provisions of 4.2.1.9.2 to 4.2.1.9.4.1 do not apply to portable tanks which contain substances maintained at a temperature above 50°C during transport (such as by means of a heating device). For portable tanks equipped with a heating device, a temperature regulator shall be used to ensure the maximum degree of filling is not more than 95% full at any time during transport.

4.2.1.9.5.1   The maximum degree of filling (in %) for solids transported above their melting points and for elevated temperature liquids shall be determined by the following formula:

$$\text{Degree of filling} = 95\frac{d_r}{d_f}$$

in which d_f and d_r are the densities of the liquid at the mean temperature of the liquid during filling and the maximum mean bulk temperature during transport respectively.

4.2.1.9.6   Portable tanks shall not be offered for transport:

.1   with a degree of filling, for liquids having a viscosity less than 2,680 mm^2/s at 20°C or at the maximum temperature of the substance during transport in the case of a heated substance, of more than 20% but less than 80% unless the shells of portable tanks are divided, by partitions or surge plates, into sections of not more than 7,500 ℓ capacity;

.2   with residue of substances previously transported adhering to the outside of the shell or service equipment;

.3   when leaking or damaged to such an extent that the integrity of the portable tank or its lifting or securing arrangements may be affected; and

.4   unless the service equipment has been examined and found to be in good working order.

For certain dangerous substances, a lower degree of filling may be required.

4.2.1.9.7   Forklift pockets of portable tanks shall be closed off where the tank is filled. This provision does not apply to portable tanks which, according to 6.7.2.17.4, need not be provided with a means of closing off the forklift pockets.

4.2.1.9.8   Portable tanks shall not be filled or discharged while they remain on board.

4.2.1.10   **Additional provisions applicable to the transport of class 3 substances in portable tanks**

All portable tanks intended for the transport of flammable liquids shall be closed and be fitted with relief devices in accordance with 6.7.2.8 to 6.7.2.15.

4.2.1.11   **Additional provisions applicable to the transport of class 4 substances (other than class 4.1 self-reactive substances) in portable tanks**

[Reserved]

Note: For class 4.1 self-reactive substances, see 4.2.1.13.

4.2.1.12   **Additional provisions applicable to the transport of class 5.1 substances in portable tanks**

[Reserved]

4.2.1.13   **Additional provisions applicable to the transport of class 5.2 substances and class 4.1 self-reactive substances in portable tanks**

4.2.1.13.1   Each substance shall have been tested and a report submitted to the competent authority of the country of origin for approval. Notification thereof shall be sent to the competent authority of the country of destination. The notification shall contain relevant transport information and the report with test results. The tests undertaken shall include those necessary:

.1   to prove the compatibility of all materials normally in contact with the substance during transport;

.2   to provide data for the design of the pressure and emergency relief devices, taking into account the design characteristics of the portable tank.

Any additional provisions necessary for safe transport of the substance shall be clearly described in the report.

4.2.1.13.2 The following provisions apply to portable tanks intended for the transport of type F organic peroxides or type F self-reactive substances with a self-accelerating decomposition temperature (SADT) of 55°C or more. In case of conflict, these provisions prevail over those specified in 6.7.2. Emergencies to be taken into account are self-accelerating decomposition of the substance and fire-engulfment as described in 4.2.1.13.8.

4.2.1.13.3 The additional provisions for transport of organic peroxides or self-reactive substances with an SADT less than 55°C in portable tanks shall be specified by the competent authority of the country of origin. Notification thereof shall be sent to the competent authority of the country of destination.

4.2.1.13.4 The portable tank shall be designed for a test pressure of at least 0.4 MPa (4 bar).

4.2.1.13.5 Portable tanks shall be fitted with temperature-sensing devices.

4.2.1.13.6 Portable tanks shall be fitted with pressure-relief devices and emergency relief devices. Vacuum-relief devices may also be used. Pressure-relief devices shall operate at pressures determined according to both the properties of the substance and the construction characteristics of the portable tank. Fusible elements are not allowed in the shell.

4.2.1.13.7 The pressure-relief devices shall consist of spring-loaded valves fitted to prevent significant build-up within the portable tank of the decomposition products and vapours released at a temperature of 50°C. The capacity and start-to-discharge pressure of the relief valves shall be based on the results of the tests specified in 4.2.1.13.1. The start-to-discharge pressure shall, however, in no case be such that liquid would escape from the valve(s) if the portable tank were overturned.

4.2.1.13.8 The emergency relief devices may be of the spring-loaded or frangible types, or a combination of the two, designed to vent all the decomposition products and vapours evolved during a period of not less than one hour of complete fire-engulfment as calculated by the following formula:

$$q = 70961\, FA^{0.82}$$

where:

q = heat absorption (W)
A = wetted area (m^2)
F = insulation factor;

F = 1 for non-insulated vessels, or

$F = \dfrac{U(923 - T)}{47032}$ for insulated shells

where:

K = heat conductivity of insulation layer   (W·m^{-1}·K^{-1})
L = thickness of insulation layer   (m)
U = K/L = heat transfer coefficient of the insulation   (W·m^{-2}·K^{-1})
T = temperature of substance at relieving conditions   (K)

The start-to-discharge pressure of the emergency relief device(s) shall be higher than that specified in 4.2.1.13.7 and based on the results of the tests referred to in 4.2.1.13.1. The emergency relief devices shall be dimensioned in such a way that the maximum pressure in the tank never exceeds the test pressure of the portable tank.

Note: An example of a method to determine the size of emergency relief devices is given in Appendix 5 of the *Manual of Tests and Criteria*.

4.2.1.13.9 For insulated portable tanks, the capacity and setting of emergency relief device(s) shall be determined assuming a loss of insulation from 1% of the surface area.

4.2.1.13.10 Vacuum-relief devices and spring-loaded valves shall be provided with flame arresters. Due attention shall be paid to the reduction of the relief capacity caused by the flame arrester.

4.2.1.13.11 Service equipment such as valves and external piping shall be so arranged that no substance remains in them after filling the portable tank.

4.2.1.13.12 Portable tanks may be either insulated or protected by a sunshield. If the SADT of the substance in the portable tank is 55°C or less, or the portable tank is constructed of aluminium, the portable tank shall be completely insulated. The outer surface shall be finished in white or bright metal.

4.2.1.13.13 The degree of filling shall not exceed 90% at 15°C.

4.2.1.13.14 The marking as required in 6.7.2.20.2 shall include the UN Number and the technical name with the approved concentration of the substance concerned.

4.2.1.13.15   Organic peroxides and self-reactive substances specifically listed in portable tank instruction T23 in 4.2.5.2.6 may be transported in portable tanks.

4.2.1.14   **Additional provisions applicable to the transport of class 6.1 substances in portable tanks**

[Reserved]

4.2.1.15   **Additional provisions applicable to the transport of class 6.2 substances in portable tanks**

[Reserved]

4.2.1.16   **Additional provisions applicable to the transport of class 7 substances in portable tanks**

4.2.1.16.1   Portable tanks used for the transport of radioactive material shall not be used for the transport of other goods.

4.2.1.16.2   The degree of filling for portable tanks shall not exceed 90% or, alternatively, any other value approved by the competent authority.

4.2.1.17   **Additional provisions applicable to the transport of class 8 substances in portable tanks**

4.2.1.17.1   Pressure-relief devices of portable tanks used for the transport of class 8 substances shall be inspected at intervals not exceeding one year.

4.2.1.18   **Additional provisions applicable to the transport of class 9 substances in portable tanks**

[Reserved]

4.2.1.19   **Additional provisions applicable to the transport of solid substances transported above their melting point**

4.2.1.19.1   Solid substances transported or offered for transport above their melting point which are not assigned a portable tank instruction in column 13 of the Dangerous Goods List of chapter 3.2 or when the assigned portable tank instruction does not apply to transport at temperatures above their melting point may be transported in portable tanks provided that the solid substances are classified in classes 4.1, 4.2, 4.3, 5.1, 6.1, 8 or 9 and have no subsidiary risk other than that of class 6.1 or class 8 and are in packing group II or III.

4.2.1.19.2   Unless otherwise indicated in the Dangerous Goods List, portable tanks used for the transport of these solid substances above their melting point shall conform to the provisions of portable tank instruction T4 for solid substances of packing group III or T7 for solid substances of packing group II. A portable tank that affords an equivalent or greater level of safety may be selected in accordance with 4.2.5.2.5. The maximum degree of filling (in %) shall be determined according to 4.2.1.9.5 (TP3).

4.2.2   General provisions for the use of portable tanks for the transport of non-refrigerated liquefied gases

4.2.2.1   This section provides general provisions applicable to the use of portable tanks for the transport of non-refrigerated liquefied gases of class 2.

4.2.2.2   Portable tanks shall conform to the design, construction, inspection and testing provisions detailed in 6.7.3. Non-refrigerated liquefied gases shall be transported in portable tanks conforming to portable tank instruction T50 as described in 4.2.5.2.6 and any portable tank special provisions assigned to specific non-refrigerated liquefied gases in the Dangerous Goods List and described in 4.2.5.3.

4.2.2.3   During transport, portable tanks shall be adequately protected against damage to the shell and service equipment resulting from lateral and longitudinal impact and overturning. If the shell and service equipment are so constructed as to withstand impact or overturning, it need not be protected in this way. Examples of such protection are given in 6.7.3.13.5.

4.2.2.4   Certain non-refrigerated liquefied gases are chemically unstable. They are accepted for transport only when the necessary steps have been taken to prevent their dangerous decomposition, transformation or polymerization during transport. To this end, care shall be taken to ensure that portable tanks do not contain any non-refrigerated liquefied gases liable to promote these reactions.

4.2.2.5   Unless the name of the gas(es) being transported appears on the metal plate described in 6.7.3.16.2, a copy of the certificate specified in 6.7.3.14.1 shall be made available upon a competent authority request and readily provided by the consignor, consignee or agent, as appropriate.

4.2.2.6   Empty portable tanks not cleaned and not gas-free shall comply with the same provisions as portable tanks filled with the previous non-refrigerated liquefied gas.

4.2.2.7   **Filling**

4.2.2.7.1   Prior to filling, the shipper shall ensure that the portable tank is approved for the non-refrigerated liquefied gas to be transported and that the portable tank is not loaded with non-refrigerated liquefied gases which, in contact with the materials of the shell, gaskets and service equipment, are likely to react dangerously with them to form dangerous products or appreciably weaken these materials. During filling, the temperature of the non-refrigerated liquefied gas shall fall within the limits of the design temperature range.

4.2.2.7.2   The maximum mass of non-refrigerated liquefied gas per litre of shell capacity (kg/ℓ) shall not exceed the density of the non-refrigerated liquefied gas at 50°C multiplied by 0.95. Furthermore, the shell shall not be liquid-full at 60°C.

4.2.2.7.3   Portable tanks shall not be filled above their maximum permissible gross mass and the maximum permissible load mass specified for each gas to be transported.

4.2.2.7.4   Portable tanks shall not be filled or discharged while they remain on board.

4.2.2.8   Portable tanks shall not be offered for transport:

.1   in an ullage condition liable to produce an unacceptable hydraulic force due to surge within the portable tank;

.2   when leaking;

.3   when damaged to such an extent that the integrity of the tank or its lifting or securing arrangements may be affected; and

.4   unless the service equipment has been examined and found to be in good working order.

4.2.2.9   Forklift pockets of portable tanks shall be closed off when the tank is filled. This provision does not apply to portable tanks which, according to 6.7.3.13.4, need not be provided with a means of closing off the forklift pockets.

4.2.3   General provisions for the use of portable tanks for the transport of refrigerated liquefied gases of class 2

4.2.3.1   This section provides general provisions applicable to the use of portable tanks for the transport of refrigerated liquefied gases.

4.2.3.2   Portable tanks shall conform to the design, construction, inspection and testing provisions detailed in 6.7.4. Refrigerated liquefied gases shall be transported in portable tanks conforming to portable tank instruction T75 as described in 4.2.5.2.6 and the portable tank special provisions assigned to each substance in columns 12 and 14 of the Dangerous Goods List and described in 4.2.5.3.

4.2.3.3   During transport, portable tanks shall be adequately protected against damage to the shell and service equipment resulting from lateral and longitudinal impact and overturning. If the shell and service equipment are so constructed as to withstand impact or overturning, it need not be protected in this way. Examples of such protection are provided in 6.7.4.12.5.

4.2.3.4   Unless the name of the gas(es) being transported appears on the metal plate described in 6.7.4.15.2, a copy of the certificate specified in 6.7.4.13.1 shall be made available upon a competent authority request and readily provided by the consignor, consignee or agent, as appropriate.

4.2.3.5   Empty portable tanks not cleaned and not gas-free shall comply with the same provisions as portable tanks filled with the previous substance.

4.2.3.6   **Filling**

4.2.3.6.1   Prior to filling, the shipper shall ensure that the portable tank is approved for the refrigerated liquefied gas to be transported and that the portable tank is not loaded with refrigerated liquefied gases which, in contact with the materials of the shell, gaskets and service equipment, are likely to react dangerously with them to form dangerous products or appreciably weaken these materials. During filling, the temperature of the refrigerated liquefied gas shall be within the limits of the design temperature range.

4.2.3.6.2 In estimating the initial degree of filling, the necessary holding time for the intended journey, including any delays which might be encountered, shall be taken into consideration. The initial degree of filling of the shell, except as provided for in 4.2.3.6.3 and 4.2.3.6.4, shall be such that if the contents, except helium, were to be raised to a temperature at which the vapour pressure is equal to the maximum allowable working pressure (MAWP) the volume occupied by liquid would not exceed 98%.

4.2.3.6.3 Shells intended for the transport of helium can be filled up to but not above the inlet of the pressure-relief device.

4.2.3.6.4 A higher initial degree of filling may be allowed, subject to approval by the competent authority, when the intended duration of transport is considerably shorter than the holding time.

4.2.3.6.5 Portable tanks shall not be filled or discharged while they remain on board.

4.2.3.7 Actual holding time

4.2.3.7.1 The actual holding time shall be calculated for each journey in accordance with a procedure recognized by the competent authority, on the basis of the following:

.1 the reference holding time for the refrigerated liquefied gas to be transported (see 6.7.4.2.8.1) (as indicated on the plate referred to in 6.7.4.15.1);

.2 the actual filling density;

.3 the actual filling pressure;

.4 the lowest set pressure of the pressure-limiting device(s).

4.2.3.7.2 The actual holding time shall be marked either on the portable tank itself or on a metal plate firmly secured to the portable tank, in accordance with 6.7.4.15.2.

4.2.3.8 Portable tanks shall not be offered for transport:

.1 in an ullage condition liable to produce an unacceptable hydraulic force due to surge within the shell;

.2 when leaking;

.3 when damaged to such an extent that the integrity of the portable tank or its lifting or securing arrangements may be affected;

.4 unless the service equipment has been examined and found to be in good working order;

.5 unless the actual holding time for the refrigerated liquefied gas being transported has been determined in accordance with 4.2.3.7 and the portable tank is marked in accordance with 6.7.4.15.2; and

.6 unless the duration of transport, after taking into consideration any delays which might be encountered, does not exceed the actual holding time.

4.2.3.9 Forklift pockets of portable tanks shall be closed off when the tank is filled. This provision does not apply to portable tanks which, according to 6.7.4.12.4, need not be provided with a means of closing off the forklift pockets.

4.2.4 General provisions for the use of multiple-element gas containers (MEGCs)

4.2.4.1 This section provides general requirements applicable to the use of multiple-element gas containers (MEGCs) for the transport of non-refrigerated gases.

4.2.4.2 MEGCs shall conform to the design, construction, inspection and testing requirements detailed in 6.7.5. The elements of MEGCs shall be periodically inspected according to the provisions set out in packing instruction P200 and in 6.2.1.6.

4.2.4.3 During transport, MEGCs shall be protected against damage to the elements and service equipment resulting from lateral and longitudinal impact and overturning. If the elements and service equipment are so constructed as to withstand impact or overturning, they need not be protected in this way. Examples of such protection are given in 6.7.5.10.4.

4.2.4.4 The periodic testing and inspection requirements for MEGCs are specified in 6.7.5.12. MEGCs or their elements shall not be charged or filled after they become due for periodic inspection but may be transported after the expiry of the time limit.

4.2.4.5 **Filling**

4.2.4.5.1 Prior to filling, the MEGC shall be inspected to ensure that it is authorized for the gas to be transported and that the applicable provisions of this Code have been met.

4.2.4.5.2 Elements of MEGCs shall be filled according to the working pressures, filling ratios and filling provisions specified in packing instruction P200 for the specific gas being filled into each element. In no case shall a MEGC or group of elements be filled as a unit in excess of the lowest working pressure of any given element.

4.2.4.5.3 MEGCs shall not be filled above their maximum permissible gross mass.

4.2.4.5.4 Isolation valves shall be closed after filling and remain closed during transport. Toxic gases of class 2.3 shall only be transported in MEGCs where each element is equipped with an isolation valve.

4.2.4.5.5 The opening(s) for filling shall be closed by caps or plugs. The leakproofness of the closures and equipment shall be verified by the shipper after filling.

4.2.4.5.6 MEGCs shall not be offered for filling:

.1 when damaged to such an extent that the integrity of the pressure receptacles or their structural or service equipment may be affected;

.2 unless the pressure receptacles and their structural and service equipment have been examined and found to be in good working order; and

.3 unless the required certification, retest, and filling markings are legible.

4.2.4.6 Filled MEGCs shall not be offered for transport;

.1 when leaking;

.2 when damaged to such an extent that the integrity of the pressure receptacles or their structural or service equipment may be affected;

.3 unless the pressure receptacles and their structural and service equipment have been examined and found to be in good working order; and

.4 unless the required certification, retest, and filling markings are legible.

4.2.4.7 Empty MEGCs that have not been cleaned and purged shall comply with the same requirements as MEGCs filled with the previous substance.

4.2.5 **Portable tank instructions and special provisions**

4.2.5.1 **General**

4.2.5.1.1 This section includes the portable tank instructions and special provisions applicable to dangerous goods authorized to be transported in portable tanks. Each portable tank instruction is identified by an alpha-numeric designation (T1 to T75). The Dangerous Goods List in chapter 3.2 indicates the portable tank instruction that shall be used for each substance permitted for transport in a portable tank. When no portable tank instruction appears in the Dangerous Goods List, transport of the substance in portable tanks is not permitted unless a competent authority approval is granted as set out in 6.7.1.3. Portable tank special provisions are assigned to specific dangerous goods in the Dangerous Goods List in chapter 3.2. Each portable tank special provision is identified by an alpha-numeric designation (such as TP1). A listing of the portable tank special provisions is provided in 4.2.5.3.

Note: The gases authorized for transport in MEGCs are indicated in the column "MEGC" in Tables 1 and 2 of packing instruction P200 in 4.1.4.1.

4.2.5.2 **Portable tank instructions**

4.2.5.2.1 Portable tank instructions apply to dangerous goods of classes 1 to 9. Portable tank instructions provide specific information relevant to portable tank provisions applicable to specific substances. These provisions shall be met in addition to the general provisions in this chapter and chapter 6.7.

4.2.5.2.2 For substances of class 1 and classes 3 to 9, the portable tank instructions indicate the applicable minimum test pressure, the minimum shell thickness (in reference steel), bottom opening provisions and pressure-relief provisions. In T23, self-reactive substances of class 4.1 and class 5.2 organic peroxides permitted to be transported in portable tanks are listed along with applicable control and emergency temperatures.

4.2.5.2.3   Non-refrigerated liquefied gases are assigned to portable tank instruction T50. T50 provides the maximum allowable working pressures, bottom opening provisions, pressure-relief provisions and degree of filling provisions for non-refrigerated liquefied gases permitted for transport in portable tanks.

4.2.5.2.4   Refrigerated liquefied gases are assigned to portable tank instruction T75.

4.2.5.2.5   *Determination of the appropriate portable tank instructions*

When a specific portable tank instruction is specified in the Dangerous Goods List, additional portable tanks which possess higher test pressures, greater shell thicknesses, more stringent bottom opening and pressure-relief device arrangements may be used. The following guidelines apply to determining the appropriate portable tanks which may be used for transport of particular substances:

Portable tank instruction specified	Portable tank instructions also permitted
T1	T2, T3, T4, T5, T6, T7, T8, T9, T10, T11, T12, T13, T14, T15, T16, T17, T18, T19, T20, T21, T22
T2	T4, T5, T7, T8, T9, T10, T11, T12, T13, T14, T15, T16, T17, T18, T19, T20, T21, T22
T3	T4, T5, T6, T7, T8, T9, T10, T11, T12, T13, T14, T15, T16, T17, T18, T19, T20, T21, T22
T4	T5, T7, T8, T9, T10, T11, T12, T13, T14, T15, T16, T17, T18, T19, T20, T21, T22
T5	T10, T14, T19, T20, T22
T6	T7, T8, T9, T10, T11, T12, T13, T14, T15, T16, T17, T18, T19, T20, T21, T22
T7	T8, T9, T10, T11, T12, T13, T14, T15, T16, T17, T18, T19, T20, T21, T22
T8	T9, T10, T13, T14, T19, T20, T21, T22
T9	T10, T13, T14, T19, T20, T21, T22
T10	T14, T19, T20, T22
T11	T12, T13, T14, T15, T16, T17, T18, T19, T20, T21, T22
T12	T14, T16, T18, T19, T20, T22
T13	T14, T19, T20, T21, T22
T14	T19, T20, T22
T15	T16, T17, T18, T19, T20, T21, T22
T16	T18, T19, T20, T22
T17	T18 , T19, T20, T21, T22
T18	T19, T20, T22
T19	T20, T22
T20	T22
T21	T22
T22	None
T23	None
T50	None

4.2.5.2.6   Portable tank instructions

Portable tank instructions specify the provisions applicable to a portable tank when used for the transport of specific substances. Portable tank instructions T1 to T22 specify the applicable minimum test pressure, the minimum shell thickness (in mm of reference steel), and the pressure relief and bottom-opening provisions.

T1 – T22	PORTABLE TANK INSTRUCTIONS			T1 – T22
These portable tank instructions apply to liquid and solid substances of classes 3 tc 9. The general provisions of 6.7.2 shall be met.				
Portable tank instruction	Minimum test pressure (bar)	Minimum shell thickness (in mm – reference steel) (see 6.7.2.4)	Pressure relief provisions[a] (see 6.7.2.8)	Bottom opening provisions (see 6.7.2.6)
T1	1.5	See 6.7.2.4.2	Normal	See 6.7.2.6.2
T2	1.5	See 6.7.2.4.2	Normal	See 6.7.2.6.3
T3	2.65	See 6.7.2.4.2	Normal	See 6.7.2.6.2
T4	2.65	See 6.7.2.4.2	Normal	See 6.7.2.6.3
T5	2.65	See 6.7.2.4.2	See 6.7.2.8.3	Not allowed
T6	4	See 6.7.2.4.2	Normal	See 6.7.2.6.2
T7	4	See 6.7.2.4.2	Normal	See 6.7.2.6.3
T8	4	See 6.7.2.4.2	Normal	Not allowed
T9	4	6 mm	Normal	Not allowed
T10	4	6 mm	See 6.7.2.8.3	Not allowed
T11	6	See 6.7.2.4.2	Normal	See 6.7.2.6.3
T12	6	See 6.7.2.4.2	See 6.7.2.8.3	See 6.7.2.6.3
T13	6	6 mm	Normal	Not allowed
T14	6	6 mm	See 6.7.2.8.3	Not allowed
T15	10	See 6.7.2.4.2	Normal	See 6.7.2.6.3
T16	10	See 6.7.2.4.2	See 6.7.2.8.3	See 6.7.2.6.3
T17	10	6 mm	Normal	See 6.7.2.6.3
T18	10	6 mm	See 6.7.2.8.3	See 6.7.2.6.3
T19	10	6 mm	See 6.7.2.8.3	Not allowed
T20	10	8 mm	See 6.7.2.8.3	Not allowed
T21	10	10 mm	Normal	Not allowed
T22	10	10 mm	See 6.7.2.8.3	Not allowed

[a] When the word "Normal" is indicated, all the provisions of 6.7.2.8 apply except for 6.7.2.8.3.

| T23 | PORTABLE TANK INSTRUCTION | | | | | | T23 |

This portable tank instruction applies to substances of class 4.1 and class 5.2, organic peroxides. The general provisions of 4.2.1 and the provisions of 6.7.2 shall be met. The provisions specific to self-reactive substances of class 4.1 and organic peroxides of class 5.2 in 4.2.1.13 shall also be met.

UN No.	Substance	Minimum test pressure (bar)	Minimum shell thickness (mm – reference steel)	Bottom opening require-ments	Pressure relief require-ments	Degree of filling	Control tempe-rature	Emergency temperature
3109	**ORGANIC PEROXIDE TYPE F, LIQUID** *tert*-Butyl hydroperoxide,* not more than 72% with water Cumyl hydroperoxide, not more than 90% in diluent type A Di-*tert*-butyl peroxide, not more than 32% in diluent type A Isopropyl cumyl hydroperoxide, not more than 72% in diluent type A *p*-Menthyl hydroperoxide, not more than 72% in diluent type A Pinanyl hydroperoxide, not more than 56% in diluent type A	4	See 6.7.2.4.2	See 6.7.2.6.3	See 6.7.2.8.2, 4.2.1.13.6, 4.2.1.13.7, 4.2.1.13.8	See 4.2.1.13.13		
3110	**ORGANIC PEROXIDE TYPE F, SOLID** Dicumyl peroxide[†]	4	See 6.7.2.4.2	See 6.7.2.6.3	See 6.7.2.8.2, 4.2.1.13.6, 4.2.1.13.7, 4.2.1.13.8	See 4.2.1.13.13		
3119	**ORGANIC PEROXIDE TYPE F, LIQUID, TEMPERATURE CONTROLLED** *tert*-Amyl peroxyneodecanoate, not more than 47% in diluent type A *tert*-Butyl peroxyacetate, not more than 32% in diluent type B *tert*-Butyl peroxy-2-ethyl-hexanoate, not more than 32% in diluent type B *tert*-Butyl peroxypivalate, not more than 27% in diluent type B *tert*-Butyl peroxy-3,5,5-trimethylhexanoate, not more than 32% in diluent type B Di-(3,5,5-trimethylhexanoyl) peroxide, not more than 38% in diluent type A or type B Peroxyacetic acid, distilled, stabilized[§]	4	See 6.7.2.4.2	See 6.7.2.6.3	See 6.7.2.8.2, 4.2.1.13.6, 4.2.1.13.7, 4.2.1.13.8	See 4.2.1.13.13	[‡] –10°C +30°C +15°C +5°C +35°C 0°C +30°C	[‡] –5°C +35°C +20°C +10°C +40°C +5°C +35°C
3120	**ORGANIC PEROXIDE TYPE F, SOLID, TEMPERATURE CONTROLLED**	4	See 6.7.2.4.2	See 6.7.2.6.3	See 6.7.2.8.2, 4.2.1.13.6, 4.2.1.13.7, 4.2.1.13.8	See 4.2.1.13.13	[‡]	[‡]

* Provided that steps have been taken to achieve the safety equivalence of 65% *tert*-butyl hydroperoxide and 35% water.

[†] Maximum quantity per portable tank: 2000 kg.

[‡] As approved by the competent authority.

[§] Formulation derived from distillation of peroxyacetic acid originating from peroxyacetic acid in concentration of not more than 41% with water, total active oxygen (peroxyacetic acid + H_2O_2) $\leqslant$ 9.5%, which fulfils the criteria of 2.5.3.3.2.6.

T23	PORTABLE TANK INSTRUCTION (continued)							T23
UN No.	Substance	Minimum test pressure (bar)	Minimum shell thickness (mm – reference steel)	Bottom opening require-ments	Pressure relief require-ments	Degree of filling	Control tempe-rature	Emergency temperature
3229	SELF-REACTIVE LIQUID TYPE F	4	See 6.7.2.4.2	See 6.7.2.6.3	See 6.7.2.8.2, 4.2.1.13.6, 4.2.1.13.7, 4.2.1.13.8	See 4.2.1.13.13		
3230	SELF-REACTIVE SOLID TYPE F	4	See 6.7.2.4.2	See 6.7.2.6.3	See 6.7.2.8.2, 4.2.1.13.6, 4.2.1.13.7, 4.2.1.13.8	See 4.2.1.13.13		
3239	SELF-REACTIVE LIQUID TYPE F, TEMPERATURE CONTROLLED	4	See 6.7.2.4.2	See 6.7.2.6.3	See 6.7.2.8.2, 4.2.1.13.6, 4.2.1.13.7, 4.2.1.13.8	See 4.2.1.13.13	*	*
3240	SELF-REACTIVE SOLID TYPE F, TEMPERATURE CONTROLLED	4	See 6.7.2.4.2	See 6.7.2.6.3	See 6.7.2.8.2, 4.2.1.13.6, 4.2.1.13.7, 4.2.1.13.8	See 4.2.1.13.13	*	*

* As approved by the competent authority.

T50	PORTABLE TANK INSTRUCTION				T50
This portable tank instruction applies to non-refrigerated liquefied gases. The general provisions of 4.2.2 and the provisions of 6.7.3 shall be met.					
UN No.	Non-refrigerated liquefied gases	Maximum allowable working pressure (bar) Small; Bare; Sunshield; Insulated respectively[a]	Openings below liquid level	Pressure relief provisions[b] (see 6.7.3.7)	Maximum filling density (kg/ℓ)
1005	Ammonia, anhydrous	29.0 25.7 22.0 19.7	Allowed	See 6.7.3.7.3	0.53
1009	Bromotrifluoromethane (Refrigerant gas R 13B1)	38.0 34.0 30.0 27.5	Allowed	Normal	1.13
1010	Butadienes, stabilized	7.5 7.0 7.0 7.0	Allowed	Normal	0.55
1010	Butadienes and hydrocarbon mixture, stabilized with more than 40% butadienes	See MAWP definition in 6.7.3.1	Allowed	Normal	See 4.2.2.7
1011	Butane	7.0 7.0 7.0 7.0	Allowed	Normal	0.51

[a] "Small" means tanks having a shell with a diameter of 1.5 metres or less; "Bare" means tanks having a shell with a diameter of more than 1.5 metres without insulation or sun shield (see 6.7.3.2.12); "Sunshield" means tanks having a shell with a diameter of more than 1.5 metres with sun shield (see 6.7.3.2.12); "Insulated" means tanks having a shell with a diameter of more than 1.5 metres with insulation (see 6.7.3.2.12); (See definition of "Design reference temperature" in 6.7.3.1).

[b] The word "Normal" in the pressure relief column indicates that a frangible disc as specified in 6.7.3.7.3 is not required.

T50	PORTABLE TANK INSTRUCTION *(continued)*				T50
UN No.	Non-refrigerated liquefied gases	Maximum allowable working pressure (bar) Small; Bare; Sunshield; Insulated respectively[a]	Openings below liquid level	Pressure relief provisions[b] (see 6.7.3.7)	Maximum filling density (kg/ℓ)
1012	Butylene	8.0 7.0 7.0 7.0	Allowed	Normal	0.53
1017	Chlorine	19.0 17.0 15.0 13.5	Not allowed	See 6.7.3.7.3	1.25
1018	Chlorodifluoromethane (Refrigerant gas R 22)	26.0 24.0 21.0 19.0	Allowed	Normal	1.03
1020	Chloropentafluoroethane (Refrigerant gas R 115)	23.0 20.0 18.0 16.0	Allowed	Normal	1.06
1021	1-Chloro-1,2,2,2-tetrafluoroethane (Refrigerant gas R 124)	10.3 9.8 7.9 7.0	Allowed	Normal	1.20
1027	Cyclopropane	18.0 16.0 14.5 13.0	Allowed	Normal	0.53
1028	Dichlorodifluoromethane (Refrigerant gas R 12)	16.0 15.0 13.0 11.5	Allowed	Normal	1.15
1029	Dichlorofluoromethane (Refrigerant gas R 21)	7.0 7.0 7.0 7.0	Allowed	Normal	1.23
1030	1,1-Difluoroethane (Refrigerant gas R 152a)	16.0 14.0 12.4 11.0	Allowed	Normal	0.79
1032	Dimethylamine, anhydrous	7.0 7.0 7.0 7.0	Allowed	Normal	0.59
1033	Dimethyl ether	15.5 13.8 12.0 10.6	Allowed	Normal	0.58
1036	Ethylamine	7.0 7.0 7.0 7.0	Allowed	Normal	0.61
1037	Ethyl chloride	7.0 7.0 7.0 7.0	Allowed	Normal	0.80

[a] "Small" means tanks having a shell with a diameter of 1.5 metres or less; "Bare" means tanks having a shell with a diameter of more than 1.5 metres without insulation or sun shield (see 6.7.3.2.12); "Sunshield" means tanks having a shell with a diameter of more than 1.5 metres with sun shield (see 6.7.3.2.12); "Insulated" means tanks having a shell with a diameter of more than 1.5 metres with insulation (see 6.7.3.2.12); (See definition of "Design reference temperature" in 6.7.3.1).

[b] The word "Normal" in the pressure relief column indicates that a frangible disc as specified in 6.7.3.7.3 is not required.

T50	PORTABLE TANK INSTRUCTION *(continued)*				T50
UN No.	Non-refrigerated liquefied gases	Maximum allowable working pressure (bar) Small; Bare; Sunshield; Insulated respectively[a]	Openings below liquid level	Pressure relief provisions[b] (see 6.7.3.7)	Maximum filling density (kg/ℓ)
1040	Ethylene oxide with nitrogen up to a total pressure of 1 MPa (10 bar) at 50°C	– – – 10.0	Not allowed	See 6.7.3.7.3	0.78
1041	Ethylene oxide and carbon dioxide mixture with more than 9% but not more than 87% ethylene oxide	See MAWP definition in 6.7.3.1	Allowed	Normal	See 4.2.2.7
1055	Isobutylene	8.1 7.0 7.0 7.0	Allowed	Normal	0.52
1060	Methylacetylene and propadiene mixture, stabilized	28.0 24.5 22.0 20.0	Allowed	Normal	0.43
1061	Methylamine, anhydrous	10.8 9.6 7.8 7.0	Allowed	Normal	0.58
1062	Methyl bromide with not more than 2% chloropicrin	7.0 7.0 7.0 7.0	Not allowed	See 6.7.3.7.3	1.51
1063	Methyl chloride (Refrigerant gas R40)	14.5 12.7 11.3 10.0	Allowed	Normal	0.81
1064	Methyl mercaptan	7.0 7.0 7.0 7.0	Not allowed	See 6.7.3.7.3	0.78
1067	Dinitrogen tetroxide	7.0 7.0 7.0 7.0	Not allowed	See 6.7.3.7.3	1.30
1075	Petroleum gas, liquefied	See MAWP definition in 6.7.3.1	Allowed	Normal	See 4.2.2.7
1077	Propylene	28.0 24.5 22.0 20.0	Allowed	Normal	0.43
1078	Refrigerant gas, N.O.S.	See MAWP definition in 6.7.3.1	Allowed	Normal	See 4.2.2.7
1079	Sulphur dioxide	11.6 10.3 8.5 7.6	Not allowed	See 6.7.3.7.3	1.23
1082	Trifluorochloroethylene, stabilized (Refrigerant gas R 1113)	17.0 15.0 13.1 11.6	Not allowed	See 6.7.3.7.3	1.13

[a] "Small" means tanks having a shell with a diameter of 1.5 metres or less; "Bare" means tanks having a shell with a diameter of more than 1.5 metres without insulation or sun shield (see 6.7.3.2.12); "Sunshield" means tanks having a shell with a diameter of more than 1.5 metres with sun shield (see 6.7.3.2.12); "Insulated" means tanks having a shell with a diameter of more than 1.5 metres with insulation (see 6.7.3.2.12); (See definition of "Design reference temperature" in 6.7.3.1).

[b] The word "Normal" in the pressure relief column indicates that a frangible disc as specified in 6.7.3.7.3 is not required.

T50	PORTABLE TANK INSTRUCTION *(continued)*				T50
UN No.	Non-refrigerated liquefied gases	Maximum allowable working pressure (bar) Small; Bare; Sunshield; Insulated respectively[a]	Openings below liquid level	Pressure relief provisions[b] (see 6.7.3.7)	Maximum filling density (kg/ℓ)
1083	Trimethylamine, anhydrous	7.0 7.0 7.0 7.0	Allowed	Normal	0.56
1085	Vinyl bromide, stabilized	7.0 7.0 7.0 7.0	Allowed	Normal	1.37
1086	Vinyl chloride, stabilized	10.6 9.3 8.0 7.0	Allowed	Normal	0.81
1087	Vinyl methyl ether, stabilized	7.0 7.0 7.0 7.0	Allowed	Normal	0.67
1581	Chloropicrin and methyl bromide mixture with more than 2% chloropicrin	7.0 7.0 7.0 7.0	Not allowed	See 6.7.3.7.3	1.51
1582	Chloropicrin and methyl chloride mixture	19.2 16.9 15.1 13.1	Not allowed	See 6.7.3.7.3	0.81
1858	Hexafluoropropylene (Refrigerant gas R 1216)	19.2 16.9 15.1 13.1	Allowed	Normal	1.11
1912	Methyl chloride and methylene chloride mixture	15.2 13.0 11.6 10.1	Allowed	Normal	0.81
1958	1,2-Dichloro-1,1,2,2-tetrafluoroethane (Refrigerant gas R 114)	7.0 7.0 7.0 7.0	Allowed	Normal	1.30
1965	Hydrocarbon gas, mixture liquefied, N.O.S.	See MAWP definition in 6.7.3.1	Allowed	Normal	See 4.2.2.7
1969	Isobutane	8.5 7.5 7.0 7.0	Allowed	Normal	0.49
1973	Chlorodifluoromethane and chloropentafluoroethane mixture with fixed boiling point, with approximately 49% chlorodifluoromethane (Refrigerant gas R 502)	28.3 25.3 22.8 20.3	Allowed	Normal	1.05
1974	Chlorodifluorobromomethane (Refrigerant gas R 12B1)	7.4 7.0 7.0 7.0	Allowed	Normal	1.61

[a] "Small" means tanks having a shell with a diameter of 1.5 metres or less; "Bare" means tanks having a shell with a diameter of more than 1.5 metres without insulation or sun shield (see 6.7.3.2.12); "Sunshield" means tanks having a shell with a diameter of more than 1.5 metres with sun shield (see 6.7.3.2.12); "Insulated" means tanks having a shell with a diameter of more than 1.5 metres with insulation (see 6.7.3.2.12); (See definition of "Design reference temperature" in 6.7.3.1).

[b] The word "Normal" in the pressure relief column indicates that a frangible disc as specified in 6.7.3.7.3 is not required.

T50	PORTABLE TANK INSTRUCTION *(continued)*				T50
UN No.	Non-refrigerated liquefied gases	Maximum allowable working pressure (bar) Small; Bare; Sunshield; Insulated respectively[a]	Openings below liquid level	Pressure relief provisions[b] (see 6.7.3.7)	Maximum filling density (kg/ℓ)
1976	Octafluorocyclobutane (Refrigerant gas RC 318)	8.8 7.8 7.0 7.0	Allowed	Normal	1.34
1978	Propane	22.5 20.4 18.0 16.5	Allowed	Normal	0.42
1983	1-Chloro-2,2,2-trifluoroethane (Refrigerant gas R 133a)	7.0 7.0 7.0 7.0	Allowed	Normal	1.18
2035	1,1,1-Trifluoroethane (Refrigerant gas R 143a)	31.0 27.5 24.2 21.8	Allowed	Normal	0.76
2424	Octafluoropropane (Refrigerant gas R 218)	23.1 20.8 18.6 16.6	Allowed	Normal	1.07
2517	1-Chloro-1,1-difluoroethane (Refrigerant gas R 142b)	8.9 7.8 7.0 7.0	Allowed	Normal	0.99
2602	Dichlorodifluoromethane and difluoroethane azeotropic mixture with approximately 74% dichlorodifluoro-methane (Refrigerant gas R 500)	20.0 18.0 16.0 14.5	Allowed	Normal	1.01
3057	Trifluoroacetyl chloride	14.6 12.9 11.3 9.9	Not allowed	See 6.7.3.7.3	1.17
3070	Ethylene oxide and dichlorodifluoro-methane mixture, with not more than 12.5% ethylene oxide	14.0 12.0 11.0 9.0	Allowed	See 6.7.3.7.3	1.09
3153	Perfluoro(methyl vinyl ether)	14.3 13.4 11.2 10.2	Allowed	Normal	1.14
3159	1,1,1,2-Tetrafluoroethane (Refrigerant gas R 134a)	17.7 15.7 13.8 12.1	Allowed	Normal	1.04
3161	Liquefied gas, flammable, N.O.S.	See MAWP definition in 6.7.3.1	Allowed	Normal	See 4.2.2.7
3163	Liquefied gas, N.O.S.	See MAWP definition in 6.7.3.1	Allowed	Normal	See 4.2.2.7
3220	Pentafluoroethane (Refrigerant gas R 125)	34.4 30.8 27.5 24.5	Allowed	Normal	0.95

[a] "Small" means tanks having a shell with a diameter of 1.5 metres or less; "Bare" means tanks having a shell with a diameter of more than 1.5 metres without insulation or sun shield (see 6.7.3.2.12); "Sunshield" means tanks having a shell with a diameter of more than 1.5 metres with sun shield (see 6.7.3.2.12); "Insulated" means tanks having a shell with a diameter of more than 1.5 metres with insulation (see 6.7.3.2.12); (See definition of "Design reference temperature" in 6.7.3.1).

[b] The word "Normal" in the pressure relief column indicates that a frangible disc as specified in 6.7.3.7.3 is not required.

4

T50	PORTABLE TANK INSTRUCTION *(continued)*				T50
UN No.	Non-refrigerated liquefied gases	Maximum allowable working pressure (bar) Small; Bare; Sunshield; Insulated respectively[a]	Openings below liquid level	Pressure relief provisions[b] (see 6.7.3.7)	Maximum filling density (kg/ℓ)
3252	Difluoromethane (Refrigerant gas R 32)	43.0 39.0 34.4 30.5	Allowed	Normal	0.78
3296	Heptafluoropropane (Refrigerant gas R 227)	16.0 14.0 12.5 11.0	Allowed	Normal	1.20
3297	Ethylene oxide and chlorotetrafluoro-ethane mixture, with not more than 8.8% ethylene oxide	8.1 7.0 7.0 7.0	Allowed	Normal	1.16
3298	Ethylene oxide and pentafluoroethane mixture, with not more than 7.9% ethylene oxide	25.9 23.4 20.9 18.6	Allowed	Normal	1.02
3299	Ethylene oxide and tetrafluoroethane mixture, with not more than 5.6% ethylene oxide	16.7 14.7 12.9 11.2	Allowed	Normal	1.03
3318	Ammonia solution, relative density less than 0.880 at 15°C in water, with more than 50% ammonia	See MAWP definition in 6.7.3.1	Allowed	See 6.7.3.7.3	See 4.2.2.7
3337	Refrigerant gas R 404A	31.6 28.3 25.3 22.5	Allowed	Normal	0.82
3338	Refrigerant gas R 407A	31.3 28.1 25.1 22.4	Allowed	Normal	0.94
3339	Refrigerant gas R 407B	33.0 29.6 26.5 23.6	Allowed	Normal	0.93
3340	Refrigerant gas R 407C	29.9 26.8 23.9 21.3	Allowed	Normal	0.95

[a] "Small" means tanks having a shell with a diameter of 1.5 metres or less; "Bare" means tanks having a shell with a diameter of more than 1.5 metres without insulation or sun shield (see 6.7.3.2.12); "Sunshield" means tanks having a shell with a diameter of more than 1.5 metres with sun shield (see 6.7.3.2.12); "Insulated" means tanks having a shell with a diameter of more than 1.5 metres with insulation (see 6.7.3.2.12); (See definition of "Design reference temperature" in 6.7.3.1).

[b] The word "Normal" in the pressure relief column indicates that a frangible disc as specified in 6.7.3.7.3 is not required.

T75	PORTABLE TANK INSTRUCTION	T75
	This portable tank instruction applies to refrigerated liquefied gases. The general provisions of 4.2.3 and 6.7.4 shall be met.	

4.2.5.3 Portable tank special provisions

Portable tank special provisions are assigned to certain substances to indicate provisions which are in addition to or in lieu of those provided by the portable tank instructions or the provisions in chapter 6.7. Portable tank special provisions are identified by an alpha-numeric designation beginning with the letters "TP" (tank provision) and are assigned to specific substances in column 14 of the Dangerous Goods List in chapter 3.2. The following is a list of the portable tank special provisions:

TP1   The degree of filling prescribed in 4.2.1.9.2 shall not be exceeded.

TP2   The degree of filling prescribed in 4.2.1.9.3 shall not be exceeded.

TP3   The maximum degree of filling (in %) for solids transported above their melting points and for elevated temperature liquids shall be determined in accordance with 4.2.1.9.5.

TP4   The degree of filling shall not exceed 90% or, alternatively, any other value approved by the competent authority (see 4.2.1.16.2).

TP5   The degree of filling prescribed in 4.2.3.6 shall be met.

TP6   To prevent the tank bursting in any event, including fire engulfment, it shall be provided with pressure-relief devices which are adequate in relation to the capacity of the tank and to the nature of the substance transported. The device shall also be compatible with the substance.

TP7   Air shall be eliminated from the vapour space by nitrogen or other means.

TP8   The test pressure for the portable tank may be reduced to 1.5 bar when the flashpoint of the substances transported is greater than 0°C.

TP9   A substance under this description shall only be transported in a portable tank under an approval granted by the competent authority.

TP10   A lead lining, not less than 5 mm thick, which shall be tested annually, or another suitable lining material approved by the competent authority is required.

TP11   [Reserved].

TP12   [Reserved].

TP13   Self-contained breathing apparatus shall be provided when this substance is transported, unless no self-contained breathing apparatus, as required by SOLAS regulation II-2/19 (II-2/54), is on board.

TP14   [Reserved].

TP15   [Reserved].

TP16   The tank shall be fitted with a special device to prevent under-pressure and excess pressure during normal transport conditions. This device shall be approved by the competent authority. Pressure-relief provisions are as indicated in 6.7.2.8.3 to prevent crystallization of the product in the pressure-relief valve.

TP17   Only inorganic non-combustible materials shall be used for thermal insulation of the tank.

TP18   Temperature shall be maintained between 18°C and 40°C. Portable tanks containing solidified methacrylic acid shall not be reheated during transport.

TP19   The calculated shell thickness shall be increased by 3 mm. Shell thickness shall be verified ultrasonically at intervals midway between periodic hydraulic tests.

TP20   This substance shall only be transported in insulated tanks under a nitrogen blanket.

TP21   The shell thickness shall be not less than 8 mm. Tanks shall be hydraulically tested and internally inspected at intervals not exceeding 2.5 years.

TP22   Lubricant for joints or other devices shall be oxygen-compatible.

TP23   Transport permitted under special conditions prescribed by the competent authorities.

TP24   The portable tank may be fitted with a device located, under maximum filling conditions, in the vapour space of the shell to prevent the build-up of excess pressure due to the slow decomposition of the substance transported. This device shall also prevent an unacceptable amount of leakage of liquid in the case of overturning or entry of foreign matter into the tank. This device shall be approved by the competent authority or its authorized body.

TP25   Sulphur trioxide 99.95% pure and above may be transported in tanks without an inhibitor provided that it is maintained at a temperature equal to or above 32.5°C.

TP26   When transported under heated conditions, the heating device shall be fitted outside the shell. For UN 3176, this provision only applies when the substance reacts dangerously with water.

TP27   A portable tank having a minimum test pressure of 4 bar may be used if it is shown that a test pressure of 4 bar or less is acceptable according to the test pressure definition in 6.7.2.1.

TP28   A portable tank having a minimum test pressure of 2.65 bar may be used if it is shown that a test pressure of 2.65 bar or less is acceptable according to the test pressure definition in 6.7.2.1.

TP29   A portable tank having a minimum test pressure of 1.5 bar may be used if it is shown that a test pressure of 1.5 bar or less is acceptable according to the test pressure definition in 6.7.2.1.

TP30   This substance shall be transported in insulated tanks.

TP31   This substance shall be transported in tanks in solid state.

TP32   For UN 0331, UN 0332 and UN 3375, portable tanks may be used subject to the following conditions:

   (a)   To avoid unnecessary confinement, each portable tank constructed of metal shall be fitted with a pressure relief device that may be of the re-closing spring-loaded type, a frangible disc or a fusible element. The set-to-discharge or burst pressure, as applicable, shall not be greater than 2.65 bar for portable tanks with minimum test pressures greater than 4 bar.

   (b)   Suitability for transport in tanks shall be demonstrated. One method to evaluate this suitability is test 8 (d) in Test Series 8 (see United Nations *Manual of Tests and Criteria*, Part 1, sub-section 18.7).

   (c)   Substances shall not be allowed to remain in the portable tank for any period that could result in caking. Appropriate measures shall be taken to avoid accumulation and packing of substances in the tank (e.g. cleaning, etc).

TP33   The portable tank instruction assigned for this substance applies for granular and powdered solids and for solids which are filled and discharged at temperatures above their melting point and which are cooled and transported as a solid mass. For solids which are transported above their melting point, see 4.2.1.19.

TP34   Portable tanks need not be subjected to the impact test in 6.7.4.14.1 if the portable tank is marked "NOT FOR RAIL TRANSPORT" on the plate specified in 6.7.4.15.1 and also in letters at least 10 cm high on both sides of the outer jacket.

TP35   Portable tank instruction T14 may continue to be applied until 31 December 2014.

TP90   Tanks with bottom openings may be used on short international voyages.

TP91   Portable tanks with bottom openings may also be used on long international voyages.

4.2.6   Additional provisions for the use of road tank vehicles

4.2.6.1   The tank of a road tank vehicle shall be attached to the vehicle during normal operations of filling, discharge and transport. IMO type 4 tanks shall be attached to the chassis when transported on board ships. Road tank vehicles shall not be filled or discharged while they remain on board. A road tank vehicle shall be driven on board on its own wheels and be fitted with permanent tie-down attachments for securing on board the ship.

4.2.6.2   Road tank vehicles shall comply with the provisions of chapter 6.8. IMO type 4, 6 and 8 tanks may be used according to the provisions of chapter 6.8 for short international voyages only.

Chapter 4.3

Use of bulk containers

Note: Sheeted bulk containers shall not be used for sea transport.

4.3.1   General provisions

4.3.1.1   These general provisions are applicable to the use of containers for the transport of solid substances in bulk. Substances shall be transported in closed bulk containers conforming to the applicable bulk container instruction identified by the code BK2 in column 13 of the Dangerous Goods List in chapter 3.2. The closed bulk container used shall conform to the requirements of chapter 6.9.

4.3.1.2   Except as provided in 4.3.1.3, bulk containers shall only be used when a substance is assigned a bulk container code in column 13 of the Dangerous Goods List.

4.3.1.3   When a substance is not assigned a bulk container code in column 13 of the Dangerous Goods List, interim approval for transport may be issued by the competent authority of the country of origin. The approval shall be included in the documentation of the consignment and contain, as a minimum, the information normally provided in the bulk container instruction and the conditions under which the substance shall be transported. Appropriate measures should be initiated by the competent authority to have the assignment included in the Dangerous Goods List.

4.3.1.4   Substances which may become liquid at temperatures likely to be encountered during transport are not permitted in bulk containers.

4.3.1.5   Bulk containers shall be siftproof and shall be so closed that none of the contents can escape under normal conditions of transport, including the effect of vibration, or by changes of temperature, humidity or pressure.

4.3.1.6   Bulk solids shall be loaded into bulk containers and evenly distributed in a manner that minimizes movement that could result in damage to the container or leakage of the dangerous goods.

4.3.1.7   Where venting devices are fitted, they shall be kept clear and operable.

4.3.1.8   Bulk solids shall not react dangerously with the material of the bulk container, gaskets, equipment including lids and tarpaulins, or with protective coatings which are in contact with the contents, or significantly weaken them. Bulk containers shall be so constructed or adapted that the goods cannot penetrate between wooden floor coverings or come into contact with those parts of the bulk containers that may be affected by the dangerous goods or residues thereof.

4.3.1.9   Before being filled and offered for transport, each bulk container shall be inspected and cleaned to ensure that it does not contain any residue on the interior or exterior that could:

–   cause a dangerous reaction with the substance intended for transport;

–   detrimentally affect the structural integrity of the bulk container; or

–   affect the dangerous goods retention capabilities of the bulk container.

4.3.1.10   During transport, no dangerous residues shall adhere to the outer surfaces of a bulk container.

4.3.1.11   If several closure systems are fitted in series, the system which is located nearest to the dangerous goods to be transported shall be closed first before filling.

4.3.1.12   Empty bulk containers that have contained dangerous goods shall be treated in the same manner as is prescribed in this Code for a filled bulk container, unless adequate measures have been taken to nullify any hazard.

4.3.1.13   If bulk containers are used for the carriage of bulk goods liable to cause a dust explosion, or evolve flammable vapours (e.g., for certain wastes), measures shall be taken to exclude sources of ignition and to prevent dangerous electrostatic discharge during transport, loading or unloading of the goods.

4.3.1.14 Substances, for example wastes, which may react dangerously with one another and substances of different classes and goods not subject to this Code, which are liable to react dangerously with one another, shall not be mixed together in the same bulk container. Dangerous reactions are:

.1 combustion and/or evolution of considerable heat;

.2 emission of flammable and/or toxic gases;

.3 formation of corrosive liquids; or

.4 formation of unstable substances.

4.3.1.15 Before a bulk container is filled, it shall be visually examined to ensure it is structurally serviceable, its interior walls, ceiling and floors are free from protrusions or damage and that any inner liners or substance retaining equipment are free from rips, tears or any damage that would compromise its cargo retention capabilities. "Structurally serviceable" means the bulk container does not have major defects in its structural components, such as top and bottom side rails, top and bottom end rails, door sill and header, floor cross members, corner posts, and corner fittings in a freight container. Major defects include:

.1 bends, cracks or breaks in the structural or supporting members that affect the integrity of the container;

.2 more than one splice or an improper splice (such as a lapped splice) in top or bottom end rails or door headers;

.3 more than two splices in any one top or bottom side rail;

.4 any splice in a door sill or corner post;

.5 door hinges and hardware that are seized, twisted, broken, missing, or otherwise inoperative;

.6 gaskets and seals that do not seal;

.7 any distortion of the overall configuration great enough to prevent proper alignment of handling equipment, mounting and securing chassis or vehicle, or insertion into ships' cargo spaces;

.8 any damage to lifting attachments or handling equipment interface features; or

.9 any damage to service or operational equipment.

4.3.2 Additional provisions applicable to bulk goods of classes 4.2, 4.3, 5.1, 6.2, 7 and 8

4.3.2.1 Bulk goods of class 4.2

The total mass carried in a bulk container shall be such that its spontaneous ignition temperature is greater than 55°C.

4.3.2.2 Bulk goods of class 4.3

Such goods shall be transported in bulk containers which are watertight.

4.3.2.3 Bulk goods of class 5.1

Bulk containers shall be so constructed or adapted that the goods cannot come into contact with wood or any other incompatible material.

4.3.2.4 Bulk goods of class 6.2

4.3.2.4.1 *Transport in bulk containers of animal material of class 6.2*

Animal material containing infectious substances (UN Nos. 2814, 2900 and 3373) is authorized for transport in bulk containers provided the following conditions are met:

.1 Closed bulk containers, and their openings, shall be leakproof by design or by the fitting of a suitable liner.

.2 The animal material shall be thoroughly treated with an appropriate disinfectant before loading prior to transport.

.3 Closed bulk containers shall not be re-used until they have been thoroughly cleaned and disinfected.

Note: Additional provisions may be required by appropriate national health authorities.

4.3.2.4.2 *Bulk wastes of class 6.2 (UN 3291)*

.1 only closed bulk containers (BK2) shall be permitted;

.2 closed bulk containers, and their openings, shall be leakproof by design. These bulk containers shall have non-porous interior surfaces and shall be free from cracks or other features that could damage packagings inside, impede disinfection or permit inadvertent release;

.3 wastes of UN 3291 shall be contained within the closed bulk container in UN type tested and approved sealed leakproof plastics bags tested for solids of packing group II and marked in accordance with 6.1.3.1. Such plastics bags shall be capable of passing the tests for tear and impact resistance according to ISO 7765-1:1988 "Plastics film and sheeting – Determination of impact resistance by the free-falling dart method – Part 1: Staircase methods" and ISO 6383-2:1983 "Plastics – Film and sheeting – Determination of tear resistance – Part 2: Elmendorf method". Each bag shall have an impact resistance of at least 165 g and a tear resistance of at least 480 g in both parallel and perpendicular planes with respect to the length of the bag. The maximum net mass of each plastics bag shall be 30 kg;

.4 single articles exceeding 30 kg such as soiled mattresses may be transported without the need for a plastics bag when authorized by the competent authority;

.5 wastes of UN 3291 which contain liquids shall only be transported in plastics bags containing sufficient absorbent material to absorb the entire amount of liquid without it spilling in the bulk container;

.6 wastes of UN 3291 containing sharp objects shall only be transported in UN type tested and approved rigid packagings meeting the provisions of packing instructions P621, IBC620 or LP621.

.7 rigid packagings specified in packing instructions P621, IBC620 or LP621 may also be used. They shall be properly secured to prevent damage during normal conditions of transport. Wastes transported in rigid packagings and plastics bags together in the same closed bulk container shall be adequately segregated from each other, e.g., by suitable rigid barriers or dividers, mesh nets or otherwise securing the packagings, such that they prevent damage to the packagings during normal conditions of transport;

.8 wastes of UN 3291 in plastics bags shall not be compressed in a closed bulk container in such a way that bags may be rendered no longer leakproof;

.9 the closed bulk container shall be inspected for leakage or spillage after each journey. If any wastes of UN 3291 have leaked or been spilled in the closed bulk container, it shall not be re-used until after it has been thoroughly cleaned and, if necessary, disinfected or decontaminated with an appropriate agent. No other goods shall be transported together with UN 3291 other than medical or veterinary wastes. Any such other wastes transported in the same closed bulk container shall be inspected for possible contamination.

4.3.2.5 **Bulk material of class 7**

For the transport of unpackaged radioactive material, see 4.1.9.2.3.

4.3.2.6 **Bulk goods of class 8**

Such goods shall be transported in closed bulk containers which are watertight.

4

4.

PART 5

CONSIGNMENT PROCEDURES

5

5

Chapter 5.1

General provisions

5.1.1   Application and general provisions

5.1.1.1   This part sets forth the provisions for dangerous goods consignments relative to authorization of consignments and advance notifications, marking, labelling, documentation (by manual, electronic data processing (EDP) or electronic data interchange (EDI) techniques) and placarding.

5.1.1.2   Except as otherwise provided in this Code, no person may offer dangerous goods for transport unless those goods are properly marked, labelled, placarded, described and certified on a transport document, and otherwise in a condition for transport as required by this part.

5.1.1.3   The purpose of indicating the Proper Shipping Name (see 3.1.2.1 and 3.1.2.2) and the UN Number of a substance, material or article offered for transport and, in the case of a marine pollutant, of the addition of "marine pollutant" on documentation accompanying the consignment, and of marking the Proper Shipping Name in accordance with 5.2.1 on the package, including IBCs containing the goods, is to ensure that the substance, material or article can be readily identified during transport. This ready identification is particularly important in the case of an accident involving these goods, in order to determine what emergency procedures are necessary to deal properly with the situation and, in the case of marine pollutants, for the master to comply with the reporting requirements of Protocol I of MARPOL 73/78.

5.1.2   Use of overpacks and unit loads

5.1.2.1   An overpack and unit load shall be marked with the Proper Shipping Name and the UN Number and marked and labelled, as required for packages by chapter 5.2, for each item of dangerous goods contained in the overpack or unit load unless markings and labels representative of all dangerous goods in the overpack or unit load are visible. An overpack, in addition, shall be marked with the word "OVERPACK" unless markings and labels representative of all dangerous goods, as required by chapter 5.2, in the overpack are visible, except as required in 5.2.2.1.12.

5.1.2.2   The individual packages comprising a unit load or an overpack shall be marked and labelled in accordance with chapter 5.2. Each package of dangerous goods contained in the unit load or overpack shall comply with all applicable provisions of the Code. The "OVERPACK" marking on an overpack is an indication of compliance with this provision. The intended function of each package shall not be impaired by the unit load or overpack.

5.1.2.3   Each package bearing package orientation markings as prescribed in 5.2.1.7 of this Code and which is overpacked, placed in a unit load or used as an inner packaging in a large packaging shall be oriented in accordance with such markings.

5.1.3   Empty uncleaned packagings or units

5.1.3.1   Other than for class 7, a packaging, including an IBC, which previously contained dangerous goods shall be identified, marked, labelled and placarded as required for those dangerous goods unless steps such as cleaning, purging of vapours or refilling with a non-dangerous substance are taken to nullify any hazard.

5.1.3.2   Packagings, including IBCs, and tanks used for the transport of radioactive material shall not be used for the transport of other goods unless decontaminated below the level of 0.4 Bq/cm^2 for beta and gamma emitters and low-toxicity alpha emitters and 0.04 Bq/cm^2 for all other alpha emitters.

5.1.3.3   Empty cargo transport units still containing residues of dangerous goods, or loaded with empty uncleaned packages or empty uncleaned bulk containers, shall comply with the provisions applicable to the goods last contained in the unit, packagings or bulk container.

5.1.4 Mixed packing

When two or more dangerous goods are packed within the same outer packaging, the package shall be labelled and marked as required for each substance. Subsidiary risk labels need not be applied if the hazard is already represented by a primary risk label.

5.1.5 General provisions for class 7

5.1.5.1 Approval of shipments and notification

5.1.5.1.1 *General*

In addition to the approval for package designs described in chapter 6.4, multilateral shipment approval is also required in certain circumstances (5.1.5.1.2 and 5.1.5.1.3). In some circumstances it is also necessary to notify competent authorities of a shipment (5.1.5.1.4).

5.1.5.1.2 *Shipment approvals*

Multilateral approval shall be required for:

.1 the shipment of Type B(M) packages not conforming with the provisions of 6.4.7.5 or designed to allow controlled intermittent venting;

.2 the shipment of Type B(M) packages containing radioactive material with an activity greater than $3000A_1$ or $3000A_2$, as appropriate, or 1000 TBq, whichever is the lower;

.3 the shipment of packages containing fissile materials if the sum of the criticality safety indexes of the packages in a single freight container or in a single conveyance exceeds 50. Excluded from this requirement shall be shipments by seagoing vessels, if the sum of the criticality safety indexes does not exceed 50 for any hold, compartment or defined deck area and the distance of 6 m between groups of packages or overpacks as required in table 7.1.14.5.4 is met; and

.4 radiation protection programmes for shipments by special use vessels according to 7.1.14.9

except that a competent authority may authorize transport into or through its country without shipment approval, by a specific provision in its design approval (see 5.1.5.2.1).

5.1.5.1.3 *Shipment approval by special arrangement*

Provisions may be approved by a competent authority under which a consignment which does not satisfy all of the applicable provisions of this Code may be transported under special arrangement (see 1.5.4).

5.1.5.1.4 *Notifications*

Notification to competent authorities is required as follows:

.1 Before the first shipment of any package requiring competent authority approval, the consignor shall ensure that copies of each applicable competent authority certificate applying to that package design have been submitted to the competent authority of each country through or into which the consignment is to be transported. The consignor is not required to await an acknowledgement from the competent authority, nor is the competent authority required to make such acknowledgement of receipt of the certificate.

.2 For each of the following types of shipments:

.1 Type C packages containing radioactive material with an activity greater than $3000A_1$ or $3000A_2$, as appropriate, or 1000 TBq, whichever is the lower;

.2 Type B(U) packages containing radioactive material with an activity greater than $3000A_1$ or $3000A_2$, as appropriate, or 1000 TBq, whichever is the lower;

.3 Type B(M) packages;

.4 shipment under special arrangement

the consignor shall notify the competent authority of each country through or into which the consignment is to be transported. This notification shall be in the hands of each competent authority prior to the commencement of the shipment, and preferably at least 7 days in advance.

.3 The consignor is not required to send a separate notification if the required information has been included in the application for shipment approval.

.4 The consignment notification shall include:

.1 sufficient information to enable the identification of the package or packages, including all applicable certificate numbers and identification marks;

.2 information on the date of shipment, the expected date of arrival and proposed routeing;

.3 the names of the radioactive material or nuclides;

.4 descriptions of the physical and chemical forms of the radioactive material, or whether it is special form radioactive material or low dispersible radioactive material; and

.5 the maximum activity of the radioactive contents during transport, expressed in units of becquerels (Bq) with an appropriate SI prefix symbol (see 1.2.2.1). For fissile material, the mass of fissile material in units of grams (g), or multiples thereof, may be used in place of activity.

5.1.5.2 Certificates issued by competent authority

5.1.5.2.1 Certificates issued by the competent authority are required for the following:

.1 Designs for:

   .1 special form radioactive material;

   .2 low dispersible radioactive material;

   .3 packages containing 0.1 kg or more of uranium hexafluoride;

   .4 all packages containing fissile material unless excepted by 6.4.11.2;

   .5 Type B(U) packages and Type B(M) packages;

   .6 Type C packages;

.2 Special arrangements;

.3 Certain shipments (see 5.1.5.1.2).

The certificates shall confirm that the applicable provisions are met, and for design approvals shall attribute to the design an identification mark.

The package design and shipment approval certificates may be combined into a single certificate.

Certificates and applications for these certificates shall be in accordance with the provisions in 6.4.23.

5.1.5.2.2 The consignor shall be in possession of a copy of each applicable certificate.

5.1.5.2.3 For package designs where a competent authority issued certificate is not required, the consignor shall, on request, make available, for inspection by the relevant competent authority, documentary evidence of the compliance of the package design with all the applicable provisions.

5.1.5.3 Determination of transport index (TI) and criticality safety index (CSI)

5.1.5.3.1 The transport index (TI) for a package, overpack or freight container, or for unpackaged LSA-I or SCO-I, shall be the number derived in accordance with the following procedure:

.1 Determine the maximum radiation level in units of millisieverts per hour (mSv/h) at a distance of 1 m from the external surfaces of the package, overpack, freight container, or unpackaged LSA-I and SCO-I. The value determined shall be multiplied by 100 and the resulting number is the transport index. For uranium and thorium ores and their concentrates, the maximum radiation level at any point 1 m from the external surface of the load may be taken as:

  0.4 mSv/h   for ores and physical concentrates of uranium and thorium;

  0.3 mSv/h   for chemical concentrates of thorium;

  0.02 mSv/h   for chemical concentrates of uranium, other than uranium hexafluoride;

.2 For tanks, freight containers and unpackaged LSA-I and SCO-I, the value determined in 5.1.5.3.1.1 above shall be multiplied by the appropriate factor from table 5.1.5.3.1;

.3 The value obtained in 5.1.5.3.1.1 and 5.1.5.3.1.2 above shall be rounded up to the first decimal place (e.g., 1.13 becomes 1.2), except that a value of 0.05 or less may be considered as zero.

Table 5.1.5.3.1 – Multiplication factors for tanks, freight containers and unpackaged LSA-I and SCO-I

Size of load[a]	Multiplication factor
size of load $\leqslant$ 1 m^2	1
1 m^2 < size of load $\leqslant$ 5 m^2	2
5 m^2 < size of load $\leqslant$ 20 m^2	3
20 m^2 < size of load	10

[a] Largest cross-sectional area of the load being measured.

5.1.5.3.2 The transport index for each overpack, freight container or conveyance shall be determined as either the sum of the TIs of all the packages contained, or by direct measurement of radiation level, except in the case of non-rigid overpacks for which the transport index shall be determined only as the sum of the TIs of all the packages.

5.1.5.3.3 The criticality safety index for each overpack or freight container shall be determined as the sum of the CSIs of all the packages contained. The same procedure shall be followed for determining the total sum of the CSIs in a consignment or aboard a conveyance.

5.1.5.3.4 Packages and overpacks shall be assigned to either category I – WHITE, II – YELLOW or III – YELLOW in accordance with the conditions specified in table 5.1.5.3.4 and with the following requirements:

.1 For a package or overpack, both the transport index and the surface radiation level conditions shall be taken into account in determining which is the appropriate category. Where the transport index satisfies the condition for one category but the surface radiation level satisfies the condition for a different category, the package or overpack shall be assigned to the higher category. For this purpose, category I – WHITE shall be regarded as the lowest category;

.2 The transport index shall be determined following the procedures specified in 5.1.5.3.1 and 5.1.5.3.2;

.3 If the surface radiation level is greater than 2 mSv/h, the package or overpack shall be transported under exclusive use and under the provisions of 7.2.3.1.3, 7.2.3.2.1, or 7.2.3.3.3, as appropriate;

.4 A package transported under a special arrangement shall be assigned to category III – YELLOW except when otherwise specified in the competent authority approval certificate of the country of origin of design (see 2.7.2.4.6);

.5 An overpack which contains packages transported under special arrangement shall be assigned to category III – YELLOW except when otherwise specified in the competent authority approval certificate of the country of origin of design (see 2.7.2.4.6).

Table 5.1.5.3.4 – Categories of packages and overpacks

Conditions		Category
Transport index	Maximum radiation level at any point on external surface	
0[a]	Not more than 0.005 mSv/h	I – WHITE
More than 0 but not more than 1[a]	More than 0.005 mSv/h but not more than 0.5 mSv/h	II – YELLOW
More than 1 but not more than 10	More than 0.5 mSv/h but not more than 2 mSv/h	III – YELLOW
More than 10	More than 2 mSv/h but not more than 10 mSv/h	III – YELLOW[b]

[a] If the measured TI is not greater than 0.05, the value quoted may be zero in accordance with 5.1.5.3.1.3.

[b] Shall also be transported under "exclusive use".

5.1.6 Packages packed into a cargo transport unit

5.1.6.1 Regardless of the placarding and marking provisions for cargo transport units, each package containing dangerous goods packed into a cargo transport unit shall be marked and labelled in accordance with the requirements of chapter 5.2.

Chapter 5.2

Marking and labelling of packages including IBCs

Note: These provisions relate essentially to the marking and labelling of dangerous goods according to their properties. However, additional markings or symbols indicating precautions to be taken in handling or storing a package (such as a symbol representing an umbrella, indicating that a package shall be kept dry) may be displayed on a package if appropriate.

5.2.1   Marking of packages including IBCs

5.2.1.1   Unless provided otherwise in this Code, the Proper Shipping Name for the dangerous goods as determined in accordance with 3.1.2 and the corresponding UN Number, preceded by the letters "UN", shall be displayed on each package. In the case of unpackaged articles, the marking shall be displayed on the article, on its cradle or on its handling, storage or launching device. For goods of division 1.4, compatibility group S, the division and compatibility group letter shall also be marked unless the label for 1.4S is displayed. A typical package marking is:

        CORROSIVE LIQUID, ACIDIC, ORGANIC, N.O.S. (caprylyl chloride) UN 3265.

5.2.1.2   All package markings required by 5.2.1.1:

    .1   shall be readily visible and legible;

    .2   shall be such that this information will still be identifiable on packages surviving at least three months' immersion in the sea. In considering suitable marking methods, account shall be taken of the durability of the packaging materials used and the surface of the package;

    .3   shall be displayed on a background of contrasting colour on the external surface of the package; and

    .4   shall not be located with other package markings that could substantially reduce their effectiveness.

5.2.1.3   Salvage packagings shall additionally be marked with the word "SALVAGE".

5.2.1.4   Intermediate bulk containers of more than 450 ℓ capacity and large packagings shall be marked on two opposing sides.

5.2.1.5   **Special marking provisions for class 7**

5.2.1.5.1   Each package shall be legibly and durably marked on the outside of the packaging with an identification of either the consignor or consignee, or both.

5.2.1.5.2   In the case of excepted packages, marking the Proper Shipping Name is not required.

5.2.1.5.3   Each package of gross mass exceeding 50 kg shall have its permissible gross mass legibly and durably marked on the outside of the packaging.

5.2.1.5.4   Each package which conforms to:

    .1   a Type IP-1 package, a Type IP-2 package or a Type IP-3 package design shall be legibly and durably marked on the outside of the packaging with "TYPE IP-1", "TYPE IP-2" or "TYPE IP-3" as appropriate;

    .2   a Type A package design shall be legibly and durably marked on the outside of the packaging with "TYPE A";

    .3   a Type IP-2 package, a Type IP-3 package or a Type A package design shall be legibly and durably marked on the outside of the packaging with the international vehicle registration code (VRI code) of the country of origin of design and either the name of the manufacturer or other identification of the packaging specified by the competent authority of the country of origin of design.

5.2.1.5.5   Each package which conforms to a design approved by the competent authority under 6.4.22.1–6.4.22.5 or 6.4.24.2–6.4.24.3 shall be legibly and durably marked on the outside of the packaging with:

.1   the identification mark allocated to that design by the competent authority;

.2   a serial number to uniquely identify each packaging which conforms to that design;

.3   in the case of a Type B(U) or Type B(M) package design, with ''TYPE B(U)'' or ''TYPE B(M)''; and

.4   in the case of a Type C package design, with ''TYPE C''.

5.2.1.5.6   Each package which conforms to a Type B(U), Type B(M) or Type C package design shall have the outside of the outermost receptacle which is resistant to the effects of fire and water plainly marked by embossing, stamping or other means resistant to the effects of fire and water with the trefoil symbol shown below.

Basic trefoil symbol with proportions based on a central circle of radius *X*.
The minimum allowable size of *X* shall be 4 mm.

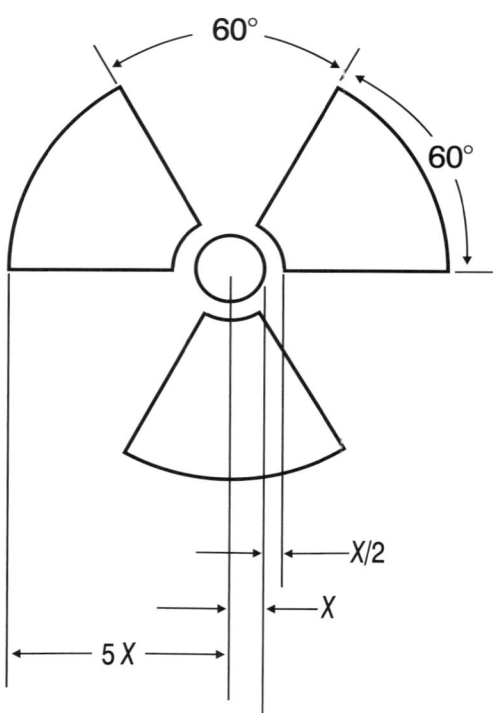

5.2.1.5.7   Where LSA-I or SCO-I material is contained in receptacles or wrapping materials and is transported under exclusive use as permitted by 4.1.9.2.3, the outer surface of these receptacles or wrapping materials may bear the marking ''RADIOACTIVE LSA-I'' or ''RADIOACTIVE SCO-I'', as appropriate.

5.2.1.5.8   In case of international transport of packages requiring competent authority design or shipment approval, for which different approval types apply in the different countries concerned, marking shall be in accordance with the certificate of the country of origin of the design.

5.2.1.6   **Special marking provisions for marine pollutants**

5.2.1.6.1   Packages containing marine pollutants meeting the criteria of 2.10.3 shall be durably marked with the marine pollutant mark with the exception of single packagings and combination packagings containing inner packagings with:

– contents of 5 ℓ or less for liquids; or

– contents of 5 kg or less for solids.

5.2.1.6.2   The marine pollutant mark shall be located adjacent to the markings required by 5.2.1.1. The provisions of 5.2.1.2 and 5.2.1.4 shall be met.

5.2.1.6.3 The marine pollutant mark shall be as shown below. For packagings, the dimensions shall be at least 100 mm × 100 mm, except in the case of packages of such dimensions that they can only bear smaller marks.

Marine pollutant mark

Symbol (fish and tree): black on white or suitable contrasting background

5.2.1.7 Except as provided in 5.2.1.7.1:

– combination packagings having inner packagings containing liquid dangerous goods;

– single packagings fitted with vents; and

– cryogenic receptacles intended for the transport of refrigerated liquefied gases

shall be legibly marked with package orientation arrows which are similar to the illustration shown below or with those meeting the specifications of ISO 780:1985. The orientation arrows shall appear on two opposite vertical sides of the package with the arrows pointing in the correct upright direction. They shall be rectangular and of a size that is clearly visible commensurate with the size of the package. Depicting a rectangular border around the arrows is optional.

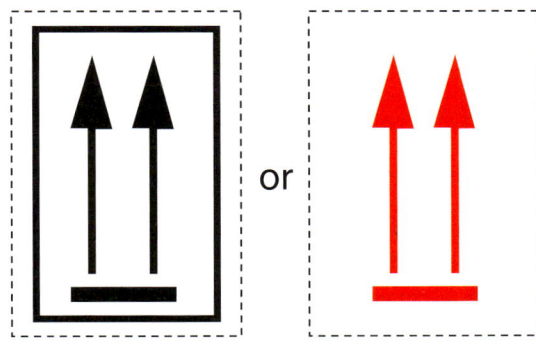

Two black or red arrows on white or suitable contrasting background.
The rectangular border is optional

5.2.1.7.1 Orientation arrows are not required on packages containing:

(a) pressure receptacles except for cryogenic receptacles;

(b) dangerous goods in inner packagings of not more than 120 mℓ which are prepared with sufficient absorbent material between the inner and outer packagings to completely absorb the liquid contents;

(c) class 6.2 infectious substances in primary receptacles of not more than 50 mℓ;

(d) class 7 radioactive material in Type IP-2, IP-3, A, B(U), B(M) or C packages; or

(e) articles which are leak-tight in all orientations (e.g. alcohol or mercury in thermometers, aerosols, etc.).

5.2.1.7.2 Arrows for purposes other than indicating proper package orientation shall not be displayed on a package marked in accordance with this sub-section.

5.2.1.8 **Excepted quantity mark**

5.2.1.8.1 Packages containing excepted quantities of dangerous goods shall be marked according to 3.5.4.

5.2.2 Labelling of packages including IBCs

5.2.2.1 Labelling provisions

These provisions are related essentially to danger labels. However, additional markings or symbols indicating precautions to be taken in handling or storing a package (such as a symbol representing an umbrella, indicating that a package shall be kept dry) may be displayed on a package if appropriate.

5.2.2.1.1 Labels identifying primary and subsidiary risks shall conform to models Nos. 1 to 9 illustrated in 5.2.2.2.2. The "EXPLOSIVE" subsidiary risk label is model No. 1.

5.2.2.1.2 Where articles or substances are specifically listed in the Dangerous Goods List, a danger class label shall be affixed for the hazard shown in column 3. A subsidiary risk label shall also be affixed for any risk indicated by a class or division number in column 4 of the Dangerous Goods List. However, special provisions indicated in column 6 may also require a subsidiary risk label where no subsidiary risk is indicated in column 4 or may exempt from the requirement for a subsidiary risk label where such a risk is indicated in the Dangerous Goods List.

5.2.2.1.2.1 A package containing a dangerous substance which has a low degree of danger may be exempt from these labelling requirements. In this case, a special provision specifying that no hazard label is required appears in column 6 of the Dangerous Goods List for the relevant substance. However, for certain substances the package shall be marked with the appropriate text as it appears in the special provision, e.g.:

Substance	UN No.	Class	Mark required on bales
Baled hay in cargo transport unit	UN 1327	4.1	None
Baled hay not in cargo transport unit	UN 1327	4.1	Class 4.1
Baled dry vegetable fibres in cargo transport unit	UN 3360	4.1	None

Substance	UN No.	Class	Mark required on packages in addition to the Proper Shipping Name and UN Number
Fishmeal*	UN 1374	4.2	Class 4.2[†]
Batteries, wet, non-spillable	UN 2800	8	Class 8[‡]

* only applicable to fishmeal in packing group III.

[†] exempt from class marking when loaded in a cargo transport unit containing only fishmeal under UN 1374.

[‡] exempt from class marking when loaded in a cargo transport unit containing only batteries under UN 2800.

5.2.2.1.3 Except as provided in 5.2.2.1.3.1, if a substance which meets the definition of more than one class is not specifically listed by name in the Dangerous Goods List in chapter 3.2, the provisions in chapter 2.0 shall be used to determine the primary risk class of the goods. In addition to the label required for that primary risk class, subsidiary risk labels shall also be applied as specified in the Dangerous Goods List.

5.2.2.1.3.1 Packagings containing substances of class 8 need not bear subsidiary risk label model No. 6.1 if the toxicity arises solely from the destructive effect on tissue. Substances of class 4.2 need not bear subsidiary risk label model No. 4.1.

5.2.2.1.4 *Labels for class 2 gases with subsidiary risk(s)*

Class	Subsidiary risk(s) shown in chapter 2.2	Primary risk label	Subsidiary risk label(s)
2.1	None	2.1	None
2.2	None	2.2	None
	5.1	2.2	5.1
2.3	None	2.3	None
	2.1	2.3	2.1
	5.1	2.3	5.1
	5.1, 8	2.3	5.1, 8
	8	2.3	8
	2.1, 8	2.3	2.1, 8

5.2.2.1.5 Three separate labels have been provided for class 2, one for flammable gases of class 2.1 (red), one for non-flammable, non-toxic gases of class 2.2 (green) and one for toxic gases of class 2.3 (white). Where the Dangerous Goods List indicates that a class 2 gas possesses single or multiple subsidiary risks, labels shall be used in accordance with the table in 5.2.2.1.4.

5.2.2.1.6 Except as provided in 5.2.2.2.1.2, each label shall:

.1 be located on the same surface of the package near the Proper Shipping Name marking, if the package dimensions are adequate;

.2 be so placed on the packaging that it is not covered or obscured by any part or attachment to the packaging or any other label or marking; and

.3 when primary and subsidiary risk labels are required, be displayed next to each other.

Where a package is of such an irregular shape or small size that a label cannot be satisfactorily affixed, the label may be attached to the package by a securely affixed tag or other suitable means.

5.2.2.1.7 Intermediate bulk containers of more than 450 ℓ capacity and large packagings shall be labelled on two opposing sides.

5.2.2.1.8 Labels shall be affixed on a surface of contrasting colour.

5.2.2.1.9 *Special provisions for the labelling of self-reactive substances*

An "EXPLOSIVE" subsidiary risk label (No. 1) shall be applied for type B self-reactive substances, unless the competent authority has permitted this label to be dispensed with for a specific packaging because test data have proved that the self-reactive substance in such a packaging does not exhibit explosive behaviour.

5.2.2.1.10 *Special provisions for the labelling of organic peroxides*

The class 5.2 label (model No. 5.2) shall be affixed to packages containing organic peroxides classified as types B, C, D, E or F. This label also implies that the product may be flammable and hence no "FLAMMABLE LIQUID" subsidiary risk label (model No. 3) is required. In addition, the following subsidiary risk labels shall be applied:

.1 An "EXPLOSIVE" subsidiary risk label (model No. 1) for organic peroxides type B, unless the competent authority has permitted this label to be dispensed with for a specific packaging because test data have proved that the organic peroxide in such a packaging does not exhibit explosive behaviour.

.2 A "CORROSIVE" subsidiary risk label (model No. 8) is required when packing group I or II criteria of class 8 are met.

5.2.2.1.11 *Special provisions for the labelling of infectious substances packages*

In addition to the primary risk label (model No. 6.2), infectious substances packages shall bear any other label required by the nature of the contents.

5.2.2.1.12 *Special provisions for the labelling of radioactive material*

5.2.2.1.12.1 Except when enlarged labels are used in accordance with 5.3.1.1.5.1, each package, overpack and freight container containing radioactive material shall bear at least two labels which conform to the models Nos. 7A, 7B, and 7C, as appropriate, according to the category (see 5.1.5.3.4) of that package, overpack or freight container. Labels shall be affixed to two opposite sides on the outside of the package or on the outside of all four sides of the freight container. Each overpack containing radioactive material shall bear at least two labels on opposite sides of the outside of the overpack. In addition, each package, overpack and freight container containing fissile material, other than fissile material excepted under the provisions of 6.4.11.2, shall bear labels which conform to model No. 7E; such labels, where applicable, shall be affixed adjacent to the labels for radioactive material. Labels shall not cover the markings specified in this chapter. Any labels which do not relate to the contents shall be removed or covered.

5.2.2.1.12.2 Each label conforming to the models Nos. 7A, 7B and 7C shall be completed with the following information:

.1 *Contents:*

.1 Except for LSA-I material, the name(s) of the radionuclide(s) as taken from the table under 2.7.2.2.1, using the symbols prescribed therein. For mixtures of radionuclides, the most restrictive nuclides must be listed to the extent the space on the line permits. The group of LSA or SCO shall be shown following the name(s) of the radionuclide(s). The terms "LSA-II", "LSA-III", "SCO-I" and "SCO-II" shall be used for this purpose.

.2 For LSA-I material, the term "LSA-I" is all that is necessary; the name of the radionuclide is not necessary.

.2 *Activity:* The maximum activity of the radioactive contents during transport, expressed in units of becquerels (Bq) with the appropriate SI prefix symbol (see 1.2.2.1). For fissile material, the mass of fissile material in units of grams (g), or multiples thereof, may be used in place of activity.

.3 For overpacks and freight containers, the "contents" and "activity" entries on the label shall bear the information required in 5.2.2.1.12.2.1 and 5.2.2.1.12.2.2, respectively, totalled together for the entire contents of the overpack or freight container except that, on labels for overpacks or freight containers containing mixed loads of packages containing different radionuclides, such entries may read "See transport documents".

.4 *Transport index:* The number determined in accordance with 5.1.5.3.1 and 5.1.5.3.2. (No transport index entry is required for category I – WHITE.)

5.2.2.1.12.3 Each label conforming to the model No. 7E shall be completed with the criticality safety index (CSI) as stated in the certificate of approval for special arrangement or the certificate of approval for the package design issued by the competent authority.

5.2.2.1.12.4 For overpacks and freight containers, the criticality safety index (CSI) on the label shall bear the information required in 5.2.2.1.12.3 totalled together for the fissile contents of the overpack or freight container.

5.2.2.1.12.5 In case of international transport of packages requiring competent authority design or shipment approval, for which different approval types apply in the different countries concerned, labelling shall be in accordance with the certificate of the country of origin of design.

5.2.2.2   Provisions for labels

5.2.2.2.1 Labels shall satisfy the provisions of this section and conform, in terms of colour, symbols, numbers and general format, to the specimen labels shown in 5.2.2.2.2.

Note: Where appropriate, labels in 5.2.2.2.2 are shown with a dotted outer boundary as provided for in 5.2.2.2.1.1. This is not required when the label is applied on a background of contrasting colour.

5.2.2.2.1.1 Labels shall be in the form of a square set at an angle of 45° (diamond-shaped) with minimum dimensions of 100 mm by 100 mm, except in the case of packages of such dimensions that they can only bear smaller labels and as provided in 5.2.2.2.1.2. They shall have a line 5 mm inside the edge and running parallel with it. In the upper half of a label the line shall have the same colour as the symbol and in the lower half it shall have the same colour as the figure in the bottom corner. Labels shall be displayed on a background of contrasting colour, or shall have either a dotted or solid outer boundary line.*

5.2.2.2.1.2 Cylinders for class 2 may, on account of their shape, orientation and securing mechanisms for transport, bear labels representative of those specified in this section, which have been reduced in size, according to ISO 7225:2005, for display on the non-cylindrical part (shoulder) of such cylinders. Labels may overlap to the extent provided for by ISO 7225:2005 "Gas cylinders – Precautionary labels"; however, in all cases, the labels representing the primary hazard and the numbers appearing on any label shall remain fully visible and the symbols recognizable.

5.2.2.2.1.3 With the exception of divisions 1.4, 1.5 and 1.6 of class 1, the upper half of the label shall contain the pictorial symbol and the lower half shall contain the class number 1, 2, 3, 4, 5.1, 5.2, 6, 7, 8 or 9 as appropriate. The label may include text such as the UN Number, or words describing the hazard class (e.g., "flammable") in accordance with 5.2.2.2.1.5 provided the text does not obscure or detract from the other label elements.

5.2.2.2.1.4 In addition, except for divisions 1.4, 1.5 and 1.6, labels for class 1 shall show in the lower half, above the class number, the division number and compatibility group letter for the substance or article. Labels for divisions 1.4, 1.5 and 1.6 shall show in the upper half the division number and in the lower half the class number and the compatibility group letter. For division 1.4, compatibility group S, no label is generally required. However, in cases where a label is considered necessary for such goods, it shall be based on model No. 1.4.

5.2.2.2.1.5 On labels other than those for material of class 7, the insertion of any text (other than the class or division number) in the space below the symbol shall be confined to particulars indicating the nature of the risk and precautions to be taken in handling.

5.2.2.2.1.6 The symbols, text and numbers shall be shown in black on all labels except for:
.1 the class 8 label, where the text (if any) and class number shall appear in white;
.2 labels with entirely green, red or blue backgrounds, where they may be shown in white;
.3 the class 5.2 label, where the symbol may be shown in white; and
.4 class 2.1 labels displayed on cylinders and gas cartridges for liquefied petroleum gases, where they may be shown in the background colour of the receptacle if adequate contrast is provided

5.2.2.2.1.7 The method of affixing the label(s) or applying stencil(s) of label(s) on packages containing dangerous goods shall be such that the label(s) or stencil(s) will still be identifiable on packages surviving at least three months' immersion in the sea. In considering suitable labelling methods, account shall be taken of the durability of the packaging materials used and the surface of the package.

---

* The use of existing labels with the white symbol and black line may be permitted until 1 January 2010.

5.2.2.2.2   *Specimen labels*

Class 1 – Explosive substances or articles

(No. 1)
Divisions 1.1, 1.2 and 1.3

Symbol (exploding bomb): black. Background: orange. Figure '**1**' in bottom corner.

(No. 1.4)
Division 1.4

(No. 1.5)
Division 1.5

(No. 1.6)
Division 1.6

Background: orange. Figures: black. Numerals shall be about 30 mm in height and be about 5 mm thick (for a label measuring 100 mm × 100 mm). Figure '**1**' in bottom corner.

** Place for division – to be left blank if explosive is the subsidiary risk.
* Place for compatibility group – to be left blank if explosive is the subsidiary risk.

Class 2 – Gases

(No. 2.1)
Class 2.1
Flammable gases

Symbol (flame): black or white
(except as provided for in 5.2.2.2.1.6.4).
Background: red. Figure '**2**' in bottom corner.

(No. 2.2)
Class 2.2
Non-flammable, non-toxic gases

Symbol (gas cylinder): black or white.
Background: green. Figure '**2**' in bottom corner.

Class 3 – Flammable liquids

(No. 2.3)
Class 2.3
Toxic gases

Symbol (skull and crossbones): black.
Background: white. Figure '**2**' in bottom corner.

(No. 3)

Symbol (flame): black or white.
Background: red. Figure '**3**' in bottom corner.

Class 4

(No. 4.1)
Class 4.1
Flammable solids

Symbol (flame): black.
Background: white with
seven vertical red stripes.
Figure '**4**' in bottom corner.

(No. 4.2)
Class 4.2
*Substances liable to
spontaneous combustion*

Symbol (flame): black.
Background: upper half white,
lower half red.
Figure '**4**' in bottom corner.

(No. 4.3)
Class 4.3
*Substances which, in contact with water,
emit flammable gases*

Symbol (flame): black or white.
Background: blue.
Figure '**4**' in bottom corner.

Class 5

(No. 5.1)
Class 5.1
Oxidizing substances

(No. 5.2(a)*)
Class 5.2
Organic peroxides

Symbol (flame over circle): black; Background: yellow.
Figure '**5.1**' in bottom corner.   Figure '**5.2**' in bottom corner.

(No. 5.2(b))
Class 5.2
Organic peroxides

Symbol (flame): black or white;
Background: upper half red; lower half yellow;
Figure '**5.2**' in bottom corner

*   May be used until 1 January 2011.

Class 6

(No. 6.1)
Class 6.1
Toxic substances

Symbol (skull and crossbones): black.
Background: white. Figure '**6**' in bottom corner.

(No. 6.2)
Class 6.2
Infectious substances

The lower half of the label may bear the inscriptions **INFECTIOUS SUBSTANCE** and
In case of damage or leakage immediately notify Public Health Authority.
Symbol (three crescents superimposed on a circle) and inscriptions: black.
Background: white. Figure '**6**' in bottom corner.

Class 7 – Radioactive material

(No. 7A)
Category I – White

Symbol (trefoil): black.
Background: white.
Text (mandatory): black in
lower half of label:
RADIOACTIVE
CONTENTS ...
ACTIVITY ...

One red bar shall follow the word
RADIOACTIVE.
Figure '**7**' in bottom corner.

(No. 7B)
Category II – Yellow

(No. 7C)
Category III – Yellow

Symbol (trefoil): black.
Background: upper half yellow with white border, lower half white.
Text (mandatory): black in lower half of label:
RADIOACTIVE
CONTENTS ...
ACTIVITY ...
IN A BLACK OUTLINED BOX: **TRANSPORT INDEX ...**

Two red vertical bars shall
follow the word **RADIOACTIVE**.

Three red vertical bars shall
follow the word **RADIOACTIVE**.

Figure '**7**' in bottom corner.

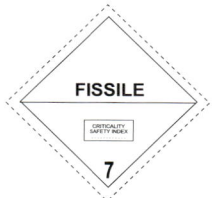

(No. 7E)
Class 7 fissile material

Background: white.
Text (mandatory): black in upper half of label: **FISSILE**.
In a black outlined box in the lower half of the label: **CRITICALITY SAFETY INDEX ...**
Figure '**7**' in bottom corner.

Class 8 – Corrosive substances

(No. 8)

Symbol (liquids, spilling from two glass vessels
and attacking a hand and a metal): black.
Background: upper half white;
lower half black with white border.
Figure '**8**' in bottom corner.*

* A class 8 label with a shaded hand may also be used.

Class 9 – Miscellaneous dangerous substances and articles

(No. 9)

Symbol (seven vertical stripes in upper half): black.
Background: white.
Figure '**9**' underlined in bottom corner.

Chapter 5.3

Placarding and marking of cargo transport units

5.3.1 Placarding

5.3.1.1 Placarding provisions

5.3.1.1.1 *General provisions*

.1 Enlarged labels (placards) and marks and signs shall be affixed to the exterior surfaces of a cargo transport unit to provide a warning that the contents of the unit are dangerous goods and present risks, unless the labels and/or marks affixed to the packages are clearly visible from the exterior of the cargo transport unit;

.2 the methods of placarding and marking as required in 5.3.1.1.4 and 5.3.2 on cargo transport units shall be such that this information will still be identifiable on cargo transport units surviving at least three months' immersion in the sea. In considering suitable marking methods, account shall be taken of the ease with which the surface of the cargo transport unit can be marked; and

.3 all placards, orange panels, marks and signs shall be removed from cargo transport units or masked as soon as both the dangerous goods or their residues which led to the application of those placards, orange panels, marks or signs are discharged.

5.3.1.1.2
Placards shall be affixed to the exterior surface of cargo transport units to provide a warning that the contents of the unit are dangerous goods and present risks. Placards shall correspond to the primary risk of the goods contained in the cargo transport unit except that:

.1 placards are not required on cargo transport units carrying any quantity of explosives of division 1.4, compatibility group S, dangerous goods packed in limited quantities, or excepted packages of radioactive material (class 7); and

.2 placards indicating the highest risk only need be affixed on cargo transport units carrying substances and articles of more than one division in class 1.

Placards shall be displayed on a background of contrasting colour, or shall have either a dotted or solid outer boundary line.

5.3.1.1.3
Placards shall also be displayed for those subsidiary risks for which a subsidiary risk label is required according to 5.2.2.1.2. However, cargo transport units containing goods of more than one class need not bear a subsidiary risk placard if the hazard represented by that placard is already indicated by a primary risk placard.

5.3.1.1.4 *Placarding requirements*

5.3.1.1.4.1
A cargo transport unit containing dangerous goods or residues of dangerous goods shall clearly display placards as follows:

.1 *a freight container, semi-trailer or portable tank:* one on each side and one on each end of the unit;

.2 *a railway wagon:* at least on each side;

.3 *a multiple-compartment tank containing more than one dangerous substance or their residues:* along each side at the positions of the relevant compartments; and

.4 *any other cargo transport unit:* at least on both sides and on the back of the unit.

5.3.1.1.5 *Special provisions for class 7*

5.3.1.1.5.1
Large freight containers carrying packages other than excepted packages, and tanks, shall bear four placards which conform with the model No. 7D given in the figure. The placards shall be affixed in a vertical orientation to each side wall and each end wall of the large freight container or tank. Any placards which do not relate to the contents shall be removed. Instead of using both labels and placards, it is permitted as an alternative to use enlarged labels only, as shown in label model Nos. 7A, 7B and 7C, and where appropriate 7E, with dimensions as required for the placard in the figure.

5.3.1.1.5.2
Rail and road vehicles carrying packages, overpacks or freight containers labelled with any of the labels shown in 5.2.2.2.2 as models Nos. 7A, 7B, 7C or 7E, or carrying consignments under exclusive use, shall display the placard shown in the figure (model No. 7D) on each of:

.1 the two external lateral walls, in the case of a rail vehicle;

.2 the two external lateral walls and the external rear wall, in the case of a road vehicle.

In the case of a vehicle without sides, the placards may be affixed directly on the cargo-carrying unit provided that they are readily visible; in the case of physically large tanks or freight containers, the placards on the tanks or freight containers shall suffice. In the case of vehicles which have insufficient area to allow the fixing of larger placards, the dimensions of the placard as described in the figure may be reduced to 100 mm. Any placards which do not relate to the contents shall be removed.

5.3.1.2   Specifications for placards

5.3.1.2.1   Except as provided in 5.3.1.2.2 for the class 7 placard, a placard shall:

.1   be not less than 250 mm by 250 mm, with a line running 12.5 mm inside the edge and parallel with it. In the upper half of the placard the line shall have the same colour as the symbol and in the lower half it shall have the same colour as the figure in the bottom corner.

.2   correspond to the label for the class of the dangerous goods in question with respect to colour and symbol; and

.3   display the number of the class or division (and, for goods in class 1, the compatibility group letter) of the dangerous goods in question in the manner prescribed in 5.2.2.2 for the corresponding label, in digits not less than 25 mm high.

5.3.1.2.2   For class 7, the placard shall have minimum overall dimensions of 250 mm by 250 mm (except as permitted by 5.3.1.1.5.2) with a black line running 5 mm inside the edge and parallel with it, and shall be otherwise as shown in the figure below. When different dimensions are used, the relative proportions shall be maintained. The number "7" shall not be less than 25 mm high. The background colour of the upper half of the placard shall be yellow and of the lower half white; the colour of the trefoil and the printing shall be black. The use of the word "RADIOACTIVE" in the bottom half is optional to allow the use of this placard to display the appropriate United Nations Number for the consignment.

Placard for radioactive material of class 7

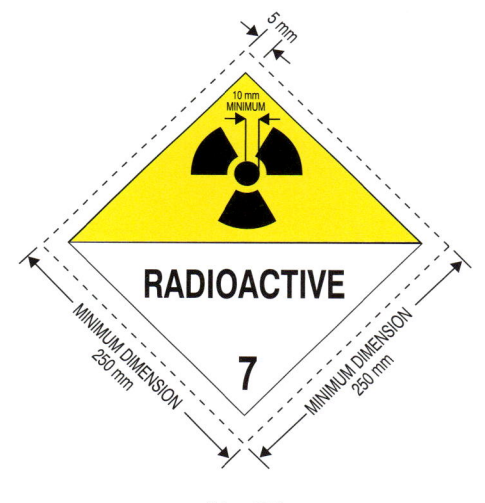

(No. 7D)
Symbol (trefoil): black.
Background: upper half yellow with white border, lower half white.
The lower half shall show the word **RADIOACTIVE** or
alternatively, when required (see 5.3.2.1), the appropriate UN Number
and the figure '**7**' in the bottom corner.

5.3.1.3   Fumigated units

Class 9 placards shall not be affixed to a fumigated unit except as required for other class 9 substances or articles packed therein.

5.3.2   Marking of cargo transport units

5.3.2.0   Display of Proper Shipping Name

The Proper Shipping Name of the contents shall be durably marked on at least both sides of:

.1   tank transport units containing dangerous goods;

.2   bulk containers containing dangerous goods; or

.3   any other cargo transport unit containing packaged dangerous goods of a single commodity for which no placard, UN Number or marine pollutant mark is required. Alternatively, the UN Number may be displayed.

5.3.2.1   Display of UN Numbers

5.3.2.1.1   Except for goods of class 1, the UN Number shall be displayed as required by this chapter on consignments of:

.1   solids, liquids or gases transported in tank cargo transport units, including on each compartment of a multi-compartment tank cargo transport unit;

.2   packaged dangerous goods loaded in excess of 4000 kg gross mass, to which only one UN Number has been assigned and which are the only dangerous goods in the cargo transport unit;

.3   unpackaged LSA-I or SCO-I material of class 7 in or on a vehicle, or in a freight container, or in a tank;

.4   packaged radioactive material with a single UN Number under exclusive use in or on a vehicle, or in a freight container;

.5   solid dangerous goods in bulk containers.

5.3.2.1.2   The UN Number for the goods shall be displayed in black digits not less then 65 mm high, either:

.1   against a white background in the area below the pictorial symbol and above the class number and the compatibility group letter in a manner that does not obscure or detract from the other required label elements (see 5.3.2.1.3); or

.2   on an orange rectangular panel not less than 120 mm high and 300 mm wide, with a 10 mm black border, to be placed immediately adjacent to each placard or marine pollutant mark (see 5.3.2.1.3). When no placard or marine pollutant mark is required, the UN Number shall be displayed immediately adjacent to the Proper Shipping Name.

5.3.2.1.3   *Examples of display of UN Numbers*

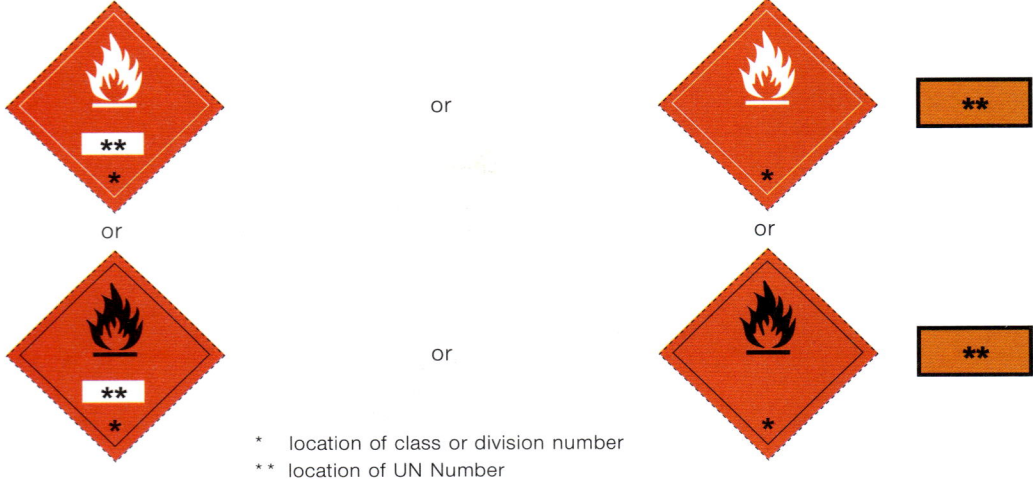

*   location of class or division number
**   location of UN Number

5.3.2.2   Elevated temperature substances

5.3.2.2.1   Cargo transport units containing a substance that is transported or offered for transport in a liquid state at a temperature equal to or exceeding 100°C or in a solid state at a temperature equal to or exceeding 240°C shall bear on each side and on each end the mark shown in the figure. The triangular shaped mark shall have sides of at least 250 mm and shall be shown in red.

Mark for transport at elevated temperature

5.3.2.2.2   In addition to the elevated temperature mark, the maximum temperature of the substance expected to be reached during transport shall be durably marked on both sides of the portable tank or insulation jacket, immediately adjacent to the elevated temperature mark, in characters at least 100 mm high.

5.3.2.3   **Marine pollutant mark**

Cargo transport units containing marine pollutants shall clearly display the marine pollutant mark in locations indicated in 5.3.1.1.4.1, even if the cargo transport unit contains packages not required to bear the marine pollutant mark. The mark shall conform to the specifications given in 5.2.1.6.3, and shall have minimum dimensions of 250 mm × 250 mm.

5.3.2.4   **Limited quantities**

Cargo transport units containing dangerous goods in only limited quantities need not be placarded nor marked according to 5.3.2.0 and 5.3.2.1. They shall, however, be suitably marked on the exterior as "LIMITED QUANTITIES" or "LTD QTY" not less than 65 mm high in locations indicated in 5.3.1.1.4.1.

5.3.2.5   **Fumigated units**

.1   The marking of the Proper Shipping Name (FUMIGATED UNIT) and the UN Number (UN 3359) is not required on fumigated units. However, if a fumigated unit is loaded with dangerous goods, any mark required by the provisions in 5.3.2.0 to 5.3.2.4 shall be marked on the fumigated unit.

.2   A fumigated unit shall be marked with the warning sign, as specified in .3, affixed in a location where it will be easily seen by persons attempting to enter the interior of the unit. The marking, as required by this paragraph, shall remain on the unit until the following provisions are met:

.1   the fumigated unit has been ventilated to remove harmful concentrations of fumigant gas; and

.2   the fumigated goods or materials have been unloaded.

.3   The fumigation warning sign shall be rectangular and shall be not less than 300 mm wide and 250 mm high. The markings shall be in black print on a white background with lettering not less than 25 mm high. An illustration of this sign is given below:

Fumigation warning sign

Chapter 5.4

Documentation

Note 1 The provisions of this Code do not preclude the use of electronic data processing (EDP) and electronic data interchange (EDI) transmission techniques as an aid to paper documentation.

Note 2 When dangerous goods are offered for transport, similar documents to those required for other categories of goods have to be prepared. The form of these documents, the particulars to be entered on them and the obligations they entail may be fixed by international conventions applying to certain modes of transport and by national legislation.

Note 3 One of the primary requirements of a transport document for dangerous goods is to convey the fundamental information relative to the hazards of the goods. It is, therefore, necessary to include certain basic information on the document for a consignment of dangerous goods unless otherwise exempted or required in this Code.

Note 4 In addition to the provisions of this chapter, other elements of information may be required by the competent authority.

5.4.1 Dangerous goods transport documentation

5.4.1.1 General

Except as otherwise provided, the consignor who offers dangerous goods for transport shall describe the dangerous goods on a transport document and provide additional information and documentation as specified in this Code.

5.4.1.2 Form of the transport document

5.4.1.2.1 A dangerous goods transport document may be in any form, provided it contains all of the information required by the provisions of this Code.

5.4.1.2.2 If both dangerous and non-dangerous goods are listed in one document, the dangerous goods shall be listed first, or otherwise be emphasized.

5.4.1.2.3 *Continuation page*

A dangerous goods transport document may consist of more than one page, provided pages are consecutively numbered.

5.4.1.2.4 The information on a dangerous goods transport document shall be easy to identify, legible and durable.

5.4.1.2.5 *Example of a dangerous goods transport document*

The form shown in figure 5.4.5 is an example of a dangerous goods transport document.*

5.4.1.3 Consignor, consignee and date

The name and address of the consignor and the consignee of the dangerous goods shall be included on the dangerous goods transport document. The date the dangerous goods transport document or an electronic copy of it was prepared or given to the initial carrier shall be included.

---

* For standardized formats, see also the relevant recommendations of the UNECE United Nations Centre for Trade Facilitation and Electronic Business (UN/CEFACT), in particular Recommendation No. 1 (United Nations Layout Key for Trade Documents) (ECE/TRADE/137, edition 81.3), UN Layout Key for Trade Documents – Guidelines for Applications (ECE/TRADE/270, edition 2002), Recommendation No. 11 (Documentary Aspects of the International Transport of Dangerous Goods) (ECE/TRADE/204, edition 96.1 – currently under revision) and Recommendation No. 22 (Layout Key for Standard Consignment Instructions) (ECE/TRADE/168, edition 1989). Refer also to the UN/CEFACT Summary of Trade Facilitation Recommendations (ECE/TRADE/346, edition 2006) and the United Nations Trade Data Elements Directory (UNTDED) (ECE/TRADE/362, edition 2005).

5.4.1.4   **Information required on the dangerous goods transport document**

5.4.1.4.1   *Dangerous goods description*

The dangerous goods transport document shall contain the following information for each dangerous substance, material or article offered for transport:

.1   The UN Number preceded by the letters "UN";

.2   The Proper Shipping Name, as determined according to 3.1.2, including the technical name enclosed in parenthesis, as applicable (see 3.1.2.8);

.3   The primary hazard class or, when assigned, the division of the goods, including, for class 1, the compatibility group letter. The words "Class" or "Division" may be included preceding the primary hazard class or division numbers;

.4   Subsidiary hazard class or division number(s) corresponding to the subsidiary risk label(s) required to be applied, when assigned, shall be entered following the primary hazard class or division and shall be enclosed in parenthesis. The words "Class" or "Division" may be included preceding the subsidiary hazard class or division numbers;

.5   Where assigned, the packing group for the substance or article, which may be preceded by "PG" (e.g. "PG II").

5.4.1.4.2   *Sequence of the dangerous goods description*

The five elements of the dangerous goods description specified in 5.4.1.4.1 shall be shown in the order listed above (i.e. .1, .2, .3, .4, and .5) with no information interspersed, except as provided in this Code. Unless permitted or required by this Code, additional information shall be placed after the dangerous goods description.

5.4.1.4.3   *Information which supplements the Proper Shipping Name in the dangerous goods description*

The Proper Shipping Name (see 3.1.2) in the dangerous goods description shall be supplemented as follows:

.1   *Technical names for "n.o.s." and other generic descriptions*: Proper Shipping Names that are assigned special provision 274 in column 6 of the Dangerous Goods List shall be supplemented with their technical or chemical group names as described in 3.1.2.8;

.2   *Empty uncleaned packagings, bulk containers and tanks*: Empty means of containment (including packagings, IBCs, bulk containers, portable tanks, road tank vehicles and railway tank wagons) which contain the residue of dangerous goods of classes other than class 7 shall be described as such by, for example, placing the words "EMPTY UNCLEANED" or "RESIDUE LAST CONTAINED" before or after the Proper Shipping Name;

.3   *Wastes*: For waste dangerous goods (other than radioactive wastes) which are being transported for disposal, or for processing for disposal, the Proper Shipping Name shall be preceded by the word "WASTE", unless this is already a part of the Proper Shipping Name;

.4   *Elevated temperature substances*: If the Proper Shipping Name of a substance which is transported or offered for transport in a liquid state at a temperature equal to or exceeding 100°C, or in a solid state at a temperature equal to or exceeding 240°C, does not convey the elevated temperature condition (for example, by using the term "MOLTEN" or "ELEVATED TEMPERATURE" as part of the Proper Shipping Name), the word "HOT" shall immediately precede the Proper Shipping Name.

.5   *Marine pollutants:* If the goods to be transported are marine pollutants, the goods shall be identified as "MARINE POLLUTANT", and for generic or "not otherwise specified" (N.O.S.) entries the Proper Shipping Name shall be supplemented with the recognized chemical name of the marine pollutant (see 3.1.2.9);

.6   *Flashpoint:* If the dangerous goods to be transported have a flashpoint of 60°C or below (in °C closed-cup (c.c.)), the minimum flashpoint shall be indicated. Because of the presence of impurities, the flashpoint may be lower or higher than the reference temperature indicated in the Dangerous Goods List for the substance. For class 5.2 organic peroxides which are also flammable, the flashpoint need not be declared.

5.4.1.4.4   *Examples of dangerous goods descriptions:*

UN1098 ALLYL ALCOHOL 6.1 (3) I (21°C c.c.)

UN1098, ALLYL ALCOHOL, class 6.1, (class 3), PG I, (21°C c.c.)

UN 1092, Acrolein, stabilized, class 6.1 (3), PG I, (–24°C c.c.) MARINE POLLUTANT

UN 2761, Organochlorine pesticide, solid, toxic, (Aldrin 19%), class 6.1, PG III, MARINE POLLUTANT

5.4.1.5   **Information required in addition to the dangerous goods description**

In addition to the dangerous goods description, the following information shall be included after the dangerous goods description on the dangerous goods transport document.

5.4.1.5.1   *Total quantity of dangerous goods*

Except for empty uncleaned packagings, the total quantity of dangerous goods covered by the description (by volume or mass as appropriate) of each item of dangerous goods bearing a different Proper Shipping Name, UN Number or packing group shall be included. For class 1 dangerous goods, the quantity shall be the net explosive mass. For dangerous goods transported in salvage packagings, an estimate of the quantity of dangerous goods shall be given. The number and kind (e.g. drum, box, etc.) of packages shall also be indicated. UN packaging codes may only be used to supplement the description of the kind of package (e.g., one box (4G)). Abbreviations may be used to specify the unit of measurement for the total quantity.

5.4.1.5.2   *Limited quantities*

5.4.1.5.2.1   When dangerous goods are transported according to the exceptions for dangerous goods packed in limited quantities provided for in column 7a of the Dangerous Goods List and chapter 3.4, the words "limited quantity" or "LTD QTY" shall be included.

5.4.1.5.2.2   Where a shipment is offered in accordance with 3.4.4.1.2, the following statement shall be included in the transport document: "Transport in accordance with 3.4.4.1.2 of the IMDG Code".

5.4.1.5.3   *Salvage packagings*

For dangerous goods transported in salvage packagings, the words "SALVAGE PACKAGE" shall be included.

5.4.1.5.4   *Substances stabilized by temperature control*

If the word "STABILIZED" is part of the Proper Shipping Name (see also 3.1.2.6), when stabilization is by means of temperature control, the control and emergency temperatures (see 7.7.2) shall be indicated in the transport document, as follows:

"Control temperature: . . .°C   Emergency temperature: . . . °C".

5.4.1.5.5   *Self-reactive substances and organic peroxides*

For self-reactive substances of class 4.1 and for organic peroxides which require temperature control during transport, the control and emergency temperatures (see 7.7.2) shall be indicated on the dangerous goods transport document, as follows:

"Control temperature: . . .°C   Emergency temperature: . . . °C".

5.4.1.5.5.1   When, for certain self-reactive substances of class 4.1 and organic peroxides of class 5.2, the competent authority has permitted the "EXPLOSIVE" subsidiary risk label (model No. 1) to be dispensed with for the specific package, a statement to this effect shall be included.

5.4.1.5.5.2   When organic peroxides and self-reactive substances are transported under conditions where approval is required (for organic peroxides, see 2.5.3.2.5, 4.1.7.2.2, 4.2.1.13.1 and 4.2.1.13.3; for self-reactive substances, see 2.4.2.3.2.4 and 4.1.7.2.2), a statement to this effect shall be included in the dangerous goods transport document. A copy of the classification approval and conditions of transport for non-listed organic peroxides and self-reactive substances shall be attached to the dangerous goods transport document.

5.4.1.5.5.3   When a sample of an organic peroxide (see 2.5.3.2.5.1) or a self-reactive substance (see 2.4.2.3.2.4.2) is transported, a statement to this effect shall be included in the dangerous goods transport document.

5.4.1.5.6   *Infectious substances*

The full address of the consignee shall be shown on the document, together with the name of a responsible person and his telephone number.

5.4.1.5.7   *Radioactive material*

5.4.1.5.7.1   The following information shall be included for each consignment of class 7 material, as applicable, in the order given:

.1   The name or symbol of each radionuclide or, for mixtures of radionuclides, an appropriate general description or a list of the most restrictive nuclides;

.2   A description of the physical and chemical form of the material, or a notation that the material is special form radioactive material or low dispersible radioactive material. A generic chemical description is acceptable for chemical form;

.3 The maximum activity of the radioactive contents during transport expressed in units of becquerels (Bq) with an appropriate SI prefix symbol (see 1.2.2.1). For fissile material, the mass of fissile material in units of grams (g), or appropriate multiples thereof, may be used in place of activity;

.4 The category of the package, i.e. I – WHITE, II – YELLOW, III – YELLOW;

.5 The transport index (categories II – YELLOW and III – YELLOW only);

.6 For consignments including fissile material other than consignments excepted under 6.4.11.2, the criticality safety index;

.7 The identification mark for each competent authority approval certificate (special form radioactive material, low dispersible radioactive material, special arrangement, package design, or shipment) applicable to the consignment;

.8 For consignments of more than one package, the information contained in 5.4.1.4.1.1 to .3 and 5.4.1.5.7.1.1 to .7 shall be given for each package. For packages in an overpack, freight container, or conveyance, a detailed statement of the contents of each package within the overpack, freight container, or conveyance and, where appropriate, of each overpack, freight container, or conveyance shall be included. If packages are to be removed from the overpack, freight container, or conveyance at a point of intermediate unloading, appropriate transport documents shall be made available;

.9 Where a consignment is required to be shipped under exclusive use, the statement "EXCLUSIVE USE SHIPMENT"; and

.10 For LSA-II, LSA-III, SCO-I and SCO-II, the total activity of the consignment as a multiple of A_2.

5.4.1.5.7.2 The transport document shall include a statement regarding actions, if any, that are required to be taken by the carrier. The statement shall be in the languages deemed necessary by the carrier or the authorities concerned, and shall include at least the following points:

.1 Supplementary requirements for loading, stowage, transport, handling and unloading of the package, overpack or freight container, including any special stowage provisions for the safe dissipation of heat (see 7.1.14.4), or a statement that no such requirements are necessary;

.2 Restrictions on the mode of transport or conveyance and any necessary routeing instructions;

.3 Emergency arrangements appropriate to the consignment.

5.4.1.5.7.3 In case of international transport of packages requiring competent authorities design or shipment approval, for which different approval types apply in the different countries concerned, the UN Number and Proper Shipping Name required in 5.4.1.4.1 shall be in accordance with the certificate of the country of origin of design.

5.4.1.5.7.4 The applicable competent authority certificates need not necessarily accompany the consignment. The consignor shall make them available to the carrier(s) before loading and unloading.

5.4.1.5.8 *Aerosols*

If the capacity of an aerosol is above 1000 mℓ, this shall be declared in the transport document.

5.4.1.5.9 *Explosives*

The following information shall be included for each consignment of class 1 goods, as applicable:

.1 Entries have been included for "SUBSTANCES, EXPLOSIVE, N.O.S.", "ARTICLES, EXPLOSIVE, N.O.S.", and "COMPONENTS, EXPLOSIVE TRAIN, N.O.S.". When a specific entry does not exist, the competent authority of the country of origin shall use the entry appropriate to the hazard division and compatibility group. The transport document shall contain the statement: "Transport under this entry approved by the competent authority of . . ." followed by the State's distinguishing sign for motor vehicles in international traffic of the country of the competent authority.

.2 The transport of explosive substances for which a minimum water or phlegmatizer content is specified in the individual entry is prohibited when containing less water or phlegmatizer than the specified minimum. Such substances shall only be transported with special authorization granted by the competent authority of the country of origin. The transport document shall contain the statement "Transport under this entry approved by the competent authority of . . ." followed by the State's distinguishing sign for motor vehicles in international traffic of the country of the competent authority.

.3 When explosive substances or articles are packaged "as approved by the competent authority", the transport document shall contain the statement "Packaging approved by the competent authority of . . ." followed by the State's distinguishing sign for motor vehicles in international traffic of the country of the competent authority.

.4 There are some hazards which are not indicated by the hazard division and compatibility group of a substance. The shipper shall provide an indication of any such hazards on the dangerous goods documentation.

5.4.1.5.10 *Viscous substances*

When viscous substances are transported in accordance with 2.3.2.5, the following statement shall be included in the transport document: "Transport in accordance with 2.3.2.5 of the IMDG Code.".

5.4.1.5.11 *Special provisions for segregation*

5.4.1.5.11.1 For substances, mixtures, solutions or preparations classified under N.O.S. entries not included in the segregation groups listed in 3.1.4.4 but belonging, in the opinion of the consignor, to one of these groups (see 3.1.4.2), the appropriate segregation group name preceded by the phrase "IMDG Code segregation group" shall be included in the transport document after the dangerous goods description. For example:

"UN 1760 CORROSIVE LIQUID, N.O.S. (Phosphoric acid) 8 III IMDG Code segregation group 1 – Acids".

5.4.1.5.11.2 When substances are loaded together in a cargo transport unit in accordance with 7.2.1.13.1.2, the following statement shall be included in the transport document: "Transport in accordance with 7.2.1.13.1.2 of the IMDG Code".

5.4.1.5.11.3 When acid and alkali substances of class 8 are transported in the same cargo transport unit, whether in the same packaging or not, in accordance with 7.2.1.13.2, the following statement shall be included in the transport document: "Transport in accordance with 7.2.1.13.2 of the IMDG Code".

5.4.1.5.12 *Transport of solid dangerous goods in bulk containers*

For bulk containers other than freight containers, the following statement shall be included on the transport document (see 6.9.4.6):

"Bulk container BK2 approved by the competent authority of ...".

5.4.1.5.13 *Transport of IBCs or portable tanks after the date of expiry of the last periodic test or inspection*

For transport in accordance with 4.1.2.2.2.2, 6.7.2.19.6.2, 6.7.3.15.6.2 or 6.7.4.14.6.2, a statement to this effect shall be included in the transport document, as follows: "Transport in accordance with 4.1.2.2.2.2", "Transport in accordance with 6.7.2.19.6.2", "Transport in accordance with 6.7.3.15.6.2" or "Transport in accordance with 6.7.4.14.6.2" as appropriate.

5.4.1.5.14 *Dangerous goods in excepted quantities*

5.4.1.5.14.1 When dangerous goods are transported according to the exceptions for dangerous goods packed in excepted quantities provided for in column 7b of the Dangerous Goods List and chapter 3.5, the words "dangerous goods in excepted quantities" shall be included.

5.4.1.6 **Certification**

5.4.1.6.1 The dangerous goods transport document shall include a certification or declaration that the consignment is acceptable for transport and that the goods are properly packaged, marked and labelled, and in proper condition for transport in accordance with the applicable regulations. The text for this certification is:

"I hereby declare that the contents of this consignment are fully and accurately described above by the Proper Shipping Name, and are classified, packaged, marked and labelled/placarded, and are in all respects in proper condition for transport according to applicable international and national government regulations."

The certification shall be signed and dated by the consignor. Facsimile signatures are acceptable where applicable laws and regulations recognize the legal validity of facsimile signatures.

5.4.1.6.2 If the dangerous goods documentation is presented to the carrier by means of electronic data processing (EDP) or electronic data interchange (EDI) transmission techniques, the signature(s) may be replaced by the name(s) (in capitals) of the person authorized to sign.

5.4.2 Container/vehicle packing certificate

5.4.2.1 When dangerous goods are packed or loaded into any container* or vehicle, those responsible for packing the container or vehicle shall provide a "container/vehicle packing certificate" specifying the container/vehicle identification number(s) and certifying that the operation has been carried out in accordance with the following conditions:

.1 The container/vehicle was clean, dry and apparently fit to receive the goods;

.2 Packages which need to be segregated in accordance with applicable segregation requirements have not been packed together onto or in the container/vehicle (unless approved by the competent authority concerned in accordance with 7.2.2.3);

.3 All packages have been externally inspected for damage, and only sound packages have been loaded;

.4 Drums have been stowed in an upright position, unless otherwise authorized by the competent authority, and all goods have been properly loaded and, where necessary, adequately braced with securing material to suit the mode(s)† of transport for the intended journey;

.5 Goods loaded in bulk have been evenly distributed within the container/vehicle;

.6 For consignments including goods of class 1 other than division 1.4, the container/vehicle is structurally serviceable in accordance with 7.4.6;

.7 The container/vehicle and packages are properly marked, labelled and placarded, as appropriate;

.8 When solid carbon dioxide (CO_2 – dry ice) is used for cooling purposes, the container/vehicle is externally marked or labelled in a conspicuous place, such as, at the door end, with the words: "DANGEROUS CO_2 (DRY ICE) INSIDE. VENTILATE THOROUGHLY BEFORE ENTERING"; and

.9 A dangerous goods transport document, as indicated in 5.4.1, has been received for each dangerous goods consignment loaded in the container/vehicle.

Note: The container/vehicle packing certificate is not required for portable tanks.

5.4.2.2 The information required in the dangerous goods transport document and the container/vehicle packing certificate may be incorporated into a single document; if not, these documents shall be attached one to the other. If the information is incorporated into a single document, the document shall include a signed declaration such as "It is declared that the packing of the goods into the container/vehicle has been carried out in accordance with the applicable provisions". This declaration shall be dated and the person signing this declaration shall be identified on the document. Facsimile signatures are acceptable where applicable laws and regulations recognize the legal validity of facsimile signatures.

5.4.2.3 If the dangerous goods documentation is presented to the carrier by means of electronic data processing (EDP) or electronic data interchange (EDI) transmission techniques, the signature(s) may be replaced by the name(s) (in capitals) of the person authorized to sign.

5.4.3 Documentation required aboard the ship

5.4.3.1 Each ship carrying dangerous goods and marine pollutants shall have a special list or manifest‡ setting out, in accordance with regulation 4.5 of chapter VII of SOLAS 1974, as amended, and with regulation 4(3) of Annex III of MARPOL 73/78, the dangerous goods and marine pollutants and the location thereof. A detailed stowage plan, which identifies by class and sets out the location of all dangerous goods and marine pollutants, may be used in place of such a special list or manifest. This dangerous goods or marine pollutants list or manifest shall be based on the documentation and certification required in this Code and shall at least contain, in addition to the information in 5.4.1.4 and 5.4.1.5, the stowage location and the total quantity of the dangerous goods and marine pollutants. A copy of one of these documents shall be made available before departure to the person or organization designated by the port State authority.

5.4.3.2 Emergency response information

5.4.3.2.1 For consignments of dangerous goods, appropriate information shall be immediately available at all times for use in emergency response to accidents and incidents involving dangerous goods in transport. The information shall be available away from packages containing the dangerous goods and immediately accessible in the event of an incident. Methods of compliance include:

.1 appropriate entries in the special list, manifest or dangerous goods declaration; or

---

* See definition of "freight container" in 1.2.1.
† See IMO/ILO/UN ECE Guidelines for Packing of Cargo Transport Units.
‡ FAL.2/Circ.51/Rev.1 may be used for this purpose.

.2 provision of a separate document such as a safety data sheet; or

.3 provision of separate documentation, such as the *Emergency Response Procedures for Ships Carrying Dangerous Goods (EmS Guide)* for use in conjunction with the transport document and the *Medical First Aid Guide for Use in Accidents Involving Dangerous Goods (MFAG)*.

5.4.4 Other required information and documentation

5.4.4.1 In certain circumstances, special certificates or other documents are required such as:

.1 a weathering certificate; as required in the individual entries of the Dangerous Goods List;

.2 a certificate exempting a substance, material or article from the provisions of the IMDG Code (such as, see individual entries for charcoal, fishmeal, seedcake);

.3 for new self-reactive substances and organic peroxides or new formulation of currently assigned self-reactive substances and organic peroxides, a statement by the competent authority of the country of origin of the approved classification and conditions of transport.

5.4.4.2 Fumigated units

The transport document for a fumigated unit shall show the type and amount of fumigant used and the date and time of fumigation. In addition, instructions for disposal of any residual fumigant, including fumigation devices, if used, shall be provided.

5.4.5 Multimodal Dangerous Goods Form

5.4.5.1 This form meets the requirements of SOLAS 74, chapter VII, regulation 4, MARPOL 73/78, Annex III, regulation 4 and the provisions of this chapter. The information required by the provisions of this chapter is mandatory; however, the layout of this form is not mandatory.

MULTIMODAL DANGEROUS GOODS FORM

This form may be used as a dangerous goods declaration as it meets the requirements of SOLAS 74, chapter VII, regulation 4; MARPOL 73/78, Annex III, regulation 4.

1 Shipper/Consignor/Sender	2 Transport document number		
	3 Page 1 of pages	4 Shipper's reference	
		5 Freight forwarder's reference	
6 Consignee	7 Carrier (to be completed by the carrier)		

SHIPPER'S DECLARATION
I hereby declare that the contents of this consignment are fully and accurately described below by the Proper Shipping Name, and are classified, packaged, marked and labelled/placarded and are in all respects in proper condition for transport according to the applicable international and national governmental regulations.

8 This shipment is within the limitations prescribed for: (Delete non-applicable)		9 Additional handling information
PASSENGER AND CARGO AIRCRAFT	CARGO AIRCRAFT ONLY	
10 Vessel/flight No. and date	11 Port/place of loading	
12 Port/place of discharge	13 Destination	

14 Shipping marks	* Number and kind of packages; description of goods	Gross mass (kg)	Net mass (kg)	Cube (m³)

15 Container identification No./ vehicle registration No.	16 Seal number(s)	17 Container/vehicle size & type	18 Tare mass (kg)	19 Total gross mass (including tare) (kg)

CONTAINER/VEHICLE PACKING CERTIFICATE

I hereby declare that the goods described above have been packed/loaded into the container/vehicle identified above in accordance with the applicable provisions.†
MUST BE COMPLETED AND SIGNED FOR ALL CONTAINER/VEHICLE LOADS BY PERSON RESPONSIBLE FOR PACKING/LOADING

21 RECEIVING ORGANISATION RECEIPT

Received the above number of packages/containers/trailers in apparent good order and condition, unless stated hereon: RECEIVING ORGANISATION REMARKS:

20 Name of company	Haulier's name	22 Name of company (OF SHIPPER PREPARING THIS NOTE)
	Vehicle reg. no.	
Name/status of declarant	Signature and date	Name/status of declarant
Place and date		Place and date
Signature of declarant	DRIVER'S SIGNATURE	Signature of declarant

* **DANGEROUS GOODS:**
You must specify: UN No., Proper Shipping Name, hazard class, packing group, (where assigned) marine pollutant and observe the mandatory requirements under applicable national and international governmental regulations. For the purposes of the IMDG Code, see 5.4.1.4.
† For the purposes of the IMDG Code, see 5.4.2.

Documentary Aspects of the International Transport of Dangerous Goods

Container/Vehicle Packing Certificate

The signature given overleaf in Box 20 must be that of the person controlling the container/vehicle operation.

It is certified that:

The container/vehicle was clean, dry and apparently fit to receive the goods.

If the consignments include goods of class 1, other than division 1.4, the container is structurally serviceable.

No incompatible goods have been packed into the container/vehicle unless specially authorized by the Competent Authority.

All packages have been externally inspected for damage and only sound packages packed.

Drums have been stowed in an upright position unless otherwise authorized by the Competent Authority.

All packages have been properly packed and secured in the container/vehicle.

When materials are transported in bulk packagings, the cargo has been evenly distributed in the container/vehicle.

The packages and the container/vehicle have been properly marked, labelled and placarded. Any irrelevant mark, labels and placards have been removed.

When solid carbon dioxide (CO_2 – dry ice) is used for cooling purposes, the vehicle or freight container is externally marked or labelled in a conspicuous place, e.g. at the door end, with the words: DANGEROUS CO_2 GAS (DRY ICE) INSIDE – VENTILATE THOROUGHLY BEFORE ENTERING.

When this Dangerous Goods Form is used as a container/vehicle packing certificate only, not a combined document, a dangerous goods declaration signed by the shipper or supplier must have been issued/received to cover each dangerous goods consignment packed in the container.

Note: The container packing certificate is not required for tanks.

1 Shipper/Consignor/Sender	2 Transport document number		
	3 Page of pages	4 Shipper's reference	
		5 Freight forwarder's reference	
14 Shipping marks * Number and kind of packages; description of goods	Gross mass (kg)	Net mass (kg)	Cube (m³)

PART 6

CONSTRUCTION AND TESTING OF PACKAGINGS,
INTERMEDIATE BULK CONTAINERS (IBCs),
LARGE PACKAGINGS, PORTABLE TANKS,
MULTIPLE-ELEMENT GAS CONTAINERS (MEGCs)
AND ROAD TANK VEHICLES

6

Chapter 6.1

Provisions for the construction and testing of packagings (other than for class 6.2 substances)

6.1.1 Applicability and general provisions

6.1.1.1 Applicability

The provisions in this chapter do not apply to:

.1 pressure receptacles;

.2 packages containing radioactive material, which shall comply with the Regulations of the International Atomic Energy Agency (IAEA), except that:

   (i) radioactive material possessing other dangerous properties (subsidiary risks) shall also comply with special provision 172 in chapter 3.3; and

   (ii) low specific activity (LSA) material and surface contaminated objects (SCO) may be carried in certain packagings defined in this Code provided that the supplementary provisions set out in the IAEA Regulations are also met;

.3 packages whose net mass exceeds 400 kg; and

.4 packages with a capacity exceeding 450 ℓ.

6.1.1.2 General provisions

6.1.1.2.1 The provisions for packagings in 6.1.4 are based on packagings currently used. In order to take into account progress in science and technology, there is no objection to the use of packagings having specifications different from those in 6.1.4, provided that they are equally effective, acceptable to the competent authority and able successfully to withstand the tests described in 6.1.1.2 and 6.1.5. Methods of testing other than those described in this chapter are acceptable, provided that they are equivalent.

6.1.1.2.2 Every packaging intended to contain liquids shall successfully undergo a suitable leakproofness test and be capable of meeting the appropriate test level indicated in 6.1.5.4.4:

.1 before it is first used for transport;

.2 after remanufacturing or reconditioning, before it is re-used for transport.

For this test, packagings need not have their own closures fitted.

The inner receptacle of a composite packaging may be tested without the outer packaging provided the test results are not affected. This test is not necessary for an inner packaging of a combination packaging.

6.1.1.2.3 Receptacles, parts of receptacles and closures (stoppers) made of plastics which may be directly in contact with a dangerous substance shall be resistant to it and shall not incorporate materials which may react dangerously or form hazardous compounds or lead to softening, weakening or failure of the receptacle or closure.

6.1.1.2.4 Plastics packagings shall be adequately resistant to ageing and to degradation caused either by the substance contained or by ultraviolet radiation. Any permeation of the substance contained shall not constitute a danger under normal conditions of transport.

6.1.1.3 Packagings shall be manufactured, reconditioned and tested under a quality-assurance programme which satisfies the competent authority in order to ensure that each packaging meets the provisions of this chapter.

Note: ISO 16106:2006 "Packaging – Transport packages for dangerous goods – Dangerous goods packagings, intermediate bulk containers (IBCs) and large packagings – Guidelines for the application of ISO 9001" provides acceptable guidance on procedures which may be followed.

6.1.1.4 Manufacturers and subsequent distributors of packagings shall provide information regarding procedures to be followed and a description of the types and dimensions of closures (including required gaskets) and any other components needed to ensure that packages as presented for transport are capable of passing the applicable performance tests of this chapter.

6.1.2 Code for designating types of packagings

6.1.2.1 The code consists of:

.1 an Arabic numeral indicating the kind of packaging, such as drum, jerrican, etc., followed by

.2 one or more capital letters in Latin characters indicating the nature of the material, such as steel, wood, etc., followed where necessary by

.3 an Arabic numeral indicating the category of packaging within the type to which the packaging belongs.

6.1.2.2 In the case of composite packagings, two capital letters in Latin characters shall be used in sequence in the second position of the code. The first indicates the material of the inner receptacle and the second that of the outer packaging.

6.1.2.3 In the case of combination packagings, only the code number for the outer packaging shall be used.

6.1.2.4 The letters 'T', 'V' or 'W' may follow the packaging code. The letter 'T' signifies a salvage packaging conforming to the provisions of 6.1.5.1.11. The letter 'V' signifies a special packaging conforming to the provisions of 6.1.5.1.7. The letter 'W' signifies that the packaging, although of the same type as that indicated by the code, is manufactured to a specification different to that in 6.1.4 but is considered equivalent under the provisions of 6.1.1.2.

6.1.2.5 The following numerals shall be used for the kinds of packaging:

1 Drum

2 (Reserved)

3 Jerrican

4 Box

5 Bag

6 Composite packaging

6.1.2.6 The following capital letters shall be used for the types of material:

A Steel (all types and surface treatments)

B Aluminium

C Natural wood

D Plywood

F Reconstituted wood

G Fibreboard

H Plastics material

L Textile

M Paper, multiwall

N Metal (other than steel or aluminium)

P Glass, porcelain or stoneware

Note: "Plastics material" is taken to include other polymeric materials such as rubber.

6.1.2.7 The following table indicates the codes to be used for designating types of packagings depending on the kind of packagings, the material used for their construction and their category; it also refers to the paragraphs to be consulted for the appropriate provisions:

Kind	Material	Category	Code	Paragraph
1 Drums	A Steel	non-removable head	1A1	6.1.4.1
		removable head	1A2	
	B Aluminium	non-removable head	1B1	6.1.4.2
		removable head	1B2	
	D Plywood	–	1D	6.1.4.5
	G Fibre	–	1G	6.1.4.7
	H Plastics	non-removable head	1H1	6.1.4.8
		removable head	1H2	
	N Metal, other than steel or aluminium	non-removable head	1N1	6.1.4.3
		removable head	1N2	

Kind		Material		Category	Code	Paragraph
2	(Reserved)					
3	Jerricans	A	Steel	non-removable head	3A1	6.1.4.4
				removable head	3A2	
		B	Aluminium	non-removable head	3B1	6.1.4.4
				removable head	3B2	
		H	Plastics	non-removable head	3H1	6.1.4.8
				removable head	3H2	
4	Boxes	A	Steel	–	4A	6.1.4.14
		B	Aluminium	–	4B	6.1.4.14
		C	Natural wood	ordinary	4C1	6.1.4.9
				with sift-proof walls	4C2	
		D	Plywood	–	4D	6.1.4.10
		F	Reconstituted wood	–	4F	6.1.4.11
		G	Fibreboard	–	4G	6.1.4.12
		H	Plastics	expanded	4H1	6.1.4.13
				solid	4H2	
5	Bags	H	Woven plastics	without inner lining or coating	5H1	6.1.4.16
				sift-proof	5H2	
				water-resistant	5H3	
		H	Plastics film	–	5H4	6.1.4.17
		L	Textile	without inner lining or coating	5L1	6.1.4.15
				sift-proof	5L2	
				water-resistant	5L3	
		M	Paper	multiwall	5M1	6.1.4.18
				multiwall, water-resistant	5M2	
6	Composite packagings	H	Plastics receptacle	in steel drum	6HA1	6.1.4.19
				in steel crate or box	6HA2	6.1.4.19
				in aluminium drum	6HB1	6.1.4.19
				in aluminium crate or box	6HB2	6.1.4.19
				in wooden box	6HC	6.1.4.19
				in plywood drum	6HD1	6.1.4.19
				in plywood box	6HD2	6.1.4.19
				in fibre drum	6HG1	6.1.4.19
				in fibreboard box	6HG2	6.1.4.19
				in plastics drum	6HH1	6.1.4.19
				in solid plastics box	6HH2	6.1.4.19
		P	Glass, porcelain or stoneware receptacle	in steel drum	6PA1	6.1.4.20
				in steel crate or box	6PA2	6.1.4.20
				in aluminium drum	6PB1	6.1.4.20
				in aluminium crate or box	6PB2	6.1.4.20
				in wooden box	6PC	6.1.4.20
				in plywood drum	6PD1	6.1.4.20
				in wickerwork hamper	6PD2	6.1.4.20
				in fibre drum	6PG1	6.1.4.20
				in fibreboard box	6PG2	6.1.4.20
				in expanded plastics packaging	6PH1	6.1.4.20
				in solid plastics packaging	6PH2	6.1.4.20

6

6.1.3 Marking

Note 1: The marking indicates that the packaging which bears it corresponds to a successfully tested design type and that it complies with the provisions of this chapter which are related to the manufacture, but not to the use, of the packaging. In itself, therefore, the mark does not necessarily confirm that the packaging may be used for any substance. The type of packaging (such as steel drum), its maximum capacity or mass, and any special provisions are specified for each substance or article n part 3 of this Code.

Note 2: The marking is intended to be of assistance to packaging manufacturers, reconditioners, packaging users, carriers and regulatory authorities. In relation to the use of a new packaging, the original marking is a means for its manufacturer to identify the type and to indicate those performance test provisions that have been met.

Note 3: The marking does not always provide full details of the test levels, etc. and these may need to be taken further into account, such as by reference to a test certificate, test reports or register of successfully tested packagings. For example, a packaging having an X or Y marking may be used for substances to which a packing group having a lesser degree of danger has been assigned, with the relevant maximum permissible value of the relative density* determined by taking into account the factor 1.5 or 2.25 indicated in the packaging test provisions in 6.1.5 as appropriate, i.e. packing group I packaging tested for products of relative density 1.2 could be used as a packing group II packaging for products of relative density 1.8 or packing group III packaging of relative density 2.7, provided, of course, that all the performance criteria can still be met with the product having the higher relative density.

6.1.3.1 Each packaging intended for use according to this Code shall bear markings which are durable, legible and placed in such a location and of such a size relative to the packaging as to be readily visible. For packages with a gross mass of more than 30 kg, the markings or a duplicate thereof shall appear on the top or on a side of the packaging. Letters, numerals and symbols shall be at least 12 mm high, except for packagings of 30 ℓ or 30 kg capacity or less, when they shall be at least 6 mm in height, and for packagings of 5 ℓ or 5 kg or less, when they shall be of an appropriate size.

The marking shall show:

(a) The United Nations packaging symbol

This symbol shall not be used for any purpose other than certifying that a packaging complies with the relevant requirements in chapter 6.1, 6.2, 6.3, 6.5 or 6.6. For embossed metal packagings, the capital letters "UN" may be applied as the symbol.

(b) The code designating the type of packaging according to 6.1.2.

(c) A code in two parts:

(i) a letter designating the packing group or groups for which the design type has been successfully tested:

"X" for packing groups I, II and III

"Y" for packing groups II and III

"Z" for packing group III only;

(ii) the relative density, rounded off to the first decimal, for which the design type has been tested for packagings, without inner packagings, intended to contain liquids; this may be omitted when the relative density does not exceed 1.2. For packagings intended to contain solids or inner packagings, the maximum gross mass in kilograms.

(d) Either a letter "S", denoting that the packaging is intended for the transport of solids or inner packagings, or, for packagings (other than combination packagings) intended to contain liquids, the hydraulic test pressure which the packaging was shown to withstand in kilopascals, rounded down to the nearest 10 kPa.

---

* Relative density *(d)* is considered to be synonymous with specific gravity (SG) and will be used throughout this text.

(e) The last two digits of the year during which the packaging was manufactured. Packagings of types 1H and 3H shall also be appropriately marked with the month of manufacture; this may be marked on the packaging in a different place from the remainder of the marking. An appropriate method is:

(f) The State authorizing the allocation of the mark, indicated by the distinguishing sign for motor vehicles in international traffic.

(g) The name of the manufacturer or other identification of the packaging specified by the competent authority.

6.1.3.2 In addition to the durable markings prescribed in 6.1.3.1, every new metal drum of a capacity greater than 100 ℓ shall bear the marks described in 6.1.3.1 (a) to (e) on the bottom, with an indication of the nominal thickness of at least the metal used in the body (in millimetres, to 0.1 mm), in permanent form (such as embossed). When the nominal thickness of either head of a metal drum is thinner than that of the body, the nominal thickness of the top head, body and bottom head shall be marked on the bottom in permanent form (such as embossed), for example '1.0 – 1.2 – 1.0' or '0.9 – 1.0 – 1.0'. Nominal thicknesses of metal shall be determined according to the appropriate ISO standard, for example ISO 3574:1999 for steel. The marks indicated in 6.1.3.1 (f) and (g) shall not be applied in a permanent form (such as embossed) except as provided in 6.1.3.5.

6.1.3.3 Every packaging other than those referred to in 6.1.3.2 liable to undergo a reconditioning process shall bear the marks indicated in 6.1.3.1 (a) to (e) in a permanent form. Marks are permanent if they are able to withstand the reconditioning process (e.g., embossed). For packagings other than metal drums of a capacity greater than 100 ℓ, these permanent marks may replace the corresponding durable markings prescribed in 6.1.3.1.

6.1.3.4 For remanufactured metal drums, if there is no change to the packaging type and no replacement or removal of integral structural components, the required markings need not be permanent (such as embossed). Every other remanufactured metal drum shall bear the markings in 6.1.3.1 (a) to (e) in a permanent form (such as embossed) on the top head or side.

6.1.3.5 Metal drums made from materials (such as stainless steel) designed to be re-used repeatedly may bear the markings indicated in 6.1.3.1 (f) and (g) in a permanent (such as embossed) form.

6.1.3.6 Packagings manufactured with recycled plastics material as defined in 1.2.1 shall be marked "REC". This mark shall be placed near the mark prescribed in 6.1.3.1.

6.1.3.7 Marking shall be applied in the sequence of the subparagraphs in 6.1.3.1; each element of the marking required in these subparagraphs and when appropriate subparagraphs (h) to (j) of 6.1.3.8 shall be clearly separated, e.g., by a slash or space, so as to be easily identifiable. For examples, see 6.1.3.10. Any additional markings authorized by a competent authority shall still enable the parts of the mark to be correctly identified with reference to 6.1.3.1.

6.1.3.8 After reconditioning a packaging, the reconditioner shall apply to it, in the following sequence, a durable marking showing:

(h) the State in which the reconditioning was carried out, indicated by the distinguishing sign for motor vehicles in international traffic;

(i) the name of the reconditioner or other identification of the packaging specified by the competent authority;

(j) the year of reconditioning; the letter "R"; and, for every packaging successfully passing the leakproofness test in 6.1.1.2.2, the additional letter "L".

6.1.3.9 When, after reconditioning, the markings required by 6.1.3.1 (a) to (d) no longer appear on the top head or the side of a metal drum, the reconditioner shall apply them in a durable form followed by those required by 6.1.3.8 (h), (i) and (j). These markings shall not identify a greater performance capability than that for which the original design type has been tested and marked.

6.1.3.10   Examples of markings for NEW packagings

4G/Y145/S/02   as in 6.1.3.1 (a), (b), (c), (d) and (e)   For a new fibreboard box

NL/VL823   as in 6.1.3.1 (f) and (g)

1A1/Y1.4/150/98   as in 6.1.3.1 (a), (b), (c), (d) and (e)   For a new steel drum to contain liquids

NL/VL824   as in 6.1.3.1 (f) and (g)

1A2/Y150/S/01   as in 6.1.3.1 (a), (b), (c), (d) and (e)   For a new steel drum to contain solids or inner packagings

NL/VL825   as in 6.1.3.1 (f) and (g)

4HW/Y136/S/98   as in 6.1.3.1 (a), (b), (c), (d) and (e)   For a new plastics box of a specification equivalent to that indicated by the packaging code

NL/VL826   as in 6.1.3.1 (f) and (g)

1A2/Y/100/01   as in 6.1.3.1 (a), (b), (c), (d) and (e)   For a remanufactured steel drum to contain liquids of relative density not exceeding 1.2

USA/MM5   as in 6.1.3.1 (f) and (g)   **Note**: For liquids, the marking of relative density not exceeding 1.2 is optional; see 6.1.3.1 (c)(ii)

6.1.3.11   Examples of markings for RECONDITIONED packagings

1A1/Y1.4/150/97   as in 6.1.3.1 (a), (b), (c), (d) and (e)

NL/RB/01 RL   as in 6.1.3.8 (h), (i) and (j)

1A2/Y150/S/99   as in 6.1.3.1 (a), (b), (c), (d) and (e)

USA/RB/00 R   as in 6.1.3.8 (h), (i) and (j)

6.1.3.12   Examples of markings for SALVAGE packagings

1A2T/Y300/S/01   as in 6.1.3.1 (a), (b), (c), (d) and (e)

USA/abc   as in 6.1.3.1 (f) and (g)

Note: The markings, for which examples are given in 6.1.3.10, 6.1.3.11 and 6.1.3.12, may be applied in a single line or in multiple lines provided the correct sequence is respected.

6.1.4   Provisions for packagings

6.1.4.1   **Steel drums**

1A1   non-removable head

1A2   removable head

6.1.4.1.1   Body and heads shall be constructed of steel sheet of suitable type and adequate thickness in relation to the capacity of the drum and the intended use.

Note: For carbon steel drums, "suitable" steels are identified in ISO 3573:1999 "Hot rolled carbon steel sheet of commercial and drawing qualities" and ISO 3574:1999 "Cold-reduced carbon steel sheet of commercial and drawing qualities".
For carbon steel drums below 100 litres, "suitable" steels in addition to the above standards are also identified in ISO 11949:1995 "Cold-reduced electrolytic tinplate", ISO 11950:1995 "Cold-reduced electrolytic chromium/chromium oxide-coated steel" and ISO 11951:1995 "Cold-reduced blackplate in coil form for the production of tinplate or electrolytic chromium/chromium-oxide coated steel".

6.1.4.1.2   Body seams of drums intended to contain more than 40 ℓ of liquid shall be welded. Body seams of drums intended to contain solids or 40 ℓ or less of liquids shall be mechanically seamed or welded.

6.1.4.1.3   Chimes shall be mechanically seamed or welded. Separate reinforcing rings may be applied.

6.1.4.1.4   The body of a drum of a capacity greater than 60 ℓ shall, in general, have at least two expanded rolling hoops or, alternatively, at least two separate rolling hoops. If there are separate rolling hoops, they shall be fitted tightly on the body and so secured that they cannot shift. Rolling hoops shall not be spot-welded.

6.1.4.1.5   Openings for filling, emptying and venting in the bodies or heads of drums with a non-removable head (1A1) shall not exceed 7 cm in diameter. Drums with larger openings are considered to be of the removable-head type (1A2). Closures for openings in the bodies and heads of drums shall be so designed and applied that they will remain secure and leakproof under normal conditions of transport. Closure flanges may be mechanically seamed or welded in place. Gaskets or other sealing elements shall be used with closures, unless the closure is inherently leakproof.

6.1.4.1.6   Closure devices for removable-head drums shall be so designed and applied that they will remain secure and drums will remain leakproof under normal conditions of transport. Gaskets or other sealing elements shall be used with all removable heads.

6.1.4.1.7   If materials used for body, heads, closures and fittings are not in themselves compatible with the contents to be transported, suitable internal protective coatings or treatments shall be applied. These coatings or treatments shall retain their properties under normal conditions of transport.

6.1.4.1.8   Maximum capacity of drum: 450 ℓ.

6.1.4.1.9   Maximum net mass: 400 kg.

6.1.4.2   **Aluminium drums**

      1B1   non-removable head

      1B2   removable head

6.1.4.2.1   Body and heads shall be constructed of aluminium at least 99% pure or of an aluminium-based alloy. Material shall be of a suitable type and of adequate thickness in relation to the capacity of the drum and the intended use.

6.1.4.2.2   All seams shall be welded. Chime seams, if any, shall be reinforced by the application of separate reinforcing rings.

6.1.4.2.3   The body of a drum of a capacity greater than 60 ℓ shall, in general, have at least two expanded rolling hoops or, alternatively, at least two separate rolling hoops. If there are separate rolling hoops, they shall be fitted tightly on the body and so secured that they cannot shift. Rolling hoops shall not be spot-welded.

6.1.4.2.4   Openings for filling, emptying and venting in the bodies or heads of drums with a non-removable head (1B1) shall not exceed 7 cm in diameter. Drums with larger openings are considered to be of the removable-head type (1B2). Closures for openings in the bodies and heads of drums shall be so designed and applied that they will remain secure and leakproof under normal conditions of transport. Closure flanges shall be welded in place so that the weld provides a leakproof seam. Gaskets or other sealing elements shall be used with closures, unless the closure is inherently leakproof.

6.1.4.2.5   Closure devices for removable-head drums shall be so designed and applied that they will remain secure and drums will remain leakproof under normal conditions of transport. Gaskets or other sealing elements shall be used with all removable heads.

6.1.4.2.6   Maximum capacity of drum: 450 ℓ.

6.1.4.2.7   Maximum net mass: 400 kg.

6.1.4.3   **Drums of metal other than aluminium or steel**

      1N1   non-removable head

      1N2   removable head

6.1.4.3.1   The body and heads shall be constructed of metal or metal alloy other than steel or aluminium. Material shall be of a suitable type and of adequate thickness in relation to the capacity of the drum and to its intended use.

6.1.4.3.2   Chime seams, if any, shall be reinforced by the application of separate reinforcing rings. All seams, if any, shall be joined (welded, soldered, etc.) in accordance with the technical state of the art for the used metal or metal alloy.

6.1.4.3.3 The body of a drum of a capacity greater than 60 ℓ shall, in general, have at least two expanded rolling hoops or, alternatively, at least two separate rolling hoops. If there are separate rolling hoops, they shall be fitted tightly on the body and so secured that they cannot shift. Rolling hoops shall not be spot-welded.

6.1.4.3.4 Openings for filling, emptying and venting in the bodies or heads of non-removable-head (1N1) drums shall not exceed 7 cm in diameter. Drums with larger openings are considered to be of the removable-head type (1N2). Closures for openings in the bodies and heads of drums shall be so designed and applied that they will remain secure and leakproof under normal conditions of transport. Closure flanges shall be joined in place (welded, soldered, etc.) in accordance with the technical state of the art for the used metal or metal alloy so that the seam join is leakproof. Gaskets or other sealing elements shall be used with closures, unless the closure is inherently leakproof.

6.1.4.3.5 Closure devices for removable-head drums shall be so designed and applied that they will remain secure and drums will remain leakproof under normal conditions of transport. Gaskets or other sealing elements shall be used with all removable heads.

6.1.4.3.6 Maximum capacity of drum: 450 ℓ.

6.1.4.3.7 Maximum net mass: 400 kg.

6.1.4.4 Steel or aluminium jerricans

    3A1   steel, non-removable head

    3A2   steel, removable head

    3B1   aluminium, non-removable head

    3B2   aluminium, removable head

6.1.4.4.1 Body and heads shall be constructed of steel sheet, of aluminium at least 99% pure or of an aluminium-based alloy. Material shall be of a suitable type and of adequate thickness in relation to the capacity of the jerrican and to its intended use.

6.1.4.4.2 Chimes of steel jerricans shall be mechanically seamed or welded. Body seams of steel jerricans intended to contain more than 40 ℓ of liquid shall be welded. Body seams of steel jerricans intended to contain 40 ℓ or less shall be mechanically seamed or welded. For aluminium jerricans, all seams shall be welded. Chime seams, if any, shall be reinforced by the application of a separate reinforcing ring.

6.1.4.4.3 Openings in jerricans (3A1 and 3B1) shall not exceed 7 cm in diameter. Jerricans with larger openings are considered to be of the removable-head type (3A2 and 3B2). Closures shall be so designed that they will remain secure and leakproof under normal conditions of transport. Gaskets or other sealing elements shall be used with closures, unless the closure is inherently leakproof

6.1.4.4.4 If materials used for body, heads, closures and fittings are not in themselves compatible with the contents to be transported, suitable internal protective coatings or treatments shall be applied. These coatings or treatments shall retain their protective properties under normal conditions of transport.

6.1.4.4.5 Maximum capacity of jerrican: 60 ℓ.

6.1.4.4.6 Maximum net mass: 120 kg.

6.1.4.5 Plywood drums

    1D

6.1.4.5.1 The wood used shall be well seasoned, commercially dry and free from any defect likely to lessen the effectiveness of the drum for the purpose intended. If a material other than plywood is used for the manufacture of the heads, it shall be of a quality equivalent to the plywood.

6.1.4.5.2 At least two-ply plywood shall be used for the body and at least three-ply plywood for the heads; the plies shall be firmly glued together by a water-resistant adhesive with their grain crosswise.

6.1.4.5.3 The body and heads of the drum and their joins shall be of a design appropriate to the capacity of the drum and its intended use.

6.1.4.5.4 In order to prevent sifting of the contents, lids shall be lined with kraft paper or some other equivalent material, which shall be securely fastened to the lid and extend to the outside along its full circumference.

6.1.4.5.5 Maximum capacity of drum: 250 ℓ.

6.1.4.5.6 Maximum net mass: 400 kg.

6.1.4.6   (Deleted)

6.1.4.7   **Fibre drums**

    1G

6.1.4.7.1   The body of the drum shall consist of multiple plies of heavy paper or fibreboard (without corrugations) firmly glued or laminated together and may include one or more protective layers of bitumen, waxed kraft paper, metal foil, plastics material, etc.

6.1.4.7.2   Heads shall be of natural wood, fibreboard, metal, plywood, plastics or other suitable material and may include one or more protective layers of bitumen, waxed kraft paper, metal foil, plastics material, etc.

6.1.4.7.3   The body and heads of the drum and their joins shall be of a design appropriate to the capacity of the drum and its intended use.

6.1.4.7.4   The assembled packaging shall be sufficiently water-resistant so as not to delaminate under normal conditions of transport.

6.1.4.7.5   Maximum capacity of drum: 450 ℓ.

6.1.4.7.6   Maximum net mass: 400 kg.

6.1.4.8   **Plastics drums and jerricans**

    1H1 drums, non-removable head
    1H2 drums, removable head
    3H1 jerricans, non-removable head
    3H2 jerricans, removable head

6.1.4.8.1   The packaging shall be manufactured from suitable plastics material and be of adequate strength in relation to its capacity and intended use. Except for *recycled plastics material* as defined in.1.2.1, no used material other than production residues or regrind from the same manufacturing process may be used. The packaging shall be adequately resistant to ageing and to degradation caused by the substance contained or by ultraviolet radiation.

6.1.4.8.2   If protection against ultraviolet radiation is required, it shall be provided by the addition of carbon black or other suitable pigments or inhibitors. These additives shall be compatible with the contents and remain effective throughout the life of the packaging. Where use is made of carbon black, pigments or inhibitors other than those used in the manufacture of the tested design type, retesting may be waived if the carbon black content does not exceed 2% by mass or if the pigment content does not exceed 3% by mass; the content of inhibitors of ultraviolet radiation is not limited.

6.1.4.8.3   Additives serving purposes other than protection against ultraviolet radiation may be included in the composition of the plastics material, provided that they do not adversely affect the chemical and physical properties of the material of the packaging. In such circumstances, retesting may be waived.

6.1.4.8.4   The wall thickness at every point of the packaging shall be appropriate to its capacity and intended use, taking into account the stresses to which each point is liable to be exposed.

6.1.4.8.5   Openings for filling, emptying and venting in the bodies or heads of non-removable-head drums (1H1) and jerricans (3H1) shall not exceed 7 cm in diameter. Drums and jerricans with larger openings are considered to be of the removable-head type (1H2 and 3H2). Closures for openings in the bodies or heads of drums and jerricans shall be so designed and applied that they will remain secure and leakproof under normal conditions of transport. Gaskets or other sealing elements shall be used with closures, unless the closure is inherently leakproof.

6.1.4.8.6   Closure devices for removable-head drums and jerricans shall be so designed and applied that they will remain secure and leakproof under normal conditions of transport. Gaskets shall be used with all removable heads unless the drum or jerrican design is such that, where the removable head is properly secured, the drum or jerrican is inherently leakproof.

6.1.4.8.7   Maximum capacity of drums and jerricans: 1H1, 1H2: 450 ℓ
                        3H1, 3H2: 60 ℓ

6.1.4.8.8   Maximum net mass: 1H1, 1H2: 400 kg
               3H1, 3H2: 120 kg

6

6.1.4.9   Boxes of natural wood

       4C1 ordinary

       4C2 with sift-proof walls

6.1.4.9.1   The wood used shall be well seasoned, commercially dry and free from defects that would materially lessen the strength of any part of the box. The strength of the material used and the method of construction shall be appropriate to the capacity and intended use of the box. The tops and bottoms may be made of water-resistant reconstituted wood such as hardboard, particle board or other suitable type.

6.1.4.9.2   Fastenings shall be resistant to vibration experienced under normal conditions of transport. Nailing into the end shall be avoided whenever practicable. Joins which are likely to be highly stressed shall be made using clenched or annular ring nails or equivalent fastenings.

6.1.4.9.3   Box 4C2: each part shall consist of one piece or be equivalent thereto. Parts are considered equivalent to one piece when one of the following methods of glued assembly is used: Lindermann joint, tongue and groove joint, ship lap or rabbet joint or butt joint, all with at least two corrugated metal fasteners at each joint.

6.1.4.9.4   Maximum net mass: 400 kg.

6.1.4.10   Plywood boxes

       4D

6.1.4.10.1 Plywood used shall be at least three-ply. It shall be made from well-seasoned rotary-cut, sliced or sawn veneer, commercially dry and free from defects that would materially lessen the strength of the box. The strength of the material used and the method of construction shall be appropriate to the capacity and intended use of the box. All adjacent plies shall be glued with water-resistant adhesive. Other suitable materials may be used together with plywood in the construction of boxes. Boxes shall be firmly nailed or secured to corner posts or ends or be assembled by equally suitable devices.

6.1.4.10.2 Maximum net mass: 400 kg.

6.1.4.11   Reconstituted wood boxes

       4F

6.1.4.11.1 The walls of boxes shall be made of water-resistant reconstituted wood such as hardboard, particle board or other suitable type. The strength of the material used and the method of construction shall be appropriate to the capacity of the boxes and their intended use.

6.1.4.11.2 Other parts of the boxes may be made of other suitable material.

6.1.4.11.3 Boxes shall be securely assembled by means of suitable devices.

6.1.4.11.4 Maximum net mass: 400 kg.

6.1.4.12   Fibreboard boxes

       4G

6.1.4.12.1 Strong and good-quality solid or double-faced corrugated fibreboard (single or multiwall) shall be used, appropriate to the capacity of the box and to its intended use. The water resistance of the outer surface shall be such that the increase in mass, as determined in a test carried out over a period of 30 minutes by the Cobb method of determining water absorption, is not greater than 155 g/m^2 – see ISO 535:1991. It shall have proper bending qualities. Fibreboard shall be cut, creased without scoring, and slotted so as to permit assembly without cracking, surface breaks or undue bending. The fluting of corrugated fibreboard shall be firmly glued to the facings.

6.1.4.12.2 The ends of boxes may have a wooden frame or be entirely of wood or other suitable material. Reinforcements of wooden battens or other suitable material may be used.

6.1.4.12.3 Manufacturing joins in the body of boxes shall be taped, lapped and glued or lapped and stitched with metal staples. Lapped joins shall have an appropriate overlap.

6.1.4.12.4 Where closing is effected by gluing or taping, a water-resistant adhesive shall be used.

6.1.4.12.5 Boxes shall be designed so as to provide a good fit to the contents.

6.1.4.12.6 Maximum net mass: 400 kg.

6.1.4.13 **Plastics boxes**

> 4H1 expanded plastics boxes
>
> 4H2 solid plastics boxes

6.1.4.13.1 The box shall be manufactured from suitable plastics material and be of adequate strength in relation to its capacity and intended use. The box shall be adequately resistant to ageing and to degradation caused either by the substance contained or by ultraviolet radiation.

6.1.4.13.2 An expanded plastics box shall comprise two parts made of a moulded expanded plastics material, a bottom section containing cavities for the inner packagings and a top section covering and interlocking with the bottom section. The top and bottom sections shall be designed so that the inner packagings fit snugly. The closure cap for any inner packaging shall not be in contact with the inside of the top section of this box.

6.1.4.13.3 For dispatch, an expanded plastics box shall be closed with a self-adhesive tape having sufficient tensile strength to prevent the box from opening. The adhesive tape shall be weather-resistant and its adhesive compatible with the expanded plastics material of the box. Other closing devices at least equally effective may be used.

6.1.4.13.4 For solid plastics boxes, protection against ultraviolet radiation, if required, shall be provided by the addition of carbon black or other suitable pigments or inhibitors. These additives shall be compatible with the contents and remain effective throughout the life of the box. Where use is made of carbon black, pigments or inhibitors other than those used in the manufacture of the tested design type, retesting may be waived if the carbon black content does not exceed 2% by mass or if the pigment content does not exceed 3% by mass; the content of inhibitors of ultraviolet radiation is not limited.

6.1.4.13.5 Additives serving purposes other than protection against ultraviolet radiation may be included in the composition of the plastics material provided that they do not adversely affect the chemical and physical properties of the material of the box. In such circumstances, retesting may be waived.

6.1.4.13.6 Solid plastics boxes shall have closure devices made of a suitable material of adequate strength and be so designed as to prevent the box from unintentional opening.

6.1.4.13.7 Maximum net mass: 4H1: 60 kg.
4H2: 400 kg.

6.1.4.14 **Steel or aluminium boxes**

> 4A steel
>
> 4B aluminium

6.1.4.14.1 The strength of the metal and the construction of the box shall be appropriate to the capacity of the box and to its intended use.

6.1.4.14.2 Boxes shall be lined with fibreboard or felt packing pieces or shall have an inner liner or coating of suitable material, as required. If a double-seamed metal liner is used, steps shall be taken to prevent the ingress of substances, particularly explosives, into the recesses of the seams.

6.1.4.14.3 Closures may be of any suitable type; they shall remain secured under normal conditions of transport.

6.1.4.14.4 Maximum net mass: 400 kg.

6.1.4.15 **Textile bags**

> 5L1 without inner lining or coating
>
> 5L2 sift-proof
>
> 5L3 water-resistant

6.1.4.15.1 The textiles used shall be of good quality. The strength of the fabric and the construction of the bag shall be appropriate to the capacity of the bag and its intended use.

6.1.4.15.2 Bags, sift-proof, 5L2: the bag shall be made sift-proof, for example by the use of:

.1 paper bonded to the inner surface of the bag by a water-resistant adhesive such as bitumen; or

.2 plastics film bonded to the inner surface of the bag; or

.3 one or more inner liners made of paper or plastics material.

6.1.4.15.3   Bags, water-resistant, 5L3: to prevent the entry of moisture, the bag shall be made waterproof, for example by the use of:

.1   separate inner liners of water-resistant paper (such as waxed kraft paper, tarred paper or plastics-coated kraft paper); or

.2   plastics film bonded to the inner surface of the bag; or

.3   one or more inner liners made of plastics material.

6.1.4.15.4   Maximum net mass: 50 kg.

6.1.4.16   **Woven plastics bags**

5H1   without inner liner or coating

5H2   sift-proof

5H3   water-resistant

6.1.4.16.1   Bags shall be made from stretched tapes or monofilaments of a suitable plastics material. The strength of the material used and the construction of the bag shall be appropriate to the capacity of the bag and its intended use.

6.1.4.16.2   If the fabric is woven flat, the bags shall be made by sewing or some other method ensuring closure of the bottom and one side. If the fabric is tubular, the bag shall be closed by sewing, weaving or some other equally strong method of closure.

6.1.4.16.3   Bags, sift-proof, 5H2: the bag shall be made sift-proof, for example by means of:

.1   paper or a plastics film bonded to the inner surface of the bag; or

.2   one or more separate inner liners made of paper or plastics material.

6.1.4.16.4   Bags, water-resistant, 5H3: to prevent the entry of moisture, the bag shall be made waterproof, for example by means of:

.1   separate inner liners of water-resistant paper (such as waxed kraft paper, double-tarred kraft paper or plastics-coated kraft paper); or

.2   plastics film bonded to the inner or outer surface of the bag; or

.3   one or more inner plastics liners.

6.1.4.16.5   Maximum net mass: 50 kg.

6.1.4.17   **Plastics film bags**

5H4

6.1.4.17.1   Bags shall be made of a suitable plastics material. The strength of the material used and the construction of the bag shall be appropriate to the capacity of the bag and its intended use. Joins and closures shall withstand pressures and impacts liable to occur under normal conditions of transport.

6.1.4.17.2   Maximum net mass: 50 kg.

6.1.4.18   **Paper bags**

5M1   multiwall

5M2   multiwall, water-resistant

6.1.4.18.1   Bags shall be made of a suitable kraft paper or of an equivalent paper with at least three plies, the middle ply of which may be net-cloth with adhesive bonding to the outermost ply. The strength of the paper and the construction of the bags shall be appropriate to the capacity of the bag and its intended use. Joins and closures shall be sift-proof.

6.1.4.18.2   Bags 5M2: to prevent the entry of moisture, a bag of four plies or more shall be made waterproof by the use of either a water-resistant ply as one of the two outermost plies or a water-resistant barrier made of a suitable protective material between the two outermost plies; a bag of three plies shall be made waterproof by the use of a water-resistant ply as the outermost ply. Where there is a danger of the substance contained reacting with moisture or where it is packed damp, a waterproof ply or barrier, such as double-tarred kraft paper, plastics-coated kraft paper, plastics film bonded to the inner surface of the bag, or one or more inner plastics liners, shall also be placed next to the substance. Joins and closures shall be waterproof.

6.1.4.18.3   Maximum net mass: 50 kg.

6.1.4.19   Composite packagings (plastics material)

6HA1	plastics receptacle with outer steel drum
6HA2	plastics receptacle with outer steel crate or box
6HB1	plastics receptacle with outer aluminium drum
6HB2	plastics receptacle with outer aluminium crate or box
6HC	plastics receptacle with outer wooden box
6HD1	plastics receptacle with outer plywood drum
6HD2	plastics receptacle with outer plywood box
6HG1	plastics receptacle with outer fibre drum
6HG2	plastics receptacle with outer fibreboard box
6HH1	plastics receptacle with outer plastics drum
6HH2	plastics receptacle with outer solid plastics box

6.1.4.19.1   *Inner receptacle*

.1   The provisions of 6.1.4.8.1 and 6.1.4.8.3 to 6.1.4.8.6 shall apply to inner plastics receptacles.

.2   The inner plastics receptacle shall fit snugly inside the outer packaging, which shall be free of any projection that might abrade the plastics material.

.3   Maximum capacity of inner receptacle:

6HA1, 6HB1, 6HD1, 6HG1, 6HH1	250 ℓ
6HA2, 6HB2, 6HC, 6HD2, 6HG2, 6HH2	60 ℓ

.4   Maximum net mass:

6HA1, 6HB1, 6HD1, 6HG1, 6HH1	400 kg
6HA2, 6HB2, 6HC, 6HD2, 6HG2, 6HH2	75 kg

6.1.4.19.2   *Outer packaging*

.1   Plastics receptacle with outer steel or aluminium drum (6HA1 or 6HB1): the relevant provisions of 6.1.4.1 or 6.1.4.2, as appropriate, shall apply to the construction of the outer packaging.

.2   Plastics receptacle with outer steel or aluminium crate or box (6HA2 or 6HB2): the relevant provisions of 6.1.4.14 shall apply to the construction of the outer packaging.

.3   Plastics receptacle with outer wooden box 6HC: the relevant provisions of 6.1.4.9 shall apply to the construction of the outer packaging.

.4   Plastics receptacle with outer plywood drum 6HD1: the relevant provisions of 6.1.4.5 shall apply to the construction of the outer packaging.

.5   Plastics receptacle with outer plywood box 6HD2: the relevant provisions of 6.1.4.10 shall apply to the construction of the outer packaging.

.6   Plastics receptacle with outer fibre drum 6HG1: the provisions of 6.1.4.7.1 to 6.1.4.7.4 shall apply to the construction of the outer packaging.

.7   Plastics receptacle with outer fibreboard box 6HG2: the relevant provisions of 6.1.4.12 shall apply to the construction of the outer packaging.

.8   Plastics receptacle with outer plastics drum 6HH1: the provisions of 6.1.4.8.1 and 6.1.4.8.2 to 6.1.4.8.6 shall apply to the construction of the outer packaging.

.9   Plastics receptacle with outer solid plastics box (including corrugated plastics material) 6HH2; the provisions of 6.1.4.13.1 and 6.1.4.13.4 to 6.1.4.13.6 shall apply to the construction of the outer packaging.

6.1.4.20   Composite packagings (glass, porcelain or stoneware)

6PA1	receptacle with outer steel drum
6PA2	receptacle with outer steel crate or box
6PB1	receptacle with outer aluminium drum
6PB2	receptacle with outer aluminium crate or box
6PC	receptacle with outer wooden box
6PD1	receptacle with outer plywood drum
6PD2	receptacle with outer wickerwork hamper
6PG1	receptacle with outer fibre drum
6PG2	receptacle with outer fibreboard box
6PH1	receptacle with outer expanded plastics packaging
6PH2	receptacle with outer solid plastics packaging

6.1.4.20.1 *Inner receptacle*

.1 Receptacles shall be of a suitable form (cylindrical or pear-shaped) and be made of good-quality material free from any defect that could impair their strength. The walls shall be sufficiently thick at every point.

.2 Screw-threaded plastics closures, ground glass stoppers or closures at least equally effective shall be used as closures for receptacles. Any part of the closure likely to come into contact with the contents of the receptacle shall be resistant to those contents. Care shall be taken to ensure that the closures are so fitted as to be leakproof and are suitably secured to prevent any loosening during transport. If vented closures are necessary, they shall comply with 4.1.1.8.

.3 The receptacle shall be firmly secured in the outer packaging by means of cushioning and/or absorbent materials.

.4 Maximum capacity of receptacle: 60 ℓ.

.5 Maximum net mass: 75 kg.

6.1.4.20.2 *Outer packaging*

.1 Receptacle with outer steel drum 6PA1: the relevant provisions of 6.1.4.1 shall apply to the construction of the outer packaging. The removable lid required for this type of packaging may nevertheless be in the form of a cap.

.2 Receptacle with outer steel crate or box 6PA2: the relevant provisions of 6.1.4.14 shall apply to the construction of the outer packaging. For cylindrical receptacles, the outer packaging shall, when upright, rise above the receptacle and its closure. If the crate surrounds a pear-shaped receptacle and is of matching shape, the outer packaging shall be fitted with a protective cover (cap).

.3 Receptacle with outer aluminium drum 6PB1: the relevant provisions of 6.1.4.2 shall apply to the construction of the outer packaging.

.4 Receptacle with outer aluminium crate or box 6PB2: the relevant provisions of 6.1.4.14 shall apply to the construction of the outer packaging.

.5 Receptacle with outer wooden box 6PC: the relevant provisions of 6.1.4.9 shall apply to the construction of the outer packaging.

.6 Receptacle with outer plywood drum 6PD1: the relevant provisions of 6.1.4.5 shall apply to the construction of the outer packaging.

.7 Receptacle with outer wickerwork hamper 6PD2: the wickerwork hamper shall be properly made with material of good quality. It shall be fitted with a protective cover (cap) so as to prevent damage to the receptacle.

.8 Receptacle with outer fibre drum 6PG1: the relevant provisions of 6.1.4.7.1 to 6.1.4.7.4 shall apply to the body of the outer packaging.

.9 Receptacle with outer fibreboard box 6PG2: the relevant provisions of 6.1.4.12 shall apply to the construction of the outer packaging.

.10 Receptacle with outer expanded plastics or solid plastics packaging (6PH1 or 6PH2): the materials of both outer packagings shall meet the relevant provisions of 6.1.4.13. Solid plastics packaging shall be manufactured from high-density polyethylene or some other comparable plastics material. The removable lid for this type of packaging may nevertheless be in the form of a cap.

6.1.5 Test provisions for packagings

6.1.5.1 Performance and frequency of tests

6.1.5.1.1 The design type of each packaging shall be tested as provided in this section, in accordance with procedures established by the competent authority.

6.1.5.1.2 Each packaging design type shall successfully pass the tests prescribed in this chapter before being used. A packaging design type is defined by the design, size, material and thickness, manner of construction and packing, but may include various surface treatments. It also includes packagings which differ from the design type only in their lesser design height.

6.1.5.1.3 Tests shall be repeated on production samples at intervals established by the competent authority. For such tests on paper or fibreboard packagings, preparation at ambient conditions is considered equivalent to the provisions of 6.1.5.2.3.

6.1.5.1.4 Tests shall also be repeated after each modification which alters the design, material or manner of construction of a packaging.

6.1.5.1.5 The competent authority may permit the selective testing of packagings that differ only in minor respects from a tested type, such as smaller sizes of inner packagings or inner packagings of lower net mass; and packagings such as drums, bags and boxes which are produced with small reductions in external dimensions.

6.1.5.1.6   (Reserved)

      Note: For the conditions for assembling different inner packagings in an outer packaging and permissible variations in inner packagings, see 4.1.1.5.1.

6.1.5.1.7   Articles or inner packagings of any type for solids or liquids may be assembled and transported without testing in an outer packaging under the following conditions:

    .1   The outer packaging shall have been successfully tested in accordance with 6.1.5.3 with fragile (such as glass) inner packagings containing liquids, using the drop height for packing group I.

    .2   The total combined gross mass of inner packagings shall not exceed one half the gross mass of inner packagings used for the drop test in .1 above.

    .3   The thickness of the cushioning material between inner packagings and between inner packagings and the outside of the packaging shall not be reduced below the corresponding thicknesses in the originally tested packaging; and when a single inner packaging was used in the original test, the thicknesses of the cushioning between inner packagings shall not be less than the thickness of cushioning between the outside of the packaging and the inner packaging in the original test. When either fewer or smaller inner packagings are used (as compared to the inner packagings used in the drop test), sufficient additional cushioning material shall be used to take up void spaces.

    .4   The outer packaging shall have successfully passed the stacking test in 6.1.5.6 while empty. The total mass of identical packages shall be based on the combined mass of inner packagings used in the drop test in .1 above.

    .5   Inner packagings containing liquids shall be completely surrounded with a sufficient quantity of absorbent material to absorb the entire liquid contents of the inner packagings.

    .6   When the outer packaging is intended to contain inner packagings for liquids and is not leakproof, or is intended to contain inner packagings for solids and is not sift-proof, a means of containing any liquid or solid contents in the event of leakage shall be provided in the form of a leakproof liner, plastics bag or other equally efficient means of containment. For packagings containing liquids, the absorbent material required in .5 above shall be placed inside the means of containing the liquid contents.

    .7   Packagings shall be marked in accordance with section 6.1.3 as having been tested to packing group I performance for combination packagings. The marked gross mass, in kilograms, shall be the sum of the mass of the outer packaging plus one half of the mass of the inner packaging(s) as used for the drop test referred to in .1 above. Such a packaging mark shall also contain the letter 'V' as described in 6.1.2.4.

6.1.5.1.8   The competent authority may at any time require proof, by tests in accordance with this section, that serially produced packagings meet the provisions of the design type tests.

6.1.5.1.9   If an inner treatment or coating is required for safety reasons, it shall retain its protective properties after the tests.

6.1.5.1.10   Provided the validity of the test results is not affected, and with the approval of the competent authority, several tests may be made on one sample.

6.1.5.1.11   *Salvage packagings*

6.1.5.1.11.1   Salvage packagings (see 1.2.1) shall be tested and marked in accordance with the provisions applicable to packing group II packagings intended for the transport of solids or inner packagings, except as follows:

    .1   The test substance used in performing the tests shall be water and the packagings shall be filled to not less than 98% of their maximum capacity. It is permissible to use additives, such as bags of lead shot, to achieve the requisite total package mass so long as they are placed in such a way that the test results are not affected. Alternatively, in performing the drop test, the drop height may be varied in accordance with 6.1.5.3.5(b);

    .2   Packagings shall, in addition, have been successfully subjected to the leakproofness test at 30 kPa, with the results of this test reflected in the test report required by 6.1.5.7; and

    .3   Packagings shall be marked with the letter 'T' as described in 6.1.2.4.

6.1.5.2   **Preparation of packagings for testing**

6.1.5.2.1   Tests shall be carried out on packagings prepared as for transport, including, with respect to combination packagings, the inner packagings used. Inner or single receptacles or packagings, other than bags, shall be filled to not less than 98% of their maximum capacity for liquids or 95% for solids. Bags shall be filled to the maximum mass at which they may be used. For combination packagings where the inner packaging is designed to carry liquids and solids, separate testing is required for both solid and liquid contents. The substances or articles to be transported in the packagings may be replaced by other substances or articles except where this would invalidate the results of the tests. For solids, when another substance is used, it shall have the same physical characteristics (mass, grain size, etc.) as the substance to be carried. It is permissible to use additives, such as bags of lead shot, to achieve the requisite total package mass, so long as they are placed so that the test results are not affected.

6.1.5.2.2　In the drop tests for liquids, when another substance is used, it shall be of similar relative density and viscosity to those of the substance being transported. Water may also be used for the liquid drop test under the conditions in 6.1.5.3.5.

6.1.5.2.3　Paper or fibreboard packagings shall be conditioned for at least 24 hours in an atmosphere having controlled temperature and relative humidity (r.h.). There are three options, one of which shall be chosen. The preferred atmosphere is $23°C \pm 2°C$ and $50\% \pm 2\%$ r.h. The two other options are $20°C \pm 2°C$ and $65\% \pm 2\%$ r.h. or $27°C \pm 2°C$ and $65\% \pm 2\%$ r.h.

Note: Average values shall fall within these limits. Short-term fluctuations and measurement limitations may cause individual measurements to vary by up to $\pm 5\%$ relative humidity without significant impairment of test reproducibility.

6.1.5.2.4　Additional steps shall be taken to ascertain that the plastics material used in the manufacture of plastics drums, plastics jerricans and composite packagings (plastics material) intended to contain liquids complies with the provisions in 6.1.1.2, 6.1.4.8.1 and 6.1.4.8.3. This may be done, for example, by submitting sample receptacles or packagings to a preliminary test extending over a long period, for example six months, during which the samples would remain filled with the substances they are intended to contain and after which the samples shall be submitted to the applicable tests listed in 6.1.5.3, 6.1.5.4, 6.1.5.5, and 6.1.5.6. For substances which may cause stress cracking or weakening in plastics drums or jerricans, the sample, filled with the substance or another substance that is known to have at least as severe stress cracking influence on the plastics materials in question, shall be subjected to a superimposed load equivalent to the total mass of identical packages which might be stacked on it during transport. The minimum height of the stack including the test sample shall be 3 m.

6.1.5.3　Drop test

6.1.5.3.1　*Number of test samples (per design type and manufacturer) and drop orientation*

For other than flat drops, the centre of gravity shall be vertically over the point of impact.

Packaging	Number of test samples	Drop orientation
Steel drums Aluminium drums Metal drums, other than steel or aluminium drums Steel jerricans Aluminium jerricans Plywood drums Fibre drums Plastics drums and jerricans Composite packagings which are in the shape of a drum	Six (three for each drop)	*First drop* (using three samples): the packaging shall strike the target diagonally on the chime or, if the packaging has no chime, on a circumferential seam or an edge *Second drop* (using the other three samples): the packaging shall strike the target on the weakest part not tested by the first drop, for example a closure or, for some cylindrical drums, the welded longitudinal seam of the body.
Boxes of natural wood Plywood boxes Reconstituted wood boxes Fibreboard boxes Plastics boxes Steel or aluminium boxes Composite packagings which are in the shape of a box	Five (one for each drop)	*First drop:* flat on the bottom *Second drop:* flat on the top *Third drop:* flat on the long side *Fourth drop:* flat on the short side *Fifth drop:* on a corner
Bags – single-ply with a side seam	Three (three drops per bag)	*First drop*: flat on a wide face *Second drop*: flat on a narrow face *Third drop*: on the end of the bag
Bags – single-ply without a side seam or multi-ply	Three (two drops per bag)	*First drop*: flat on a wide face *Second drop*: on an end of the bag

Where more than one orientation is possible for a given drop test, the orientation most likely to result in failure of the packaging shall be used.

6.1.5.3.2   *Special preparation of test samples for the drop test*

The temperature of the test sample and its contents shall be reduced to −18°C or lower for the following packagings:

.1   plastics drums (see 6.1.4.8);

.2   plastics jerricans (see 6.1.4.8);

.3   plastics boxes other than expanded plastics boxes (see 6.1.4.13);

.4   composite packagings (plastics material) (see 6.1.4.19); and

.5   combination packagings with plastics inner packagings, other than plastics bags intended to contain solids or articles.

Where the test samples are prepared in this way, the conditioning in 6.1.5.2.3 may be waived. Test liquids shall be kept in the liquid state by the addition of anti-freeze if necessary.

6.1.5.3.3   Removable head packagings for liquids shall not be dropped until at least 24 hours after filling and closing to allow for any possible gasket relaxation.

6.1.5.3.4   *Target*

The target shall be a non-resilient and horizontal surface and shall be:

.1   integral and massive enough to be immovable;

.2   flat with a surface kept free from local defects capable of influencing the test results;

.3   rigid enough to be non-deformable under test conditions and not liable to become damaged by the tests; and

.4   sufficiently large to ensure that the test package falls entirely upon the surface.

6.1.5.3.5   *Drop height*

For solids and liquids, if the test is performed with the solid or liquid to be carried or with another substance having essentially the same physical characteristics:

Packing group I	Packing group II	Packing group III
1.8 m	1.2 m	0.8 m

For liquids in single packagings and for inner packagings of combination packagings, if the test is performed with water:

Note: The term "water" includes water/antifreeze solutions with a minimum specific gravity of 0.95 for testing at −18°C.

(a) where the substances to be transported have a relative density not exceeding 1.2:

Packing group I	Packing group II	Packing group III
1.8 m	1.2 m	0.8 m

(b) where the substances to be transported have a relative density exceeding 1.2, the drop height shall be calculated on the basis of the relative density (d) of the substance to be carried, rounded up to the first decimal, as follows:

Packing group I	Packing group II	Packing group III
$d \times 1.5$ m	$d \times 1.0$ m	$d \times 0.67$ m

6.1.5.3.6   *Criteria for passing the test*

.1   Each packaging containing liquid shall be leakproof when equilibrium has been reached between the internal and external pressures, except for inner packagings of combination packagings, when it is not necessary that the pressures be equalized.

.2   Where a packaging for solids undergoes a drop test and its upper face strikes the target, the test sample passes the test if the entire contents are retained by an inner packaging or inner receptacle (such as a plastics bag), even if the closure, while retaining its containment function, is no longer sift-proof.

.3 The packaging or outer packaging of a composite or combination packaging shall not exhibit any damage liable to affect safety during transport. There shall be no leakage of the filling substance from the inner receptacle or inner packagings.

.4 Neither the outermost ply of a bag nor an outer packaging shall exhibit any damage liable to affect safety during transport.

.5 A slight discharge from the closures upon impact shall not be considered to be a failure of the packaging provided that no further leakage occurs.

.6 No rupture is permitted in packagings for goods of class 1 which would permit the spillage of loose explosive substances or articles from the outer packaging

6.1.5.4   Leakproofness test

6.1.5.4.1   The leakproofness test shall be performed on all design types of packagings intended to contain liquids; however, this test is not required for the inner packagings of combination packagings.

6.1.5.4.2   Number of test samples: three test samples per design type and manufacturer.

6.1.5.4.3   Special preparation of test samples for the test: vented closures shall either be replaced by similar non-vented closures or the vent shall be sealed.

6.1.5.4.4   Test method and pressure to be applied: the packagings, including their closures, shall be restrained under water for 5 minutes while an internal air pressure is applied. The method of restraint shall not affect the results of the test.

The air pressure (gauge) to be applied shall be:

Packing group I	Packing group II	Packing group III
Not less than 30 kPa (0.3 bar)	Not less than 20 kPa (0.2 bar)	Not less than 20 kPa (0.2 bar)

Other methods at least equally as effective may be used.

6.1.5.4.5   Criterion for passing the test: there shall be no leakage.

6.1.5.5   Internal pressure (hydraulic) test

6.1.5.5.1   Packagings to be tested: the internal pressure (hydraulic) test shall be carried out on all design types of metal, plastics and composite packagings intended to contain liquids. This test is not required for inner packagings of combination packagings.

6.1.5.5.2   Number of test samples: three test samples per design type and manufacture.

6.1.5.5.3   Special preparation of packagings for testing: vented closures shall either be replaced by similar non-vented closures or the vent shall be sealed.

6.1.5.5.4   Test method and pressure to be applied: metal packagings and composite packagings (glass, porcelain or stoneware), including their closures, shall be subjected to the test pressure for 5 minutes. Plastics packagings and composite packagings (plastics material), including their closures, shall be subjected to the test pressure for 30 minutes. This pressure is the one to be included in the marking required by 6.1.3.1(d). The manner in which the packagings are supported shall not invalidate the test. The test pressure shall be applied continuously and evenly; it shall be kept constant throughout the test period. The hydraulic pressure (gauge) applied, as determined by any one of the following methods, shall be:

.1 not less than the total gauge pressure measured in the packaging (i.e. the vapour pressure of the filling liquid and the partial pressure of the air or other inert gases, minus 100 kPa) at 55°C, multiplied by a safety factor of 1.5; this total gauge pressure shall be determined on the basis of a maximum degree of filling in accordance with 4.1.1.4 and a filling temperature of 15°C

.2 not less than 1.75 times the vapour pressure at 50°C of the liquid to be transported, minus 100 kPa, but with a minimum test pressure of 100 kPa;

.3 not less than 1.5 times the vapour pressure at 55°C of the liquid to be transported minus 100 kPa, but with a minimum test pressure of 100 kPa.

6.1.5.5.5   In addition, packagings intended to contain liquids of packing group I shall be tested to a minimum test pressure of 250 kPa (gauge) for a test period of 5 or 30 minutes, depending upon the material of construction of the packaging.

6.1.5.5.6   Criterion for passing the test: no packaging shall leak.

6.1.5.6   **Stacking test**

All design types of packagings other than bags shall be subjected to a stacking test.

6.1.5.6.1   Number of test samples: three test samples per design type and manufacturer.

6.1.5.6.2   Test method: the test sample shall be subjected to a force applied to the top surface of the test sample equivalent to the total mass of identical packages which might be stacked on it during transport: where the contents of the test sample are liquids with relative density different from that of the liquid to be transported, the force shall be calculated in relation to the latter. The minimum height of the stack including the test sample shall be 3 m. The duration of the test shall be 24 hours except that plastics drums, jerricans, and composite packagings 6HH1 and 6HH2 intended for liquids shall be subjected to the stacking test for a period of 28 days at a temperature of not less than 40°C.

6.1.5.6.3   Criteria for passing the test: no test sample shall leak. In composite packagings or combination packagings, there shall be no leakage of the filling substance from the inner receptacle or inner packaging. No test sample shall show any deterioration which could adversely affect transport safety or any distortion liable to reduce its strength or cause instability in stacks of packages. Plastics packagings shall be cooled to ambient temperature before the assessment.

6.1.5.7   **Test report**

6.1.5.7.1   A test report containing at least the following particulars shall be drawn up and shall be available to the users of the packaging:

.1   name and address of the test facility;

.2   name and address of applicant (where applicable);

.3   a unique test report identification;

.4   date of the test report;

.5   manufacturer of the packaging;

.6   description of the packaging design type (such as dimensions, materials, closures, thickness, etc.), including method of manufacture (such as blow-moulding), and which may include drawing(s) and/or photograph(s);

.7   maximum capacity;

.8   characteristics of test contents, such as viscosity and relative density for liquids and particle size for solids;

.9   test descriptions and results;

.10 signature, with the name and status of the signatory.

6.1.5.7.2   The test report shall contain statements that the packaging prepared as for transport was tested in accordance with the appropriate provisions of this chapter and that the use of other packaging methods or components may render it invalid. A copy of the test report shall be available to the competent authority.

6

Chapter 6.2

*Provisions for the construction and testing
of pressure receptacles, aerosol dispensers,
small receptacles containing gas
(gas cartridges) and fuel cell cartridges
containing liquefied flammable gas*

6.2.1 General provisions

Note: For aerosol dispensers, small receptacles containing gas (gas cartridges) and fuel cell cartridges containing liquefied flammable gas, see 6.2.4.

6.2.1.1 Design and construction

6.2.1.1.1 Pressure receptacles and their closures shall be designed, manufactured, tested and equipped in such a way as to withstand all conditions, including fatigue, to which they will be subjected during normal conditions of transport.

6.2.1.1.2 In recognition of scientific and technological advances, and recognizing that pressure receptacles other than those that are marked with a UN certification marking may be used on a national or regional basis, pressure receptacles conforming to requirements other than those specified in this Code may be used if approved by the competent authorities in the countries of transport and use.

6.2.1.1.3 In no case shall the minimum wall thickness be less than that specified in the design and construction technical standards.

6.2.1.1.4 For welded pressure receptacles, only metals of weldable quality shall be used.

6.2.1.1.5 The test pressure of cylinders, tubes, pressure drums and bundles of cylinders shall be in accordance with packing instruction P200. The test pressure for closed cryogenic receptacles shall be in accordance with packing instruction P203.

6.2.1.1.6 Pressure receptacles assembled in bundles shall be structurally supported and held together as a unit. Pressure receptacles shall be secured in a manner that prevents movement in relation to the structural assembly and movement that would result in the concentration of harmful local stresses. Manifold assemblies (e.g., manifold, valves, and pressure gauges) shall be designed and constructed such that they are protected from impact damage and forces normally encountered in transport. Manifolds shall have at least the same test pressure as the cylinders. For toxic liquefied gases, each pressure receptacle shall have an isolation valve to ensure that each pressure receptacle can be filled separately and that no interchange of pressure receptacle contents can occur during transport.

6.2.1.1.7 Contact between dissimilar metals which could result in damage by galvanic action shall be avoided.

6.2.1.1.8 The following additional provisions apply to the construction of closed cryogenic receptacles for refrigerated liquefied gases:

.1 The mechanical properties of the metal used shall be established for each pressure receptacle, including the impact strength and the bending coefficient;

.2 The pressure receptacles shall be thermally insulated. The thermal insulation shall be protected against impact by means of a jacket. If the space between the pressure receptacle and the jacket is evacuated of air (vacuum insulation), the jacket shall be designed to withstand, without permanent deformation, an external pressure of at least 100 kPa (1 bar) calculated in accordance with a recognized technical code or a calculated critical collapsing pressure of not less than 200 kPa (2 bar) gauge pressure. If the jacket is so closed as to be gas-tight (e.g., in the case of vacuum insulation), a device shall be provided to prevent any dangerous pressure from developing in the insulating layer in the event of inadequate gas-tightness of the pressure receptacle or its fittings. The device shall prevent moisture from penetrating into the insulation.

.3   Closed cryogenic receptacles intended for the transport of refrigerated liquefied gases having a boiling point below –182°C at atmospheric pressure shall not include materials which may react with oxygen or oxygen-enriched atmospheres in a dangerous manner, when located in parts of the thermal insulation where there is a risk of contact with oxygen or with oxygen-enriched liquid.

.4   Closed cryogenic receptacles shall be designed and constructed with suitable lifting and securing arrangements.

6.2.1.1.9   *Additional requirements for the construction of pressure receptacles for acetylene*

Pressure receptacles for UN 1001 acetylene, dissolved, and UN 3374 acetylene, solvent free, shall be filled with a porous material, uniformly distributed, of a type that conforms to the requirements and testing specified by the competent authority and which:

.1   is compatible with the pressure receptacle and does not form harmful or dangerous compounds either with the acetylene or with the solvent in the case of UN 1001; and

.2   is capable of preventing the spread of decomposition of the acetylene in the porous material.

In the case of UN 1001, the solvent shall be compatible with the pressure receptacles.

6.2.1.2   Materials

6.2.1.2.1   Construction materials of pressure receptacles and their closures which are in direct contact with dangerous goods shall not be affected or weakened by the dangerous goods intended and shall not cause a dangerous effect, e.g., catalysing a reaction or reacting with the dangerous goods.

6.2.1.2.2   Pressure receptacles and their closures shall be made of the materials specified in the design and construction technical standards and the applicable packing instruction for the substances intended for transport in the pressure receptacle. The materials shall be resistant to brittle fracture and to stress corrosion cracking as indicated in the design and construction technical standards.

6.2.1.3   Service equipment

6.2.1.3.1   Valves, piping and other fittings subjected to pressure, excluding pressure relief devices, shall be designed and constructed so that the burst pressure is at least 1.5 times the test pressure of the pressure receptacle.

6.2.1.3.2   Service equipment shall be configured or designed to prevent damage that could result in the release of the pressure receptacle contents during normal conditions of handling and transport. Manifold piping leading to shut-off valves shall be sufficiently flexible to protect the valves and the piping from shearing or releasing the pressure receptacle contents. The filling and discharge valves and any protective caps shall be capable of being secured against unintended opening. Valves shall be protected as specified in 4.1.6.1.8.

6.2.1.3.3   Pressure receptacles which are not capable of being handled manually or rolled shall be fitted with devices (skids, rings, straps) ensuring that they can be safely handled by mechanical means and so arranged as not to impair the strength of, nor cause undue stresses in, the pressure receptacle.

6.2.1.3.4   Individual pressure receptacles shall be equipped with pressure relief devices as specified in packing instruction P200(1) or in 6.2.1.3.6.4 and 6.2.1.3.6.5. Pressure relief devices shall be designed to prevent the entry of foreign matter, the leakage of gas and the development of any dangerous excess pressure. When fitted, pressure relief devices on manifolded horizontal pressure receptacles filled with flammable gas shall be arranged to discharge freely to the open air in such a manner as to prevent any impingement of escaping gas upon the pressure receptacle itself under normal conditions of transport.

6.2.1.3.5   Pressure receptacles where filling is measured by volume shall be provided with a level indicator.

6.2.1.3.6   *Additional provisions for closed cryogenic receptacles*

6.2.1.3.6.1   Each filling and discharge opening in a closed cryogenic receptacle used for the transport of flammable refrigerated liquefied gases shall be fitted with at least two mutually independent shut-off devices in series, the first being a stop-valve, the second being a cap or equivalent device.

6.2.1.3.6.2   For sections of piping which can be closed at both ends and where liquid product can be trapped, a method of automatic pressure relief shall be provided to prevent excess pressure build-up within the piping.

6.2.1.3.6.3   Each connection on a closed cryogenic receptacle shall be clearly marked to indicate its function (e.g., vapour or liquid phase).

6.2.1.3.6.4 *Pressure relief devices*

6.2.1.3.6.4.1 Each closed cryogenic receptacle shall be provided with at least one pressure relief device. The pressure relief device shall be of the type that will resist dynamic forces, including surge.

6.2.1.3.6.4.2 Closed cryogenic receptacles may, in addition, have a frangible disc in parallel with the spring-loaded device(s) in order to meet the provisions of 6.2.1.3.6.5.

6.2.1.3.6.4.3 Connections to pressure relief devices shall be of sufficient size to enable the required discharge to pass unrestricted to the pressure relief device.

6.2.1.3.6.4.4 All pressure relief device inlets shall, under maximum filling conditions, be situated in the vapour space of the closed cryogenic receptacle and the devices shall be so arranged as to ensure that the escaping vapour is discharged unrestrictedly.

6.2.1.3.6.5 *Capacity and setting of pressure relief devices*

Note: In relation to pressure relief devices of closed cryogenic receptacles, "MAWP" means the maximum effective gauge pressure permissible at the top of a loaded closed cryogenic receptacle in its operating position, including the highest effective pressure during filling and discharge.

6.2.1.3.6.5.1 The pressure relief device shall open automatically at a pressure not less than the MAWP and be fully open at a pressure equal to 110% of the MAWP. It shall, after discharge, close at a pressure not lower than 10% below the pressure at which discharge starts and shall remain closed at all lower pressures.

6.2.1.3.6.5.2 Frangible discs shall be set to rupture at a nominal pressure which is the lower of either the test pressure or 150% of the MAWP.

6.2.1.3.6.5.3 In the case of the loss of vacuum in a vacuum-insulated closed cryogenic receptacle, the combined capacity of all pressure relief devices installed shall be sufficient so that the pressure (including accumulation) inside the closed cryogenic receptacle does not exceed 120% of the MAWP.

6.2.1.3.6.5.4 The required capacity of the pressure relief devices shall be calculated in accordance with an established technical code recognized by the competent authority.*

6.2.1.4 **Approval of pressure receptacles**

6.2.1.4.1 The conformity of pressure receptacles shall be assessed at time of manufacture as required by the competent authority. Pressure receptacles shall be inspected, tested and approved by an inspection body. The technical documentation shall include full specifications on design and construction, and full documentation on the manufacturing and testing.

6.2.1.4.2 Quality assurance systems shall conform to the requirements of the competent authority.

6.2.1.5 **Initial inspection and test**

6.2.1.5.1 New pressure receptacles, other than closed cryogenic receptacles, shall be subjected to testing and inspection during and after manufacture in accordance with the applicable design standards including the following:

On an adequate sample of pressure receptacles:

.1 testing of the mechanical characteristics of the material of construction;

.2 verification of the minimum wall thickness;

.3 verification of the homogeneity of the material for each manufacturing batch;

.4 inspection of the external and internal conditions of the pressure receptacles;

.5 inspection of the neck threads;

.6 verification of the conformance with the design standard;

For all pressure receptacles:

.7 a hydraulic pressure test. Pressure receptacles shall withstand the test pressure without expansion greater than that allowed in the design specification;

Note: With the agreement of the competent authority, the hydraulic pressure test may be replaced by a test using a gas, where such an operation does not entail any danger.

---

* See, for example, CGA Publications S-1.2-2003 ''Pressure Relief Device Standards – Part 2 – Cargo and Portable Tanks for Compressed Gases'' and S-1.1-2003 ''Pressure Relief Device Standards – Part 1 – Cylinders for Compressed Gases''.

    .8 inspection and assessment of manufacturing defects and either repairing them or rendering the pressure receptacles unserviceable. In the case of welded pressure receptacles, particular attention shall be paid to the quality of the welds;

    .9 an inspection of the markings on the pressure receptacles;

    .10 in addition, pressure receptacles intended for the transport of UN 1001 acetylene, dissolved and UN 3374 acetylene, solvent free shall be inspected to ensure proper installation and condition of the porous material and, if applicable, the quantity of solvent.

6.2.1.5.2   On an adequate sample of closed cryogenic receptacles, the inspections and tests specified in 6.2.1.5.1.1, .2, .4, and .6 shall be performed. In addition, welds shall be inspected by radiographic, ultrasonic or another suitable non-destructive test method on a sample of closed cryogenic receptacles, according to the applicable design and construction standard. This weld inspection does not apply to the jacket.

    Additionally, all closed cryogenic receptacles shall undergo the inspections and tests specified in 6.2.1.5.1, .7, .8, and .9, as well as a leakproofness test and a test of the satisfactory operation of the service equipment after assembly.

6.2.1.6   **Periodic inspection and test**

6.2.1.6.1   Refillable pressure receptacles, other than cryogenic receptacles, shall be subjected to periodic inspections and tests, by a body authorized by the competent authority, in accordance with the following:

    .1 Check of the external conditions of the pressure receptacle and verification of the equipment and the external markings;

    .2 Check of the internal conditions of the pressure receptacle (e.g., internal inspection, verification of minimum wall thickness);

    .3 Check of the threads if there is evidence of corrosion or if the fittings are removed;

    .4 A hydraulic pressure test and, if necessary, verification of the characteristics of the material by suitable tests;

    .5 Check of service equipment, other accessories and pressure-relief devices, if to be reintroduced into service.

    Note 1: With the agreement of the competent authority, the hydraulic pressure test may be replaced by a test using a gas, where such an operation does not entail any danger.

    Note 2: With the agreement of the competent authority, the hydraulic pressure test of cylinders or tubes may be replaced by an equivalent method based on acoustic emission testing, ultrasonic examination or a combination of acoustic emission testing and ultrasonic examination.

6.2.1.6.2   Pressure receptacles intended for the transport of UN 1001 acetylene, dissolved and UN 3374 acetylene, solvent free shall be examined only as specified in 6.2.1.6.1.1, 6.2.1.6.3 and 6.2.1.6.1.5. In addition, the condition of the porous material (e.g., cracks, top clearance, loosening, or settlement) shall be examined.

6.2.1.7   **Requirements for manufacturers**

6.2.1.7.1   The manufacturer shall be technically able and shall possess all resources required for the satisfactory manufacture of pressure receptacles; this relates in particular to qualified personnel:

    .1 to supervise the entire manufacturing process;

    .2 to carry out joining of materials; and

    .3 to carry out the relevant tests.

6.2.1.7.2   The proficiency test of a manufacturer shall in all instances be carried out by an inspection body approved by the competent authority of the country of approval.

6.2.1.8   **Requirements for inspection bodies**

6.2.1.8.1   Inspection bodies shall be independent from manufacturing enterprises and competent to perform the tests, inspections and approvals required.

6.2.2   **Provisions for UN pressure receptacles**

    In addition to the general requirements of 6.2.1, UN pressure receptacles shall comply with the provisions of this section, including the standards, as applicable.

    Note: With the agreement of the competent authority, more recently published versions of the standards, if available, may be used.

6.2.2.1   Design, construction and initial inspection and test

6.2.2.1.1   The following standards apply for the design, construction, and initial inspection and test of UN cylinders, except that inspection requirements related to the conformity assessment system and approval shall be in accordance with 6.2.2.5:

ISO 9809-1:1999	Gas cylinders – Refillable seamless steel gas cylinders – Design, construction and testing – Part 1: Quenched and tempered steel cylinders with tensile strength less than 1100 MPa. **Note:** The note concerning the *F* factor in section 7.3 of this standard shall not be applied for UN cylinders.
ISO 9809-2:2000	Gas cylinders – Refillable seamless steel gas cylinders – Design, construction and testing – Part 2: Quenched and tempered steel cylinders with tensile strength greater than or equal to 1100 MPa
ISO 9809-3:2000	Gas cylinders – Refillable seamless steel gas cylinders – Design, construction and testing – Part 3: Normalized steel cylinders
ISO 7866:1999	Gas cylinders – Refillable seamless aluminium alloy gas cylinders – Design, construction and testing **Note:** The note concerning the *F* factor in section 7.2 of this standard shall not be applied for UN cylinders. Aluminium alloy 6351A-T6 or equivalent shall not be authorized.
ISO 11118:1999	Gas cylinders – Non-refillable metallic gas cylinders — Specification and test methods
ISO 11119-1:2002	Gas cylinders of composite construction – Specification and test methods – Part 1: Hoop wrapped composite gas cylinders
ISO 11119-2:2002	Gas cylinders of composite construction – Specification and test methods – Part 2: Fully wrapped fibre reinforced composite gas cylinders with load-sharing metal liners
ISO 11119-3:2002	Gas cylinders of composite construction – Specification and test methods – Part 3: Fully wrapped fibre reinforced composite gas cylinders with non-load-sharing metallic or non-metallic liners

Note 1: In the above referenced standards, composite cylinders shall be designed for unlimited service life.

Note 2: After the first 15 years of service, composite cylinders manufactured according to these standards may be approved for extended service by the competent authority which was responsible for the original approval of the cylinders and which will base its decision on the test information supplied by the manufacturer or owner or user.

6.2.2.1.2   The following standards apply for the design, construction, and initial inspection and test of UN tubes, except that inspection requirements related to the conformity assessment system and approval shall be in accordance with 6.2.2.5:

ISO 11120:1999	Gas cylinders – Refillable seamless steel tubes for compressed gas transport, of water capacity between 150 ℓ and 3000 ℓ – Design, construction and testing **Note:** The note concerning the *F* factor in section 7.1 of this standard shall not be applied for UN tubes

6.2.2.1.3   The following standards apply for the design, construction and initial inspection and test of UN acetylene cylinders, except that inspection requirements related to the conformity assessment system and approval shall be in accordance with 6.2.2.5:

For the cylinder shell:

ISO 9809-1:1999	Gas cylinders – Refillable seamless steel gas cylinders – Design, construction and testing – Part 1: Quenched and tempered steel cylinders with tensile strength less than 1100 MPa **Note:** The note concerning the *F* factor in section 7.3 of this standard shall not be applied for UN cylinders.
ISO 9809-3:2000	Gas cylinders – Refillable seamless steel gas cylinders – Design, construction and testing – Part 3: Normalized steel cylinders

For the porous material in the cylinder:

ISO 3807-1:2000	Cylinders for acetylene – Basic requirements – Part 1: Cylinders without fusible plugs
ISO 3807-2:2000	Cylinders for acetylene – Basic requirements – Part 2: Cylinders with fusible plugs

6.2.2.1.4   The following standard applies for the design, construction and initial inspection and test of UN cryogenic receptacles, except that inspection requirements related to the conformity assessment system and approval shall be in accordance with 6.2.2.5:

ISO 21029-1:2004	Cryogenic vessels – Transportable vacuum insulated vessels of not more than 1000 litres volume – Part 1: Design, fabrication, inspection and tests

6.2.2.2 Materials

In addition to the material requirements specified in the pressure receptacle design and construction standards, and any restrictions specified in the applicable packing instruction for the gas(es) to be transported (e.g., packing instruction P200), the following standards apply to material compatibility:

ISO 11114-1:1997	Transportable gas cylinders – Compatibility of cylinder and valve materials with gas contents – Part 1: Metallic materials
ISO 11114-2:2000	Transportable gas cylinders – Compatibility of cylinder and valve materials with gas contents – Part 2: Non-metallic materials

Note: The limitations imposed in ISO 11114-1 on high strength steel alloys at ultimate tensile strength levels up to 1100 MPa do not apply to SILANE (UN 2203).

6.2.2.3 Service equipment

The following standards apply to closures and their protection:

ISO 11117:1998	Gas cylinders – Valve protection caps and valve guards for industrial and medical gas cylinders – Design, construction and tests
ISO 10297:1999	Gas cylinders – Refillable gas cylinder valves – Specification and type testing.

6.2.2.4 Periodic inspection and test

The following standards apply to the periodic inspection and testing of UN cylinders:

ISO 6406:2005	Seamless steel gas cylinders – Periodic inspection and testing
ISO 10461:2005/ Amd 1:2006	Seamless aluminium-alloy gas cylinders – Periodic inspection and testing
ISO 10462:2005	Transportable cylinders for dissolved acetylene – Periodic inspection and maintenance
ISO 11623:2002	Transportable gas cylinders – Periodic inspection and testing of composite gas cylinders

6.2.2.5 Conformity assessment system and approval for manufacture of pressure receptacles

6.2.2.5.1 *Definitions*

For the purposes of this section:

Conformity assessment system means a system for competent authority approval of a manufacturer, by pressure receptacle design type approval, approval of manufacturer's quality system and approval of inspection bodies;

Design type means a pressure receptacle design as specified by a particular pressure receptacle standard;

Verify means confirm by examination or provision of objective evidence that specified requirements have been fulfilled.

6.2.2.5.2 *General requirements*

Competent authority

6.2.2.5.2.1 The competent authority that approves the pressure receptacle shall approve the conformity assessment system for the purpose of ensuring that pressure receptacles conform to the provisions of this Code. In instances where the competent authority that approves a pressure receptacle is not the competent authority in the country of manufacture, the marks of the approval country and the country of manufacture shall be indicated in the pressure receptacle marking (see 6.2.2.7 and 6.2.2.8).

The competent authority of the country of approval shall supply, upon request, evidence demonstrating compliance of this conformity assessment system to its counterpart in a country of use.

6.2.2.5.2.2 The competent authority may delegate its functions in this conformity assessment system in whole or in part.

6.2.2.5.2.3 The competent authority shall ensure that a current list of approved inspection bodies and their identity marks and approved manufacturers and their identity marks is available.

Inspection body

6.2.2.5.2.4 The inspection body shall be approved by the competent authority for the inspection of pressure receptacles and shall:

.1   have a staff with an organizational structure, capable, trained, competent, and skilled, to satisfactorily perform its technical functions;

.2   have access to suitable and adequate facilities and equipment;

.3   operate in an impartial manner and be free from any influence which could prevent it from doing so;

.4   ensure commercial confidentiality of the commercial and proprietary activities of the manufacturer and other bodies;

.5   maintain clear demarcation between actual inspection body functions and unrelated functions;

.6   operate a documented quality system;

.7   ensure that the tests and inspections specified in the relevant pressure receptacle standard and in this Code are performed; and

.8   maintain an effective and appropriate report and record system in accordance with 6.2.2.5.6.

6.2.2.5.2.5 The inspection body shall perform design type approval, pressure receptacle production testing and inspection, and certification to verify conformity with the relevant pressure receptacle standard (see 6.2.2.5.4 and 6.2.2.5.5).

Manufacturer

6.2.2.5.2.6 The manufacturer shall:

.1   operate a documented quality system in accordance with 6.2.2.5.3;

.2   apply for design type approvals in accordance with 6.2.2.5.4;

.3   select an inspection body from the list of approved inspection bodies maintained by the competent authority in the country of approval; and

.4   maintain records in accordance with 6.2.2.5.6.

Testing laboratory

6.2.2.5.2.7 The testing laboratory shall have:

.1   staff with an organizational structure, sufficient in number, competence, and skill; and

.2   suitable and adequate facilities and equipment to perform the tests required by the manufacturing standard to the satisfaction of the inspection body.

6.2.2.5.3 **Manufacturer's quality system**

6.2.2.5.3.1 The quality system shall contain all the elements, requirements and provisions adopted by the manufacturer. It shall be documented in a systematic and orderly manner in the form of written policies, procedures and instructions.

The contents shall in particular include adequate descriptions of:

.1   the organizational structure and responsibilities of personnel with regard to design and product quality;

.2   the design control and design verification techniques, processes, and procedures that will be used when designing the pressure receptacles;

.3   the relevant pressure receptacle manufacturing, quality control, quality assurance and process operation instructions that will be used;

.4   quality records, such as inspection reports, test data and calibration data;

.5   management reviews to ensure the effective operation of the quality system arising from the audits in accordance with 6.2.2.5.3.2;

.6   the process describing how customer requirements are met;

.7   the process for control of documents and their revision;

.8   the means for control of non-conforming pressure receptacles, purchased components, in-process and final materials; and

.9   training programmes and qualification procedures for relevant personnel.

6.2.2.5.3.2 *Audit of the quality system*

The quality system shall be initially assessed to determine whether it meets the requirements in 6.2.2.5.3.1 to the satisfaction of the competent authority.

The manufacturer shall be notified of the results of the audit. The notification shall contain the conclusions of the audit and any corrective actions required.

Periodic audits shall be carried out, to the satisfaction of the competent authority, to ensure that the manufacturer maintains and applies the quality system. Reports of the periodic audits shall be provided to the manufacturer.

6.2.2.5.3.3 *Maintenance of the quality system*

The manufacturer shall maintain the quality system as approved in order that it remains adequate and efficient.

The manufacturer shall notify the competent authority that approved the quality system of any intended changes. The proposed changes shall be evaluated in order to determine whether the amended quality system will still satisfy the requirements in 6.2.2.5.3.1.

6.2.2.5.4 **Approval process**

Initial design type approval

6.2.2.5.4.1 The initial design type approval shall consist of approval of the manufacturer's quality system and approval of the pressure receptacle design to be produced. An application for an initial design type approval shall meet the requirements cf 6.2.2.5.3, 6.2.2.5.4.2 to 6.2.2.5.4.6 and 6.2.2.5.4.9.

6.2.2.5.4.2 A manufacturer desiring to produce pressure receptacles in accordance with a pressure receptacle standard and this Code shall apply for, obtain, and retain a Design Type Approval Certificate issued by the competent authority in the country of approval for at least one pressure receptacle design type in accordance with the procedure given in 6.2.2.5.4.9. This certificate shall, on request, be submitted to the competent authority of the country of use.

6.2.2.5.4.3 An application shall be made for each manufacturing facility and shall include:

.1 the name and registered address of the manufacturer and in addition, if the application is submitted by an authorized representative, its name and address;

.2 the address of the manufacturing facility (if different from the above);

.3 the name and title of the person(s) responsible for the quality system;

.4 the designation of the pressure receptacle and the relevant pressure receptacle standard;

.5 details of any refusal of approval of a similar application by any other competent authority;

.6 the identity of the inspection body for design type approval;

.7 documentation on the manufacturing facility as specified under 6.2.2.5.3.1; and

.8 the technical documentation required for design type approval, which shall enable verification of the conformity of the pressure receptacles with the requirements of the relevant pressure receptacle design standard. The technical documentation shall cover the design and method of manufacture and shall contain, as far as is relevant for assessment, at least the following:

.1 pressure receptacle design standard, design and manufacturing drawings, showing components and sub-assemblies, if any;

.2 descriptions and explanations necessary for the understanding of the drawings and intended use of the pressure receptacles;

.3 a list of the standards necessary to fully define the manufacturing process;

.4 design calculations and material specifications; and

.5 design type approval test reports, describing the results of examinations and tests carried out in accordance with 6.2.2.5.4.9.

6.2.2.5.4.4 An initial audit in accordance with 6.2.2.5.3.2 shall be performed to the satisfaction of the competent authority.

6.2.2.5.4.5 If the manufacturer is denied approval, the competent authority shall provide written detailed reasons for such denial.

6.2.2.5.4.6 Following approval, changes to the information submitted under 6.2.2.5.4.3 relating to the initial approval shall be provided to the competent authority.

Subsequent design type approvals

6.2.2.5.4.7 An application for a subsequent design type approval shall encompass the requirements of 6.2.2.5.4.8 and 6.2.2.5.4.9, provided a manufacturer is in the possession of an initial design type approval. In such a case, the manufacturer's quality system according to 6.2.2.5.3 shall have been approved during the initial design type approval and shall be applicable for the new design.

6.2.2.5.4.8 The application shall include:

.1 the name and address of the manufacturer and in addition, if the application is submitted by an authorized representative, its name and address;

.2 details of any refusal of approval of a similar application by any other competent authority;

.3 evidence that initial design type approval has been granted; and

.4 the technical documentation, as described in 6.2.2.5.4.3.8.

Procedure for design type approval

6.2.2.5.4.9 The inspection body shall:

.1 examine the technical documentation to verify that:

.1 the design is in accordance with the relevant provisions of the standard, and

.2 the prototype lot has been manufactured in conformity with the technical documentation and is representative of the design;

.2 verify that the production inspections have been carried out as required in accordance with 6.2.2.5.5;

.3 select pressure receptacles from a prototype production lot and supervise the tests of these pressure receptacles as required for design type approval;

.4 perform or have performed the examinations and tests specified in the pressure receptacle standard to determine that:

.1 the standard has been applied and fulfilled, and

.2 the procedures adopted by the manufacturer meet the requirements of the standard; and

.5 ensure that the various type approval examinations and tests are correctly and competently carried out.

After prototype testing has been carried out with satisfactory results and all applicable requirements of 6.2.2.5.4 have been satisfied, a Design Type Approval Certificate shall be issued which shall include the name and address of the manufacturer, results and conclusions of the examination, and the necessary data for identification of the design type.

If the manufacturer is denied a design type approval, the competent authority shall provide written detailed reasons for such denial.

6.2.2.5.4.10 *Modifications to approved design types*

The manufacturer shall either:

(a) inform the issuing competent authority of modifications to the approved design type, where such modifications do not constitute a new design, as specified in the pressure receptacle standard; or

(b) request a subsequent design type approval where such modifications constitute a new design according to the relevant pressure receptacle standard. This additional approval shall be given in the form of an amendment to the original design type approval certificate.

6.2.2.5.4.11 Upon request, the competent authority shall communicate to any other competent authority information concerning design type approval, modifications of approvals, and withdrawn approvals.

6.2.2.5.5 *Production inspection and certification*

An inspection body, or its delegate, shall carry out the inspection and certification of each pressure receptacle. The inspection body selected by the manufacturer for inspection and testing during production may be different from the inspection body used for the design type approval testing.

Where it can be demonstrated to the satisfaction of the inspection body that the manufacturer has trained and competent inspectors, independent of the manufacturing operations, inspection may be performed by those inspectors. In such a case, the manufacturer shall maintain training records of the inspectors.

The inspection body shall verify that the inspections by the manufacturer and tests performed on those pressure receptacles fully conform to the standard and the provisions of this Code. Should non-conformance in conjunction with this inspection and testing be determined, the permission to have inspection performed by the manufacturer's inspectors may be withdrawn.

The manufacturer shall, after approval by the inspection body, make a declaration of conformity with the certified design type. The application of the pressure receptacle certification marking shall be considered a declaration that the pressure receptacle complies with the applicable pressure receptacle standards and the requirements of this conformity assessment system and with the provisions of this Code. The inspection body shall affix or delegate the manufacturer to affix the pressure receptacle certification marking and the registered mark of the inspection body to each approved pressure receptacle.

A certificate of compliance, signed by the inspection body and the manufacturer, shall be issued before the pressure receptacles are filled.

6.2.2.5.6 *Records*

Design type approval and certificate of compliance records shall be retained by the manufacturer and the inspection body for not less than 20 years.

6.2.2.6 **Approval system for periodic inspection and testing of pressure receptacles**

6.2.2.6.1 *Definitions*

For the purposes of this section:

Approval system means a system for competent authority approval of a body performing periodic inspection and testing of pressure receptacles (hereinafter referred to as "periodic inspection and testing body"), including approval of that body's quality system.

6.2.2.6.2 *General provisions*

Competent authority

6.2.2.6.2.1 The competent authority shall establish an approval system for the purpose of ensuring that the periodic inspection and testing of pressure receptacles conform to the provisions of this Code. In instances where the competent authority that approves a body performing periodic inspection and testing of a pressure receptacle is not the competent authority of the country approving the manufacture of the pressure receptacle, the marks of the approval country of periodic inspection and testing shall be indicated in the pressure receptacle marking (see 6.2.2.7). The competent authority of the country of approval for the periodic inspection and testing shall supply, upon request, evidence demonstrating compliance with this approval system, including the records of the periodic inspection and testing, to its counterpart in a country of use. The competent authority of the country of approval may terminate the approval certificate referred to in 6.2.2.6.4.1, upon evidence demonstrating non-compliance with the approval system.

6.2.2.6.2.2 The competent authority may delegate its functions in this approval system, in whole or in part.

6.2.2.6.2.3 The competent authority shall ensure that a current list of approved periodic inspection and testing bodies and their identity marks is available.

Periodic inspection and testing body

6.2.2.6.2.4 The periodic inspection and testing body shall be approved by the competent authority and shall:

.1   have a staff with an organizational structure, capable, trained, competent, and skilled, satisfactorily to perform its technical functions;

.2   have access to suitable and adequate facilities and equipment;

.3   operate in an impartial manner and be free from any influence which could prevent it from doing so;

.4   ensure commercial confidentiality;

.5   maintain clear demarcation between actual periodic inspection and testing body functions and unrelated functions;

.6   operate a documented quality system in accordance with 6.2.2.6.3;

.7   apply for approval in accordance with 6.2.2.6.4;

.8   ensure that the periodic inspections and tests are performed in accordance with 6.2.2.6.5; and

.9   maintain an effective and appropriate report and record system in accordance with 6.2.2.6.6.

6.2.2.6.3 *Quality system and audit of the periodic inspection and testing body*

6.2.2.6.3.1 *Quality system.* The quality system shall contain all the elements, requirements, and provisions adopted by the periodic inspection and testing body. It shall be documented in a systematic and orderly manner in the form of written policies, procedures, and instructions. The quality system shall include:

.1   a description of the organizational structure and responsibilities;

.2 the relevant inspection and test, quality control, quality assurance, and process operation instructions that will be used;

.3 quality records, such as inspection reports, test data, calibration data and certificates;

.4 management reviews to ensure the effective operation of the quality system arising from the audits performed in accordance with 6.2.2.6.3.2;

.5 a process for control of documents and their revision;

.6 a means for control of non-conforming pressure receptacles; and

.7 training programmes and qualification procedures for relevant personnel.

6.2.2.6.3.2 *Audit*. The periodic inspection and testing body and its quality system shall be audited in order to determine whether it meets the requirements of this Code to the satisfaction of the competent authority. An audit shall be conducted as part of the initial approval process (see 6.2.2.6.4.3). An audit may be required as part of the process to modify an approval (see 6.2.2.6.4.6). Periodic audits shall be conducted, to the satisfaction of the competent authority, to ensure that the periodic inspection and testing body continues to meet the provisions of this Code. The periodic inspection and testing body shall be notified of the results of any audit. The notification shall contain the conclusions of the audit and any corrective actions required.

6.2.2.6.3.3 *Maintenance of the quality system*. The periodic inspection and testing body shall maintain the quality system as approved in order that it remains adequate and efficient. The periodic inspection and testing body shall notify the competent authority that approved the quality system of any intended changes, in accordance with the process for modification of an approval in 6.2.2.6.4.6.

6.2.2.6.4 **Approval process for periodic inspection and testing bodies**

Initial approval

6.2.2.6.4.1 A body desiring to perform periodic inspection and testing of pressure receptacles in accordance with a pressure receptacle standard and with this Code shall apply for, obtain, and retain an Approval Certificate issued by the competent authority. This written approval shall, on request, be submitted to the competent authority of a country of use.

6.2.2.6.4.2 An application shall be made for each periodic inspection and testing body and shall include:

.1 the name and address of the periodic inspection and testing body and, if the application is submitted by an authorized representative, its name and address;

.2 the address of each facility performing periodic inspection and testing;

.3 the name and title of the person(s) responsible for the quality system;

.4 the designation of the pressure receptacles, the periodic inspection and test methods, and the relevant pressure receptacle standards met by the quality system;

.5 documentation on each facility, the equipment, and the quality system as specified under 6.2.2.6.3.1;

.6 the qualifications and training records of the periodic inspection and test personnel; and

.7 details of any refusal of approval of a similar application by any other competent authority.

6.2.2.6.4.3 The competent authority shall:

.1 examine the documentation to verify that the procedures are in accordance with the requirements of the relevant pressure receptacle standards and of this Code; and

.2 conduct an audit in accordance with 6.2.2.6.3.2 to verify that the inspections and tests are carried out as required by the relevant pressure receptacle standards and by this Code, to the satisfaction of the competent authority.

6.2.2.6.4.4 After the audit has been carried out with satisfactory results and all applicable requirements of 6.2.2.6.4 have been satisfied, an Approval Certificate shall be issued. It shall include the name of the periodic inspection and testing body, the registered mark, the address of each facility, and the necessary data for identification of its approved activities (e.g., designation of pressure receptacles, periodic inspection and test method and pressure receptacle standards).

6.2.2.6.4.5 If the periodic inspection and testing body is denied approval, the competent authority shall provide written detailed reasons for such denial.

Modifications to periodic inspection and testing body approvals

6.2.2.6.4.6 Following approval, the periodic inspection and testing body shall notify the issuing competent authority of any modifications to the information submitted under 6.2.2.6.4.2 relating to the initial approval. The modifications shall be evaluated in order to determine whether the requirements of the relevant pressure receptacle standards and of this Code will be satisfied. An audit in accordance with 6.2.2.6.3.2 may be required. The competent authority shall accept or reject these modifications in writing, and an amended Approval Certificate shall be issued as necessary.

6.2.2.6.4.7 Upon request, the competent authority shall communicate to any other competent authority, information concerning initial approvals, modifications of approvals, and withdrawn approvals.

6.2.2.6.5 *Periodic inspection and test and certification*

The application of the periodic inspection and test marking to a pressure receptacle shall be considered a declaration that the pressure receptacle complies with the applicable pressure receptacle standards and with the provisions of this Code. The periodic inspection and testing body shall affix the periodic inspection and test marking, including its registered mark, to each approved pressure receptacle (see 6.2.2.7.6). A record certifying that a pressure receptacle has passed the periodic inspection and test shall be issued by the periodic inspection and testing body, before the pressure receptacle is filled.

6.2.2.6.6 *Records*

The periodic inspection and testing body shall retain records of pressure receptacle periodic inspection and tests (both passed and failed), including the location of the test facility, for not less than 15 years. The owner of the pressure receptacle shall retain an identical record until the next periodic inspection and test unless the pressure receptacle is permanently removed from service.

6.2.2.7 **Marking of refillable UN pressure receptacles**

Refillable UN pressure receptacles shall be marked clearly and legibly with certification, operational and manufacturing marks. These marks shall be permanently affixed (e.g., stamped, engraved, or etched) on the pressure receptacle. The marks shall be on the shoulder, top end or neck of the pressure receptacle or on a permanently affixed component of the pressure receptacle (e.g., welded collar or corrosion-resistant plate welded on the outer jacket of a closed cryogenic receptacle). Except for the UN packaging symbol, the minimum size of the marks shall be 5 mm for pressure receptacles with a diameter greater than or equal to 140 mm and 2.5 mm for pressure receptacles with a diameter less than 140 mm. The minimum size of the UN packaging symbol shall be 10 mm for pressure receptacles with a diameter greater than or equal to 140 mm and 5 mm for pressure receptacles with a diameter less than 140 mm.

6.2.2.7.1 The following certification marks shall be applied:

(a) The UN packaging symbol

This symbol shall not be used for any purpose other than certifying that a packaging complies with the relevant requirements in chapter 6.1, 6.2, 6.3, 6.5 or 6.6.

(b) The technical standard (e.g., ISO 9809-1) used for design, construction and testing;

(c) The character(s) identifying the country of approval as indicated by the distinguishing signs of motor vehicles in international traffic;

(d) The identity mark or stamp of the inspection body that is registered with the competent authority of the country authorizing the marking;

(e) The date of the initial inspection, the year (four digits) followed by the month (two digits) separated by a slash (i.e. "/").

6.2.2.7.2 The following operational marks shall be applied:

(f) The test pressure in bar, preceded by the letters "PH" and followed by the letters "BAR";

(g) The mass of the empty pressure receptacle including all permanently attached integral parts (e.g., neck ring, foot ring, etc.) in kilograms, followed by the letters "KG". This mass shall not include the mass of valve, valve cap or valve guard, any coating, or porous mass for acetylene. The mass shall be expressed to three significant figures rounded up to the last digit. For cylinders of less than 1 kg, the mass shall be expressed to two significant figures rounded up to the last digit. In the case of pressure receptacles for UN 1001 acetylene, dissolved and UN 3374 acetylene, solvent free, at least one decimal shall be shown after the decimal point and two digits for pressure receptacles of less than 1 kg;

(h) The minimum guaranteed wall thickness of the pressure receptacle in millimetres followed by the letters "MM". This mark is not required for pressure receptacles with a water capacity less than or equal to 1 litre or for composite cylinders or for closed cryogenic receptacles;

(i) In the case of pressure receptacles for compressed gases, UN 1001 acetylene, dissolved, and UN 3374 acetylene, solvent free, the working pressure in bar, preceded by the letters "PW". In the case of closed cryogenic receptacles, the maximum allowable working pressure preceded by the letters "MAWP";

(j) In the case of pressure receptacles for liquefied gases and refrigerated liquefied gases, the water capacity in litres expressed to three significant figures rounded down to the last digit, followed by the letter "L". If the value of the minimum or nominal water capacity is an integer, the digits after the decimal point may be neglected;

(k) In the case of pressure receptacles for UN 1001 acetylene, dissolved, the total of the mass of the empty receptacle, the fittings and accessories not removed during filling, any coating, the porous material, the solvent and the saturation gas expressed to three significant figures rounded down to the last digit followed by the letters "KG". At least one decimal shall be shown after the decimal point. For pressure receptacles of less than 1 kg, the mass shall be expressed to two significant figures rounded down to the last digit;

(l) In the case of pressure receptacles for UN 3374 acetylene, solvent free, the total of the mass of the empty receptacle, the fittings and accessories not removed during filling, any coating and the porous material expressed to three significant figures rounded down to the last digit followed by the letters "KG". At least one decimal shall be shown after the decimal point. For pressure receptacles of less than 1 kg, the mass shall be expressed to two significant figures rounded down to the last digit.

6.2.2.7.3   The following manufacturing marks shall be applied:

(m) Identification of the cylinder thread (e.g., 25E). This mark is not required for closed cryogenic receptacles;

(n) The manufacturer's mark registered by the competent authority. When the country of manufacture is not the same as the country of approval, then the manufacturer's mark shall be preceded by the character(s) identifying the country of manufacture as indicated by the distinguishing signs of motor vehicles in international traffic. The country mark and the manufacturer's mark shall be separated by a space or slash;

(o) The serial number assigned by the manufacturer;

(p) In the case of steel pressure receptacles and composite pressure receptacles with steel liner intended for the transport of gases with a risk of hydrogen embrittlement, the letter "H" showing compatibility of the steel (see ISO 11114-1:1997).

6.2.2.7.4   The above marks shall be placed in three groups.

–   Manufacturing marks shall be the top grouping and shall appear consecutively in the sequence given in 6.2.2.7.3.

–   The operational marks in 6.2.2.7.2 shall be the middle grouping and the test pressure (f) shall be immediately preceded by the working pressure (i) when the latter is required.

–   Certification marks shall be the bottom grouping and shall appear in the sequence given in 6.2.2.7.1.

The following is an example of the markings applied to a cylinder.

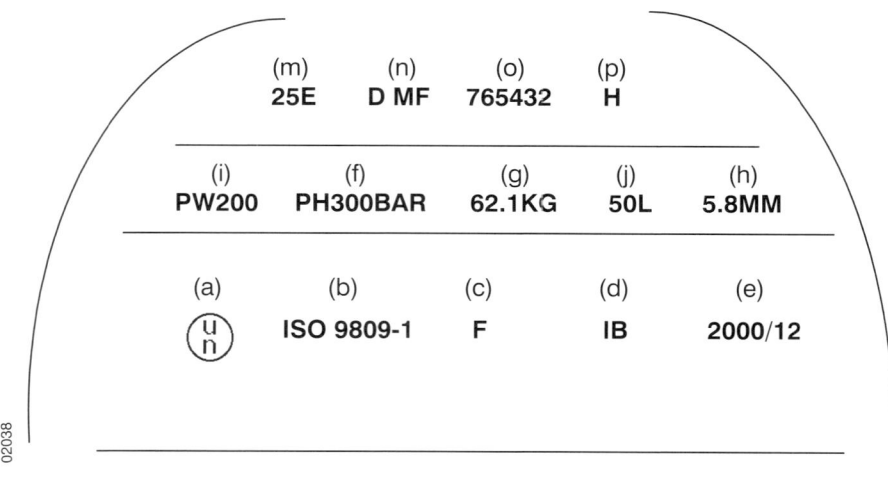

6.2.2.7.5   Other marks are allowed in areas other than the side wall, provided they are made in low-stress areas and are not of a size and depth that will create harmful stress concentrations. In the case of closed cryogenic receptacles, such marks may be on a separate plate attached to the outer jacket. Such marks shall not conflict with required marks.

6.2.2.7.6   In addition to the preceding marks, each refillable pressure receptacle that meets the periodic and test requirements of 6.2.2.4 shall be marked in sequence as follows:

     (a) the character(s) identifying the country authorizing the body performing the periodic inspection and test. This marking is not required if this body is approved by the competent authority of the country approving manufacture;

     (b) the registered mark of the body authorized by the competent authority for performing periodic inspection and test;

     (c) the date of the periodic inspection and test, the year (two digits) followed by the month (two digits) separated by a slash (i.e. "/"). Four digits may be used to indicate the year.

6.2.2.7.7   For acetylene cylinders, with the agreement of the competent authority, the date of the most recent periodic inspection and the stamp of the body performing the periodic inspection and test may be engraved on a ring held on the cylinder by the valve. The ring shall be configured so that it can only be removed by disconnecting the valve from the cylinder.

6.2.2.8   **Marking of non-refillable UN pressure receptacles**

      Non-refillable UN pressure receptacles shall be marked clearly and legibly with certification and gas or pressure receptacle specific marks. These marks shall be permanently affixed (e.g., stencilled, stamped, engraved, or etched) on the pressure receptacle. Except when stencilled, the marks shall be on the shoulder, top end or neck of the pressure receptacle or on a permanently affixed component of the pressure receptacle (e.g., welded collar). Except for the "UN" mark and the "DO NOT REFILL" mark, the minimum size of the marks shall be 5 mm for pressure receptacles with a diameter greater than or equal to 140 mm and 2.5 mm for pressure receptacles with a diameter less than 140 mm. The minimum size of the "UN" mark shall be 10 mm for pressure receptacles with a diameter greater than or equal to 140 mm and 5 mm for pressure receptacles with a diameter less than 140 mm. The minimum size of the "DO NOT REFILL" mark shall be 5 mm.

6.2.2.8.1   The marks listed in 6.2.2.7.1 to 6.2.2.7.3 shall be applied with the exception of (g), (h) and (m). The serial number (o) may be replaced by the batch number. In addition, the words "DO NOT REFILL" in letters of at least 5 mm in height are required.

6.2.2.8.2   The requirements of 6.2.2.7.4 shall apply.

      Note: Non-refillable pressure receptacles may, on account of their size, substitute this marking by a label.

6.2.2.8.3   Other marks are allowed provided they are made in low-stress areas other than the side wall and are not of a size and depth that will create harmful stress concentrations. Such marks shall not conflict with required marks.

6.2.3   **Provisions for non-UN pressure receptacles**

6.2.3.1   Pressure receptacles not designed, constructed, inspected, tested and approved according to 6.2.2 shall be designed, constructed, inspected, tested and approved in accordance with a technical code recognized by the competent authority and the general provisions of 6.2.1.

6.2.3.2   Pressure receptacles designed, constructed, inspected, tested and approved under the provisions of this section shall not be marked with the UN packaging symbol.

6.2.3.3   For metallic cylinders, tubes, pressure drums and bundles of cylinders, the construction shall be such that the minimum burst ratio (burst pressure divided by test pressure) is:

        1.50 for refillable pressure receptacles,

        2.00 for non-refillable pressure receptacles.

6.2.3.4   Marking shall be in accordance with the requirements of the competent authority of the country of use.

6

6.2.4 Provisions for aerosol dispensers, small receptacles containing gas (gas cartridges) and fuel cell cartridges containing liquefied flammable gas

6.2.4.1 Small receptacles containing gas (gas cartridges) and fuel cell cartridges containing liquefied flammable gas

6.2.4.1.1 Each receptacle or fuel cell cartridge shall be subjected to a test performed in a hot water bath. The temperature of the bath and the duration of the test shall be such that the internal pressure reaches that which would be reached at 55°C (50°C if the liquid phase does not exceed 95% of the capacity of the receptacle or the fuel cell cartridge at 50°C). If the contents are sensitive to heat or if the receptacles or the fuel cell cartridges are made of plastics material which softens at this test temperature, the temperature of the bath shall be set at between 20°C and 30°C, but in addition one receptacle or fuel cell cartridge in 2000 shall be tested at the higher temperature.

6.2.4.1.2 No leakage or permanent deformation of a receptacle or fuel cell cartridge shall occur, except that a plastics receptacle or fuel cell cartridge may be deformed through softening provided that it does not leak.

6.2.4.2 Aerosol dispensers

Each filled aerosol dispenser shall be subjected to a test performed in a hot water bath or an approved water bath alternative.

6.2.4.2.1 *Hot water bath test*

6.2.4.2.1.1 The temperature of the water bath and the duration of the test shall be such that the internal pressure reaches that which would be reached at 55°C (50°C if the liquid phase does not exceed 95% of the capacity of the aerosol dispenser at 50°C). If the contents are sensitive to heat or if the aerosol dispensers are made of plastics material which softens at this test temperature, the temperature of the bath shall be set at between 20°C and 30°C but, in addition, one aerosol dispenser in 2000 shall be tested at the higher temperature.

6.2.4.2.1.2 No leakage or permanent deformation of an aerosol dispenser may occur, except that a plastic aerosol dispenser may be deformed through softening provided that it does not leak.

6.2.4.2.2 *Alternative methods*

With the approval of the competent authority, alternative methods which provide an equivalent level of safety may be used provided that the requirements of 6.2.4.2.2.1, 6.2.4.2.2.2 and 6.2.4.2.2.3 are met.

6.2.4.2.2.1 *Quality system*

Aerosol dispenser fillers and component manufacturers shall have a quality system. The quality system shall implement procedures to ensure that all aerosol dispensers that leak or that are deformed are rejected and not offered for transport.

The quality system shall include:

(a) a description of the organizational structure and responsibilities;

(b) the relevant inspection and test, quality control, quality assurance, and process operation instructions that will be used;

(c) quality records, such as inspection reports, test data, calibration data and certificates;

(d) management reviews to ensure the effective operation of the quality system;

(e) a process for control of documents and their revision;

(f) a means for control of non-conforming aerosol dispensers;

(g) training programmes and qualification procedures for relevant personnel; and

(h) procedures to ensure that there is no damage to the final product.

An initial audit and periodic audits shall be conducted to the satisfaction of the competent authority. These audits shall ensure the approved system is and remains adequate and efficient. Any proposed changes to the approved system shall be notified to the competent authority in advance.

6.2.4.2.2.2 *Pressure and leak testing of aerosol dispensers before filling*

Every empty aerosol dispenser shall be subjected to a pressure equal to or in excess of the maximum expected in the filled aerosol dispensers at 55°C (50°C if the liquid phase does not exceed 95% of the capacity of the receptacle at 50°C). This shall be at least two-thirds of the design pressure of the aerosol dispenser. If any aerosol dispenser shows evidence of leakage at a rate equal to or greater than 3.3×10^{-2} mbar·ℓ·s^{-1} at the test pressure, distortion or other defect, it shall be rejected.

6.2.4.2.2.3   *Testing of the aerosol dispensers after filling*

Prior to filling, the filler shall ensure that the crimping equipment is set appropriately and the specified propellant is used.

Each filled aerosol dispenser shall be weighed and leak tested. The leak detection equipment shall be sufficiently sensitive to detect at least a leak rate of 2.0×10^{-3} mbar·ℓ·s^{-1} at 20°C.

Any filled aerosol dispenser which shows evidence of leakage, deformation or excessive mass shall be rejected.

6.2.4.3   With the approval of the competent authority, aerosols and receptacles, small, containing pharmaceutical products and non-flammable gases which are required to be sterile, but may be adversely affected by water bath testing, are not subject to 6.2.4.1 and 6.2.4.2 if:

(a)   They are manufactured under the authority of a national health administration and, if required by the competent authority, follow the principles of Good Manufacturing Practice (GMP) established by the World Health Organization (WHO);* and

(b)   An equivalent level of safety is achieved by the manufacturer's use of alternative methods for leak detection and pressure resistance, such as helium detection and water bathing a statistical sample of at least 1 in 2000 from each production batch.

---

* WHO Publication: ''Quality assurance of pharmaceuticals. A compendium of guidelines and related materials. Volume 2: Good manufacturing practices and inspection''.

Chapter 6.3

Provisions for the construction and testing of packagings for class 6.2 infectious substances of category A

6.3.1 General

6.3.1.1 The provisions of this chapter apply to packagings intended for the transport of infectious substances of category A.

6.3.2 Provisions for packagings

6.3.2.1 The provisions for packagings in this section are based on packagings, as specified in 6.1.4, currently used. In order to take into account progress in science and technology, there is no objection to the use of packagings having specifications different from those in this chapter provided that they are equally effective, acceptable to the competent authority and able successfully to withstand the tests described in 6.3.5. Methods of testing other than those described in the provisions of this Code are acceptable provided they are equivalent.

6.3.2.2 Packagings shall be manufactured and tested under a quality assurance programme which satisfies the competent authority in order to ensure that each packaging meets the provisions of this chapter.

Note: ISO 16106:2006 "Packaging – Transport packages for dangerous goods – Dangerous goods packagings, intermediate bulk containers (IBCs) and large packagings – Guidelines for the application of ISO 9001" provides acceptable guidance on procedures which may be followed.

6.3.2.3 Manufacturers and subsequent distributors of packagings shall provide information regarding procedures to be followed and a description of the types and dimensions of closures (including required gaskets) and any other components needed to ensure that packages as presented for transport are capable of passing the applicable performance tests of this chapter.

6.3.3 Code for designating types of packagings

6.3.3.1 The codes for designating types of packagings are set out in 6.1.2.7.

6.3.3.2 The letters "U" or "W" may follow the packaging code. The letter "U" signifies a special packaging conforming to the provisions of 6.3.5.1.6. The letter "W" signifies that the packaging, although of the same type as indicated by the code, is manufactured to a specification different from that in 6.1.4 and is considered equivalent under the provisions of 6.3.2.1.

6.3.4 Marking

Note 1: The marking indicates that the packaging which bears it corresponds to a successfully tested design type and that it complies with the provisions of this chapter which are related to the manufacture, but not to the use, of the packaging.

Note 2: The marking is intended to be of assistance to packaging manufacturers, reconditioners, packaging users, carriers and regulatory authorities.

Note 3: The marking does not always provide full details of the test levels, etc., and these may need to be taken further into account, e.g., by reference to a test certificate, to test reports or to a register of successfully tested packagings.

6.3.4.1 Each packaging intended for use according to the provisions of this Code shall bear markings which are durable, legible and placed in a location and of such a size relative to the packaging as to be readily visible. For packages with a gross mass of more than 30 kg, the markings or a duplicate thereof shall appear on the top or on a side of the packaging. Letters, numerals and symbols shall be at least 12 mm high, except for packagings of 30 litres or 30 kg capacity or less, when they shall be at least 6 mm in height, and for packagings of 5 litres or 5 kg or less, when they shall be of an appropriate size.

6.3.4.2   A packaging that meets the provisions of this section and of 6.3.5 shall be marked with:

(a) the United Nations packaging symbol. This symbol shall not be used for any purpose other than certifying that a packaging complies with the relevant provisions in chapter 6.1, 6.2, 6.3, 6.5 or 6.6;

(b) the code designating the type of packaging according to the provisions of 6.1.2;

(c) the text "CLASS 6.2";

(d) the last two digits of the year of manufacture of the packaging;

(e) the State authorizing the allocation of the mark, indicated by the distinguishing sign for motor vehicles in international traffic;

(f) the name of the manufacturer or other identification of the packaging specified by the competent authority;

(g) for packagings meeting the provisions of 6.3.5.1.6, the letter "U" shall be inserted immediately following the marking required in (b) above; and

(h) each element of the marking applied in accordance with subparagraphs (a) to (g).

6.3.4.3   Marking shall be applied in the sequence shown in 6.3.4.2 (a) to (g); each element of the marking required in these subparagraphs shall be clearly separated, e.g., by a slash or space, so as to be easily identifiable. For examples, see 6.3.4.4

Any additional markings authorized by a competent authority shall still enable the parts of the mark to be correctly identified with reference to 6.3.4.1

6.3.4.4   **Example of marking**

   **4G/CLASS 6.2/06**   as in 6.3.4.2 (a), (b), (c) and (d)

S/SP-9989-ERIKSSON as in 6.3.4.2 (e) and (f)

6.3.5   Test provisions for packagings

6.3.5.1   Performance and frequency of tests

6.3.5.1.1   The design type of each packaging shall be tested as provided in this section in accordance with procedures established by the competent authority.

6.3.5.1.2   Each packaging design type shall successfully pass the tests prescribed in this chapter before being used. A packaging design type is defined by the design, size, material and thickness, manner of construction and packing, but may include various surface treatments. It also includes packagings which differ from the design type only in their lesser design height.

6.3.5.1.3   Tests shall be repeated on production samples at intervals established by the competent authority.

6.3.5.1.4   Tests shall also be repeated after each modification which alters the design, material or manner of construction of a packaging.

6.3.5.1.5   The competent authority may permit the selective testing of packagings that differ only in minor respects from a tested type, such as smaller sizes or lower net mass of primary receptacles; and packagings such as drums and boxes which are produced with small reductions in external dimension(s).

6.3.5.1.6   Primary receptacles of any type may be assembled within an secondary packaging and transported without testing in the rigid outer packaging under the following conditions:

.1 the rigid outer packaging shall have been successfully tested in accordance with 6.3.5.2.2 with fragile (such as glass) primary receptacles;

.2 the total combined gross mass of primary receptacles shall not exceed one half of the gross mass of primary receptacles used for the drop test in .1 above;

.3 the thickness of cushioning between primary receptacles and between primary receptacles and the outside of the secondary packaging shall not be reduced below the corresponding thicknesses in the originally tested packaging; and if a single primary receptacle was used in the original test, the thickness of cushioning between primary receptacles shall not be less than the thickness of cushioning between the outside of the secondary packaging and the primary receptacle in the original test. When either fewer or smaller primary receptacles are used (as compared to the primary receptacles used in the drop test), sufficient additional cushioning material shall be used to take up the void spaces;

.4 the rigid outer packaging shall have successfully passed the stacking test in 6.1.5.6 while empty. The total mass of identical packages shall be based on the combined mass of packagings used in the drop test in .1 above;

.5 for primary receptacles containing liquids, an adequate quantity of absorbent material to absorb the entire liquid content of the primary receptacles shall be present;

.6 if the rigid outer packaging is intended to contain primary receptacles for liquids and is not leakproof, or is intended to contain primary receptacles for solids and is not sift-proof, a means of containing any liquid or solid contents in the event of leakage shall be provided in the form of a leakproof liner, plastics bag or other equally effective means of containment; and

.7 in addition to the markings prescribed in 6.3.4.2 (a) to (f), packagings shall be marked in accordance with 6.3.4.2(g).

6.3.5.1.7 The competent authority may at any time require proof, by tests in accordance with this section, that serially produced packagings meet the provisions of the design type tests.

6.3.5.1.8 Provided the validity of the test results is not affected and with the approval of the competent authority, several tests may be made on one sample.

6.3.5.2 Preparation of packagings for testing

6.3.5.2.1 Samples of each packaging shall be prepared as for transport except that a liquid or solid infectious substance shall be replaced by water or, where conditioning at $-18^{\circ}C$ is specified, by water containing antifreeze. Each primary receptacle shall be filled to not less than 98% of its capacity.

Note: The term "water" includes water/antifreeze solution with a minimum specific gravity of 0.95 for testing at $-18^{\circ}C$.

6.3.5.2.2 *Tests and number of samples required*

Tests required for packaging types

Type of packaging[a]			Tests required					
Rigid outer packaging	Primary receptacle		Water spray 6.3.5.3.6.1	Cold conditioning 6.3.5.3.6.2	Drop 6.3.5.3	Additional drop 6.3.5.3.6.3	Puncture 6.3.5.4	Stack 6.1.5.6
	Plastics	Other	Number of samples	Number of samples	Number of samples	Number of samples	Number of samples	Number of samples
Fibreboard box	x		5	5	10	Required on one sample when the packaging is intended to contain dry ice.	2	Required on three samples when testing a "U"-marked packaging as defined in 6.3.5.1.6 for specific provisions.
		x	5	0	5		2	
Fibreboard drum	x		3	3	6		2	
		x	3	0	3		2	
Plastics box	x		0	5	5		2	
		x	0	5	5		2	
Plastics drum/ jerrican	x		0	3	3		2	
		x	0	3	3		2	
Boxes of other material	x		0	5	5		2	
		x	0	0	5		2	
Drums/jerricans of other material	x		0	3	3		2	
		x	0	0	3		2	

[a] "Type of packaging" categorizes packagings for test purposes according to the kind of packaging and its material characteristics.

Note 1: In instances where a primary receptacle is made of two or more materials, the material most liable to damage determines the appropriate test.

Note 2: The materials of the secondary packagings are not taken into consideration when selecting the test or conditioning for the test.

Explanation for use of the table:

If the packaging to be tested consists of a fibreboard outer box with a plastics primary receptacle, five samples must undergo the water spray test (see 6.3.5.3.6.1) prior to dropping and another five must be conditioned to −18°C (see 6.3.5.3.6.2) prior to dropping. If the packaging is to contain dry ice then one further single sample shall be dropped five times after conditioning in accordance with 6.3.5.3.6.3.

Packagings prepared as for transport shall be subjected to the tests in 6.3.5.3 and 6.3.5.4. For outer packagings, the headings in the table relate to fibreboard or similar materials whose performance may be rapidly affected by moisture; plastics which may embrittle at low temperature; and other materials such as metal whose performance is not affected by moisture or temperature.

6.3.5.3   Drop test

6.3.5.3.1   Samples shall be subjected to free-fall drops from a height of 9 m onto a non-resilient, horizontal, flat, massive and rigid surface in conformity with 6.1.5.3.4.

6.3.5.3.2   Where the samples are in the shape of a box, five shall be dropped, one in each of the following orientations:

.1   flat on the base;

.2   flat on the top;

.3   flat on the longest side;

.4   flat on the shortest side; and

.5   on a corner.

6.3.5.3.3   Where the samples are in the shape of a drum, three shall be dropped, one in each of the following orientations:

.1   diagonally on the top chime, with the centre of gravity directly above the point of impact;

.2   diagonally on the base chime; and

.3   flat on the side.

6.3.5.3.4   While the sample shall be released in the required orientation, it is accepted that, for aerodynamic reasons, the impact may not take place in that orientation.

6.3.5.3.5   Following the appropriate drop sequence, there shall be no leakage from the primary receptacle(s), which shall remain protected by cushioning/absorbent material in the secondary packaging.

6.3.5.3.6   *Special preparation of test sample for the drop test*

6.3.5.3.6.1   *Fibreboard – Water spray test*

Fibreboard outer packagings: The sample shall be subjected to a water spray that simulates exposure to rainfall of approximately 5 cm per hour for at least one hour. It shall then be subjected to the test described in 6.3.5.3.1.

6.3.5.3.6.2   *Plastics material – Cold conditioning*

Plastics primary receptacles or outer packagings: The temperature of the test sample and its contents shall be reduced to −18°C or lower for a period of at least 24 hours and within 15 minutes of removal from that atmosphere the test sample shall be subjected to the test described in 6.3.5.3.1. Where the sample contains dry ice, the conditioning period shall be reduced to 4 hours.

6.3.5.3.6.3   *Packagings intended to contain dry ice – Additional drop test*

Where the packaging is intended to contain dry ice, a test additional to that specified in 6.3.5.3.1 and, when appropriate, in 6.3.5.3.6.1 or 6.3.5.3.6.2 shall be carried out. One sample shall be stored so that all the dry ice dissipates and then that sample shall be dropped in one of the orientations described in 6.3.5.3.2 which shall be that most likely to result in failure of the packaging.

6.3.5.4   Puncture test

6.3.5.4.1   *Packagings with a gross mass of 7 kg or less*

Samples shall be placed on a level hard surface. A cylindrical steel rod with a mass of at least 7 kg, a diameter of 38 mm and the impact end edges having a radius not exceeding 6 mm shall be dropped in a vertical free fall from a height of 1 m, measured from the impact end to the impact surface of a sample. One sample shall be placed on its base. A second sample shall be placed in an orientation perpendicular to that used for the first. In each instance, the steel rod shall be aimed to impact the primary receptacle. Following each impact, penetration of the secondary packaging is acceptable, provided that there is no leakage from the primary receptacle(s).

6.3.5.4.2   *Packagings with a gross mass exceeding 7 kg*

Samples shall be dropped on to the end of a cylindrical steel rod. The rod shall be set vertically in a level hard surface. It shall have a diameter of 38 mm and the edges of the upper end a radius not exceeding 6 mm. The rod shall protrude from the surface a distance at least equal to that between the centre of the primary receptacle(s) and the outer surface of the outer packaging with a minimum of 200 mm. One sample shall be dropped with its top face lowermost in a vertical free fall from a height of 1 m, measured from the top of the steel rod. A second sample shall be dropped from the same height in an orientation perpendicular to that used for the first. In each instance, the packaging shall be so orientated that the steel rod would be capable of penetrating the primary receptacle(s). Following each impact, penetration of the secondary packaging is acceptable, provided that there is no leakage from the primary receptacle(s).

6.3.5.5   **Test report**

6.3.5.5.1   A written test report containing at least the following particulars shall be drawn up and shall be available to the users of the packaging:

.1   Name and address of the test facility;

.2   Name and address of applicant (where appropriate);

.3   A unique test report identification;

.4   Date of the test and of the report;

.5   Manufacturer of the packaging;

.6   Description of the packaging design type (e.g., dimensions, materials, closures, thickness, etc.), including method of manufacture (e.g., blow moulding) and which may include drawing(s) and/or photograph(s);

.7   Maximum capacity;

.8   Test contents;

.9   Test descriptions and results;

.10 The test report shall be signed with the name and status of the signatory.

6.3.5.5.2   The test report shall contain statements that the packaging prepared as for transport was tested in accordance with the appropriate requirements of this chapter and that the use of other packaging methods or components may render it invalid. A copy of the test report shall be available to the competent authority.

Chapter 6.4

Provisions for the construction, testing and approval of packages and material of class 7

Note:	This chapter includes provisions which apply to the construction, testing and approval of certain packages and material only when transported by air. Whilst these provisions do not apply to packages/material transported by sea, the provisions are reproduced for information/identification purposes, since such packages/material, designed, tested and approved for air transport, may also be transported by sea.

6.4.1   [reserved]

6.4.2   General provisions

6.4.2.1   The package shall be so designed in relation to its mass, volume and shape that it can be easily and safely transported. In addition, the package shall be so designed that it can be properly secured in or on the conveyance during transport.

6.4.2.2   The design shall be such that any lifting attachments on the package will not fail when used in the intended manner and that, if failure of the attachments shall occur, the ability of the package to meet other provisions of this Code would not be impaired. The design shall take account of appropriate safety factors to cover snatch lifting.

6.4.2.3   Attachments and any other features on the outer surface of the package which could be used to lift it shall be designed either to support its mass in accordance with the provisions of 6.4.2.2 or shall be removable or otherwise rendered incapable of being used during transport.

6.4.2.4   As far as practicable, the packaging shall be so designed and finished that the external surfaces are free from protruding features and can be easily decontaminated.

6.4.2.5   As far as practicable, the outer layer of the package shall be so designed as to prevent the collection and the retention of water.

6.4.2.6   Any features added to the package at the time of transport which are not part of the package shall not reduce its safety.

6.4.2.7   The package shall be capable of withstanding the effects of any acceleration, vibration or vibration resonance which may arise under routine conditions of transport without any deterioration in the effectiveness of the closing devices on the various receptacles or in the integrity of the package as a whole. In particular, nuts, bolts and other securing devices shall be so designed as to prevent them from becoming loose or being released unintentionally, even after repeated use.

6.4.2.8   The materials of the packaging and any components or structures shall be physically and chemically compatible with each other and with the radioactive contents. Account shall be taken of their behaviour under irradiation.

6.4.2.9   All valves through which the radioactive contents could otherwise escape shall be protected against unauthorized operation.

6.4.2.10   The design of the package shall take into account ambient temperatures and pressures that are likely to be encountered in routine conditions of transport.

6.4.2.11   For radioactive material having other dangerous properties, the package design shall take into account those properties; see 4.1.9.1.5, 2.0.3.1 and 2.0.3.2.

6.4.2.12   Manufacturers and subsequent distributors of packagings shall provide information regarding procedures to be followed and a description of the types and dimensions of closures (including required gaskets) and any other components needed to ensure that packages as presented for transport are capable of passing the applicable performance tests of this chapter.

6

6.4.3 Additional provisions for packages transported by air

6.4.3.1 For packages to be transported by air, the temperature of the accessible surfaces shall not exceed 50°C at an ambient temperature of 38°C with no account taken for insolation.

6.4.3.2 Packages to be transported by air shall be so designed that, if they were exposed to ambient temperatures ranging from –40°C to +55°C, the integrity of containment would not be impaired.

6.4.3.3 Packages containing radioactive material, to be transported by air, shall be capable of withstanding, without leakage, an internal pressure which produces a pressure differential of not less than maximum normal operating pressure plus 95 kPa.

6.4.4 Provisions for excepted packages

An excepted package shall be designed to meet the provisions specified in 6.4.2 and, in addition, shall meet the provisions of 6.4.3 if carried by air.

6.4.5 Provisions for industrial packages

6.4.5.1 A Type IP-1 package shall be designed to meet the provisions specified in 6.4.2 and 6.4.7.2, and, in addition, shall meet the provisions of 6.4.3 if carried by air.

6.4.5.2 A package, to be qualified as a Type IP-2 package, shall be designed to meet the provisions for Type IP-1 as specified in 6.4.5.1 and, in addition, if it were subjected to the tests specified in 6.4.15.4 and 6.4.15.5, it would prevent:

.1 loss or dispersal of the radioactive contents, and

.2 more than a 20% increase in the maximum radiation level at any external surface of the package.

6.4.5.3 A package, to be qualified as a Type IP-3 package, shall be designed to meet the provisions for Type IP-1 as specified in 6.4.5.1 and, in addition, the provisions specified in 6.4.7.2–6.4.7.15.

6.4.5.4 Alternative provisions for Type IP-2 and Type IP-3 packages

6.4.5.4.1 Packages may be used as Type IP-2 package provided that:

.1 they satisfy the provisions for Type IP-1 specified in 6.4.5.1;

.2 they are designed to satisfy the provisions for packing group I or II in chapter 6.1 of this Code; and

.3 when subjected to the tests for UN packing group I or II in chapter 6.1, they would prevent:

    (i) loss or dispersal of the radioactive contents; and

    (ii) more than a 20% increase in the maximum radiation level at any external surface of the package.

6.4.5.4.2 Portable tanks may also be used as Type IP-2 or Type IP-3 packages provided that:

.1 they satisfy the provisions for Type IP-1 specified in 6.4.5.1;

.2 they are designed to satisfy the provisions of chapter 6.7 of this Code, and are capable of withstanding a test pressure of 265 kPa; and

.3 they are designed so that any shielding which is provided shall be capable of withstanding the static and dynamic stresses resulting from handling and routine conditions of transport and of preventing an increase of more than 20% in the maximum radiation level at any external surface of the portable tanks.

6.4.5.4.3 Tanks, other than portable tanks, may also be used as Type IP-2 or Type IP-3 packages for transporting LSA-I and LSA-II liquids and gases as prescribed in the table under 4.1.9.2.4, provided that:

.1 they satisfy the provisions of 6.4.5.1;

.2 they are designed to satisfy the provisions prescribed in regional or national regulations for the transport of dangerous goods and are capable of withstanding a test pressure of 265 kPa; and

.3 they are designed so that any additional shielding which is provided shall be capable of withstanding the static and dynamic stresses resulting from handling and routine conditions of transport and of preventing an increase of more than 20% in the maximum radiation level at any external surface of the tanks.

6.4.5.4.4 Freight containers of a permanent enclosed character may also be used as Type IP-2 or Type IP-3 packages provided that:

.1 the radioactive contents are restricted to solid materials;

.2 they satisfy the provisions for Type IP-1 specified in 6.4.5.1; and

.3 they are designed to conform to the standards prescribed in the International Organization for Standardization document ISO 1496-1:1990(E), "Series 1 Freight Containers – Specifications and Testing – Part 1: General Cargo Containers", excluding dimensions and ratings. They shall be designed such that, if subjected to the tests prescribed in that document and the accelerations occurring during routine conditions of transport, they would prevent:

.1 loss or dispersal of the radioactive contents; and

.2 more than a 20% increase in the maximum radiation level at any external surface of the package.

6.4.5.4.5 Metal intermediate bulk containers may also be used as Type IP-2 or Type IP-3 packages provided that:

.1 they satisfy the provisions for Type IP-1 specified in 6.4.5.1; and

.2 they are designed to satisfy the provisions of chapter 6.5 of this Code for packing group I or II, and if they were subjected to the tests prescribed in that chapter, but with the drop test conducted in the most damaging orientation, they would prevent:

.1 loss or dispersal of the radioactive contents; and

.2 more than a 20% increase in the maximum radiation level at any external surface of the package.

6.4.6 Provisions for packages containing uranium hexafluoride

6.4.6.1 Packages designed to contain uranium hexafluoride shall meet the requirements prescribed elsewhere in this Code which pertain to the radioactive and fissile properties of the material. Except as allowed in 6.4.6.4, uranium hexafluoride in quantities of 0.1 kg or more shall also be packaged and transported in accordance with ISO 7195:1993(E), "Packaging of uranium hexafluoride (UF_6) for transport", and the provisions of 6.4.6.2–6.4.6.3.

6.4.6.2 Each package designed to contain 0.1 kg or more of uranium hexafluoride shall be designed so that it would meet the following provisions:

.1 withstand, without leakage and without unacceptable stress, as specified in ISO 7195:1993(E), the structural test as specified in 6.4.21;

.2 withstand, without loss or dispersal of the uranium hexafluoride, the free drop test specified in 6.4.15.4; and

.3 withstand, without rupture of the containment system, the thermal test specified in 6.4.17.3.

6.4.6.3 Packages designed to contain 0.1 kg or more of uranium hexafluoride shall not be provided with pressure relief devices.

6.4.6.4 Subject to the approval of the competent authority, packages designed to contain 0.1 kg or more of uranium hexafluoride may be transported if:

(a) the packages are designed to international or national standards other than ISO 7195:1993, provided an equivalent level of safety is maintained;

(b) the packages are designed to withstand, without leakage and without unacceptable stress, a test pressure of less than 2.76 MPa as specified in 6.4.21; or

(c) for packages designed to contain 9000 kg or more of uranium hexafluoride, the packages do not meet the requirement of 6.4.6.2.3.

In all other respects, the provisions of 6.4.6.1 to 6.4.6.3 shall be satisfied.

6.4.7 Provisions for Type A packages

6.4.7.1 Type A packages shall be designed to meet the general provisions of 6.4.2, shall meet the provisions of 6.4.3 if carried by air, and shall meet the provisions of 6.4.7.2–6.4.7.17.

6.4.7.2 The smallest overall external dimension of the package shall not be less than 10 cm.

6.4.7.3 The outside of the package shall incorporate a feature, such as a seal, which is not readily breakable and which, while intact, will be evidence that it has not been opened.

6

6.4.7.4   Any tie-down attachments on the package shall be so designed that, under normal and accident conditions of transport, the forces in those attachments shall not impair the ability of the package to meet the provisions of this Code.

6.4.7.5   The design of the package shall take into account temperatures ranging from –40°C to +70°C for the components of the packaging. Attention shall be given to freezing temperatures for liquids and to the potential degradation of packaging materials within the given temperature range.

6.4.7.6   The design and manufacturing techniques shall be in accordance with national or international standards, or other provisions, acceptable to the competent authority.

6.4.7.7   The design shall include a containment system securely closed by a positive fastening device which cannot be opened unintentionally or by a pressure which may arise within the package.

6.4.7.8   Special form radioactive material may be considered as a component of the containment system.

6.4.7.9   If the containment system forms a separate unit of the package, it shall be capable of being securely closed by a positive fastening device which is independent of any other part of the packaging.

6.4.7.10   The design of any component of the containment system shall take into account, where applicable, the radiolytic decomposition of liquids and other vulnerable materials and the generation of gas by chemical reaction and radiolysis.

6.4.7.11   The containment system shall retain its radioactive contents under a reduction of ambient pressure to 60 kPa.

6.4.7.12   All valves, other than pressure relief valves, shall be provided with an enclosure to retain any leakage from the valve.

6.4.7.13   A radiation shield which encloses a component of the package specified as a part of the containment system shall be so designed as to prevent the unintentional release of that component from the shield. Where the radiation shield and such component within it form a separate unit, the radiation shield shall be capable of being securely closed by a positive fastening device which is independent of any other packaging structure.

6.4.7.14   A package shall be so designed that, if it were subjected to the tests specified in 6.4.15, it would prevent:

.(a) loss or dispersal of the radioactive contents; and

.(b) more than a 20% increase in the maximum radiation level at any external surface of the package.

6.4.7.15   The design of a package intended for liquid radioactive material shall make provision for ullage to accommodate variations in the temperature of the contents, dynamic effects and filling dynamics.

Type A packages to contain liquids

6.4.7.16   A Type A package designed to contain liquid radioactive material shall, in addition:

.1 be adequate to meet the conditions specified in 6.4.7.14(a) above if the package is subjected to the tests specified in 6.4.16; and

.2 either

(i) be provided with sufficient absorbent material to absorb twice the volume of the liquid contents. Such absorbent material must be suitably positioned so as to contact the liquid in the event of leakage; or

(ii) be provided with a containment system composed of primary inner and secondary outer containment components designed to ensure retention of the liquid contents within the secondary outer containment components even if the primary inner components leak.

Type A packages to contain gas

6.4.7.17   A package designed for gases shall prevent loss or dispersal of the radioactive contents if the package were subjected to the tests specified in 6.4.16. A Type A package designed for tritium gas or for noble gases shall be excepted from this requirement.

6.4.8   Provisions for Type B(U) packages

6.4.8.1   Type B(U) packages shall be designed to meet the provisions specified in 6.4.2, shall also meet the provisions of 6.4.3 if carried by air, and shall meet the provisions of 6.4.7–6.4.8, except as specified in 6.4.7.14(a), and, in addition, the provisions specified in 6.4.8.2–6.4.8.15.

6.4.8.2   A package shall be so designed that, under the ambient conditions specified in 6.4.8.5 and 6.4.8.6, heat generated within the package by the radioactive contents shall not, under normal conditions of transport, as demonstrated by the tests in 6.4.15, adversely affect the package in such a way that it would fail to meet the applicable provisions for containment and shielding if left unattended for a period of one week. Particular attention shall be paid to the effects of heat, which may:

(a)   alter the arrangement, the geometrical form or the physical state of the radioactive contents or, if the radioactive material is enclosed in a can or receptacle (for example, clad fuel elements), cause the can, receptacle or radioactive material to deform or melt; or

(b)   lessen the efficiency of the packaging through differential thermal expansion or cracking or melting of the radiation shielding material; or

(c)   in combination with moisture, accelerate corrosion.

6.4.8.3   A package shall be so designed that, under the ambient condition specified in 6.4.8.5 and in the absence of insolation, the temperature of the accessible surfaces of a package shall not exceed 50°C, unless the package is transported under exclusive use.

6.4.8.4   Except as required in 6.4.3.1 for a package transported by air, the maximum temperature of any surface readily accessible during transport of a package under exclusive use shall not exceed 85°C in the absence of insolation under the ambient conditions specified in 6.4.8.5. Account may be taken of barriers or screens intended to give protection to persons without the need for the barriers or screens being subject to any test.

6.4.8.5   The ambient temperature shall be assumed to be 38°C.

6.4.8.6   The solar insolation conditions shall be assumed to be as specified in the table hereunder.

Insolation data

Case	Form and location of surface	Insolation for 12 hours per day (W/m^2)
1	Flat surfaces transported horizontally – downward facing	0
2	Flat surfaces transported horizontally – upward facing	800
3	Surfaces transported vertically	200*
4	Other downward-facing (not horizontal) surfaces	200*
5	All other surfaces	400*

* Alternatively, a sine function may be used, with an absorption coefficient adopted and the effects of possible reflection from neighbouring objects neglected.

6.4.8.7   A package which includes thermal protection for the purpose of satisfying the provisions of the thermal test specified in 6.4.17.3 shall be so designed that such protection will remain effective if the package is subjected to the tests specified in 6.4.15 and 6.4.17.2(a) and (b) or 6.4.17.2(b) and (c), as appropriate. Any such protection on the exterior of the package shall not be rendered ineffective by ripping, cutting, skidding, abrasion or rough handling.

6.4.8.8   A package shall be so designed that, if it were subjected to:

.1   the tests specified in 6.4.15, it would restrict the loss of radioactive contents to not more than $10^{-6}A_2$ per hour; and

.2   the tests specified in 6.4.17.1, 6.4.17.2(b), 6.4.17.3 and 6.4.17.4 and the tests in:

(i)   6.4.17.2(c), when the package has a mass not greater than 500 kg, an overall density not greater than 1000 kg/m^3 based on the external dimensions, and radioactive contents greater than $1000A_2$ not as special form radioactive material, or

(ii)   6.4.17.2(a), for all other packages,

it would meet the following provisions:

• retain sufficient shielding to ensure that the radiation level at 1 m from the surface of the package would not exceed 10 mSv/h with the maximum radioactive contents which the package is designed to contain; and

• restrict the accumulated loss of radioactive contents in a period of one week to not more than $10A_2$ for krypton-85 and not more than A_2 for all other radionuclides.

6

Where mixtures of different radionuclides are present, the provisions of 2.7.2.2.4–2.7.2.2.6 shall apply except that for krypton-85 an effective $A_2(i)$ value equal to $10A_2$ may be used. For case (.1) above, the assessment shall take into account the external contamination limits of 4.ˉ.9.1.2.

6.4.8.9   A package for radioactive contents with activity greater than 10^5A_2 shall be so designed that, if it were subjected to the enhanced water immersion test specified in 6.4.18, there would be no rupture of the containment system.

6.4.8.10   Compliance with the permitted activity release limits shall depend neither upon filters nor upon a mechanical cooling system.

6.4.8.11   A package shall not include a pressure relief system from the containment system which would allow the release of radioactive material to the environment under the conditions of the tests specified in 6.4.15 and 6.4.17.

6.4.8.12   A package shall be so designed that, if it were at the maximum normal operating pressure and it were subjected to the tests specified in 6.4.15 and 6.4.17, the level of strains in the containment system would not attain values which would adversely affect the package in such a way that it would fail to meet the applicable provisions.

6.4.8.13   A package shall not have a maximum normal operating pressure in excess of a gauge pressure of 700 kPa.

6.4.8.14   A package containing low dispersible radioactive material shall be so designed that any features added to the low dispersible radioactive material that are not part of it, or any internal components of the packaging, shall not adversely affect the performance of the low dispersible radioactive material.

6.4.8.15   A package shall be designed for an ambient temperature range from -40°C to $+38^\circ$C.

6.4.9   Provisions for Type B(M) packages

6.4.9.1   Type B(M) packages shall meet the provisions for Type B(U) packages specified in 6.4.8.1, except that, for packages to be transported solely within a specified country or solely between specified countries, conditions other than those given in 6.4.7.5, 6.4.8.4, 6.4.8.5 and 6.4.8.8–6.4.8.15 above may be assumed, with the approval of the competent authorities of these countries. Notwithstanding, the provisions for Type B(U) packages specified in 6.4.8.8–6.4.8.15 shall be met as far as practicable.

6.4.9.2   Intermittent venting of Type B(M) packages may be permitted during transport, provided that the operational controls for venting are acceptable to the relevant competent authorities.

6.4.10   Provisions for Type C packages

6.4.10.1   Type C packages shall be designed to meet the provisions specified in 6.4.2 and 6.4.3, and of 6.4.7.2–6.4.7.15, except as specified in 6.4.7.14, and of the provisions specified in 6.4.8.2–6.4.8.5, 6.4.8.9–6.4.8.15, and, in addition, of 6.4.10.2–6.4.10.4.

6.4.10.2   A package shall be capable of meeting the assessment criteria prescribed for tests in 6.4.8.7.2 and 6.4.8.11 after burial in an environment defined by a thermal conductivity of 0.33 W/m·K and a temperature of 38°C in the steady state. Initial conditions for the assessment shall assume that any thermal insulation of the package remains intact, the package is at the maximum normal operating pressure and the ambient temperature is 38°C.

6.4.10.3   A package shall be so designed that, if it were at the maximum normal operating pressure and subjected to:

(a) the tests specified in 6.4.15, it would restrict the loss of radioactive contents to not more than $10^{-6}A_2$ per hour; and

(b) the test sequences in 6.4.20.1, it would meet the following provisions:

     (i) retain sufficient shielding to ensure that the radiation level at 1 m from the surface of the package would not exceed 10 mSv/h with the maximum radioactive contents which the package is designed to contain; and

     (ii) restrict the accumulated loss of radioactive contents in a period of 1 week to not more than $10A_2$ for krypton-85 and not more than A_2 for all other radionuclides.

Where mixtures of different radionuclides are present, the provisions of 2.7.2.2.4–2.7.2.2.6 shall apply except that for krypton-85 an effective $A_2(i)$ value equal to $10A_2$ may be used. For case (a) above, the assessment shall take into account the external contamination limits of 4.1.9.1.2.

6.4.10.4   A package shall be so designed that there will be no rupture of the containment system following performance of the enhanced water immersion test specified in 6.4.18.

6.4.11   Provisions for packages containing fissile material

6.4.11.1   Fissile material shall be transported so as to:

(a) maintain subcriticality during normal and accident conditions of transport; in particular, the following contingencies shall be considered:

(i)   water leaking into or out of packages;

(ii) the loss of efficiency of built-in neutron absorbers or moderators;

(iii) rearrangement of the contents either within the package or as a result of loss from the package;

(iv) reduction of spaces within or between packages;

(v)   packages becoming immersed in water or buried in snow; and

(vi) temperature changes; and

(b) meet the provisions:

(i)   of 6.4.7.2 for packages containing fissile material;

(ii) prescribed elsewhere in this Code which pertain to the radioactive properties of the material; and

(iii) specified in 6.4.11.3–6.4.11.12, unless excepted by 6.4.11.2.

6.4.11.2   Fissile material meeting one of the provisions .1 to .4 of 2.7.2.3.5 is excepted from the requirement to be transported in packages that comply with 6.4.11.3–6.4.11.12 as well as the other provisions of this Code that apply to fissile material. Only one type of exception is allowed per consignment.

6.4.11.3   Where the chemical or physical form, isotopic composition, mass or concentration, moderation ratio or density, or geometric configuration is not known, the assessments of 6.4.11.7–6.4.11.12 shall be performed assuming that each parameter that is not known has the value which gives the maximum neutron multiplication consistent with the known conditions and parameters in these assessments.

6.4.11.4   For irradiated nuclear fuel, the assessments of 6.4.11.7–6.4.11.12 shall be based on an isotopic composition demonstrated to provide:

(a) the maximum neutron multiplication during the irradiation history, or

(b) a conservative estimate of the neutron multiplication for the package assessments. After irradiation, but prior to shipment, a measurement shall be performed to confirm the conservatism of the isotopic composition.

6.4.11.5   The package, after being subjected to the tests specified in 6.4.15, shall prevent the entry of a 10 cm cube.

6.4.11.6   The package shall be designed for an ambient temperature range of –40°C to +38°C unless the competent authority specifies ctherwise in the certificate of approval for the package design.

6.4.11.7   For a package in isolation, it shall be assumed that water can leak into or out of all void spaces of the package, including those within the containment system. However, if the design incorporates special features to prevent such leakage of water into or out of certain void spaces, even as a result of error, absence of leakage may be assumed in respect of those void spaces. Special features shall include the following:

(a) Multiple high-standard water barriers, each of which would remain watertight if the package were subject to the tests prescribed in 6.4.11.12(b), a high degree of quality control in the manufacture, maintenance and repair of packagings and tests to demonstrate the closure of each package before each shipment; or

(b) For packages containing uranium hexafluoride only, with maximum enrichment of 5 mass percent uranium-235:

(i)   packages where, following the tests prescribed in 6.4.11.12(b), there is no physical contact between the valve and any other component of the packaging other than at its original point of attachment and where, in addition, following the test prescribed in 6.4.17.3, the valves remain leaktight; and

(ii) a high degree of quality control in the manufacture, maintenance and repair of packagings coupled with tests to demonstrate closure of each package before each shipment.

6.4.11.8   It shall be assumed that the confinement system is closely reflected by at least 20 cm of water or such greater reflection as may additionally be provided by the surrounding material of the packaging. However, when it can be demonstrated that the confinement system remains within the packaging following the tests prescribed in 6.4.11.12(b), close reflection of the package by at least 20 cm of water may be assumed in 6.4.11.9(c).

6

6.4.11.9 The package shall be subcritical under the conditions of 6.4.11.7 and 6.4.11.8 and with the package conditions that result in the maximum neutron multiplication consistent with:

(a) routine conditions of transport (incident-free);

(b) the tests specified in 6.4.11.11(b);

(c) the tests specified in 6.4.11.12(b).

6.4.11.10 For packages to be transported by air:

(a) the package shall be subcritical under conditions consistent with the Type C package tests specified in 6.4.20.1 assuming reflection by at least 20 cm of water but no water in-leakage; and

(b) in the assessment of 6.4.11.9, allowance shall not be made for special features of 6.4.11.7 unless, following the Type C package tests specified in 6.4.20.1 and, subsequently, the water in-leakage test of 6.4.19.3, leakage of water into or out of the void spaces is prevented.

6.4.11.11 A number "*N*" shall be derived, such that five times "*N*" packages shall be subcritical for the arrangement and package conditions that provide the maximum neutron multiplication consistent with the following:

(a) there shall not be anything between the packages, and the package arrangement shall be reflected on all sides by at least 20 cm of water; and

(b) the state of the packages shall be their assessed or demonstrated condition if they had been subjected to the tests specified in 6.4.15.

6.4.11.12 A number "*N*" shall be derived, such that two times "*N*" packages shall be subcritical for the arrangement and package conditions that provide the maximum neutron multiplication consistent with the following:

(a) hydrogenous moderation between packages, and the package arrangement reflected on all sides by at least 20 cm of water; and

(b) the tests specified in 6.4.15 followed by whichever of the following is the more limiting:

(i) the tests specified in 6.4.17.2(b) and either 6.4.17.2(c), for packages having a mass not greater than 500 kg and an overall density not greater than 1000 kg/m^3 based on the external dimensions, or 6.4.17.2(a), for all other packages; followed by the test specified in 6.4.17.3 and completed by the tests specified in 6.4.19.1–6.4.19.3; or

(ii) the test specified in 6.4.17.4; and

(c) where any part of the fissile material escapes from the containment system following the tests specified in 6.4.11.12(b), it shall be assumed that fissile material escapes from each package in the array and all of the fissile material shall be arranged in the configuration and moderation that results in the maximum neutron multiplication with close reflection by at least 20 cm of water.

6.4.11.13 The criticality safety index (CSI) for packages containing fissile material shall be obtained by dividing the number 50 by the smaller of the two values of *N* derived in 6.4.11.11 and 6.4.11.12 (i.e. CSI = 50/*N*). The value of the criticality safety index may be zero, provided that an unlimited number of packages is subcritical (i.e. *N* is effectively equal to infinity in both cases).

6.4.12 Test procedures and demonstration of compliance

6.4.12.1 Demonstration of compliance with the performance standards required in 2.7.2.3.1.3, 2.7.2.3.1.4, 2.7.2.3.3.1, 2.7.2.3.3.2, 2.7.2.3.4.1, 2.7.2.3.4.2 and 6.4.2–6.4.11 shall be accomplished by any of the methods listed below or by a combination thereof.

(a) Performance of tests with specimens representing LSA-III material, or special form radioactive material, or low dispersible radioactive material or with prototypes or samples of the packaging, where the contents of the specimen or the packaging for the tests shall simulate as closely as practicable the expected range of radioactive contents and the specimen or packaging to be tested shall be prepared as presented for transport.

(b) Reference to previous satisfactory demonstrations of a sufficiently similar nature.

(c) Performance of tests with models of appropriate scale incorporating those features which are significant with respect to the item under investigation when engineering experience has shown results of such tests to be suitable for design purposes. When a scale model is used, the need for adjusting certain test parameters, such as penetrator diameter or compressive load, shall be taken into account.

(d) Calculation, or reasoned argument, when the calculation procedures and parameters are generally agreed to be reliable or conservative.

6.4.12.2   After the specimen, prototype or sample has been subjected to the tests, appropriate methods of assessment shall be used to assure that the provisions of this chapter have been fulfilled in compliance with the performance and acceptance standards prescribed in this chapter (see 2.7.2.3.1.3, 2.7.2.3.1.4, 2.7.2.3.3.1, 2.7.2.3.3.2, 2.7.2.3.4.1, 2.7.2.3.4.2 and 6.4.2–6.4.11).

6.4.12.3   All specimens shall be inspected before testing in order to identify and record faults or damage, including the following:

(a) divergence from the design;

(b) defects in manufacture;

(c) corrosion or other deterioration; and

(d) distortion of features.

The containment system of the package shall be clearly specified. The external features of the specimen shall be clearly identified so that reference may be made simply and clearly to any part of such specimen.

6.4.13   Testing the integrity of the containment system and shielding and evaluating criticality safety

After each of the applicable tests specified in 6.4.15–6.4.21:

(a) faults and damage shall be identified and recorded;

(b) it shall be determined whether the integrity of the containment system and shielding has been retained to the extent required in this chapter for the package under test; and

(c) for packages containing fissile material, it shall be determined whether the assumptions and conditions used in the assessments required by 6.4.11.1–6.4.11.12 for one or more packages are valid.

6.4.14   Target for drop tests

The target for the drop tests specified in 2.7.2.3.3.5, 6.4.15.4, 6.4.16(a), 6.4.17.2 and 6.4.20.2 shall be a flat, horizontal surface of such a character that any increase in its resistance to displacement or deformation upon impact by the specimen would not significantly increase the damage to the specimen.

6.4.15   Test for demonstrating ability to withstand normal conditions of transport

6.4.15.1   The tests are: the water spray test, the free drop test, the stacking test and the penetration test. Specimens of the package shall be subjected to the free drop test, the stacking test and the penetration test, preceded in each case by the water spray test. One specimen may be used for all the tests, provided that the provisions of 6.4.15.2 are fulfilled.

6.4.15.2   The time interval between the conclusion of the water spray test and the succeeding test shall be such that the water has soaked in to the maximum extent, without appreciable drying of the exterior of the specimen. In the absence of any evidence to the contrary, this interval shall be taken to be two hours if the water spray is applied from four directions simultaneously. No time interval shall elapse, however, if the water spray is applied from each of the four directions consecutively.

6.4.15.3   Water spray test: The specimen shall be subjected to a water spray test that simulates exposure to rainfall of approximately 5 cm per hour for at least one hour.

6.4.15.4   Free drop test: The specimen shall drop onto the target so as to suffer maximum damage in respect of the safety features to be tested.

(a) The height of drop measured from the lowest point of the specimen to the upper surface of the target shall be not less than the distance specified in the table hereunder for the applicable mass. The target shall be as defined in 6.4.14.

(b) For rectangular fibreboard or wood packages not exceeding a mass of 50 kg, a separate specimen shall be subjected to a free drop onto each corner from a height of 0.3 m.

(c) For cylindrical fibreboard packages not exceeding a mass of 100 kg, a separate specimen shall be subjected to a free drop onto each of the quarters of each rim from a height of 0.3 m.

6

Free drop distance for testing packages to normal conditions of transport

Package mass (kg)	Free drop distance (m)
Package mass < 5 000	1.2
5 000 ≤ Package mass < 10 000	0.9
10 000 ≤ Package mass < 15 000	0.6
15 000 ≤ Package mass	0.3

6.4.15.5 Stacking test: Unless the shape of the packaging effectively prevents stacking, the specimen shall be subjected, for a period of 24 hours, to a compressive load equal to the greater of the following:

(a) The equivalent of 5 times the mass of the actual package;

(b) The equivalent of 13 kPa multiplied by the vertically projected area of the package.

The load shall be applied uniformly to two opposite sides of the specimen, one of which shall be the base on which the package would typically rest.

6.4.15.6 Penetration test: The specimen shall be placed on a rigid, flat, horizontal surface which will not move significantly while the test is being carried out.

(a) A bar of 3.2 cm in diameter with a hemispherical end and a mass of 6 kg shall be dropped and directed to fall, with its longitudinal axis vertical, onto the centre of the weakest part of the specimen, so that, if it penetrates sufficiently far, it will hit the containment system. The bar shall not be significantly deformed by the test performance.

(b) The height of drop of the bar measured from its lower end to the intended point of impact on the upper surface of the specimen shall be 1 m.

6.4.16 Additional tests for Type A packages designed for liquids and gases

A specimen or separate specimens shall be subjected to each of the following tests unless it can be demonstrated that one test is more severe for the specimen in question than the other, in which case one specimen shall be subjected to the more severe test.

(a) Free drop test: The specimen shall drop onto the target so as to suffer the maximum damage in respect of containment. The height of the drop measured from the lowest part of the specimen to the upper surface of the target shall be 9 m. The target shall be as defined in 6.4.14.

(b) Penetration test: The specimen shall be subjected to the test specified in 6.4.15.6 except that the height of drop shall be increased to 1.7 m from the 1 m specified in 6.4.15.6(b).

6.4.17 Tests for demonstrating ability to withstand accident conditions of transport

6.4.17.1 The specimen shall be subjected to the cumulative effects of the tests specified in 6.4.17.2 and 6.4.17.3, in that order. Following these tests, either this specimen or a separate specimen shall be subjected to the effect(s) of the water immersion test(s) as specified in 6.4.17.4 and, if applicable, 6.4.18.

6.4.17.2 Mechanical test: The mechanical test consists of three different drop tests. Each specimen shall be subjected to the applicable drops as specified in 6.4.8.7 or 6.4.11.12. The order in which the specimen is subjected to the drops shall be such that, on completion of the mechanical test, the specimen shall have suffered such damage as will lead to the maximum damage in the thermal test which follows.

(a) For drop I, the specimen shall drop onto the target so as to suffer the maximum damage, and the height of the drop measured from the lowest point of the specimen to the upper surface of the target shall be 9 m. The target shall be as defined in 6.4.14.

(b) For drop II, the specimen shall drop so as to suffer the maximum damage onto a bar rigidly mounted perpendicularly on the target. The height of the drop measured from the intended point of impact of the specimen to the upper surface of the bar shall be 1 m. The bar shall be of solid mild steel of circular section, (15.0 ± 0.5) cm in diameter and 20 cm long unless a longer bar would cause greater damage, in which case a bar of sufficient length to cause maximum damage shall be used. The upper end of the bar shall be flat and horizontal with its edge rounded off to a radius of not more than 6 mm. The target on which the bar is mounted shall be as described in 6.4.14.

(c) For drop III, the specimen shall be subjected to a dynamic crush test by positioning the specimen on the target so as to suffer maximum damage by the drop of a 500 kg mass from 9 m onto the specimen. The mass shall consist of a solid mild steel plate 1 m by 1 m and shall fall in a horizontal attitude. The height of the drop shall be measured from the underside of the plate to the highest point of the specimen. The target on which the specimen rests shall be as defined in 6.4.14.

6.4.17.3 Thermal test: The specimen shall be in thermal equilibrium under conditions of an ambient temperature of 38°C, subject to the solar insolation conditions specified in the table under 6.4.8.6 and subject to the design maximum rate of internal heat generation within the package from the radioactive contents. Alternatively, any of these parameters are allowed to have different values prior to and during the test, providing due account is taken of them in the subsequent assessment of package response.

The thermal test shall then consist of:

(a) exposure of a specimen for a period of 30 minutes to a thermal environment which provides a heat flux at least equivalent to that of a hydrocarbon fuel/air fire in sufficiently quiescent ambient conditions to give a minimum average flame emissivity coefficient of 0.9 and an average temperature of at least 800°C, fully engulfing the specimen, with a surface absorptivity coefficient of 0.8 or that value which the package may be demonstrated to possess if exposed to the fire specified, followed by;

(b) exposure of the specimen to an ambient temperature of 38°C, subject to the solar insolation conditions specified in the table under 6.4.8.6 and subject to the design maximum rate of internal heat generation within the package by the radioactive contents, for a sufficient period to ensure that temperatures in the specimen are everywhere decreasing and/or are approaching initial steady-state conditions. Alternatively, any of these parameters are allowed to have different values following cessation of heating, providing due account is taken of them in the subsequent assessment of package response.

During and following the test, the specimen shall not be artificially cooled and any combustion of materials of the specimen shall be permitted to proceed naturally.

6.4.17.4 Water immersion test: The specimen shall be immersed under a head of water of at least 15 m for a period of not less than eight hours in the attitude which will lead to maximum damage. For demonstration purposes, an external gauge pressure of at least 150 kPa shall be considered to meet these conditions.

6.4.18 **Enhanced water immersion test for Type B(U) and Type B(M) packages containing more than $10^5 A_2$ and Type C packages**

Enhanced water immersion test: The specimen shall be immersed under a head of water of at least 200 m for a period of not less than one hour. For demonstration purposes, an external gauge pressure of at least 2 MPa shall be considered to meet these conditions.

6.4.19 **Water leakage test for packages containing fissile material**

6.4.19.1 Packages for which water in-leakage or out-leakage to the extent which results in greatest reactivity has been assumed for purposes of assessment under 6.4.11.7–6.4.11.12 shall be excepted from the test.

6.4.19.2 Before the specimen is subjected to the water leakage test specified below, it shall be subjected to the tests in 6.4.17.2(b), and either 6.4.17.2(a) or (c) as required by 6.4.11.12, and the test specified in 6.4.17.3.

6.4.19.3 The specimen shall be immersed under a head of water of at least 0.9 m for a period of not less than eight hours and in the attitude for which maximum leakage is expected.

6.4.20 **Tests for Type C packages**

6.4.20.1 Specimens shall be subjected to the effects of each of the following test sequences in the orders specified:

(a) the tests specified in 6.4.17.2(a), 6.4.17.2(c), 6.4.20.2 and 6.4.20.3; and

(b) the test specified in 6.4.20.4.

Separate specimens are allowed to be used for each of the sequences (a) and (b).

6.4.20.2 Puncture/tearing test: The specimen shall be subjected to the damaging effects of a solid probe made of mild steel. The orientation of the probe to the surface of the specimen shall be as to cause maximum damage at the conclusion of the test sequence specified in 6.4.20.1(a).

(a) The specimen, representing a package having a mass less than 250 kg, shall be placed on a target and subjected to a probe having a mass of 250 kg falling from a height of 3 m above the intended impact point. For this test, the probe shall be a 20 cm diameter cylindrical bar with the striking end forming a frustum of a right circular cone with the following dimensions: 30 cm height and 2.5 cm in diameter at the top with its edge rounded off to a radius of not more than 6 mm. The target on which the specimen is placed shall be as specified in 6.4.14.

(b) For packages having a mass of 250 kg or more, the base of the probe shall be placed on a target and the specimen dropped onto the probe. The height of the drop, measured from the point of impact with the specimen to the upper surface of the probe, shall be 3 m. For this test, the probe shall have the same properties and dimensions as specified in (a) above, except that the length and mass of the probe shall be such as to incur maximum damage to the specimen. The target on which the base of the probe is placed shall be as specified in 6.4.14.

6.4.20.3   Enhanced thermal test: The conditions for this test shall be as specified in 6.4.17.3, except that the exposure to the thermal environment shall be for a period of 60 minutes.

6.4.20.4   Impact test: The specimen shall be subject to an impact on a target at a velocity of not less than 90 m/s, at such an orientation as to suffer maximum damage. The target shall be as defined in 6.4.14, except that the target surface may be at any orientation provided that the surface is normal to the specimen path.

6.4.21   Tests for packagings designed to contain uranium hexafluoride

Specimens that comprise or simulate packagings designed to contain 0.1 kg or more of uranium hexafluoride shall be tested hydraulically at an internal pressure of at least 1 38 MPa but, when the test pressure is less than 2.76 MPa, the design will require multilateral approval. For retesting packagings, any other equivalent non-destructive testing may be applied, subject to multilateral approval.

6.4.22   Approvals of package designs and materials

6.4.22.1   The approval of designs for packages containing 0.1 kg or more of uranium hexafluoride requires that:

(a) Each design that meets the provisions of 6.4.6.4 shall require multilateral approval;

(b) Each design that meets the provisions of 6.4.6.1 to 6.4.6.3 shall require unilateral approval by the competent authority of the country of origin of the design, unless multilateral approval is otherwise required by this Code.

6.4.22.2   Each Type B(U) and Type C package design will require unilateral approval, except that:

(a) a package design for fissile material which is also subject to 6.4.22.4, 6.4.23.7 and 5.1.5.2.1 will require multilateral approval; and

(b) a Type B(U) package design for low dispersible radioactive material will require multilateral approval.

6.4.22.3   Each Type B(M) package design, including those for fissile material which are also subject to 6.4.22.4, 6.4.23.7 and 5.1.5.2.1 and those for low dispersible radioactive material, will require multilateral approval.

6.4.22.4   Each package design for fissile material which is not excepted according to 6.4.11.2 from the provisions that apply specifically to packages containing fissile material will require multilateral approval.

6.4.22.5   The design for special form radioactive material will require unilateral approval. The design for low dispersible radioactive material will require multilateral approval (see also 6.4.23.8).

6.4.23   Applications for approval and approvals for radioactive material transport

6.4.23.1   [reserved]

6.4.23.2   An application for shipment approval shall include:

(a) the period of time, related to the shipment, for which the approval is sought;

(b) the actual radioactive contents, the expected modes of transport, the type of conveyance, and the probable or proposed route; and

(c) the details of how the precautions and administrative or operational controls referred to in the package design approval certificates issued under 5.1.5.2.1 are to be put into effect.

6.4.23.3   An application for approval of shipments under special arrangement shall include all the information necessary to satisfy the competent authority that the overall level of safety in transport is at least equivalent to that which would be provided if all the applicable provisions of this Code had been met. The application shall also include:

(a) a statement of the respects in which, and of the reasons why, the shipment cannot be made in full accordance with the applicable provisions; and

(b) a statement of any special precautions or special administrative or operational controls which are to be employed during transport to compensate for the failure to meet the applicable provisions.

6.4.23.4 An application for approval of Type B(U) or Type C package design shall include:

(a) a detailed description of the proposed radioactive contents with reference to their physical and chemical states and the nature of the radiation emitted;

(b) a detailed statement of the design, including complete engineering drawings and schedules of materials and methods of manufacture;

(c) a statement of the tests which have been done and their results, or evidence based on calculative methods or other evidence that the design is adequate to meet the applicable provisions;

(d) the proposed operating and maintenance instructions for the use of the packaging;

(e) if the package is designed to have a maximum normal operating pressure in excess of 100 kPa gauge, a specification of the materials of manufacture of the containment system, the samples to be taken, and the tests to be made;

(f) where the proposed radioactive contents are irradiated fuel, a statement and a justification of any assumption in the safety analysis relating to the characteristics of the fuel and a description of any pre-shipment measurement required by 6.4.11.4(b);

(g) any special stowage provisions necessary to ensure the safe dissipation of heat from the package, considering the various modes of transport to be used and type of conveyance or freight container;

(h) a reproducible illustration, not larger than 21 cm by 30 cm, showing the make-up of the package; and

(i) a specification of the applicable quality-assurance programme as required in 1.5.3.1.

6.4.23.5 An application for approval of a Type B(M) package design shall include, in addition to the information required in 6.4.23.4 for Type B(U) packages:

(a) a list of the provisions specified in 6.4.7.5, 6.4.8.4, 6.4.8.5 and 6.4.8.8–6.4.8.15 with which the package does not conform;

(b) any proposed supplementary operational controls to be applied during transport not regularly provided for in this Code, but which are necessary to ensure the safety of the package or to compensate for the deficiencies listed in (a) above;

(c) a statement relative to any restrictions on the mode of transport and to any special loading, carriage, unloading or handling procedures; and

(d) the range of ambient conditions (temperature, solar radiation) which are expected to be encountered during transport and which have been taken into account in the design.

6.4.23.6 The application for approval of designs for packages containing 0.1 kg or more of uranium hexafluoride shall include all information necessary to satisfy the competent authority that the design meets the provisions of 6.4.6.1, and a specification of the applicable quality-assurance programme as required by 1.5.3.1.

6.4.23.7 An application for a fissile package approval shall include all information necessary to satisfy the competent authority that the design meets the provisions of 6.4.11.1, and a specification of the applicable quality-assurance programme as required in 1.5.3.1.

6.4.23.8 An application for approval of design for special form radioactive material and design for low dispersible radioactive material shall include:

(a) a detailed description of the radioactive material or, if a capsule, the contents; particular reference shall be made to both physical and chemical states;

(b) a detailed statement of the design of any capsule to be used;

(c) a statement of the tests which have been done and their results, or evidence based on calculative methods to show that the radioactive material is capable of meeting the performance standards, or other evidence that the special form radioactive material or low dispersible radioactive material meets the applicable provisions of this Code;

(d) a specification of the applicable quality-assurance programme as required in 1.5.3.1; and

(e) any proposed pre-shipment actions for use in the consignment of special form radioactive material or low dispersible radioactive material.

6

6.4.23.9   Each approval certificate issued by a competent authority shall be assigned an identification mark. The mark shall be of the following generalized type:

VRI/number/type code

(a) Except as provided in 6.4.23.10(b), "VRI" represents the international vehicle registration identification code of the country issuing the certificate.*

(b) The number shall be assigned by the competent authority, and shall be unique and specific with regard to the particular design or shipment. The shipment approval identification mark shall be clearly related to the design approval identification mark.

(c) The following type codes shall be used, in the order listed, to indicate the types of approval certificates issued:

AF	Type A package design for fissile material
B(U)	Type B(U) package design ("B(U)F" if for fissile material)
B(M)	Type B(M) package design ("B(M)F" if for fissile material)
C	Type C package design ("CF" if for fissile material)
IF	industrial package design for fissile material
S	special form radioactive material
LD	low dispersible radioactive material
T	shipment
X	special arrangement.

In the case of package designs for non-fissile or fissile excepted uranium hexafluoride, where none of the above codes apply, then the following type codes shall be used:

H(U)	unilateral approval
H(M)	multilateral approval

(d) For package design and special form radioactive material approval certificates, other than those issued under the provisions of 6.4.24.2–6.4.24.4, and for low dispersible radioactive material approval certificates, the symbols "-96" shall be added to the type code.

6.4.23.10   These type codes shall be applied as follows:

(a) Each certificate and each package shall bear the appropriate identification mark, comprising the symbols prescribed in 6.4.23.9(a), (b), (c) and (d) above, except that, for packages, only the applicable design type codes, including, if applicable, the symbols '-96', shall appear following the second stroke; that is, the 'T' or 'X' shall not appear in the identification marking on the package. Where the design approval and shipment approval are combined, the applicable type codes do not need to be repeated. For example:

A/132/B(M)F-96: A Type B(M) package design approved for fissile material, requiring multilateral approval, for which the competent authority of Austria has assigned the design number 132 (to be marked on both the package and on the package design approval certificate);

A/132/B(M)F-96T: The shipment approval issued for a package bearing the identification mark elaborated above (to be marked on the certificate only);

A/137/X: A special arrangement approval issued by the competent authority of Austria, to which the number 137 has been assigned (to be marked on the certificate only);

A/139/IF-96: An Industrial package design for fissile material approved by the competent authority of Austria, to which package design number 139 has been assigned (to be marked on both the package and on the package design approval certificate); and

A/145/H(U)-96: A package design for fissile excepted uranium hexafluoride approved by the competent authority of Austria, to which package design number 145 has been assigned (to be marked on both the package and on the package design approval certificate);

---

* See Convention on Road Traffic, Vienna, 1968.

(b) Where multilateral approval is effected by validation according to 6.4.23.16, only the identification mark issued by the country of origin of the design or shipment shall be used. Where multilateral approval is effected by issue of certificates by successive countries, each certificate shall bear the appropriate identification mark and the package whose design was so approved shall bear all appropriate identification marks. For example:

A/132/B(M)F-96

CH/28/B(M)F-96

would be the identification mark of a package which was originally approved by Austria and was subsequently approved, by separate certificate, by Switzerland. Additional identification marks would be tabulated in a similar manner on the package;

(c) The revision of a certificate shall be indicated by a parenthetical expression following the identification mark on the certificate. For example, **A/132/B(M)F-96(Rev.2)** would indicate revision 2 of the Austrian package design approval certificate; or **A/132/B(M)F-96(Rev.0)** would indicate the original issuance of the Austrian package design approval certificate. For original issuances, the parenthetical entry is optional and other words such as 'original issuance' may also be used in place of 'Rev.0'. Certificate revision numbers may only be issued by the country issuing the original approval certificate;

(d) Additional symbols (as may be necessitated by national provisions) may be added in parentheses to the end of the identification mark. For example, **A/132/B(M)F-96(SP503)**; and

(e) It is not necessary to alter the identification mark on the packaging each time that a revision to the design certificate is made. Such re-marking shall be required only in those cases where the revision to the package design certificate involves a change in the letter type codes for the package design following the second stroke.

6.4.23.11   Each approval certificate issued by a competent authority for special form radioactive material or low dispersible radioactive material shall include the following information:

(a) Type of certificate.

(b) The competent authority identification mark.

(c) The issue date and an expiry date.

(d) List of applicable national and international regulations, including the edition of the IAEA Regulations for the Safe Transport of Radioactive Material under which the special form radioactive material or low dispersible radioactive material is approved.

(e) The identification of the special form radioactive material or low dispersible radioactive material.

(f) A description of the special form radioactive material or low dispersible radioactive material.

(g) Design specifications for the special form radioactive material or low dispersible radioactive material, which may include references to drawings.

(h) A specification of the radioactive contents which includes the activities involved and which may include the physical and chemical form.

(i) A specification of the applicable quality-assurance programme as required in 1.5.3.1.

(j) Reference to information provided by the applicant relating to specific actions to be taken prior to shipment.

(k) If deemed appropriate by the competent authority, reference to the identity of the applicant.

(l) Signature and identification of the certifying official.

6.4.23.12   Each approval certificate issued by a competent authority for a special arrangement shall include the following information:

(a) Type of certificate.

(b) The competent authority identification mark.

(c) The issue date and an expiry date.

(d) Mode(s) of transport.

(e) Any restrictions on the modes of transport, type of conveyance, freight container, and any necessary routeing instructions.

(f) List of applicable national and international regulations, including the edition of the IAEA Regulations for the Safe Transport of Radioactive Material under which the special arrangement is approved.

(g) The following statement: "This certificate does not relieve the consignor from compliance with any requirement of the government of any country through or into which the package will be transported."

    (h) References to certificates for alternative radioactive contents, other competent authority validation, or additional technical data or information, as deemed appropriate by the competent authority.

    (i) Description of the packaging by a reference to the drawings or a specification of the design. If deemed appropriate by the competent authority, a reproducible illustration, not larger than 21 cm by 30 cm, showing the make-up of the package shall also be provided, accompanied by a brief description of the packaging, including materials of manufacture, gross mass, general outside dimensions and appearance.

    (j) A specification of the authorized radioactive contents, including any restrictions on the radioactive contents which might not be obvious from the nature of the packaging. This shall include the physical and chemical forms, the activities involved (including those of the various isotopes, if appropriate), amounts in grams (for fissile material), and whether special form radioactive material or low dispersible radioactive material, if applicable.

    (k) Additionally, for packages containing fissile material:

        (i) a detailed description of the authorized radioactive contents;

        (ii) the value of the criticality safety index;

        (iii) reference to the documentation that demonstrates the criticality safety of the contents;

        (iv) any special features, on the basis of which the absence of water from certain void spaces has been assumed in the criticality assessment;

        (v) any allowance (based on 6.4.11.4(b)) for a change in neutron multiplication assumed in the criticality assessment as a result of actual irradiation experience; and

        (vi) the ambient temperature range for which the special arrangement has been approved.

    (l) A detailed listing of any supplementary operational controls required for preparation, loading, carriage, unloading and handling of the consignment, including any special stowage provisions for the safe dissipation of heat.

    (m) If deemed appropriate by the competent authority, reasons for the special arrangement.

    (n) Description of the compensatory measures to be applied as a result of the shipment being under special arrangement.

    (o) Reference to information provided by the applicant relating to the use of the packaging or specific actions to be taken prior to the shipment.

    (p) A statement regarding the ambient conditions assumed for purposes of design if these are not in accordance with those specified in 6.4.8.4, 6.4.8.5 and 6.4.8.15, as applicable.

    (q) Any emergency arrangements deemed necessary by the competent authority.

    (r) A specification of the applicable quality-assurance programme as required in 1.5.3.1.

    (s) If deemed appropriate by the competent authority, reference to the identity of the applicant and to the identity of the carrier.

    (t) Signature and identification of the certifying official.

6.4.23.13 Each approval certificate for a shipment issued by a competent authority shall include the following information:

    (a) Type of certificate.

    (b) The competent authority identification mark(s).

    (c) The issue date and an expiry date.

    (d) List of applicable national and international regulations, including the edition of the IAEA Regulations for the Safe Transport of Radioactive Material under which the shipment is approved.

    (e) Any restrictions on the modes of transport, type of conveyance, freight container, and any necessary routeing instructions.

    (f) The following statement: "This certificate does not relieve the consignor from compliance with any requirement of the government of any country through or into which the package will be transported."

    (g) A detailed listing of any supplementary operational controls required for preparation, loading, carriage, unloading and handling of the consignment, including any special stowage provisions for the safe dissipation of heat or maintenance of criticality safety.

    (h) Reference to information provided by the applicant relating to specific actions to be taken prior to shipment.

    (i) Reference to the applicable design approval certificate(s).

(j) A specification of the actual radioactive contents, including any restrictions on the radioactive contents which might not be obvious from the nature of the packaging. This shall include the physical and chemical forms, the total activities involved (including those of the various isotopes, if appropriate), amounts in grams (for fissile material), and whether special form radioactive material or low dispersible radioactive material, if applicable.

(k) Any emergency arrangements deemed necessary by the competent authority.

(l) A specification of the applicable quality-assurance programme as required in 1.5.3.1.

(m) If deemed appropriate by the competent authority, reference to the identity of the applicant.

(n) Signature and identification of the certifying official.

6.4.23.14 Each approval certificate of the design of a package issued by a competent authority shall include the following information:

(a) Type of certificate.

(b) The competent authority identification mark.

(c) The issue date and an expiry date.

(d) Any restriction on the modes of transport, if appropriate.

(e) List of applicable national and international regulations, including the edition of the IAEA Regulations for the Safe Transport of Radioactive Material under which the design is approved.

(f) The following statement: "This certificate does not relieve the consignor from compliance with any requirement of the government of any country through or into which the package will be transported."

(g) References to certificates for alternative radioactive contents, other competent authority validation, or additional technical data or information, as deemed appropriate by the competent authority.

(h) A statement authorizing shipment where shipment approval is required under 5.1.5.1.2, if deemed appropriate.

(i) Identification of the packaging.

(j) Description of the packaging by a reference to the drawings or specification of the design. If deemed appropriate by the competent authority, a reproducible illustration, not larger than 21 cm by 30 cm, showing the make-up of the package shall also be provided, accompanied by a brief description of the packaging, including materials of manufacture, gross mass, general outside dimensions and appearance.

(k) Specification of the design by reference to the drawings.

(l) A specification of the authorized radioactive content, including any restrictions on the radioactive contents which might not be obvious from the nature of the packaging. This shall include the physical and chemical forms, the activities involved (including those of the various isotopes, if appropriate), amounts in grams (for fissile material), and whether special form radioactive material or low dispersible radioactive material, if applicable.

(m) A description of the containment system;

(n) Additionally, for packages containing fissile material:

  (i) a detailed description of the authorized radioactive contents;

  (ii) A description of the confinement system;

  (iii) the value of the criticality safety index;

  (iv) reference to the documentation that demonstrates the criticality safety of the contents;

  (v) any special features, on the basis of which the absence of water from certain void spaces has been assumed in the criticality assessment;

  (vi) any allowance (based on 6.4.11.4(b)) for a change in neutron multiplication assumed in the criticality assessment as a result of actual irradiation experience; and

  (vii) the ambient temperature range for which the package design has been approved.

(o) For Type B(M) packages, a statement specifying those prescriptions of 6.4.7.5, 6.4.8.4, 6.4.8.5, 6.4.8.6 and 6.4.8.9–6.4.8.15 with which the package does not conform and any amplifying information which may be useful to other competent authorities.

(p) For packages containing more than 0.1 kg of uranium hexafluoride, a statement specifying those prescriptions of 6.4.6.4 that apply, if any, and any amplifying information which may be useful to other competent authorities.

(q) A detailed listing of any supplementary operational controls required for preparation, loading, carriage, unloading and handling of the consignment, including any special stowage provisions for the safe dissipation of heat.

(r) Reference to information provided by the applicant relating to the use of the packaging or specific actions to be taken prior to shipment.

(s) A statement regarding the ambient conditions assumed for purposes of design if these are not in accordance with those specified in 6.4.8.5, 6.4.8.6 and 6.4.8.15, as applicable.

(t) A specification of the applicable quality-assurance programme as required in 1.5.3.1.

(u) Any emergency arrangements deemed necessary by the competent authority.

(v) If deemed appropriate by the competent authority, reference to the identity of the applicant.

(w) Signature and identification of the certifying official.

6.4.23.15 The competent authority shall be informed of the serial number of each packaging manufactured to a design approved under 6.4.22.2, 6.4.22.3, 6.4.22.4, 6.4.24.2 and 6.4.24.3.

6.4.23.16 Multilateral approval may be by validation of the original certificate issued by the competent authority of the country of origin of the design or shipment. Such validation may take the form of an endorsement on the original certificate or the issuance of a separate endorsement, annex, supplement, etc., by the competent authority of the country through or into which the shipment is made.

6.4.24   Transitional measures for class 7

Packages not requiring competent authority approval of design under the 1985 and 1985 (as amended 1990) editions of IAEA Safety Series No. 6

6.4.24.1 Excepted packages, Type IP-1, Type IP-2 and Type IP-3 and Type A packages that did not require approval of design by the competent authority and which meet the provisions of the 1985 or 1985 (as amended 1990) editions of IAEA Regulations for the Safe Transport of Radioactive Material (IAEA Safety Series No. 6) may continue to be used, subject to the mandatory programme of quality assurance in accordance with the provisions of 1.5.3.1 and the activity limits and material restrictions of 2.7.2.2, 2.7.2.4.1, 2.7.2.4.4, 2.7.2.4.5, 2.7.2.4.6 and 4.1.9.3. Any packaging modified, unless to improve safety, or manufactured after 31 December 2003 shall meet the provisions of this Code in full. Packages prepared for transport not later than 31 December 2003 under the 1985 or 1985 (as amended 1990) editions of IAEA Safety Series No. 6 may continue in transport. Packages prepared for transport after this date shall meet the provisions of this Code in full.

Packages approved under the 1973, 1973 (as amended), 1985 and 1985 (as amended 1990) editions of IAEA Safety Series No. 6

6.4.24.2 Packagings manufactured to a package design approved by the competent authority under the provisions of the 1973 or 1973 (as amended) editions of IAEA Safety Series No. 6 may continue to be used, subject to: multilateral approval of package design; the mandatory programme of quality assurance in accordance with the applicable provisions of 1.5.3.1; the activity limits and material restrictions of 2.7.2.2, 2.7.2.4.1, 2.7.2.4.4, 2.7.2.4.5, 2.7.2.4.6 and 4.1.9.3; and, for a package containing fissile material and transported by air, the requirement of 6.4.11.10 shall be met. No new manufacture of such packaging shall be permitted to commence. Changes in the design of the packaging or in the nature or quantity of the authorized radioactive contents which, as determined by the competent authority, would significantly affect safety shall require that the provisions of this Code be met in full. A serial number according to the provision of 5.2.1.5.5 shall be assigned to and marked on the outside of each packaging.

6.4.24.3 Packagings manufactured to a package design approved by the competent authority under the provisions of the 1985 or 1985 (as amended 1990) editions of IAEA Safety Series No. 6 may continue to be used, subject to: the multilateral approval of package design; the mandatory programme of quality assurance in accordance with the provisions of 1.5.3.1; the activity limits and material restrictions of 2.7.2.2, 2.7.2.4.1, 2.7.2.4.4, 2.7.2.4.5, 2.7.2.4.6 and 4.1.9.3; and, for a package containing fissile material and transported by air, the requirement of 6.4.11.10 shall be met. Changes in the design of the packaging or in the nature or quantity of the authorized radioactive contents which, as determined by the competent authority, would significantly affect safety shall require that the provisions of this Code be met in full. All packagings for which manufacture begins after 31 December 2006 shall meet the provisions of this Code in full.

Special form radioactive material approved under the 1973, 1973 (as amended), 1985 and 1985 (as amended 1990) editions of IAEA Safety Series No. 6

6.4.24.4   Special form radioactive material manufactured to a design which had received unilateral approval by the competent authority under the 1973, 1973 (as amended), 1985 or 1985 (as amended 1990) editions of IAEA Safety Series No. 6 may continue to be used when in compliance with the mandatory programme of quality assurance in accordance with the applicable provisions of 1.5.3.1. All special form radioactive material manufactured after 31 December 2003 shall meet the provisions of this Code in full.

Chapter 6.5

Provisions for the construction and testing of intermediate bulk containers (IBCs)

6.5.1 General requirements

6.5.1.1 Scope

6.5.1.1.1 The provisions of this chapter apply to IBCs intended for the transport of certain dangerous substances and materials.

6.5.1.1.2 IBCs and their service equipment not conforming strictly to the provisions herein, but conforming to acceptable alternatives, may be considered by the competent authority concerned for approval. In order to take into account progress in science and technology, the use of alternative arrangements which offer at least an equivalent degree of safety in transport in respect of compatibility with the substances to be loaded therein and an equivalent or superior resistance to handling impact, and fire, may be considered by the competent authority concerned.

6.5.1.1.3 The construction, equipment, testing, marking and operation of IBCs shall be subject to acceptance by the competent authority of the country in which the IBCs are approved.

6.5.1.1.4 Manufacturers and subsequent distributors of IBCs shall provide information regarding procedures to be followed and a description of the types and dimensions of closures (including required gaskets) and any other components needed to ensure that IBCs as presented for transport are capable of passing the applicable performance tests of this chapter.

6.5.1.2 Definitions

Body (for all categories of IBCs other than composite IBCs) means the receptacle proper, including openings and their closures, but does not include service equipment;

Handling device (for flexible IBCs) means any sling, loop, eye or frame attached to the body of the IBC or formed from a continuation of the IBC body material;

Maximum permissible gross mass means the mass of the IBC and any service or structural equipment together with the maximum net mass;

Plastics material, when used in connection with inner receptacles for composite IBCs, is taken to include other polymeric materials such as rubber;

Protected (for metal IBCs) means the IBC being provided with additional protection against impact, the protection taking the form of, for example, a multi-layer (sandwich) or double-wall construction or a frame with a metal latticework packaging;

Service equipment means filling and discharge devices and, according to the category of IBC, pressure relief or venting, safety, heating and heat-insulating devices and measuring instruments;

Structural equipment (for all categories of IBCs other than flexible IBCs) means the reinforcing, fastening, handling, protective or stabilizing members of the body, including the base pallet for composite IBCs with plastics inner receptacle, fibreboard and wooden IBCs;

Woven plastics (for flexible IBCs) means a material made from stretched tapes or monofilaments of a suitable plastics material.

6.5.1.3 Categories of IBCs

6.5.1.3.1 *Metal* IBCs consist of a metal body together with appropriate service and structural equipment.

6.5.1.3.2 *Flexible* IBCs consist of a body constituted of film, woven fabric or any other flexible material or combinations thereof, and if necessary an inner coating or liner, together with any appropriate service equipment and handling devices.

6.5.1.3.3 *Rigid plastics* IBCs consist of a rigid plastics body, which may have structural equipment together with appropriate service equipment.

6.5.1.3.4   *Composite* IBCs consist of structural equipment in the form of a rigid outer packaging enclosing a plastics inner receptacle together with any service or other structural equipment. The IBC is so constructed that the inner receptacle and outer packaging, once assembled, form, and are used as, an integrated single unit to be filled, stored, transported or emptied as such.

6.5.1.3.5   *Fibreboard* IBCs consist of a fibreboard body with or without separate top and bottom caps, if necessary, an inner liner (but no inner packagings) and appropriate service and structural equipment.

6.5.1.3.6   *Wooden* IBCs consist of a rigid or collapsible wooden body together with an inner liner (but no inner packagings) and appropriate service and structural equipment.

6.5.1.4   **Designatory code system for IBCs**

6.5.1.4.1   The code shall consist of two Arabic numerals as specified in .1 followed by one or more capital letters as specified in .2; followed, when specified in an individual section, by an Arabic numeral indicating the category of IBC.

.1

Type	For solids, filled or discharged		For liquids
	by gravity	under pressure of more than 10 kPa (0.1 bar)	
Rigid	11	21	31
Flexible	13	–	–

.2   A   Steel (all types and surface treatments)
B   Aluminium
C   Natural wood
D   Plywood
F   Reconstituted wood
G   Fibreboard
H   Plastics material
L   Textile
M   Paper, multiwall
N   Metal (other than steel or aluminium)

6.5.1.4.2   For a composite IBC, two capital letters in Latin characters shall be used in sequence in the second position of the code. The first shall indicate the material of the inner receptacle of the IBC and the second that of the outer packaging of the IBC.

6.5.1.4.3   The following types and codes of IBCs are assigned:

Material		Category	Code	Paragraph
Metal				6.5.5.1
A	Steel	for solids, filled or discharged by gravity	11A	
		for solids, filled or discharged under pressure	21A	
		for liquids	31A	
B	Aluminium	for solids, filled or discharged by gravity	11B	
		for solids, filled or discharged under pressure	21B	
		for liquids	31B	
N	Other than steel or aluminium	for solids, filled or discharged by gravity	11N	
		for solids, filled or discharged under pressure	21N	
		for liquids	31N	
Flexible				6.5.5.2
H	Plastics	woven plastics without coating or liner	13H1	
		woven plastics, coated	13H2	
		woven plastics with liner	13H3	
		woven plastics, coated and with liner	13H4	
		plastics film	13H5	
L	Textile	without coating or liner	13L1	
		coated	13L2	
		with liner	13L3	
		coated and with liner	13L4	
M	Paper	multiwall	13M1	
		multiwall, water-resistant	13M2	

6

Material	Category	Code	Paragraph
H Rigid plastics	for solids, filled or discharged by gravity, fitted with structural equipment	11H1	6.5.5.3
	for solids, filled or discharged by gravity, freestanding	11H2	
	for solids, filled or discharged under pressure, fitted with structural equipment	21H1	
	for solids, filled or discharged under pressure, freestanding	21H2	
	for liquids, fitted with structural equipment	31H1	
	for liquids, freestanding	31H2	
HZ Composite with plastics inner receptacle*	for solids, filled or discharged by gravity, with rigid plastics inner receptacle	11HZ1	6.5.5.4
	for solids, filled or discharged by gravity, with flexible plastics inner receptacle	11HZ2	
	for solids, filled or discharged under pressure, with rigid plastics inner receptacle	21HZ1	
	for solids, filled or discharged under pressure, with flexible plastics inner receptacle	21HZ2	
	for liquids, with rigid plastics inner receptacle	31HZ1	
	for liquids, with flexible plastics inner receptacle	31HZ2	
G Fibreboard	for solids, filled or discharged by gravity	11G	6.5.5.5
Wooden C Natural wood	for solids, filled or discharged by gravity, with inner liner	11C	6.5.5.6
D Plywood	for solids, filled or discharged by gravity, with inner liner	11D	
F Reconstituted wood	for solids, filled or discharged by gravity, with inner liner	11F	

*The code shall be completed by replacing the letter 'Z' by a capital letter in accordance with 6.5.1.4.1.2 to indicate the nature of the material used for the outer packaging.

6.5.1.4.4   The letter 'W' may follow the IBC code. The letter 'W' signifies that the IBC, although of the same type as indicated by the code, is manufactured to a specification different from those in section 6.5.3 and is considered equivalent in accordance with the provisions in 6.5.1.1.2.

6.5.2   Marking

6.5.2.1   Primary marking

6.5.2.1.1   Each IBC manufactured and intended for use according to these provisions shall bear durable markings which are legible and placed in a location so as to be readily visible. Letters, numbers and symbols shall be at least 12 mm high and shall show:

.1   the United Nations packaging symbol

This symbol shall not be used for any purpose other than certifying that a packaging complies with the relevant requirements in chapter 6.1, 6.2, 6.3, 6.5 or 6.6. For metal IBCs on which the marking is stamped or embossed, the capital letters "UN" may be applied instead of the symbol;

.2   the code designating the type of IBC according to 6.5.1.4;

.3   a capital letter designating the packing group(s) for which the design type has been approved:

"X" for packing groups I, II and III (IBCs for solids only);

"Y" for packing groups II and III; or

"Z" for packing group III only;

.4   the month and year (last two digits) of manufacture;

.5   the State authorizing the allocation of the mark, indicated by the distinguishing sign for motor vehicles in international traffic;

.6   the name or symbol of the manufacturer and other identifications of the IBC as specified by the competent authority;

.7   the stacking test load* in kilograms. For IBCs not designed for stacking, the figure "0" shall be shown;

.8   the maximum permissible gross mass in kilograms.

---

* The stacking test load in kilograms to be placed on the IBC shall be 1.8 times the combined maximum permissible gross mass of the number of similar IBC that may be stacked on top of the IBC during transport (see 6.5.4.6.4).

The primary marking required above shall be applied in the sequence indicated in the subparagraphs .1 to .8 above. The additional marking required by 6.5.2.2 and any further marking authorized by a competent authority shall still enable the various parts of the mark to be correctly identified.

6.5.2.1.2 Examples of markings for various types of IBCs in accordance with .1 to .8 above:

 11A/Y/02 99/ NL/...* 007/ 5500/1500 — For a metal IBC for solids discharged by gravity and made from steel/ for packing groups II and III/ manufactured in February 1999/ authorized by the Netherlands/ manufactured by . . . *(name of manufacturer) and of a design type to which the competent authority has allocated serial number 007/ the stacking test load in kilograms/ and the maximum permissible gross mass in kilograms.

 13H3/Z/03 01/ F/...* 1713/ 0/1500 — For a flexible IBC for solids discharged by gravity and made from woven plastics with a liner/ not designed to be stacked.

 31H1/Y/04 99/ GB/...* 9099/ 10800/1200 — For a rigid plastics IBC for liquids made from plastics with structural equipment withstanding the stack load.

 31HA1/Y/05 01/ D/...* 1683/ 10800/1200 — For a composite IBC for liquids with a rigid plastics inner receptacle and steel outer packaging.

 11C/X/01 02/ S/...* 9876/ 3000/910 — For a wooden IBC for solids with an inner liner and authorized for packing group I solids.

 11G/Z/06 02/ I/...* 962/ 0/500 — For a fibreboard IBC/ not designed to be stacked.

 11D/Y/07 02/ E/...* 261/ 3240/600 — For a plywood IBC with inner liner.

Each element of the marking applied in accordance with subparagraphs .1 to .8 and with 6.5.2.2 shall be clearly separated, such as by a slash or space, so as to be easily identifiable.

6.5.2.2 Additional marking

6.5.2.2.1 Each IBC shall bear the markings required in 6.5.2.1 and, in addition, the following information, which may appear on a corrosion-resistant plate permanently attached in a place readily accessible for inspection:

Note: For metal IBCs, this plate shall be a corrosion-resistant metal plate.

Additional marking	Category of IBC				
	Metal	Rigid plastics	Composite	Fibreboard	Wooden
Capacity in litres[a] at 20°C	X	X	X		
Tare mass in kg[a]	X	X	X	X	X
Test (gauge) pressure, in kPa or bar,[a] if applicable		X	X		
Maximum filling/discharge pressure in kPa or bar,[a] if applicable	X	X	X		
Body material and its minimum thickness in mm	X				
Date of last leakproofness test, if applicable (month and year)	X	X	X		
Date of last inspection (month and year)	X	X	X		
Serial number of the manufacturer	X				
Maximum permitted stacking load[b]	X	X	X	X	X

[a] The unit used shall be indicated.
[b] See 6.5.2.2.2. This additional marking shall apply to all IBCs manufactured, repaired or remanufactured as from 1 January 2011.

6.5.2.2.2 The maximum permitted stacking load applicable when the IBC is in use shall be displayed on a symbol as follows:

IBCs capable of being stacked   IBCs NOT capable of being stacked

The symbol shall be not less than 100 mm × 100 mm, be durable and clearly visible. The letters and numbers indicating the mass shall be at least 12 mm high.

The mass marked above the symbol shall not exceed the load imposed during the design type test (see 6.5.6.6.4) divided by 1.8.

Note: The provisions of 6.5.2.2.2 shall apply to all IBCs manufactured, repaired or remanufactured as from 1 January 2011.

6.5.2.2.3 Each flexible IBC may also bear a pictogram or pictograms indicating the recommended lifting methods.

6.5.2.2.4 The inner receptacle of composite IBCs shall be marked with at least the following information:

.1 the name or symbol of the manufacturer and other identification of the IBC as specified by the competent authority, as in 6.5.2.1.1.6;

.2 the date of manufacture, as in 6.5.2.1.1.4; and

.3 the distinguishing sign of the State authorizing the allocation of the mark, as in 6.5.2.1.1.5.

6.5.2.2.5 Where a composite IBC is designed in such a manner that the outer packaging is intended to be dismantled for transport when empty (such as for return of the IBC for re-use to the original consignor), each of the parts intended to be detached when so dismantled shall be marked with the month and year of manufacture and the name or symbol of the manufacturer and other identification of the IBC as specified by the competent authority (see 6.5.2.1.1.6).

6.5.2.3 **Conformity to design type**

The marking indicates that the IBCs correspond to a successfully tested design type and that the provisions referred to in the certificate have been met.

6.5.3 **Construction requirements**

6.5.3.1 **General requirements**

6.5.3.1.1 IBCs shall be resistant to or adequately protected from deterioration due to the external environment.

6.5.3.1.2 IBCs shall be so constructed and closed that none of the contents can escape under normal conditions of transport, including the effects of vibration, or by changes in temperature, humidity or pressure.

6.5.3.1.3 IBCs and their closures shall be constructed of materials compatible with their contents, or be protected internally, so that they are not liable:

.1 to be attacked by the contents so as to make their use dangerous;

.2 to cause the contents to react or decompose, or form harmful or dangerous compounds with the IBCs.

6.5.3.1.4 Gaskets, where used, shall be made of materials not subject to attack by the contents of an IBC.

6.5.3.1.5 All service equipment shall be so positioned or protected as to minimize the risk of escape of the contents owing to damage during handling and transport.

6.5.3.1.6   IBCs, their attachments and their service and structural equipment shall be designed to withstand, without loss of contents, the internal pressure of the contents and the stresses of normal handling and transport. IBCs intended for stacking shall be designed for stacking. Any lifting or securing features of IBCs shall be of sufficient strength to withstand the normal conditions of handling and transport without gross distortion or failure and shall be so positioned that no undue stress is caused in any part of the IBC.

6.5.3.1.7   Where an IBC consists of a body within a framework, it shall be so constructed that:

.1   the body does not chafe or rub against the framework so as to cause material damage to the body,

.2   the body is retained within the framework at all times,

.3   the items of equipment are fixed in such a way that they cannot be damaged if the connections between body and frame allow relative expansion or movement.

6.5.3.1.8   Where a bottom discharge valve is fitted, it shall be capable of being made secure in the closed position and the whole discharge system shall be suitably protected from damage. Valves having lever closures shall be able to be secured against accidental opening and the open or closed position shall be readily apparent. For IBCs containing liquids, a secondary means of sealing the discharge aperture shall also be provided, such as by a blank flange or equivalent device.

6.5.4   Testing, certification and inspection

6.5.4.1   Quality assurance

IBCs shall be manufactured and tested under a quality-assurance programme which satisfies the competent authority, in order to ensure that each manufactured IBC meets the provisions of this chapter.

Note: ISO 16106:2006 "Packaging – Transport packages for dangerous goods – Dangerous goods packagings, intermediate bulk containers (IBCs) and large packagings – Guidelines for the application of ISO 9001" provides acceptable guidance on procedures which may be followed.

6.5.4.2   Test provisions

IBCs shall be subjected to design type tests and, if applicable, to initial and periodic inspections and tests in accordance with 6.5.4.4.

6.5.4.3   Certification

In respect of each design type of IBC, a certificate and mark (as in 6.5.2) shall be issued attesting that the design type, including its equipment, meets the test provisions.

6.5.4.4   Inspection and testing

Note: See also 6.5.4.5 for tests and inspections on repaired IBCs.

6.5.4.4.1   Every metal, rigid plastics and composite IBC shall be inspected to the satisfaction of the competent authority:

.1   before it is put into service (including after remanufactured), and thereafter at intervals not exceeding five years, with regard to:

.1   conformity to the design type, including marking;

.2   internal and external condition; and

.3   proper functioning of service equipment.

Thermal insulation, if any, need be removed only to the extent necessary for a proper examination of the body of the IBC.

.2   at intervals of not more than two and a half years with regard to:

.1   external condition; and

.2   proper functioning of service equipment.

Thermal insulation, if any, need be removed only to the extent necessary for a proper examination of the body of the IBC.

Each IBC shall correspond in all respects to its design type.

6.5.4.4.2 Every metal, rigid plastics and composite IBC for liquids, or for solids which are filled or discharged under pressure, shall undergo a suitable leakproofness test at least equally effective as the test prescribed in 6.5.6.7.3 and be capable of meeting the test level indicated in 6.5.6.7.3:

(a) before it is first used for transport;

(b) at intervals of not more than two and a half years.

For this test the IBC shall be fitted with the primary bottom closure. The inner receptacle of a composite IBC may be tested without the outer casing, provided the test results are not affected.

6.5.4.4.3 A report of each inspection and test shall be kept by the owner of the IBC at least until the next inspection or test. The report shall include the results of the inspection and test and shall identify the party performing the inspection and test (see also the marking requirements in 6.5.2.2.1).

6.5.4.5 Repaired IBCs

6.5.4.5.1 When an IBC is impaired as a result of impact (e.g., accident) or any other cause, it shall be repaired or otherwise maintained (see definition of "Routine maintenance of IBCs" in 1.2.1), so as to conform to the design type. The bodies of rigid plastics IBCs and the inner receptacles of composite IBCs that are impaired shall be replaced.

6.5.4.5.2 In addition to any other testing and inspection requirements in this Code, an IBC shall be subjected to the full testing and inspection requirements set out in 6.5.4.4, and the required reports shall be prepared, whenever it is repaired.

6.5.4.5.3 The party performing the tests and inspections after the repair shall durably mark the IBC near the manufacturer's UN design type marking to show:

.1 the State in which the tests and inspections were carried out;

.2 the name or authorized symbol of the party performing the tests and inspections; and

.3 the date (month, year) of the tests and inspections.

6.5.4.5.4 Test and inspections performed in accordance with 6.5.4.5.2 may be considered to satisfy the requirements for the 2.5- and 5-year periodic tests and inspections.

6.5.4.5.5 The competent authority may at any time require proof, by tests in accordance with this chapter, that the IBCs meet the provisions of the design type tests.

6.5.5 Specific provisions for IBCs

6.5.5.1 Specific provisions for metal IBCs

6.5.5.1.1 These provisions apply to metal IBCs for the transport of liquids and solids. There are three categories of metal IBCs:

those for solids which are filled and discharged by gravity (11A, 11B, 11N);

those for solids which are filled and discharged at a gauge pressure greater than 10 kPa (21A, 21B, 21N); and

those for liquids (31A, 31B, 31N).

6.5.5.1.2 Bodies shall be made of suitable ductile metal in which the weldability has been fully demonstrated. Welds shall be skilfully made and afford complete safety. Low-temperature performance shall be taken into account when appropriate.

6.5.5.1.3 Care shall be taken to avoid damage by galvanic action due to the juxtaposition of dissimilar metals.

6.5.5.1.4 Aluminium IBCs intended for the transport of flammable liquids shall have no movable parts, such as covers, closures, etc., made of unprotected steel liable to rust, which might cause a dangerous reaction by coming into frictional or percussive contact with the aluminium.

6.5.5.1.5 Metal IBCs shall be made of metals which meet the following provisions:

.1 For steel, the elongation at fracture, per cent, shall not be less than $10,000/R_m$ with an absolute minimum of 20%, where R_m = guaranteed minimum tensile strength of the reference steel to be used, in N/mm^2.

.2 For aluminium and aluminium alloys, the elongation at fracture, per cent, shall not be less than $10,000/6R_m$ with an absolute minimum of 8%.

Specimens used to determine the elongation at fracture shall be taken transversely to the direction of rolling and be so secured that:

$$L_o = 5d, \text{ or}$$
$$L_o = 5.65\sqrt{A}$$

where :

L_o = gauge length of the specimen before the test;

d = diameter; and

A = cross-sectional area of the test specimen.

6.5.5.1.6   *Minimum wall thickness*

.1   For a reference steel having a product of $R_m \times A_o$ = 10,000, the wall thickness shall not be less than:

Capacity (C) in litres	Wall thickness (T) in mm			
	Types 11A, 11B, 11N		Types 21A, 21B, 21N, 31A, 31B, 31N	
	Unprotected	Protected	Unprotected	Protected
$C \leqslant 1000$	2.0	1.5	2.5	2.0
$1000 < C \leqslant 2000$	$T = C/2000 + 1.5$	$T = C/2000 + 1.0$	$T = C/2000 + 2.0$	$T = C/2000 + 1.5$
$2000 < C \leqslant 3000$	$T = C/2000 + 1.5$	$T = C/2000 + 1.0$	$T = C/1000 + 1.0$	$T = C/2000 + 1.5$

where: A_o = minimum elongation (as a percentage) of the reference steel to be used on fracture under tensile stress (see 6.5.5.1.5).

.2   For metals other than the reference steel described in .1, the minimum wall thickness is given by the following equivalence formula:

$$e_1 = \frac{21.4 \times e_0}{\sqrt[3]{R_{m1}A_1}}$$

where:

e_1 = required equivalent wall thickness of the metal to be used (in mm);

e_0 = required minimum wall thickness for the reference steel (in mm);

R_{m1} = guaranteed minimum tensile strength of the metal to be used (in N/mm^2) (see .3); and

A_1 = minimum elongation (as a percentage) of the metal to be used on fracture under tensile stress (see 6.5.5.1.5).

However, in no case shall the wall thickness be less than 1.5 mm.

.3   For purposes of the calculation described in .2, the guaranteed minimum tensile strength of the metal to be used (R_{m1}) shall be the minimum value according to national or international material standards.

However, for austenitic steels, the specified minimum value for R_m according to the material standards may be increased by up to 15% when a greater value is attested in the material inspection certificate. When no material standard exists for the material in question, the value of R_m shall be the minimum value attested in the material inspection certificate.

6.5.5.1.7   *Pressure relief provisions*

IBCs for liquids shall be capable of releasing a sufficient amount of vapour in the event of fire engulfment to ensure that no rupture of the shell will occur. This can be achieved by conventional pressure relief devices or by other constructional means. The start-to-discharge pressure shall not be higher than 65 kPa and no lower than the total gauge pressure experienced in the IBC (i.e. the vapour pressure of the filling substance plus the partial pressure of the air or other inert gases, minus 100 kPa) at 55°C, determined on the basis of a maximum degree of filling as defined in 4.1.1.4. The pressure relief devices shall be fitted in the vapour space.

6.5.5.2   **Specific provisions for flexible IBCs**

6.5.5.2.1   These provisions apply to flexible IBCs of the following types:

13H1   woven plastics without coating or liner

13H2   woven plastics, coated

13H3   woven plastics with liner

13H4   woven plastics, coated and with liner

13H5   plastics film

13L1   textile without coating or liner

6

13L2	textile, coated
13L3	textile with liner
13L4	textile, coated and with liner
13M1	paper, multiwall
13M2	paper, multiwall, water-resistant.

Flexible IBCs are intended for the transport of solids only.

6.5.5.2.2 Bodies of IBCs shall be manufactured from suitable materials. The strength of the material and the construction of a flexible IBC shall be appropriate to its capacity and its intended use.

6.5.5.2.3 All materials used in the construction of flexible IBCs of types 13M1 and 13M2 shall, after complete immersion in water for not less than 24 hours, retain at least 85% of the tensile strength as measured originally on the material conditioned to equilibrium at 67% relative humidity or less.

6.5.5.2.4 Seams of IBCs shall be formed by stitching, heat sealing, gluing or any equivalent method. All stitched seam-ends shall be secured.

6.5.5.2.5 Flexible IBCs shall provide adequate resistance to ageing and to degradation caused by ultraviolet radiation, by climatic conditions, or by the substance contained within which would thereby render them unsuitable for their intended use.

6.5.5.2.6 For plastics flexible IBCs where protection against ultraviolet radiation is required, it shall be provided by the addition of carbon black or other suitable pigments or inhibitors. These additives shall be compatible with the contents and remain effective throughout the life of the body of the IBC. Where use is made of carbon black, pigments or inhibitors other than those used in the manufacture of the tested design type, retesting may be waived if changes in the carbon black content, the pigment content or the inhibitor content do not adversely affect the physical properties of the material of construction.

6.5.5.2.7 Additives may be incorporated into the material of the body to improve the resistance to ageing or to serve other purposes, provided that these do not adversely affect the physical or chemical properties of the material.

6.5.5.2.8 No material recovered from used receptacles shall be used in the manufacture of IBC bodies. Production residues or scrap from the same manufacturing process may, however, be used. Component parts such as fittings and pallet bases may also be used provided such components have not in any way been damaged in previous use.

6.5.5.2.9 When filled, the ratio of height to width shall be not more than 2:1.

6.5.5.2.10 The liner shall be made of a suitable material. The strength of the material used and the construction of the liner shall be appropriate to the capacity of the IBC and the intended use. Joints and closures shall be sift-proof and capable of withstanding pressures and impacts liable to occur under normal conditions of handling and transport.

6.5.5.3 **Specific provisions for rigid plastics IBCs**

6.5.5.3.1 These provisions apply to rigid plastics IBCs for the transport of solids or liquids. Rigid plastics IBCs are of the following types:

11H1	fitted with structural equipment designed to withstand the whole load when IBCs are stacked, for solids which are filled or discharged by gravity
11H2	freestanding, for solids which are filled or discharged by gravity
21H1	fitted with structural equipment designed to withstand the whole load when IBCs are stacked, for solids which are filled or discharged under pressure
21H2	freestanding, for solids which are filled or discharged under pressure
31H1	fitted with structural equipment designed to withstand the whole load when IBCs are stacked, for liquids
31H2	freestanding, for liquids.

6.5.5.3.2 The body shall be manufactured from suitable plastics material of known specifications and be of adequate strength in relation to its capacity and to the service it is required to perform. The material shall be adequately resistant to ageing and to degradation caused by the substance contained within or, where relevant, by ultraviolet radiation. Low-temperature performance shall be taken into account when appropriate. Any permeation of the substance contained within shall not constitute a danger under normal conditions of transport.

6.5.5.3.3   Where protection against ultraviolet radiation is required, it shall be provided by the addition of carbon black or other suitable pigments or inhibitors. These additives shall be compatible with the contents and remain effective throughout the life of the body of the IBC. Where use is made of carbon black, pigments or inhibitors other than those used in the manufacture of the tested design type, retesting may be waived if changes in the carbon black content, the pigment content or the inhibitor content do not adversely affect the physical properties of the material of construction.

6.5.5.3.4   Additives may be incorporated in the material of the body to improve the resistance to ageing or to serve other purposes, provided that these do not adversely affect the physical or chemical properties of the material.

6.5.5.3.5   No used material other than production residues or regrind from the same manufacturing process may be used in the manufacturing of rigid plastics IBCs.

6.5.5.4   **Specific provisions for composite IBCs with plastics inner receptacles**

6.5.5.4.1   These provisions apply to composite IBCs for the transport of solids or liquids of the following types:

      11HZ1   composite IBCs with a rigid plastics inner receptacle, for solids filled or discharged by gravity

      11HZ2   composite IBCs with a flexible plastics inner receptacle, for solids filled or discharged by gravity

      21HZ1   composite IBCs with a rigid plastics inner receptacle, for solids filled or discharged under pressure

      21HZ2   composite IBCs with a flexible plastics inner receptacle, for solids filled or discharged under pressure

      31HZ1   composite IBCs with a rigid plastics inner receptacle, for liquids

      31HZ2   composite IBCs with a flexible plastics inner receptacle, for liquids.

This code shall be completed by replacing the letter 'Z' by a capital letter in accordance with 6.5.1.4.1.2 to indicate the nature of the material used for the outer packaging.

6.5.5.4.2   The inner receptacle is not intended to perform a containment function without its outer packaging. A "rigid" inner receptacle is a receptacle which retains its general shape when empty without closures in place and without the benefit of the outer packaging. Any inner receptacle that is not "rigid" is considered to be "flexible".

6.5.5.4.3   The outer packaging normally consists of rigid material formed so as to protect the inner receptacle from physical damage during handling and transport, but is not intended to perform the containment function. It includes the base pallet where appropriate.

6.5.5.4.4   A composite IBC with a fully enclosing outer packaging shall be so designed that the integrity of the inner receptacle may be readily assessed following the leakproofness and hydraulic tests.

6.5.5.4.5   IBCs of type 31HZ2 shall be limited to a capacity of not more than 1250 ℓ.

6.5.5.4.6   The inner receptacle shall be manufactured from suitable plastics material of known specifications and be of adequate strength in relation to its capacity and to the service it is required to perform. The material shall be adequately resistant to ageing and to degradation caused by the substance contained and, where relevant, by ultraviolet radiation. Low-temperature performance shall be taken into account when appropriate. Any permeation of the substance contained shall not constitute a danger under normal conditions of transport.

6.5.5.4.7   Where protection against ultraviolet radiation is required, it shall be provided by the addition of carbon black or other suitable pigments or inhibitors. These additives shall be compatible with the contents and remain effective throughout the life of the inner receptacle. Where use is made of carbon black, pigments or inhibitors other than those used in the manufacture of the tested design type, re-testing may be waived if changes in carbon black content, the pigment content or the inhibitor content do not adversely affect the physical properties of the material of construction.

6.5.5.4.8   Additives may be incorporated in the material of the inner receptacle to improve the resistance to ageing or to serve other purposes, provided that these do not adversely affect the physical or chemical properties of the material.

6.5.5.4.9   No used material other than production residues or regrind from the same manufacturing process may be used in the manufacture of inner receptacles.

6.5.5.4.10   The inner receptacle of IBCs of type 31HZ2 shall consist of at least three plies of film.

6.5.5.4.11   The strength of the material and the construction of the outer packaging shall be appropriate to the capacity of the composite IBC and its intended use.

6.5.5.4.12   The outer packaging shall be free of any projection that might damage the inner receptacle.

6.5.5.4.13   Outer packagings of steel or aluminium shall be constructed of a suitable metal of adequate thickness.

6.5.5.4.14   Outer packagings of natural wood shall be of well-seasoned wood, commercially dry and free from defects that would materially lessen the strength of any part of the packaging. The tops and bottoms may be made of water-resistant reconstituted wood such as hardboard, particle board or other suitable type.

6.5.5.4.15   Outer packagings of plywood shall be made of well-seasoned rotary-cut, sliced or sawn veneer plywood, commercially dry and free from defects that would materially lessen the strength of the packaging. All adjacent plies shall be glued with water-resistant adhesive. Other suitable materials may be used in conjunction with plywood for the construction of packagings. Packagings shall be firmly nailed or secured to corner posts or ends or be assembled by equally suitable devices.

6.5.5.4.16   The walls of outer packagings of reconstituted wood shall be made of water-resistant reconstituted wood such as hardboard, particle board or other suitable type. Other parts of the packagings may be made of other suitable material.

6.5.5.4.17   For fibreboard outer packagings, strong and good-quality solid or double-faced corrugated fibreboard (single or multiwall) shall be used appropriate to the capacity of the packaging and to its intended use. The water resistance of the outer surface shall be such that the increase in mass, as determined in a test carried out over 30 minutes by the Cobb method of determining water absorption, is not greater than 155 g/m^2 – see ISO 535:1991. It shall have proper bending qualities. Fibreboard shall be cut, creased without scoring, and slotted so as to permit assembly without cracking, surface breaks or undue bending. The fluting of corrugated fibreboard shall be firmly glued by water-resistant adhesive to the facings.

6.5.5.4.18   The ends of fibreboard outer packagings may have a wooden frame or be entirely of wood. Reinforcements of wooden battens may be used.

6.5.5.4.19   Manufacturing joins in the fibreboard outer packagings shall be taped, lapped and glued, or lapped and stitched with metal staples. Lapped joins shall have an appropriate overlap. Where closing is effected by gluing or taping, a water-resistant adhesive shall be used.

6.5.5.4.20   Where the outer packagings are of plastics material, the relevant provisions of 6.5.5.4.6 to 6.5.5.4.9 shall apply.

6.5.5.4.21   The outer packagings of IBCs of type 31HZ2 shall enclose the inner receptacle on all sides.

6.5.5.4.22   Any integral pallet base forming part of the IBC or a detachable pallet shall be suitable for mechanical handling with the IBC filled to its maximum permissible gross mass.

6.5.5.4.23   The pallet or integral base shall be designed so as to avoid any protrusion of the base of the IBC that might be liable to damage in handling.

6.5.5.4.24   The outer packagings shall be secured to a detachable pallet to ensure stability in handling and transport. Where a detachable pallet is used, its top surface shall be free from sharp protrusions that might damage the IBC.

6.5.5.4.25   Strengthening devices such as timber supports to increase stacking performance may be used but shall be external to the inner receptacle.

6.5.5.4.26   Where IBCs are intended for stacking, the bearing surfaces shall be such as to distribute the load in a safe manner. Such IBCs shall be designed so that the load is not supported by the inner receptacle.

6.5.5.5   **Specific provisions for fibreboard IBCs**

6.5.5.5.1   These provisions apply to fibreboard IBCs for the transport of solids which are filled or discharged by gravity. Fibreboard IBCs are of the following type: 11G.

6.5.5.5.2   Fibreboard IBCs shall not incorporate top lifting devices.

6.5.5.5.3 The body shall be made of strong and good-quality solid or double-faced corrugated fibreboard (single or multiwall), appropriate to the capacity of the IBC and to its intended use. The water resistance of the outer surface shall be such that the increase in mass, as determined in a test carried out over a period of 30 minutes by the Cobb method of determining water absorption, is not greater than 155 g/m^2 – see ISO 535:1991. It shall have proper bending qualities. Fibreboard shall be cut, creased without scoring, and slotted so as to permit assembly without cracking, surface breaks or undue bending. The fluting or corrugated fibreboard shall be firmly glued to the facings.

6.5.5.5.4 The walls, including top and bottom, shall have a minimum puncture resistance of 15 J, measured according to ISO 3036:1975.

6.5.5.5.5 Manufacturing joins in the body of IBCs shall be made with an appropriate overlap and shall be taped, glued, stitched with metal staples or fastened by other means at least equally effective. Where joins are effected by gluing or taping, a water-resistant adhesive shall be used. Metal staples shall pass completely through all pieces to be fastened and be formed or protected so that any inner liner cannot be abraded or punctured by them.

6.5.5.5.6 The liner shall be made of suitable material. The strength of the material used and the construction of the liner shall be appropriate to the capacity of the IBC and its intended use. Joins and closures shall be sift-proof and capable of withstanding pressure and impacts liable to occur under normal conditions of handling and transport.

6.5.5.5.7 Any integral pallet base forming part of the IBC or any detachable pallet shall be suitable for mechanical handling with the IBC filled to its maximum permissible gross mass.

6.5.5.5.8 The pallet or integral base shall be designed so as to avoid any protrusion of the base of the IBC that might be liable to damage in handling.

6.5.5.5.9 The body shall be secured to a pallet to ensure stability in handling and transport. Where a detachable pallet is used, its top surface shall be free from sharp protrusions that might damage the IBC.

6.5.5.5.10 Strengthening devices such as timber supports to increase stacking performance may be used but shall be external to the liner.

6.5.5.5.11 Where IBCs are intended for stacking, the bearing surface shall be such as to distribute the load in a safe manner.

6.5.5.6 Specific provisions for wooden IBCs

6.5.5.6.1 These provisions apply to wooden IBCs for the transport of solids which are filled or discharged by gravity. Wooden IBCs are of the following types:

> 11C natural wood with inner liner
>
> 11D plywood with inner liner
>
> 11F reconstituted wood with inner liner.

6.5.5.6.2 Wooden IBCs shall not incorporate top lifting devices.

6.5.5.6.3 The strength of the materials used and the method of construction shall be appropriate to the capacity and intended use of the IBC.

6.5.5.6.4 Natural wood shall be well seasoned, commercially dry and free from defects that would materially lessen the strength of any part of the IBC. Each part of the IBC shall consist of one piece or be equivalent thereto. Parts are considered equivalent to one piece when a suitable method of glued assembly is used (as for instance Lindermann joint, tongue and groove joint, ship lap or rabbet joint, or butt joint), with at least two corrugated metal fasteners at each joint, or when other methods at least equally effective are used.

6.5.5.6.5 Bodies of plywood shall be at least three-ply. It shall be made of well-seasoned rotary-cut, sliced or sawn veneer, commercially dry and free from defects that would materially lessen the strength of the body. All adjacent plies shall be glued with water-resistant adhesive. Other suitable materials may be used with plywood for the construction of the body.

6.5.5.6.6 Bodies of reconstituted wood shall be made of water-resistant reconstituted wood such as hardboard, particle board or other suitable type.

6.5.5.6.7 IBCs shall be firmly nailed or secured to corner posts or ends or be assembled by equally suitable devices.

6.5.5.6.8 The liner shall be made of a suitable material. The strength of the material used and the construction of the liner shall be appropriate to the capacity of the IBC and its intended use. Joins and closures shall be sift-proof and capable of withstanding pressure and impacts liable to occur under normal conditions of handling and transport.

6.5.5.6.9 Any integral pallet base forming part of the IBC or any detachable pallet shall be suitable for mechanical handling with the IBC filled to its maximum permissible gross mass.

6.5.5.6.10 The pallet or integral base shall be designed so as to avoid any protrusion of the base of the IBC that might be liable to damage in handling.

6.5.5.6.11 The body shall be secured to a pallet to ensure stability in handling and transport. Where a detachable pallet is used, its top surface shall be free from sharp protrusions that might damage the IBC.

6.5.5.6.12 Strengthening devices such as timber supports to increase stacking performance may be used but shall be external to the liner.

6.5.5.6.13 Where IBCs are intended for stacking, the bearing surface shall be such as to distribute the load in a safe manner.

6.5.6 Test provisions for IBCs

6.5.6.1 Performance and frequency of tests

6.5.6.1.1 Each IBC design type shall successfully pass the tests prescribed in this chapter before being used. An IBC design type is defined by the design, size and material and thickness, manner of construction and means of filling and discharging, but may include various surface treatments; it also includes IBCs which differ from the design type only in their lesser external dimensions.

6.5.6.1.2 Tests shall be carried out on IBCs as prepared for transport. IBCs shall be filled as indicated in the relevant section. The substances to be transported in the IBCs may be replaced by other substances except where this would invalidate the results of the tests. For solids, when another substance is used, it shall have the same physical characteristics (mass, grain size, etc.) as the substance to be transported. It is permissible to use additives, such as bags of lead shot, to achieve the requisite total package gross mass, so long as they are placed so that the test results are not affected.

6.5.6.2 Design type tests

6.5.6.2.1 One IBC of each design type, size, wall thickness and manner of construction shall be submitted to the tests in the order shown in 6.5.6.3.5 and as set out in 6.5.6.5 to 6.5.6.13. These design type tests shall be carried out as required by the competent authority.

6.5.6.2.2 The competent authority may permit the selective testing of IBCs which differ only in minor respects from the tested type, such as with small reductions in external dimensions.

6.5.6.2.3 If detachable pallets are used in the tests, the test report issued in accordance with 6.5.6.14 shall include a technical description of the pallets to be used.

6.5.6.3 Preparation of IBC for testing

6.5.6.3.1 Paper and fibreboard IBCs and composite IBCs with fibreboard outer packagings shall be conditioned for at least 24 hours in an atmosphere having a controlled temperature and relative humidity (r.h.). There are three options, one of which shall be chosen. The preferred atmosphere is $23^\circ C \pm 2^\circ C$ and $50\% \pm 2\%$ r.h. The two other options are $20^\circ C \pm 2^\circ C$ and $65\% \pm 2\%$ r.h. or $27^\circ C \pm 2^\circ C$ and $65\% \pm 2\%$ r.h.

Note: Average values shall fall within these limits. Short-term fluctuations and measurement limitations may cause individual measurements to vary by up to $\pm 5\%$ relative humidity without significant impairment of test reproducibility.

6.5.6.3.2 Additional steps shall be taken to ascertain that the plastics material used in the manufacture of rigid plastics IBCs of types 31H1 and 31H2 and composite IBCs of type 31HZ1 and 31HZ2 complies with the provisions of 6.5.5.3.2 to 6.5.5.3.4 and 6.5.5.4.6 to 6.5.5.4.9.

6.5.6.3.3 This may be done, for example, by submitting sample IBCs to a preliminary test extending over a long period, for example six months, during which the samples would remain filled with the substances they are intended to contain or with substances which are known to have at least as severe a stress-cracking, weakening or molecular degradation influence on the plastics materials in question, and after which the samples shall be submitted to the applicable tests listed in the table in 6.5.6.3.5.

6.5.6.3.4 Where the behaviour of the plastics material has been established by other means, the above compatibility test may be dispensed with.

6.5.6.3.5 Design type tests required in sequential order:

Type of IBC	Vibration[f]	Bottom lift	Top lift[a]	Stacking[b]	Leak-proofness	Hydraulic pressure	Drop	Tear	Topple	Righting[c]
Metal: 11A, 11B, 11N 21A, 21B, 21N 31A, 31B, 31N	– – 1st	1st[a] 1st[a] 2nd[a]	2nd 2nd 3rd	3rd 3rd 4th	– 4th 5th	– 5th 6th	4th[e] 6th[e] 7th[e]	– – –	– – –	– – –
Flexible[d]	–	–	x[c]	x	–	–	x	x	x	x
Rigid plastics: 11H1, 11H2 21H1, 21H2 31H1, 31H2	– – 1st	1st[a] 1st[a] 2nd[a]	2nd 2nd 3rd	3rd 3rd 4th	– 4th 5th	– 5th 6th	4th 6th 7th	– – –	– – –	– – –
Composite: 11HZ1, 11HZ2 21HZ1, 21HZ2 31HZ1, 31HZ2	– – 1st	1st[a] 1st[a] 2nd[a]	2nd 2nd 3rd	3rd 3rd 4th	– 4th 5th	– 5th 6th	4th[e] 6th[e] 7th[e]	– – –	– – –	– – –
Fibreboard	–	1st	–	2nd	–	–	3rd	–	–	–
Wooden	–	1st	–	2nd	–	–	3rd	–	–	–

[a] When IBCs are designed for this method of handling.

[b] When IBCs are designed to be stacked.

[c] When IBCs are designed to be lifted from the top or the side.

[d] Required test indicated by "x"; an IBC which has passed one test may be used for other tests, in any order.

[e] Another IBC of the same design may be used for the drop test.

[f] Another IBC of the same design may be used for the vibration test.

6.5.6.4 **Bottom lift test**

6.5.6.4.1 *Applicability*

For all fibreboard and wooden IBCs and for all types of IBCs which are fitted with means for lifting from the base, as a design type test.

6.5.6.4.2 *Preparation of the IBC for test*

The IBC shall be filled. A load shall be added and evenly distributed. The mass of filled IBC and the load shall be 1.25 times its maximum permissible gross mass.

6.5.6.4.3 *Method of testing*

The IBC shall be raised and lowered twice by a forklift truck with the forks centrally positioned so that the space between them is three quarters of the length of the side of entry (unless the points of entry are fixed). The forks shall penetrate to three quarters of the depth in the direction of entry. The test shall be repeated from each possible direction of entry.

6.5.6.4.4 *Criteria for passing the test*

No permanent deformation which renders the IBC, including the base pallet, if any, unsafe for transport and no loss of contents.

6.5.6.5 **Top lift test**

6.5.6.5.1 *Applicability*

For all types of IBCs which are designed to be lifted from the top, and for flexible IBCs designed to be lifted from the top or the side, as a design type test.

6.5.6.5.2 *Preparation of the IBC for test*

Metal, rigid plastics and composite IBCs shall be filled. A load shall be added and evenly distributed. The mass of filled IBC and the load shall be twice the maximum permissible gross mass. Flexible IBCs shall be filled with a representative material and then shall be loaded to six times their maximum permissible gross mass, the load being evenly distributed.

6.5.6.5.3   *Method of testing*

Metal and flexible IBCs shall be lifted in the manner for which they are designed until clear of the floor and maintained in that position for a period of five minutes.

Rigid plastics and composite IBCs shall be lifted:

.1   by each pair of diagonally opposite lifting devices, so that the hoisting forces are applied vertically, for a period of five minutes; and

.2   by each pair of diagonally opposite lifting devices, so that the hoisting forces are applied towards the centre at 45° to the vertical, for a period of five minutes.

6.5.6.5.4   Other methods of top-lift testing and preparation at least equally effective may be used for flexible IBCs.

6.5.6.5.5   *Criteria for passing the test*

.1   Metal, rigid plastics and composite IBCs: the IBC remains safe for normal conditions of transport, there is no observable permanent deformation of the IBC, including the base pallet, if any, and no loss of contents.

.2   Flexible IBCs: no damage to the IBC or its lifting devices which renders the IBC unsafe for transport or handling and no loss of contents.

6.5.6.6   **Stacking test**

6.5.6.6.1   *Applicability*

For all types of IBCs which are designed to be stacked on each other, as a design type test.

6.5.6.6.2   *Preparation of the IBC for test*

The IBC shall be filled to its maximum permissible gross mass. If the specific gravity of the product being used for testing makes this impracticable, the IBC shall additionally be loaded so that it is tested at its maximum permissible gross mass, the load being evenly distributed.

6.5.6.6.3   *Method of testing*

.1   The IBC shall be placed on its base on level hard ground and subjected to a uniformly distributed superimposed test load (see 6.5.6.6.4). IBCs shall be subjected to the test load for a period of at least:

– 5 minutes, for metal IBCs;

– 28 days at 40°C, for rigid plastics IBCs of types 11H2, 21H2 and 31H2 and for composite IBCs with outer packagings of plastics material which bear the stacking load (i.e., types 11HH1, 11HH2, 21HH1, 21HH2, 31HH1 and 31HH2);

– 24 hours, for all other types of IBCs.

.2   The load shall be applied by one of the following methods:

– one or more IBCs of the same type, filled to the maximum permissible gross mass, stacked on the test IBC;

– appropriate mass loaded on to either a flat plate or a reproduction of the base of the IBC, which is stacked on the test IBC.

6.5.6.6.4   *Calculation of superimposed test load*

The load to be placed on the IBC shall be 1.8 times the combined maximum permissible gross mass of the number of similar IBCs that may be stacked on top of the IBC during transport.

6.5.6.6.5   *Criteria for passing the test*

.1   All types of IBCs other than flexible IBCs: no permanent deformation which renders the IBC, including the base pallet, if any, unsafe for transport and no loss of contents.

.2   Flexible IBCs: no deterioration of the body which renders the IBC unsafe for transport and no loss of contents.

6.5.6.7   **Leakproofness test**

6.5.6.7.1   *Applicability*

For those types of IBCs used for liquids, or for solids filled or discharged under pressure, as a design type test and a periodic test.

6.5.6.7.2 *Preparation of the IBC for test*

The test shall be carried out before the fitting of any thermal insulation equipment. Vented closures shall either be replaced by similar non-vented closures or the vent shall be sealed.

6.5.6.7.3 *Method of testing and pressure to be applied*

The test shall be carried out for a period of at least 10 minutes, using air at a gauge pressure of not less than 20 kPa (0.2 bar). The airtightness of the metal IBC shall be determined by a suitable method such as coating the seams and joints with a soap solution or by air-pressure differential test or by immersing the IBC in water. In the latter case, a correction factor shall be applied for the hydrostatic pressure.

6.5.6.7.4 *Criterion for passing the test*

No leakage of air.

6.5.6.8 **Hydraulic pressure test**

6.5.6.8.1 *Applicability*

For those types of IBCs used for liquids or for solids filled or discharged under pressure, as a design type test.

6.5.6.8.2 *Preparation of the IBC for test*

The test shall be carried out before the fitting of any thermal insulation equipment. Pressure relief devices shall be removed and their apertures plugged, or shall be rendered inoperative.

6.5.6.8.3 *Method of testing*

The test shall be carried out for a period of at least ten minutes, applying a hydraulic pressure of not less than that indicated in 6.5.6.8.4. The IBC shall not be mechanically restrained during the test.

6.5.6.8.4 *Pressures to be applied*

6.5.6.8.4.1 Metal IBCs:

.1   For IBCs of types 21A, 21B and 21N, for packing group I solids, a 250 kPa (2.5 bar) gauge pressure;

.2   For IBCs of types 21A, 21B, 21N, 31A, 31B and 31N, for packing groups II or III substances, a 200 kPa (2 bar) gauge pressure;

.3   In addition, for IBCs of types 31A, 31B and 31N, a 65 kPa (0.65 bar) gauge pressure. This test shall be performed before the 200 kPa (2 bar) test.

6.5.6.8.4.2 Rigid plastics and composite IBCs:

.1   For IBCs of types 21H1, 21H2, 21HZ1 and 21HZ2: 75 kPa (0.75 bar) gauge;

.2   For IBCs of types 31H1, 31H2, 31HZ1 and 31HZ2: whichever is the greater of two values, the first as determined by one of the following methods:

–   the total gauge pressure measured in the IBC (i.e. the vapour pressure of the filling substance and the partial pressure of the air or other inert gases, minus 100 kPa) at 55°C multiplied by a safety factor of 1.5; this total gauge pressure shall be determined on the basis of a maximum degree of filling in accordance with 4.1.1.4 and a filling temperature of 15°C; or

–   1.75 times the vapour pressure at 50°C of the substance to be transported minus 100 kPa, but with a minimum test pressure of 100 kPa; or

–   1.5 times the vapour pressure at 55°C of the substance to be transported minus 100 kPa, but with a minimum test pressure of 100 kPa;

and the second as determined by the following method:

–   twice the static pressure of the substance to be transported, with a minimum of twice the static pressure of water.

6.5.6.8.5 *Criteria for passing the test(s)*

.1   For IBCs of types 21A, 21B, 21N, 31A, 31B and 31N, when subjected to the test pressure specified in 6.5.6.8.4.1.1 or .2: no leakage;

.2   For IBCs of types 31A, 31B and 31N, when subjected to the test pressure specified in 6.5.6.8.4.1.3: neither permanent deformation which would render the IBC unsafe for transport nor leakage; and

.3   For rigid plastics and composite IBCs: no permanent deformation which would render the IBC unsafe for transport and no leakage.

6.5.6.9 Drop test

6.5.6.9.1 *Applicability*

For all types of IBCs, as a design type test.

6.5.6.9.2 *Preparation of the IBC for test*

.1 Metal IBCs: the IBC shall be filled to not less than 95% of its maximum capacity for solids or 98% of its maximum capacity for liquids. Pressure relief devices shall be rendered inoperative or shall be removed and their apertures sealed.

.2 Flexible IBCs: the IBC shall be filled to the maximum permissible gross mass, the contents being evenly distributed.

.3 Rigid plastics and composite IBCs: the IBC shall be filled to not less than 95% of its maximum capacity for solids or 98% of its maximum capacity for liquids. Arrangements provided for pressure relief may be removed and sealed or rendered inoperative. Testing of IBCs shall be carried out when the temperature of the test sample and its contents has been reduced to −18°C or lower. Where test samples of composite IBCs are prepared in this way, the conditioning specified in 6.5.6.3.1 may be waived. Test liquids shall be kept in the liquid state, if necessary by the addition of anti-freeze. This conditioning may be disregarded if the materials in question are of sufficient ductility and tensile strength at low temperatures.

.4 Fibreboard and wooden IBCs: the IBC shall be filled to not less than 95% of its maximum capacity.

6.5.6.9.3 *Method of testing*

The IBC shall be dropped on its base onto a non-resilient, horizontal, flat, massive and rigid surface in conformity with the requirements of 6.1.5.3.4, in such a manner as to ensure that the point of impact is that part of the base of the IBC considered to be the most vulnerable. IBCs of 0.45 m^3 or less capacity shall also be dropped:

.1 Metal IBCs: on the most vulnerable part other than the part of the base of the IBC tested in the first drop;

.2 Flexible IBCs: on the most vulnerable side;

.3 Rigid plastics, composite, fibreboard and wooden IBCs: flat on a side, flat on the top and on a corner.

The same or different IBCs may be used for each drop.

6.5.6.9.4 *Drop height*

For solids and liquids, if the test is performed with the solid or liquid to be transported or with another substance having essentially the same physical characteristics:

Packing group I	Packing group II	Packing group III
1.8 m	1.2 m	0.8 m

For liquids, if the test is performed with water:

(a) where the substances to be transported have a relative density not exceeding 1.2:

Packing group II	Packing group III
1.2 m	0.8 m

(b) where the substances to be transported have a relative density exceeding 1.2, the drop heights shall be calculated on the basis of the relative density (*d*) of the substance to be transported rounded up to the first decimal as follows:

Packing group II	Packing group III
$d \times 1.0$ m	$d \times 0.67$ m

6.5.6.9.5 *Criterion for passing the test(s)*

.1 Metal IBCs: no loss of contents.

.2 Flexible IBCs: no loss of contents. A slight discharge, such as from closures or stitch holes, upon impact shall not be considered to be a failure of the IBC provided that no further leakage occurs after the IBC has been raised clear of the ground.

.3 Rigid plastics, composite, fibreboard and wooden IBCs: no loss of contents. A slight discharge from a closure upon impact shall not be considered to be a failure of the IBC provided that no further leakage occurs.

.4   All IBCs: no damage which renders the IBC unsafe to be transported for salvage or for disposal, and no loss of contents. In addition, the IBC shall be capable of being lifted by an appropriate means until clear of the floor for five minutes.

Note: The criterion in 6.5.6.9.5.4 applies to design types for IBCs manufactured as from 1 January 2011.

6.5.6.10   **Tear test**

6.5.6.10.1   *Applicability*

For all types of flexible IBCs, as a design type test.

6.5.6.10.2   *Preparation of the IBC for test*

The IBC shall be filled to not less than 95% of its capacity and to its maximum permissible gross mass, the contents being evenly distributed.

6.5.6.10.3   *Method of testing*

Once the IBC is placed on the ground, a 100 mm knife score, completely penetrating the wall of a wide face, is made at a 45° angle to the principal axis of the IBC, halfway between the bottom surface and the top level of the contents. The IBC shall then be subjected to a uniformly distributed superimposed load equivalent to twice the maximum permissible gross mass. The load shall be applied for at least five minutes. An IBC which is designed to be lifted from the top or the side shall then, after removal of the superimposed load, be lifted until it is clear of the floor and maintained in that position for a period of five minutes.

6.5.6.10.4   *Criterion for passing the test*

The cut shall not propagate more than 25% of its original length.

6.5.6.11   **Topple test**

6.5.6.11.1   *Applicability*

For all types of flexible IBCs, as a design type test.

6.5.6.11.2   *Preparation of the IBC for test*

The IBC shall be filled to not less than 95% of its capacity and to its maximum permissible gross mass, the contents being evenly distributed.

6.5.6.11.3   *Method of testing*

The IBC shall be caused to topple onto any part of its top onto a rigid, non-resilient, smooth, flat and horizontal surface.

6.5.6.11.4   *Topple height*

Packing group I	Packing group II	Packing group III
1.8 m	1.2 m	0.8 m

6.5.6.11.5   *Criterion for passing the test*

No loss of contents. A slight discharge, such as from closures or stitch holes, upon impact shall not be considered to be a failure of the IBC provided that no further leakage occurs.

6.5.6.12   **Righting test**

6.5.6.12.1   *Applicability*

For all flexible IBCs designed to be lifted from the top or side, as a design type test.

6.5.6.12.2   *Preparation of the IBC for test*

The IBC shall be filled to not less than 95% of its capacity and its maximum permissible gross mass, the contents being evenly distributed.

6.5.6.12.3 *Method of testing*

The IBC, lying on its side, shall be lifted at a speed of 0.1 m/s to an upright position, clear of the floor, by one lifting device or by two lifting devices when four are provided.

6.5.6.12.4 *Criterion for passing the test*

No damage to the IBC or its lifting devices which renders the IBC unsafe for transport or handling.

6.5.6.13 **Vibration test**

6.5.6.13.1 *Applicability*

For all IBCs used for liquids, as a design type test.

Note: This test applies to design types for IBCs manufactured as from 1 January 2011.

6.5.6.13.2 *Preparation of the IBC for test*

A sample IBC shall be selected at random and shall be fitted and closed as for transport. The IBC shall be filled with water to not less than 98% of its maximum capacity.

6.5.6.13.3 *Test method and duration*

6.5.6.13.3.1 The IBC shall be placed in the centre of the test machine platform with a vertical sinusoidal, double amplitude (peak-to-peak displacement) of 25 mm ± 5%. If necessary, restraining devices shall be attached to the platform to prevent the specimen from moving horizontally off the platform without restricting vertical movement.

6.5.6.13.3.2 The test shall be conducted for one hour at a frequency that causes part of the base of the IBC to be momentarily raised from the vibrating platform for part of each cycle to such a degree that a metal shim can be completely inserted intermittently at, at least, one point between the base of the IBC and the test platform. The frequency may need to be adjusted after the initial set point to prevent the packaging from going into resonance. Nevertheless, the test frequency shall continue to allow placement of the metal shim under the IBC as described in this paragraph. The continuing ability to insert the metal shim is essential to passing the test. The metal shim used for this test shall be at least 1.6 mm thick, 50 mm wide, and be of sufficient length to be inserted between the IBC and the test platform a minimum of 100 mm to perform the test.

6.5.6.13.4 *Criteria for passing the test*

No leakage or rupture shall be observed. In addition, no breakage or failure of structural components, such as broken welds or failed fastenings, shall be observed.

6.5.6.14 **Test report**

6.5.6.14.1 A test report containing at least the following particulars shall be drawn up and shall be available to the users of the IBC:

.1 name and address of the test facility;

.2 name and address of applicant (where appropriate);

.3 a unique test report identification;

.4 date of the test report;

.5 manufacturer of the IBC;

.6 description of the IBC design type (such as dimensions, materials, closures, thickness, etc.), including method of manufacture (such as blow-moulding), and which may include drawing(s) and/or photograph(s);

.7 maximum capacity;

.8 characteristics of test contents, such as viscosity and relative density for liquids and particle size for solids;

.9 test descriptions and results; and

.10 signature, with the name and status of the signatory.

6.5.6.14.2 The test report shall contain statements that the IBC, prepared as for transport, was tested in accordance with the appropriate provisions of this chapter and that the use of other packaging methods or components may render it invalid. A copy of the test report shall be available to the competent authority.

Chapter 6.6

Provisions for the construction and testing of large packagings

6.6.1 General

6.6.1.1 The provisions of this chapter do not apply to:

- class 2, except articles including aerosols;
- class 6.2, except clinical waste of UN 3291;
- class 7 packages containing radioactive material.

6.6.1.2 Large packagings shall be manufactured and tested under a quality-assurance programme which satisfies the competent authority in order to ensure that each manufactured packaging meets the provisions of this chapter.

Note: ISO 16106:2006 "Packaging – Transport packages for dangerous goods – Dangerous goods packagings, intermediate bulk containers (IBCs) and large packagings – Guidelines for the application of ISO 9001" provides acceptable guidance on procedures which may be followed.

6.6.1.3 The specific requirements for large packagings in 6.6.4 are based on large packagings currently used. In order to take into account progress in science and technology, there is no objection to the use of large packagings having specifications different from those in 6.6.4 provided they are equally effective, acceptable to the competent authority and able successfully to withstand the tests described in 6.6.5. Methods of testing other than those prescribed in this Code are acceptable provided they are equivalent.

6.6.1.4 Manufacturers and subsequent distributors of packagings shall provide information regarding procedures to be followed and a description of the types and dimensions of closures (including required gaskets) and any other components needed to ensure that packages as presented for transport are capable of passing the applicable performance tests of this chapter.

6.6.2 Code for designating types of large packagings

6.6.2.1 The code used for large packagings consists of:

(a) two Arabic numerals:

"50" for rigid large packagings; or

"51" for flexible large packagings; and

(b) capital letters in Latin characters indicating the nature of the material, such as wood, steel, etc. The capital letters used shall be those shown in 6.1.2.6.

6.6.2.2 The letter "W" may follow the large packaging code. The letter "W" signifies that the large packaging, although of the same type as indicated by the code, is manufactured to a specification different from those in 6.6.4 and is considered equivalent in accordance with the requirements in 6.6.1.3.

6.6.3 Marking

6.6.3.1 **Primary marking**

Each large packaging manufactured and intended for the use according to this Code shall bear durable and legible markings showing:

(a) The United Nations packaging symbol

This symbol shall not be used for any purpose other than certifying that a packaging complies with the relevant requirements in chapter 6.1, 6.2, 6.3, 6.5 or 6.6. For metal large packagings on which the marking is stamped or embossed, the capital letters "UN" may be applied instead of the symbol;

(b) the code "50" designating a large rigid packaging or "51" for flexible large packagings, followed by the material type in accordance with 6.5.1.4.1.2;

(c) a capital letter designating the packing group(s) for which the design type has been approved:

"X" for packing groups I, II and III

"Y" for packing groups II and III

"Z" for packing group III only;

(d) the month and year (last two digits) of manufacture;

(e) the State authorizing the allocation of the marks, indicated by the distinguishing sign for motor vehicles in international traffic;

(f) the name or symbol of the manufacturer and other identification of the large packagings as specified by the competent authority;

(g) the stacking test load* in kilograms. For large packagings not designed for stacking, the figure "0" shall be shown;

(h) the maximum permissible gross mass in kilograms.

The primary marking required above shall be applied in the sequence of the subparagraphs. Each element of the marking applied in accordance with subparagraphs (a) to (h) shall be clearly separated, such as by a slash or space, so as to be easily identifiable.

6.6.3.2 Examples of the marking

 50A/X/05 01/N/PQRS 2500/1000

For a large steel packaging suitable for stacking; stacking load: 2500 kg; maximum gross mass: 1000 kg.

 50H/Y/04 02/D/ABCD 987 0/800

For a large plastics packaging not suitable for stacking; maximum gross mass: 800 kg.

 51H/Z/06 01/S/1999 0/500

For a large flexible packaging not suitable for stacking; maximum gross mass: 500 kg.

6.6.4   Specific provisions for large packagings

6.6.4.1   Specific provisions for metal large packagings

50A   steel

50B   aluminium

50N   metal (other than steel or aluminium)

6.6.4.1.1 The large packaging shall be made of suitable ductile metal in which the weldability has been fully demonstrated. Welds shall be skillfully made and afford complete safety. Low-temperature performance shall be taken into account when appropriate.

6.6.4.1.2 Care shall be taken to avoid damage by galvanic action due to the juxtaposition of dissimilar metals.

6.6.4.2   Specific provisions for flexible material large packagings

51H   flexible plastics

51M   flexible paper

6.6.4.2.1 The large packaging shall be manufactured from suitable materials. The strength of the material and the construction of the flexible large packaging shall be appropriate to its capacity and its intended use.

6.6.4.2.2 All materials used in the construction of flexible large packagings of types 51M shall, after complete immersion in water for not less than 24 hours, retain at least 85% of the tensile strength as measured originally on the material conditioned to equilibrium at 67% relative humidity or less.

---

* The stacking test load in kilograms to be placed on the large packaging shall be 1.8 times the combined maximum permissible gross mass of the number of similar large packagings that may be stacked on top of the large packaging during transport (see 6.6.5.3.3.4).

6.6.4.2.3 Seams shall be formed by stitching, heat sealing, gluing or any equivalent method. All stitched seam-ends shall be secured.

6.6.4.2.4 Flexible large packagings shall provide adequate resistance to ageing and to degradation caused by ultraviolet radiation or the climatic conditions, or by the substance contained, thereby rendering them appropriate to their intended use.

6.6.4.2.5 For plastics flexible large packagings where protection against ultraviolet radiation is required, it shall be provided by the addition of carbon black or other suitable pigments or inhibitors. These additives shall be compatible with the contents and remain effective throughout the life of the large packaging. Where use is made of carbon black, pigments or inhibitors other than those used in the manufacture of the tested design type, re-testing may be waived if changes in the carbon black content, the pigment content or the inhibitor content do not adversely affect the physical properties of the material of construction.

6.6.4.2.6 Additives may be incorporated into the material of the large packaging to improve the resistance to ageing or to serve other purposes, provided that these do not adversely affect the physical or chemical properties of the material.

6.6.4.2.7 When filled, the ratio of height to width shall be not more than 2:1.

6.6.4.3 **Specific provisions for plastics large packagings**

50H   rigid plastics

6.6.4.3.1 The large packaging shall be manufactured from suitable plastics material of known specifications and be of adequate strength in relation to its capacity and its intended use. The material shall be adequately resistant to ageing and to degradation caused by the substance contained or, where relevant, by ultraviolet radiation. Low-temperature performance shall be taken into account when appropriate. Any permeation of the substance contained shall not constitute a danger under normal conditions of transport.

6.6.4.3.2 Where protection against ultraviolet radiation is required, it shall be provided by the addition of carbon black or other suitable pigments or inhibitors. These additives shall be compatible with the contents and remain effective throughout the life of the outer packaging. Where use is made of carbon black, pigments or inhibitors other than those used in the manufacture of the tested design type, re-testing may be waived if changes in the carbon black content, the pigment content or the inhibitor content do not adversely affect the physical properties of the material of construction.

6.6.4.3.3 Additives may be incorporated into the material of the large packaging to improve the resistance to ageing or to serve other purposes, provided that these do not adversely affect the physical or chemical properties of the material.

6.6.4.4 **Specific provisions for fibreboard large packagings**

50G   rigid fibreboard

6.6.4.4.1 Strong and good-quality solid or double-faced corrugated fibreboard (single or multiwall) shall be used, appropriate to the capacity of the large packagings and to their intended use. The water resistance of the outer surface shall be such that the increase in mass, as determined in a test carried out over a period of 30 minutes by the Cobb method of determining water absorption, is not greater than 155 g/m^2 – see ISO 535:1991. It shall have proper bending qualities. Fibreboard shall be cut, creased without scoring, and slotted so as to permit assembly without cracking, surface breaks or undue bending. The fluting of corrugated fibreboard shall be firmly glued to the facings.

6.6.4.4.2 The walls, including top and bottom, shall have a minimum puncture resistance of 15 J, measured according to ISO 3036:1975.

6.6.4.4.3 Manufacturing joins in the outer packaging of large packagings shall be made with an appropriate overlap and shall be taped, glued, stitched with metal staples or fastened by other means at least equally effective. Where joins are effected by gluing or taping, a water-resistant adhesive shall be used. Metal staples shall pass completely through all pieces to be fastened and be formed or protected so that any inner liner cannot be abraded or punctured by them.

6.6.4.4.4 Any integral pallet base forming part of a large packaging or any detachable pallet shall be suitable for mechanical handling with the large packaging filled to its maximum permissible gross mass.

6.6.4.4.5 The pallet or integral base shall be designed so as to avoid any protrusion of the base of the large packaging that might be liable to damage in handling.

6.6.4.4.6 The body shall be secured to any detachable pallet to ensure stability in handling and transport. Where a detachable pallet is used, its top surface shall be free from sharp protrusions that might damage the large packaging.

6.6.4.4.7 Strengthening devices such as timber supports to increase stacking performance may be used but shall be external to the liner.

6.6.4.4.8 Where large packagings are intended for stacking, the bearing surface shall be such as to distribute the load in a safe manner.

6.6.4.5 **Specific provisions for wooden large packagings**

    50C natural wood

    50D plywood

    50F reconstituted wood

6.6.4.5.1 The strength of the materials used and the method of construction shall be appropriate to the capacity and intended use of the large packagings.

6.6.4.5.2 Natural wood shall be well seasoned, commercially dry and free from defects that would materially lessen the strength of any part of the large packaging. Each part of the large packaging shall consist of one piece or be equivalent thereto. Parts are considered equivalent to one piece when a suitable method of glued assembly is used, as for instance Lindermann joint, tongue and groove joint, ship lap or rabbet joint, or butt joint with at least two corrugated metal fasteners at each joint, or when other methods at least equally effective are used.

6.6.4.5.3 Large packagings of plywood shall be at least three-ply. They shall be made of well-seasoned rotary-cut, sliced or sawn veneer, commercially dry and free from defects that would materially lessen the strength of the large packaging. All adjacent plies shall be glued with water-resistant adhesive. Other suitable materials may be used with plywood for the construction of the large packaging.

6.6.4.5.4 Large packagings of reconstituted wood shall be made of water-resistant reconstituted wood such as hardboard, particle board or other suitable type.

6.6.4.5.5 Large packagings shall be firmly nailed or secured to corner posts or ends or be assembled by equally suitable devices.

6.6.4.5.6 Any integral pallet base forming part of a large packaging or any detachable pallet shall be suitable for mechanical handling with the large packaging filled to its maximum permissible gross mass.

6.6.4.5.7 The pallet or integral base shall be designed so as to avoid any protrusion of the base of the large packaging that might be liable to damage in handling.

6.6.4.5.8 The body shall be secured to any detachable pallet to ensure stability in handling and transport. Where a detachable pallet is used, its top surface shall be free from sharp protrusions that might damage the large packaging.

6.6.4.5.9 Strengthening devices such as timber supports to increase stacking performance may be used but shall be external to the liner.

6.6.4.5.10 Where large packagings are intended for stacking, the bearing surface shall be such as to distribute the load in a safe manner.

6.6.5   Test provisions for large packagings

6.6.5.1   Performance and frequency of test

6.6.5.1.1 The design type of each large packaging shall be tested as provided in 6.6.5.3 in accordance with procedures established by the competent authority.

6.6.5.1.2 Each large packaging design type shall successfully pass the tests prescribed in this chapter before being used. A large packaging design type is defined by the design, size, material and thickness, manner of construction and packing, but may include various surface treatments. It also includes large packagings that differ from the design type only in their lesser design height.

6.6.5.1.3 Tests shall be repeated on production samples at intervals established by the competent authority. For such tests on fibreboard large packagings, preparation at ambient conditions is considered equivalent to the provisions of 6.6.5.2.3.

6.6.5.1.4 Tests shall also be repeated after each modification which alters the design, material or manner of construction of large packagings.

6.6.5.1.5 The competent authority may permit the selective testing of large packagings that differ only in minor respects from a tested type, such as smaller sizes of inner packagings or inner packagings of lower net mass, and large packagings which are produced with small reductions in external dimension(s).

6.6.5.1.6 (Reserved)

Note: For the conditions for assembling different inner packagings in a large packaging and permissible variations in inner packagings, see 4.1.1.5.1.

6.6.5.1.7 The competent authority may at any time require proof, by tests in accordance with this section, that serially produced large packagings meet the provisions of the design type tests.

6.6.5.1.8 Provided the validity of the test results is not affected, and with the approval of the competent authority, several tests may be made on one sample.

6.6.5.2 Preparation for testing

6.6.5.2.1 Tests shall be carried out on large packagings prepared as for transport, including the inner packagings or articles used. Inner packagings shall be filled to not less than 98% of their maximum capacity for liquids or 95% for solids. For large packagings where the inner packagings are designed to carry liquids and solids, separate testing is required for both liquid and solid contents. The substances in the inner packagings or the articles to be transported in the large packagings may be replaced by other material or articles except where this would invalidate the results of the tests. When other inner packagings or articles are used, they shall have the same physical characteristics (mass, etc.) as the inner packagings or articles to be carried. It is permissible to use additives, such as bags of lead shot, to achieve the requisite total package mass, so long as they are placed so that the test results are not affected.

6.6.5.2.2 In the drop tests for liquids, when another substance is used, its relative density and viscosity shall be similar to those of the substances being transported. Water may also be used for the drop tests for liquids under the following conditions:

.1 where the substances to be carried have a relative density not exceeding 1.2, the drop heights shall be those shown in the table in 6.6.5.3.4.4;

.2 where the substances to be transported have a relative density exceeding 1.2, the drop height shall be calculated on the basis of the relative density (d) of the substance to be transported, rounded up to the first decimal, as follows:

Packing group I	Packing group II	Packing group III
$d \times 1.5$ m	$d \times 1.0$ m	$d \times 0.67$ m

6.6.5.2.3 Large packagings made of plastics materials and large packagings containing inner packagings of plastic materials – other than bags intended to contain solids or articles – shall be drop tested when the temperature of the test sample and its contents has been reduced to -18°C or lower. This conditioning may be disregarded if the materials in question are of sufficient ductility and tensile strength at low temperatures. Where test samples are prepared in this way, the conditioning in 6.6.5.2.4 may be waived. Test liquids shall be kept in the liquid state by the addition of anti-freeze if necessary.

6.6.5.2.4 Large packagings of fibreboard shall be conditioned for at least 24 hours in an atmosphere having a controlled temperature and relative humidity (r.h.). There are three options, one of which shall be chosen. The preferred atmosphere is 23°C $\pm 2^{\circ}$C and 50% ± 2% r.h. The two other options are 20°C $\pm 2^{\circ}$C and 65% ± 2% r.h. or 27°C $\pm 2^{\circ}$C and 65% ± 2% r.h.

Note: Average values shall fall within these limits. Short-term fluctuation and measurement limitations may cause individual measurements to vary by up to ± 5% relative humidity without significant impairment of test reproducibility.

6.6.5.3 Test provisions

6.6.5.3.1 *Bottom lift test*

6.6.5.3.1.1 *Applicability*

For all types of large packagings which are fitted with means of lifting from the base, as a design type test.

6.6.5.3.1.2 *Preparation of large packaging for test*

The large packaging shall be filled to 1.25 times its maximum permissible gross mass, the load being evenly distributed.

6.6.5.3.1.3 *Method of testing*

The large packaging shall be raised and lowered twice by a lift truck with the forks centrally positioned and spaced at three quarters of the dimension of the side of entry (unless the points of entry are fixed). The forks shall penetrate to three quarters of the depth in the direction of entry. The test shall be repeated from each possible direction of entry.

6.6.5.3.1.4 *Criteria for passing the test*

No permanent deformation which renders the large packaging unsafe for transport and no loss of contents.

6.6.5.3.2 **Top lift test**

6.6.5.3.2.1 *Applicability*

For types of large packaging which are intended to be lifted from the top and fitted with means of lifting, as a design type test.

6.6.5.3.2.2 *Preparation of large packaging for test*

The large packaging shall be loaded to twice its maximum permissible gross mass. A flexible large packaging shall be loaded to six times its maximum permissible gross mass, the load being evenly distributed.

6.6.5.3.2.3 *Method of testing*

The large packaging shall be lifted in the manner for which it is designed until clear of the floor and maintained in that position for a period of five minutes.

6.6.5.3.2.4 *Criteria for passing the test*

.1   Metal, rigid plastics and composite large packagings: no permanent deformation which renders the large packaging, including the base pallet, if any, unsafe for transport and no loss of contents.

.2   Flexible large packagings: no damage to the large packaging or its lifting devices which renders the large packaging unsafe for transport or handling and no loss of contents.

6.6.5.3.3 **Stacking test**

6.6.5.3.3.1 *Applicability*

For all types of large packaging which are designed to be stacked on each other, as a design type test.

6.6.5.3.3.2 *Preparation of large packaging for test*

The large packaging shall be filled to its maximum permissible gross mass.

6.6.5.3.3.3 *Method of testing*

The large packaging shall be placed on its base on level hard ground and subjected to a uniformly distributed superimposed test load (see 6.6.5.3.3.4) for a period of at least five minutes: for large packaging of wood, fibreboard and plastics materials for a period of 24 hours.

6.6.5.3.3.4 *Calculation of superimposed test load*

The load to be placed on the large packaging shall be 1.8 times the combined maximum permissible gross mass of the number of similar large packagings that may be stacked on top of the large packaging during transport.

6.6.5.3.3.5 *Criteria for passing the test*

.1   All types of large packagings other than flexible large packagings: no permanent deformation which renders the large packaging, including the base pallet, if any, unsafe for transport and no loss of contents.

.2   Flexible large packagings: no deterioration of the body which renders the large packaging unsafe for transport and no loss of contents.

6.6.5.3.4 **Drop test**

6.6.5.3.4.1 *Applicability*

For all types of large packaging, as a design type test.

6.6.5.3.4.2 *Preparation of large packaging for testing*

The large packaging shall be filled in accordance with 6.6.5.2.1.

6.6.5.3.4.3 *Method of testing*

The large packaging shall be dropped onto a non-resilient, horizontal, flat, massive and rigid surface in conformity with the requirements of 6.1.5.3.4, in such a manner as to ensure that the point of impact is that part of the base of the large packaging considered to be the most vulnerable.

6.6.5.3.4.4 *Drop height*

Packing group I	Packing group II	Packing group III
1.8 m	1.2 m	0.8 m

Note: Packaging for substances and articles of class 1, self-reactive substances of class 4.1 and organic peroxides of class 5.2 shall be tested at the packing group II performance level.

6.6.5.3.4.5 *Criteria for passing the test*

6.6.5.3.4.5.1 The large packaging shall not exhibit any damage liable to affect safety during transport. There shall be no leakage of the filling substance from inner packaging(s) or article(s).

6.6.5.3.4.5.2 No rupture is permitted in a large packaging for articles of class 1 which would permit the spillage of loose explosive substances or articles from the large packaging.

6.6.5.3.4.5.3 Where a large packaging undergoes a drop test, the sample passes the test if the entire contents are retained even if the closure is no longer sift-proof.

6.6.5.4 **Certification and test report**

6.6.5.4.1 In respect of each design type of large packaging, a certificate and mark (as in 6.6.3) shall be issued attesting that the design type, including its equipment, meets the test provisions.

6.6.5.4.2 A test report containing at least the following particulars shall be drawn up and shall be available to the users of the large packaging:

.1 name and address of the test facility;

.2 name and address of applicant (where appropriate);

.3 a unique test report identification;

.4 date of the test report;

.5 manufacturer of the large packaging;

.6 description of the large packaging design type (such as dimensions, materials, closures, thickness, etc.) and/or photograph(s);

.7 maximum capacity/maximum permissible gross mass;

.8 characteristics of test contents, such as types and descriptions of inner packaging or articles used;

.9 test descriptions and results;

.10 the test report shall be signed with the name and status of the signatory.

6.6.5.4.3 The test report shall contain statements that the large packaging prepared as for transport was tested in accordance with the appropriate provisions of this chapter and that the use of other packaging methods or components may render it invalid. A copy of the test report shall be available to the competent authority.

6

Chapter 6.7

Provisions for the design, construction, inspection and testing of portable tanks and multiple-element gas containers (MEGCs)

Note: The provisions of this chapter also apply to road tank vehicles to the extent indicated in chapter 6.8.

6.7.1   Application and general provisions

6.7.1.1   The provisions of this chapter apply to portable tanks intended for the transport of dangerous goods, and to MEGCs intended for the transport of non-refrigerated gases of class 2, by all modes of transport. In addition to the provisions of this chapter, unless otherwise specified, the applicable provisions of the International Convention for Safe Containers (CSC) 1972, as amended, shall be fulfilled by any multimodal portable tank or MEGC which meets the definition of a "container" within the terms of that Convention. Additional provisions may apply to offshore portable tanks that are handled in open seas.

6.7.1.1.1   The International Convention for Safe Containers does not apply to offshore tank-containers that are handled in open seas. The design and testing of offshore tank-containers shall take into account the dynamic lifting and impact forces that may occur when a tank is handled in open seas in adverse weather and sea conditions. The provisions for such tanks shall be determined by the approving competent authority (see also MSC/Circ. 860 "Guidelines for the approval of offshore containers handled in open seas").

6.7.1.2   In recognition of scientific and technological advances, the technical provisions of this chapter may be varied by alternative arrangements. These alternative arrangements shall offer a level of safety not less than that given by the provisions of this chapter with respect to the compatibility with substances transported and the ability of the portable tank to withstand impact, loading and fire conditions. For international transport, alternative arrangement portable tanks or MEGCs shall be approved by the applicable competent authorities.

6.7.1.3   When a substance is not assigned a portable tank instruction (T1 to T75) in the Dangerous Goods List in chapter 3.2, interim approval for transport may be issued by the competent authority of the country of origin. The approval shall be included in the documentation of the consignment and contain, as a minimum, the information normally provided in the portable tank instructions and the conditions under which the substance shall be transported. Appropriate measures shall be initiated by the competent authority to include the assignment in the Dangerous Goods List.

6.7.2   Provisions for the design, construction, inspection and testing of portable tanks intended for the transport of substances of class 1 and classes 3 to 9

6.7.2.1   Definitions

For the purposes of this section:

Design pressure means the pressure to be used in calculations required by a recognized pressure-vessel code. The design pressure shall be not less than the highest of the following pressures:

.1   the maximum effective gauge pressure allowed in the shell during filling or discharge; or

.2   the sum of:

.1   the absolute vapour pressure (in bar) of the substance at 65°C (or at the highest temperature during filling, discharge or transport for substances which are filled, discharged or transported over 65°C), minus 1 bar;

.2   the partial pressure (in bar) of air or other gases in the ullage space, being determined by a maximum ullage temperature of 65°C and a liquid expansion due to an increase in mean bulk temperature of $t_r - t_f$ (t_f = filling temperature, usually 15°C; t_r = 50°C, maximum mean bulk temperature); and

.3   a head pressure determined on the basis of the static forces specified in 6.7.2.2.12, but not less than 0.35 bar.

.3   two thirds of the minimum test pressure specified in the applicable portable tank instruction in 4.2.5.2.6;

Design temperature range for the shell shall be –40°C to 50°C for substances transported under ambient conditions. For the other substances filled, discharged or transported above 50°C, the design temperature shall not be less than the maximum temperature of the substance during filling, discharge or transport. More severe design temperatures shall be considered for portable tanks subjected to severe climatic conditions;

Fine grain steel means steel which has a ferritic grain size of 6 or finer when determined in accordance with ASTM E 112-96 or as defined in EN 10028-3, Part 3;

Fusible element means a non-reclosable pressure relief device that is thermally actuated;

Leakproofness test means a test using gas, subjecting the shell and its service equipment to an effective internal pressure of not less than 25% of the MAWP;

Maximum allowable working pressure (MAWP) means a pressure that shall be not less than the highest of the following pressures measured at the top of the shell while in operating position:

.1   the maximum effective gauge pressure allowed in the shell during filling or discharge; or

.2   the maximum effective gauge pressure to which the shell is designed, which shall be not less than the sum of:

   .1   the absolute vapour pressure (in bar) of the substance at 65°C (or at the highest temperature during filling, discharge or transport for substances which are filled, discharged or transported over 65°C) minus 1 bar; and

   .2   the partial pressure (in bar) of air or other gases in the ullage space, being determined by a maximum ullage temperature of 65°C and a liquid expansion due to an increase in mean bulk temperature of $t_r – t_f$ (t_f = filling temperature, usually 15°C; t_r = 50°C, maximum mean bulk temperature);

Maximum permissible gross mass (MPGM) means the sum of the tare mass of the portable tank and the heaviest load authorized for transport;

Mild steel means a steel with a guaranteed minimum tensile strength of 360 N/mm^2 to 440 N/mm^2 and a guaranteed minimum elongation at fracture conforming to 6.7.2.3.3.3;

Offshore portable tank means a portable tank specially designed for repeated use for transport of dangerous goods to, from and between offshore facilities. An offshore portable tank is designed and constructed in accordance with MSC/Circ.860 "Guidelines for the approval of containers handled in open seas";

Portable tank means a multimodal tank used for the transport of substances of class 1 and classes 3 to 9. The portable tank includes a shell fitted with service equipment and structural equipment necessary for the transport of dangerous substances. The portable tank shall be capable of being filled and discharged without the removal of its structural equipment. It shall possess stabilizing members external to the shell, and shall be capable of being lifted when full. It shall be designed primarily to be loaded onto a transport vehicle or ship and shall be equipped with skids, mountings or accessories to facilitate mechanical handling. Road tank-vehicles, rail tank-wagons, non-metallic tanks and intermediate bulk containers (IBCs) are not considered to fall within the definition for portable tanks;

Reference steel means a steel with a tensile strength of 370 N/mm^2 and an elongation at fracture of 27%;

Service equipment means measuring instruments and filling, discharge, venting, safety, heating, cooling and insulating devices;

Shell means the part of the portable tank which retains the substance intended for transport (tank proper), including openings and their closures, but does not include service equipment or external structural equipment;

Structural equipment means the reinforcing, fastening, protective and stabilizing members external to the shell;

Test pressure means the maximum gauge pressure at the top of the shell during the hydraulic pressure test, equal to not less than 1.5 times the design pressure. The minimum test pressure for portable tanks intended for specific substances is specified in the applicable portable tank instruction in 4.2.5.2.6.

6.7.2.2   **General design and construction provisions**

6.7.2.2.1   Shells shall be designed and constructed in accordance with the provisions of a pressure-vessel code recognized by the competent authority. Shells shall be made of metallic materials suitable for forming. The materials shall, in principle, conform to national or international material standards. For welded shells, only a material whose weldability has been fully demonstrated shall be used. Welds shall be skillfully made and afford complete safety. When the manufacturing process or the materials make it necessary, the shells shall be suitably heat-treated to guarantee adequate toughness in the weld and in the heat-affected zones. In choosing the material, the design temperature range shall be taken into account with respect to risk of brittle fracture, to stress corrosion cracking and to resistance to impact. When fine grain-steel is used, the guaranteed value of the yield strength shall be not more than 460 N/mm^2 and the guaranteed value of the upper limit of the tensile strength shall be not more than 725 N/mm^2 according to the material specification. Aluminium may only be used as a construction material when indicated in a portable tank special provision assigned to a specific substance in the Dangerous Goods List or when approved by the competent authority. When aluminium is authorized, it shall be insulated to prevent significant loss of physical properties when subjected to a heat load of 110 kW/m^2 for a period of not less than 30 minutes. The insulation shall remain effective at all temperatures less than 649°C and shall be jacketed with a material with a melting point of not less than 700°C. Portable tank materials shall be suitable for the external environment in which they may be transported.

6.7.2.2.2 Portable tank shells, fittings, and pipework shall be constructed from materials which are:

.1 substantially immune to attack by the substance(s) intended to be transported; or

.2 properly passivated or neutralized by chemical reaction; or

.3 lined with corrosion-resistant material directly bonded to the shell or attached by equivalent means.

6.7.2.2.3 Gaskets shall be made of materials not subject to attack by the substance(s) intended to be transported.

6.7.2.2.4 When shells are lined, the lining shall be substantially immune to attack by the substance(s) intended to be transported, homogeneous, non-porous, free from perforations, sufficiently elastic and compatible with the thermal expansion characteristics of the shell. The lining of every shell, shell fittings and piping shall be continuous, and shall extend around the face of any flange. Where external fittings are welded to the tank, the lining shall be continuous through the fitting and around the face of external flanges.

6.7.2.2.5 Joints and seams in the lining shall be made by fusing the material together or by other equally effective means.

6.7.2.2.6 Contact between dissimilar metals which could result in damage by galvanic action shall be avoided.

6.7.2.2.7 The materials of the portable tank, including any devices, gaskets, linings and accessories, shall not adversely affect the substance(s) intended to be transported in the portable tank.

6.7.2.2.8 Portable tanks shall be designed and constructed with supports to provide a secure base during transport and with suitable lifting and tie-down attachments.

6.7.2.2.9 Portable tanks shall be designed to withstand, without loss of contents, at least the internal pressure due to the contents and the static, dynamic and thermal loads during normal conditions of handling and transport. The design shall demonstrate that the effects of fatigue, caused by repeated application of these loads through the expected life of the portable tank, have been taken into account.

6.7.2.2.9.1 For portable tanks that are intended for use as offshore tank-containers, the dynamic stresses imposed by handling in open seas shall be taken into account.

6.7.2.2.10 A shell which is to be equipped with a vacuum-relief device shall be designed to withstand, without permanent deformation, an external pressure of not less than 0.21 bar above the internal pressure. The vacuum-relief device shall be set to relieve at a vacuum setting not greater than –0.21 bar unless the shell is designed for a higher external overpressure, in which case the vacuum-relief pressure of the device to be fitted shall be not greater than the tank design vacuum pressure. A shell used for the transport of solid substances of packing groups II or III only which do not liquefy during transport may be designed for a lower external pressure, subject to competent authority's approval. In this case, the vacuum-relief device shall be set to relieve at this lower pressure. A shell that is not to be fitted with a vacuum-relief device shall be designed to withstand, without permanent deformation, an external pressure of not less than 0.4 bar above the internal pressure.

6.7.2.2.11 Vacuum-relief devices used on portable tanks intended for the transport of substances meeting the flashpoint criteria of class 3, including elevated-temperature substances transported at or above their flashpoint, shall prevent the immediate passage of flame into the shell, or the portable tank shall have a shell capable of withstanding, without leakage, an internal explosion resulting from the passage of flame into the shell.

6.7.2.2.12 Portable tanks and their fastenings shall, under the maximum permissible load, be capable of absorbing the following separately applied static forces:

.1 in the direction of travel: twice the MPGM multiplied by the acceleration due to gravity (*g*);*

---

* For calculation purposes, *g* = 9.81 m/s^2.

.2 horizontally at right angles to the direction of travel: the MPGM (when the direction of travel is not clearly determined, the forces shall be equal to twice the MPGM) multiplied by the acceleration due to gravity (g);*

.3 vertically upwards: the MPGM multiplied by the acceleration due to gravity (g);* and

.4 vertically downwards: twice the MPGM (total loading including the effect of gravity) multiplied by the acceleration due to gravity (g).*

6.7.2.2.13 Under each of the forces in 6.7.2.2.12, the safety factor to be observed shall be as follows:

.1 for metals having a clearly defined yield point, a safety factor of 1.5 in relation to the guaranteed yield strength; or

.2 for metals with no clearly defined yield point, a safety factor of 1.5 in relation to the guaranteed 0.2% proof strength and, for austenitic steels, the 1% proof strength.

6.7.2.2.14 The value of yield strength or proof strength shall be the value according to national or international material standards. When austenitic steels are used, the specified minimum values of yield strength or proof strength according to the material standards may be increased by up to 15% when these greater values are attested in the material inspection certificate. When no material standard exists for the metal in question, the value of yield strength or proof strength used shall be approved by the competent authority.

6.7.2.2.15 Portable tanks shall be capable of being electrically earthed when intended for the transport of substances meeting the flashpoint criteria of class 3, including elevated-temperature substances transported above their flashpoint. Measures shall be taken to prevent dangerous electrostatic discharge.

6.7.2.2.16 When required for certain substances by the applicable portable tank instruction indicated in column 12 or 13 of the Dangerous Goods List, or by a portable tank special provision indicated in column 12 or 14, portable tanks shall be provided with additional protection, which may take the form of additional shell thickness or a higher test pressure, the additional shell thickness or higher test pressure being determined in the light of the inherent risks associated with the transport of the substances concerned.

6.7.2.2.17 Thermal insulation directly in contact with the shell intended for substances transported at elevated temperature shall have an ignition temperature at least 50°C higher than the maximum designed temperature of the tank.

6.7.2.3 **Design criteria**

6.7.2.3.1 Shells shall be of a design capable of being stress-analysed mathematically or experimentally by resistance strain gauges, or by other methods approved by the competent authority.

6.7.2.3.2 Shells shall be designed and constructed to withstand a hydraulic test pressure not less than 1.5 times the design pressure. Specific provisions are laid down for certain substances in the applicable portable tank instruction indicated in the Dangerous Goods List and described in 4.2.5 or by a portable tank special provision indicated in column 13 of the Dangerous Goods List and described in 4.2.5.3. The minimum shell thickness shall not be less than that specified for these tanks in 6.7.2.4.1 to 6.7.2.4.10.

6.7.2.3.3 For metals exhibiting a clearly defined yield point or characterized by a guaranteed proof strength (0.2% proof strength, generally, or 1% proof strength for austenitic steels), the primary membrane stress σ (sigma) in the shell shall not exceed $0.75R_e$ or $0.50R_m$, whichever is lower, at the test pressure, where:

R_e = yield strength in N/mm^2, or 0.2% proof strength or, for austenitic steels, 1% proof strength;

R_m = minimum tensile strength in N/mm^2.

6.7.2.3.3.1 The values of R_e and R_m to be used shall be the specified minimum values according to national or international material standards. When austenitic steels are used, the specified minimum values for R_e and R_m according to the material standards may be increased by up to 15% when these greater values are attested in the material inspection certificate. When no material standard exists for the metal in question, the values of R_e and R_m used shall be approved by the competent authority or its authorized body.

6.7.2.3.3.2 Steels which have a R_e/R_m ratio of more than 0.85 are not allowed for the construction of welded shells. The values of R_e and R_m to be used in determining this ratio shall be the values specified in the material inspection certificate.

6

---

* For calculation purposes, g = 9.81 m/s^2.

6.7.2.3.3.3   Steels used in the construction of shells shall have an elongation at fracture, in %, of not less than $10,000/R_m$ with an absolute minimum of 16% for fine-grain steels and 20% for other steels. Aluminium and aluminium alloys used in the construction of shells shall have an elongation at fracture, in %, of not less than $10,000/6R_m$ with an absolute minimum of 12%.

6.7.2.3.3.4   For the purpose of determining actual values for materials, it shall be noted that for sheet metal, the axis of the tensile test specimen shall be at right angles (transversely) to the direction of rolling. The permanent elongation at fracture shall be measured on test specimens of rectangular cross-section in accordance with ISO 6892:1984 using a 50 mm gauge length.

6.7.2.4   **Minimum shell thickness**

6.7.2.4.1   The minimum shell thickness shall be the greater thickness based on:

.1   the minimum thickness determined in accordance with the provisions of 6.7.2.4.2 to 6.7.2.4.10;

.2   the minimum thickness determined in accordance with the recognized pressure-vessel code, including the provisions in 6.7.2.3; and

.3   the minimum thickness specified in the applicable portable tank instruction indicated in column 12 or 13 of the Dangerous Goods List, or by a portable tank special provision indicated in column 12 or 14.

6.7.2.4.2   The cylindrical portions, ends (heads) and manhole covers of shells not more than 1.80 m in diameter shall be not less than 5 mm thick in the reference steel or of equivalent thickness in the metal to be used. Shells more than 1.80 m in diameter shall be not less than 6 mm thick in the reference steel or of equivalent thickness in the metal to be used, except that for powdered or granular solid substances of packing group II or III the minimum thickness requirement may be reduced to not less than 5 mm thick in the reference steel or of equivalent thickness in the metal to be used.

6.7.2.4.3   When additional protection against shell damage is provided, portable tanks with test pressures less than 2.65 bar may have the minimum shell thickness reduced, in proportion to the protection provided, as approved by the competent authority. However, shells not more than 1.80 m in diameter shall be not less than 3 mm thick in the reference steel or of equivalent thickness in the metal to be used. Shells more than 1.80 m in diameter shall be not less than 4 mm thick in the reference steel or of equivalent thickness in the metal to be used.

6.7.2.4.4   The cylindrical portions, ends (heads) and manhole covers of all shells shall be not less than 3 mm thick regardless of the material of construction.

6.7.2.4.5   The additional protection referred to in 6.7.2.4.3 may be provided by overall external structural protection, such as suitable "sandwich" construction with the outer sheathing (jacket) secured to the shell, double-wall construction or by enclosing the shell in a complete framework with longitudinal and transverse structural members.

6.7.2.4.6   The equivalent thickness of a metal other than the thickness prescribed for the reference steel in 6.7.2.4.3 shall be determined using the following equation:

$$e_1 = \frac{21.4 \times e_0}{\sqrt[3]{R_{m1} \times A_1}}$$

where:

e_1 = required equivalent thickness (in mm) of the metal to be used;

e_0 = minimum thickness (in mm) of the reference steel specified in the applicable portable tank instruction or by a portable tank special provision indicated in column 12, 13 or 14 of the Dangerous Goods List;

R_{m1} = guaranteed minimum tensile strength (in N/mm^2) of the metal to be used (see 6.7.2.3.3);

A_1 = guaranteed minimum elongation at fracture (in %) of the metal to be used according to national or international standards.

6.7.2.4.7   When, in the applicable portable tank instruction in 4.2.5.2.6, a minimum thickness of 8 mm, 10 mm or 12 mm is specified, it shall be noted that these thicknesses are based on the properties of the reference steel and a shell diameter of 1.80 m. When a metal other than mild steel (see 6.7.2.1) is used or the shell has a diameter of more than 1.80 m, the thickness shall be determined using the following equation:

$$e_1 = \frac{21.4 \times e_0 d_1}{1.8 \sqrt[3]{R_{m1} \times A_1}}$$

where:

e_1 = required equivalent thickness (in mm) of the metal to be used;

e_0 = minimum thickness (in mm) of the reference steel specified in the applicable portable tank instruction or by a portable tank special provision indicated in column 12, 13 or 14 of the Dangerous Goods List;

d_1 = diameter of the shell (in m), but not less than 1.80 m;

R_{m1} = guaranteed minimum tensile strength (in N/mm^2) of the metal to be used (see 6.7.2.3.3);

A_1 = guaranteed minimum elongation at fracture (in %) of the metal to be used according to national or international standards.

6.7.2.4.8 In no case shall the wall thickness be less than that prescribed in 6.7.2.4.2, 6.7.2.4.3 and 6.7.2.4.4. All parts of the shell shall have a minimum thickness as determined by 6.7.2.4.2 to 6.7.2.4.4. This thickness shall be exclusive of any corrosion allowance.

6.7.2.4.9 When mild steel is used (see 6.7.2.1), calculation using the equation in 6.7.2.4.6 is not required.

6.7.2.4.10 There shall be no sudden change of plate thickness at the attachment of the ends (heads) to the cylindrical portion of the shell.

6.7.2.5 **Service equipment**

6.7.2.5.1 Service equipment shall be so arranged as to be protected against the risk of being wrenched off or damaged during handling and transport. When the connection between the frame and the shell allows relative movement between the sub-assemblies, the equipment shall be so fastened as to permit such movement without risk of damage to working parts. The external discharge fittings (pipe sockets, shut-off devices), the internal stop-valve and its seating shall be protected against the danger of being wrenched off by external forces (for example, by using shear sections). The filling and discharge devices (including flanges or threaded plugs) and any protective caps shall be capable of being secured against unintended opening.

6.7.2.5.1.1 For offshore tank-containers, where positioning of service equipment and the design and strength of protection for such equipment is concerned, the increased danger of impact damage when handling such tanks in open seas shall be taken into account.

6.7.2.5.2 All openings in the shell, intended for filling or discharging the portable tank, shall be fitted with a manually operated stop-valve located as close to the shell as reasonably practicable. Other openings, except for openings leading to venting or pressure relief devices, shall be equipped with either a stop-valve or another suitable means of closure located as close to the shell as reasonably practicable.

6.7.2.5.3 All portable tanks shall be fitted with a manhole or other inspection openings of a suitable size to allow for internal inspection and adequate access for maintenance and repair of the interior. Compartmented portable tanks shall have a manhole or other inspection openings for each compartment.

6.7.2.5.4 As far as reasonably practicable, external fittings shall be grouped together. For insulated portable tanks, top fittings shall be surrounded by a spill-collection reservoir with suitable drains.

6.7.2.5.5 Each connection to a portable tank shall be clearly marked to indicate its function.

6.7.2.5.6 Each stop-valve or other means of closure shall be designed and constructed to a rated pressure not less than the MAWP of the shell, taking into account the temperatures expected during transport. All stop-valves with screwed spindles shall close by a clockwise motion of the handwheel. For other stop-valves, the position (open and closed) and direction of closure shall be clearly indicated. All stop-valves shall be designed to prevent unintentional opening.

6.7.2.5.7 No moving parts, such as covers, components of closures, etc., shall be made of unprotected corrodible steel when they are liable to come into frictional or percussive contact with aluminium portable tanks intended for the transport of substances meeting the flashpoint criteria of class 3, including elevated-temperature substances transported above their flashpoint.

6.7.2.5.8 Piping shall be designed, constructed and installed so as to avoid the risk of damage due to thermal expansion and contraction, mechanical shock and vibration. All piping shall be of a suitable metallic material. Welded pipe joints shall be used wherever possible.

6.7.2.5.9 Joints in copper tubing shall be brazed or have an equally strong metal union. The melting point of brazing materials shall be no lower than 525°C. The joints shall not decrease the strength of the tubing, as may happen when cutting threads.

6.7.2.5.10 The burst pressure of all piping and pipe fittings shall be not less than the highest of four times the MAWP of the shell or four times the pressure to which it may be subjected in service by the action of a pump or other device (except pressure relief devices).

6.7.2.5.11   Ductile metals shall be used in the construction of valves and accessories.

6.7.2.5.12   The heating system shall be designed or controlled so that a substance cannot reach a temperature at which the pressure in the tank exceeds its MAWP or causes other hazards (e.g., dangerous thermal decomposition).

6.7.2.5.13   The heating system shall be designed or controlled so that power for internal heating elements is not available unless the heating elements are completely submerged. The temperature at the surface of the heating elements for internal heating equipment or the temperature at the shell for external heating equipment shall, in no case, exceed 80% of the auto-ignition temperature (in $^\circ$C) of the substances carried.

6.7.2.5.14   If an electrical heating system is installed inside the tank, it shall be equipped with an earth leakage circuit breaker with a releasing current of less than 100 mA.

6.7.2.5.15   Electrical switch cabinets mounted to tanks shall not have a direct connection to the tank interior and shall provide protection of at least the equivalent of IP 56 according to IEC 144 or IEC 529.

6.7.2.6   Bottom openings

6.7.2.6.1   Certain substances shall not be transported in portable tanks with bottom openings. When the applicable portable tank instruction identified in the Dangerous Goods List and described in 4.2.5.2.6 indicates that bottom openings are prohibited, there shall be no openings below the liquid level of the shell when it is filled to its maximum permissible filling limit. When an existing opening is closed, it shall be accomplished by internally and externally welding one plate to the shell.

6.7.2.6.2   Bottom discharge outlets for portable tanks carrying certain solid, crystallizable or highly viscous substances shall be equipped with not less than two serially fitted and mutually independent shut-off devices. The design of the equipment shall be to the satisfaction of the competent authority or its authorized body and shall include:

　　.1   an external stop-valve fitted as close to the shell as reasonably practicable; and

　　.2   a liquid-tight closure at the end of the discharge pipe, which may be a bolted blank flange or a screw cap.

6.7.2.6.3   Every bottom discharge outlet, except as provided in 6.7.2.6.2, shall be equipped with three serially fitted and mutually independent shut-off devices. The design of the equipment shall be to the satisfaction of the competent authority or its authorized body and include:

　　.1   a self-closing internal stop-valve, that is a stop-valve within the shell or within a welded flange or its companion flange, such that:

　　　　.1   the control devices for the operation of the valve are designed so as to prevent any unintended opening through impact or other inadvertent act;

　　　　.2   the valve may be operable from above or below;

　　　　.3   if possible, the setting of the valve (open or closed) shall be capable of being verified from the ground;

　　　　.4   except for portable tanks having a capacity of not more than 1000 ℓ, it shall be possible to close the valve from an accessible position of the portable tank that is remote from the valve itself; and

　　　　.5   the valve shall continue to be effective in the event of damage to the external device for controlling the operation of the valve;

　　.2   an external stop-valve fitted as close to the shell as reasonably practicable; and

　　.3   a liquid-tight closure at the end of the discharge pipe, which may be a bolted blank flange or a screw cap.

6.7.2.6.4   For a lined shell, the internal stop-valve required by 6.7.2.6.3.1 may be replaced by an additional external stop-valve. The manufacturer shall satisfy the provisions of the competent authority or its authorized body.

6.7.2.7   Safety relief devices

6.7.2.7.1   All portable tanks shall be fitted with at least one pressure relief device. All relief devices shall be designed, constructed and marked to the satisfaction of the competent authority or its authorized body.

6.7.2.8   Pressure relief devices

6.7.2.8.1   Every portable tank with a capacity not less than 1900 ℓ and every independent compartment of a portable tank with a similar capacity shall be provided with one or more pressure relief devices of the spring-loaded type and may in addition have a frangible disc or fusible element in parallel with the spring-loaded devices except when prohibited by reference to 6.7.2.8.3 in the applicable portable tank instruction in 4.2.5.2.6. The pressure relief devices shall have sufficient capacity to prevent rupture of the shell due to over-pressurization or vacuum resulting from filling, from discharging, or from heating of the contents.

6.7.2.8.2 Pressure relief devices shall be designed to prevent the entry of foreign matter, the leakage of liquid and the development of any dangerous excess pressure.

6.7.2.8.3 When required for certain substances by the applicable portable tank instruction identified in the Dangerous Goods List and described in 4.2.5.2.6, portable tanks shall have a pressure relief device approved by the competent authority. Unless a portable tank in dedicated service is fitted with an approved relief device constructed of materials compatible with the load, the relief device shall comprise a frangible disc preceding a spring-loaded pressure relief device. When a frangible disc is inserted in series with the required pressure relief device, the space between the frangible disc and the pressure relief device shall be provided with a pressure gauge or suitable tell-tale indicator for the detection of disc rupture, pinholing, or leakage which could cause a malfunction of the pressure relief system. The frangible disc shall rupture at a nominal pressure 10% above the start-to-discharge pressure of the relief device.

6.7.2.8.4 Every portable tank with a capacity less than 1900 ℓ shall be fitted with a pressure relief device, which may be a frangible disc when this disc complies with the provisions of 6.7.2.11.1. When no spring-loaded pressure relief device is used, the frangible disc shall be set to rupture at a nominal pressure equal to the test pressure.

6.7.2.8.5 When the shell is fitted for pressure discharge, the inlet line shall be provided with a suitable pressure relief device set to operate at a pressure not higher than the MAWP of the shell, and a stop-valve shall be fitted as close to the shell as reasonably practicable.

6.7.2.9 Setting of pressure relief devices

6.7.2.9.1 It shall be noted that the pressure relief devices shall operate only in conditions of excessive rise in temperature, since the shell shall not be subject to undue fluctuations of pressure during normal conditions of transport (see 6.7.2.12.2).

6.7.2.9.2 The required pressure relief device shall be set to start to discharge at a nominal pressure of five sixths of the test pressure for shells having a test pressure of not more than 4.5 bar and 110% of two thirds of the test pressure for shells having a test pressure of more than 4.5 bar. After discharge, the device shall close at a pressure not more than 10% below the pressure at which the discharge starts. The device shall remain closed at all lower pressures. This requirement does not prevent the use of vacuum relief or combination pressure relief and vacuum relief devices.

6.7.2.10 Fusible elements

6.7.2.10.1 Fusible elements shall operate at a temperature between 110°C and 149°C on condition that the pressure in the shell at the fusing temperature will be not more than the test pressure. They shall be placed at the top of the shell with their inlets in the vapour space, and in no case shall they be shielded from external heat. Fusible elements shall not be utilized on portable tanks with a test pressure which exceeds 2.65 bar. Fusible elements used on portable tanks intended for the transport of elevated-temperature substances shall be designed to operate at a temperature higher than the maximum temperature that will be experienced during transport and shall be to the satisfaction of the competent authority or its authorized body.

6.7.2.11 Frangible discs

6.7.2.11.1 Except as specified in 6.7.2.8.3, frangible discs shall be set to rupture at a nominal pressure equal to the test pressure throughout the design temperature range. Particular attention shall be given to the provisions of 6.7.2.5.1 and 6.7.2.8.3 if frangible discs are used.

6.7.2.11.2 Frangible discs shall be appropriate for the vacuum pressures which may be produced in the portable tank.

6.7.2.12 Capacity of pressure relief devices

6.7.2.12.1 The spring-loaded pressure relief device required by 6.7.2.8.1 shall have a minimum cross-sectional flow area equivalent to an orifice of 31.75 mm diameter. Vacuum relief devices, when used, shall have a cross-sectional flow area not less than 284 mm^2.

6.7.2.12.2 The combined delivery capacity of the pressure relief system (taking into account the reduction of the flow when the portable tank is fitted with frangible discs preceding spring-loaded pressure relief devices or when the spring-loaded pressure relief devices are provided with a device to prevent the passage of the flame), in conditions of complete fire engulfment of the portable tank shall be sufficient to limit the pressure in the shell to 20% above the start-to-discharge pressure of the pressure-limiting device. Emergency pressure relief devices may be used to achieve the full relief capacity prescribed. These devices may be fusible, spring-loaded or frangible disc components, or a combination of spring-loaded and frangible disc devices. The total required capacity of the relief devices may be determined using the formula in 6.7.2.12.2.1 or the table in 6.7.2.12.2.3.

6.7.2.12.2.1 To determine the total required capacity of the relief devices, which shall be regarded as being the sum of the individual capacities of all the contributing devices, the following formula shall be used:

$$Q = 12.4 \frac{FA^{0.82}}{LC} \sqrt{\frac{ZT}{M}}$$

where:

Q = minimum required rate of discharge in cubic metres of air per second (m³/s) at standard conditions: 1 bar and 0°C (273 K);

F = a coefficient with the following value:
for uninsulated shells, $F = 1$
for insulated shells, $F = U(649 - t)/13.6$ but in no case is less than 0.25

where:

U = thermal conductance of the insulation, in kW/m·K, at 38°C;
t = actual temperature of the substance during filling (in °C) (when this temperature is unknown, let $t = 15$°C);

The value of F given above for insulated shells may be taken provided that the insulation is in conformance with 6.7.2.12.2.4;

A = total external surface area of shell in square metres;

Z = the gas compressibility factor in the accumulating condition (when this factor is unknown, let Z equal 1.0);

T = absolute temperature in kelvin (°C + 273) above the pressure relief devices in the accumulating condition;

L = the latent heat of vaporization of the liquid, in kJ/kg, in the accumulating condition;

M = molecular mass of the discharged gas;

C = a constant which is derived from one of the following formulae as a function of the ratio k of specific heats:

$$k = \frac{C_p}{C_v}$$

where:

C_p = specific heat at constant pressure; and

C_v = specific heat at constant volume.

When $k > 1$:

$$C = \sqrt{k \left(\frac{2}{k+1} \right)^{\frac{k+1}{k-1}}}$$

When $k = 1$ or k is unknown:

$$C = \frac{1}{\sqrt{e}} = 0.607$$

where e is the mathematical constant 2.7183.

C may also be taken from the following table:

k	C	k	C	k	C
1.00	0.607	1.26	0.660	1.52	0.704
1.02	0.611	1.28	0.664	1.54	0.707
1.04	0.615	1.30	0.667	1.56	0.710
1.06	0.620	1.32	0.671	1.58	0.713
1.08	0.624	1.34	0.674	1.60	0.716
1.10	0.628	1.36	0.678	1.62	0.719
1.12	0.633	1.38	0.681	1.64	0.722
1.14	0.637	1.40	0.685	1.66	0.725
1.16	0.641	1.42	0.688	1.68	0.728
1.18	0.645	1.44	0.691	1.70	0.731
1.20	0.649	1.46	0.695	2.00	0.770
1.22	0.652	1.48	0.698	2.20	0.793
1.24	0.656	1.50	0.701		

6.7.2.12.2.2 As an alternative to the formula above, shells designed for the transport of liquids may have their relief devices sized in accordance with the table in 6.7.2.12.2.3. This table assumes an insulation value of $F = 1$ and shall be adjusted accordingly when the shell is insulated. Other values used in determining this table are:

$$M = 86.7; \qquad T = 394 \text{ K}; \qquad L = 334.94 \text{ kJ/kg}; \qquad C = 0.607; \qquad Z = 1$$

6.7.2.12.2.3 Minimum required rate of discharge, Q, in cubic metres of air per second at 1 bar and 0°C (273 K):

A Exposed area (square metres)	Q (cubic metres of air per second)	A Exposed area (square metres)	Q (cubic metres of air per second)
2	0.230	37.5	2.539
3	0.320	40	2.677
4	0.405	42.5	2.814
5	0.487	45	2.949
6	0.565	47.5	3.082
7	0.641	50	3.215
8	0.715	52.5	3.346
9	0.788	55	3.476
10	0.859	57.5	3.605
12	0.998	60	3.733
14	1.132	62.5	3.860
16	1.263	65	3.987
18	1.391	67.5	4.112
20	1.517	70	4.236
22.5	1.670	75	4.483
25	1.821	80	4.726
27.5	1.969	85	4.967
30	2.115	90	5.206
32.5	2.258	95	5.442
35	2.400	100	5.676

6.7.2.12.2.4 Insulation systems, used for the purpose of reducing venting capacity, shall be approved by the competent authority or its authorized body. In all cases, insulation systems approved for this purpose shall:

(a) remain effective at all temperatures up to 649°C; and

(b) be jacketed with a material having a melting point of 700°C or greater.

6.7.2.13 **Marking of pressure relief devices**

6.7.2.13.1 Every pressure relief device shall be clearly and permanently marked with the following:

.1 the pressure (in bar or kPa) or temperature (in °C) at which it is set to discharge;

.2 the allowable tolerance at the discharge pressure, for spring-loaded devices;

.3 the reference temperature corresponding to the rated pressure, for frangible discs;

.4 the allowable temperature tolerance, for fusible elements; and

.5 the rated flow capacity of the spring-loaded pressure relief devices, frangible discs or fusible elements in standard cubic metres of air per second (m^3/s).

When practicable, the following information shall also be shown:

.6 the manufacturer's name and relevant catalogue number.

6.7.2.13.2 The rated flow capacity marked on the spring-loaded pressure relief devices shall be determined according to ISO 4126-1:1996.

6.7.2.14 **Connections to pressure relief devices**

6.7.2.14.1 Connections to pressure relief devices shall be of sufficient size to enable the required discharge to pass unrestricted to the safety device. No stop-valve shall be installed between the shell and the pressure relief devices except where duplicate devices are provided for maintenance or other reasons and the stop-valves serving the devices actually in use are locked open or the stop-valves are interlocked so that at least one of the duplicate devices is always in use. There shall be no obstruction in an opening leading to a vent or pressure relief device which might restrict or cut off the flow from the shell to that device. Vents or pipes from the pressure relief device outlets, when used, shall deliver the relieved vapour or liquid to the atmosphere in conditions of minimum back-pressure on the relieving devices.

6

6.7.2.15 Siting of pressure relief devices

6.7.2.15.1 Each pressure relief device inlet shall be situated on top of the shell in a position as near the longitudinal and transverse centre of the shell as reasonably practicable. All pressure relief device inlets shall, under maximum filling conditions, be situated in the vapour space of the shell and the devices shall be so arranged as to ensure the escaping vapour is discharged unrestrictedly. For flammable substances, the escaping vapour shall be directed away from the shell in such a manner that it cannot impinge upon the shell. Protective devices which deflect the flow of vapour are permissible provided the required relief-device capacity is not reduced.

6.7.2.15.2 Arrangements shall be made to prevent access to the pressure relief devices by unauthorized persons and to protect the devices from damage caused by the portable tank overturning.

6.7.2.16 Gauging devices

6.7.2.16.1 Glass level-gauges and gauges made of other fragile material, which are in direct communication with the contents of the tank, shall not be used.

6.7.2.17 Portable tank supports, frameworks, lifting and tie-down attachments

6.7.2.17.1 Portable tanks shall be designed and constructed with a support structure to provide a secure base during transport. The forces specified in 6.7.2.2.12 and the safety factor specified in 6.7.2.2.13 shall be considered in this aspect of the design. Skids, frameworks, cradles or other similar structures are acceptable.

6.7.2.17.2 The combined stresses caused by portable tank mountings (such as cradles, framework, etc.) and portable tank lifting and tie-down attachments shall not cause excessive stress in any portion of the shell. Permanent lifting and tie-down attachments shall be fitted to all portable tanks. Preferably they shall be fitted to the portable tank supports but may be secured to reinforcing plates located on the shell at the points of support.

6.7.2.17.3 In the design of supports and frameworks, the effects of environmental corrosion shall be taken into account.

6.7.2.17.4 Forklift pockets shall be capable of being closed off. The means of closing forklift pockets shall be a permanent part of the framework or permanently attached to the framework. Single-compartment portable tanks with a length less than 3.65 m need not have closed-off forklift pockets provided that:

.1 the shell, including all the fittings, is well protected from being hit by the forklift blades; and

.2 the distance between the centres of the forklift pockets is at least half of the maximum length of the portable tank.

6.7.2.17.5 When portable tanks are not protected during transport, according to 4.2.1.2, the shells and service equipment shall be protected against damage to the shell and service equipment resulting from lateral or longitudinal impact or overturning. External fittings shall be protected so as to preclude the release of the shell contents upon impact or overturning of the portable tank on its fittings. Examples of protection include:

.1 protection against lateral impact, which may consist of longitudinal bars protecting the shell on both sides at the level of the median line;

.2 protection of the portable tank against overturning, which may consist of reinforcement rings or bars fixed across the frame;

.3 protection against rear impact, which may consist of a bumper or frame;

.4 protection of the shell against damage from impact or overturning by use of an ISO frame in accordance with ISO 1496-3:1995.

6.7.2.18 Design approval

6.7.2.18.1 The competent authority or its authorized body shall issue a design approval certificate for any new design of a portable tank. This certificate shall attest that a portable tank has been surveyed by that authority, is suitable for its intended purpose and meets the provisions of this chapter and, where appropriate, the provisions for substances provided in chapter 4.2 and in the Dangerous Goods List in chapter 3.2. When a series of portable tanks are manufactured without change in the design, the certificate shall be valid for the entire series. The certificate shall refer to the prototype test report, the substances or group of substances allowed to be transported, the materials of construction of the shell and lining (when applicable) and an approval number. The approval number shall consist of the distinguishing sign or mark of the State in whose territory the approval was granted, i.e. the distinguishing sign for use in international traffic as prescribed by the Convention on Road Traffic, Vienna, 1968, and a registration number. Any alternative arrangements according to 6.7.1.2 shall be indicated on the certificate. A design approval may serve for the approval of smaller portable tanks made of materials of the same kind and thickness, by the same fabrication techniques and with identical supports, equivalent closures and other appurtenances.

6.7.2.18.2 The prototype test report for the design approval shall include at least the following:

.1 the results of the applicable framework test specified in ISO 1496-3:1995;

.2 the results of the initial inspection and test in 6.7.2.19.3; and

.3 the results of the impact test in 6.7.2.19.1, when applicable.

6.7.2.19 Inspection and testing

6.7.2.19.1 Portable tanks meeting the definition of *container* in the International Convention for Safe Containers (CSC), 1972, as amended, shall not be used unless they are successfully qualified by subjecting a representative prototype of each design to the Dynamic, Longitudinal Impact Test prescribed in the United Nations *Manual of Tests and Criteria*, part IV, section 41. This provision only applies to portable tanks which are constructed according to a design approval certificate which has been issued on or after 1 January 2008.

6.7.2.19.2 The shell and items of equipment of each portable tank shall be inspected and tested before being put into service for the first time (initial inspection and test) and thereafter at not more than five-year intervals (5-year periodic inspection and test) with an intermediate periodic inspection and test (2.5-year periodic inspection and test) midway between the 5-year periodic inspections and tests. The 2.5-year periodic inspection and test may be performed within 3 months of the specified date. An exceptional inspection and test shall be performed regardless of the date of the last periodic inspection and test when necessary according to 6.7.2.19.7.

6.7.2.19.3 The initial inspection and test of a portable tank shall include a check of the design characteristics, an internal and external examination of the portable tank and its fittings with due regard to the substances to be transported, and a pressure test. Before the portable tank is placed into service, a leakproofness test and a test of the satisfactory operation of all service equipment shall also be performed. When the shell and its fittings have been pressure-tested separately, they shall be subjected together after assembly to a leakproofness test.

6.7.2.19.4 The 5-year periodic inspection and test shall include an internal and external examination and, as a general rule, a hydraulic pressure test. For tanks only used for the transport of solid substances other than toxic or corrosive substances, which do not liquefy during transport, the hydraulic pressure test may be replaced by a suitable pressure test at 1.5 times MAWP, subject to competent authority approval. Sheathing, thermal insulation and the like shall be removed only to the extent required for reliable appraisal of the condition of the portable tank. When the shell and equipment have been pressure-tested separately, they shall be subjected together after assembly to a leakproofness test.

6.7.2.19.4.1 The heating system shall be subject to inspection and tests including pressure tests on heating coils or ducts during the 5-year periodic inspection.

6.7.2.19.5 The intermediate 2.5-year periodic inspection and test shall at least include an internal and external examination of the portable tank and its fittings with due regard to the substances intended to be transported, a leakproofness test and a test of the satisfactory operation of all service equipment. Sheathing, thermal insulation and the like shall be removed only to the extent required for reliable appraisal of the condition of the portable tank. For portable tanks dedicated to the transport of a single substance, the 2.5-year internal examination may be waived or substituted by other test methods or inspection procedures specified by the competent authority or its authorized body.

6.7.2.19.6 A portable tank may not be filled and offered for transport after the date of expiry of the last 5-year or 2.5-year periodic inspection and test as required by 6.7.2.19.2. However, a portable tank filled prior to the date of expiry of the last periodic inspection and test may be transported for a period not to exceed three months beyond the date of expiry of the last periodic test or inspection. In addition, a portable tank may be transported after the date of expiry of the last periodic test and inspection:

.1 after emptying but before cleaning, for purposes of performing the next required test or inspection prior to refilling; and

.2 unless otherwise approved by the competent authority, for a period not to exceed six months beyond the date of expiry of the last periodic test or inspection, in order to allow the return of dangerous goods for proper disposal or recycling. Reference to this exemption shall be mentioned in the transport document.

6.7.2.19.7   The exceptional inspection and test is necessary when the portable tank shows evidence of damaged or corroded areas, or leakage, or other conditions that indicate a deficiency that could affect the integrity of the portable tank. The extent of the exceptional inspection and test shall depend on the amount of damage or deterioration of the portable tank. It shall include at least the 2.5-year periodic inspection and test according to 6.7.2.19.5.

6.7.2.19.8   The internal and external examinations shall ensure that:

.1   the shell is inspected for pitting, corrosion, or abrasions, dents, distortions, defects in welds or any other conditions, including leakage, that might render the portable tank unsafe for transport;

.2   the piping, valves, heating/cooling system, and gaskets are inspected for corroded areas, defects, or any other conditions, including leakage, that might render the portable tank unsafe for filling, discharge or transport;

.3   devices for tightening manhole covers are operative and there is no leakage at manhole covers or gaskets;

.4   missing or loose bolts or nuts on any flanged connection or blank flange are replaced or tightened;

.5   all emergency devices and valves are free from corrosion, distortion and any damage or defect that could prevent their normal operation. Remote closure devices and self-closing stop-valves shall be operated to demonstrate proper operation;

.6   linings, if any, are inspected in accordance with criteria outlined by the lining manufacturer;

.7   required markings on the portable tank are legible and in accordance with the applicable provisions; and

.8   the framework, supports and arrangements for lifting the portable tank are in a satisfactory condition.

6.7.2.19.9   The inspections and tests in 6.7.2.19.1, 6.7.2.19.3, 6.7.2.19.4, 6.7.2.19.5 and 6.7.2.19.7 shall be performed or witnessed by an expert approved by the competent authority or its authorized body. When the pressure test is a part of the inspection and test, the test pressure shall be the one indicated on the data plate of the portable tank. While under pressure, the portable tank shall be inspected for any leaks in the shell, piping or equipment.

6.7.2.19.10   In all cases when cutting, burning or welding operations on the shell have been effected, that work shall be to the approval of the competent authority or its authorized body, taking into account the pressure-vessel code used for the construction of the shell. A pressure test to the original test pressure shall be performed after the work is completed.

6.7.2.19.11   When evidence of any unsafe condition is discovered, the portable tank shall not be returned to service until it has been corrected and the test is repeated and passed.

6.7.2.20   **Marking**

6.7.2.20.1   Every portable tank shall be fitted with a corrosion-resistant metal plate permanently attached to the portable tank in a conspicuous place readily accessible for inspection. When, for reasons of portable tank arrangements, the plate cannot be permanently attached to the shell, the shell shall be marked with at least the information required by the pressure-vessel code. As a minimum, at least the following information shall be marked on the plate by stamping or by any other similar method:

Country of manufacture:

U	Approval	Approval	For alternative arrangements (see 6.7.1.2):
N	country	number	"AA"

Manufacturer's name or mark

Manufacturer's serial number

Authorized body for the design approval

Owner's registration number

Year of manufacture

Pressure-vessel code to which the shell is designed

Test pressure . . . . bar/kPa gauge*

MAWP . . . . bar/kPa gauge*

External design pressure[†] . . . . bar/kPa gauge*

Design temperature range . . . . $^{\circ}$C to . . . . $^{\circ}$C

Water capacity at 20°C . . . . litres

Water capacity of each compartment at 20°C . . . . litres

Initial pressure test date and witness identification

MAWP for heating/cooling system . . . . bar/kPa gauge*

Shell material(s) and material standard reference(s)

Equivalent thickness in reference steel . . . . mm

Lining material (when applicable)

Date and type of most recent periodic test(s):

Month . . . . Year . . . . Test pressure . . . . bar/kPa gauge*

Stamp of expert who performed or witnessed the most recent test.

   * The unit used shall be marked.
   † See 6.7.2.2.10.

6.7.2.20.2 The following information shall be marked either on the portable tank itself or on a metal plate firmly secured to the portable tank:

Name of the operator

Maximum permissible gross mass (MPGM) . . . . kg

Unladen (tare) mass . . . . kg

6.7.2.20.3 If a portable tank is designed and approved for handling in open seas, the words "OFFSHORE PORTABLE TANK" shall be marked on the identification plate.

6.7.3 Provisions for the design, construction, inspection and testing of portable tanks intended for the transport of non-refrigerated liquefied gases of class 2

6.7.3.1 Definitions

For the purposes of this section:

Design pressure means the pressure to be used in calculations required by a recognized pressure-vessel code. The design pressure shall be not less than the highest of the following pressures:

.1 the maximum effective gauge pressure allowed in the shell during filling or discharge; or

.2 the sum of:

   .1 the maximum effective gauge pressure to which the shell is designed, as defined in .2 of the MAWP definition (see below); and

   .2 a head pressure determined on the basis of the static forces specified in 6.7.3.2.9, but not less than 0.35 bar;

Design reference temperature means the temperature at which the vapour pressure of the contents is determined for the purpose of calculating the MAWP. The design reference temperature shall be less than the critical temperature of the non-refrigerated liquefied gas intended to be transported to ensure that the gas at all times is liquefied. This value for each portable tank type is as follows:

.1 shell with a diameter of 1.5 m or less: 65°C;

.2 shell with a diameter of more than 1.5 m:

   .1 without insulation or sunshield: 60°C;

   .2 with sunshield (see 6.7.3.2.12): 55°C; and

   .3 with insulation (see 6.7.3.2.12): 50°C;

Design temperature range for the shell shall be –40°C to 50°C for non-refrigerated liquefied gases transported under ambient conditions. More severe design temperatures shall be considered for portable tanks subjected to severe climatic conditions;

Filling density means the average mass of non-refrigerated liquefied gas per litre of shell capacity (kg/ℓ). The filling density is given in portable tank instruction T50 in 4.2.5.2.6;

Leakproofness test means a test using gas subjecting the shell and its service equipment to an effective internal pressure of not less than 25% of the MAWP;

Maximum allowable working pressure (MAWP) means a pressure that shall be not less than the highest of the following pressures measured at the top of the shell while in operating position, but in no case less than 7 bar:

.1 the maximum effective gauge pressure allowed in the shell during filling or discharge; or

.2 the maximum effective gauge pressure to which the shell is designed, which shall be:

   .1 for a non-refrigerated liquefied gas listed in the portable tank instruction T50 in 4.2.5.2.6, the MAWP (in bar) given in portable tank instruction T50 for that gas;

.2 for other non-refrigerated liquefied gases, not less than the sum of:

- the absolute vapour pressure (in bar) of the non-refrigerated liquefied gas at the design reference temperature minus 1 bar; and

- the partial pressure (in bar) of air or other gases in the ullage space, being determined by the design reference temperature and the liquid phase expansion due to an increase of the mean bulk temperature of $t_r - t_f$ (t_f = filling temperature, usually 15°C; t_r = 50°C, maximum mean bulk temperature);

Maximum permissible gross mass (MPGM) means the sum of the tare mass of the portable tank and the heaviest load authorized for transport;

Mild steel means a steel with a guaranteed minimum tensile strength of 360 N/mm^2 to 440 N/mm^2 and a guaranteed minimum elongation at fracture conforming to 6.7.3.3.3.3;

Portable tank means a multimodal tank having a capacity of more than 450 ℓ used for the transport of non-refrigerated liquefied gases of class 2. The portable tank includes a shell fitted with service equipment and structural equipment necessary for the transport of gases. The portable tank shall be capable of being filled and discharged without the removal of its structural equipment. It shall possess stabilizing members external to the shell, and shall be capable of being lifted when full. It shall be designed primarily to be loaded onto a transport vehicle or ship and shall be equipped with skids, mountings or accessories to facilitate mechanical handling. Road tank-vehicles, rail tank-wagons, non-metallic tanks, intermediate bulk containers (IBCs), gas cylinders and large receptacles are not considered to fall within the definition for portable tanks;

Reference steel means a steel with a tensile strength of 370 N/mm^2 and an elongation at fracture of 27%;

Service equipment means measuring instruments and filling, discharge, venting, safety and insulating devices;

Shell means the part of the portable tank which retains the non-refrigerated liquefied gas intended for transport (tank proper), including openings and their closures, but does not include service equipment or external structural equipment;

Structural equipment means reinforcing, fastening, protective and stabilizing members external to the shell;

Test pressure means the maximum gauge pressure at the top of the shell during the pressure test.

6.7.3.2 General design and construction provisions

6.7.3.2.1 Shells shall be designed and constructed in accordance with the provisions of a pressure-vessel code recognized by the competent authority. Shells shall be made of steel suitable for forming. The materials shall, in principle, conform to national or international material standards. For welded shells, only a material whose weldability has been fully demonstrated shall be used. Welds shall be skilfully made and afford complete safety. When the manufacturing process or the materials make it necessary, the shells shall be suitably heat-treated to guarantee adequate toughness in the weld and in the heat-affected zones. In choosing the material, the design temperature range shall be taken into account with respect to risk of brittle fracture, to stress corrosion cracking and to resistance to impact. When fine-grain steel is used, the guaranteed value of the yield strength shall be not more than 460 N/mm^2 and the guaranteed value of the upper limit of the tensile strength shall be not more than 725 N/mm^2, according to the material specification. Portable tank materials shall be suitable for the external environment in which they may be transported.

6.7.3.2.2 Portable tank shells, fittings and pipework shall be constructed of materials which are:

.1 substantially immune to attack by the non-refrigerated liquefied gas(es) intended to be transported; or

.2 properly passivated or neutralized by chemical reaction.

6.7.3.2.3 Gaskets shall be made of materials compatible with the non-refrigerated liquefied gas(es) intended to be transported.

6.7.3.2.4 Contact between dissimilar metals which could result in damage by galvanic action shall be avoided.

6.7.3.2.5 The materials of the portable tank, including any devices, gaskets, and accessories, shall not adversely affect the non-refrigerated liquefied gas(es) intended for transport in the portable tank.

6.7.3.2.6 Portable tanks shall be designed and constructed with supports to provide a secure base during transport and with suitable lifting and tie-down attachments.

6.7.3.2.7 Portable tanks shall be designed to withstand, without loss of contents, at least the internal pressure due to the contents and the static, dynamic and thermal loads during normal conditions of handling and transport. The design shall demonstrate that the effects of fatigue, caused by repeated application of these loads through the expected life of the portable tank, have been taken into account.

6.7.3.2.7.1 For portable tanks that are intended for use as offshore tank-containers, the dynamic stresses imposed by handling in open seas shall be taken into account.

6.7.3.2.8 Shells shall be designed to withstand an external pressure of at least 0.4 bar gauge above the internal pressure without permanent deformation. When the shell is to be subjected to a significant vacuum before filling or during discharge, it shall be designed to withstand an external pressure of at least 0.9 bar gauge above the internal pressure and shall be proven at that pressure.

6.7.3.2.9 Portable tanks and their fastenings shall, under the maximum permissible load, be capable of absorbing the following separately applied static forces:

.1 in the direction of travel: twice the MPGM multiplied by the acceleration due to gravity (g);*

.2 horizontally at right angles to the direction of travel: the MPGM (when the direction of travel is not clearly determined, the forces shall be equal to twice the MPGM) multiplied by the acceleration due to gravity (g);*

.3 vertically upwards: the MPGM multiplied by the acceleration due to gravity (g);* and

.4 vertically downwards: twice the MPGM (total loading including the effect of gravity) multiplied by the acceleration due to gravity (g).*

6.7.3.2.10 Under each of the forces in 6.7.3.2.9, the safety factor to be observed shall be as follows:

.1 for steels having a clearly defined yield point, a safety factor of 1.5 in relation to the guaranteed yield strength; or

.2 for steels with no clearly defined yield point, a safety factor of 1.5 in relation to the guaranteed 0.2% proof strength and, for austenitic steels, the 1% proof strength.

6.7.3.2.11 The values of yield strength or proof strength shall be the values according to national or international material standards. When austenitic steels are used, the specified minimum values of yield strength and proof strength according to the material standards may be increased by up to 15% when these greater values are attested in the material inspection certificate. When no material standard exists for the steel in question, the value of yield strength or proof strength used shall be approved by the competent authority.

6.7.3.2.12 When the shells intended for the transport of non-refrigerated liquefied gases are equipped with thermal insulation, the thermal insulation system shall satisfy the following provisions:

.1 It shall consist of a shield covering not less than the upper third but not more than the upper half of the surface of the shell and separated from the shell by an air space about 40 mm across; or

.2 It shall consist of a complete cladding of adequate thickness of insulating materials, protected so as to prevent the ingress of moisture and damage under normal conditions of transport and so as to provide a thermal conductance of not more than 0.67 W/m·K;

.3 When the protective covering is so closed as to be gas-tight, a device shall be provided to prevent any dangerous pressure from developing in the insulating layer in the event of inadequate gas-tightness of the shell or of its items of equipment;

.4 The thermal insulation shall not inhibit access to the fittings and discharge devices.

6.7.3.2.13 Portable tanks intended for the transport of flammable non-refrigerated liquefied gases shall be capable of being electrically earthed.

6.7.3.3 **Design criteria**

6.7.3.3.1 Shells shall be of a circular cross-section.

6.7.3.3.2 Shells shall be designed and constructed to withstand a test pressure not less than 1.3 times the design pressure. The shell design shall take into account the minimum MAWP values provided in portable tank instruction T50 in 4.2.5.2.6 for each non-refrigerated liquefied gas intended for transport. Attention is drawn to the minimum shell thickness provisions for these shells specified in 6.7.3.4.

6.7.3.3.3 For steels exhibiting a clearly defined yield point or characterized by a guaranteed proof strength (0.2% proof strength, generally, or 1% proof strength for austenitic steels), the primary membrane stress σ (sigma) in the shell shall not exceed $0.75R_e$ or $0.50R_m$, whichever is lower, at the test pressure, where:

R_e = yield strength in N/mm^2, or 0.2% proof strength or, for austenitic steels, 1% proof strength.

R_m = minimum tensile strength in N/mm^2.

---

* For calculation purposes, $g = 9.81$ m/s^2.

6.7.3.3.3.1   The values of R_e and R_m to be used shall be the specified minimum values according to national or international material standards. When austenitic steels are used, these specified minimum values for R_e and R_m according to the material standards may be increased by up to 15% when these greater values are attested in the material inspection certificate. When no material standard exists for the steel in question, the values of R_e and R_m used shall be approved by the competent authority or its authorized body.

6.7.3.3.3.2   Steels which have a R_e/R_m ratio of more than 0.85 are not allowed for the construction of welded shells. The values of R_e and R_m to be used in determining this ratio shall be the values specified in the material inspection certificate.

6.7.3.3.3.3   Steels used in the construction of shells shall have an elongation at fracture, in %, of not less than $10,000/R_m$ with an absolute minimum of 16% for fine-grain steels and 20% for other steels.

6.7.3.3.3.4   For the purpose of determining actual values for materials, it shall be noted that for sheet metal, the axis of the tensile test specimen shall be at right angles (transversely) to the direction of rolling. The permanent elongation at fracture shall be measured on test specimens of rectangular cross-section in accordance with ISO 6892:1984 using a 50 mm gauge length.

6.7.3.4   Minimum shell thickness

6.7.3.4.1   The minimum shell thickness shall be the greater thickness based on:

.1   the minimum thickness determined in accordance with the provisions in 6.7.3.4; and

.2   the minimum thickness determined in accordance with the recognized pressure-vessel code, including the provisions in 6.7.3.3.

6.7.3.4.2   The cylindrical portions, ends (heads) and manhole covers of shells of not more than 1.80 m in diameter shall be not less than 5 mm thick in the reference steel or of equivalent thickness in the steel to be used. Shells of more than 1.80 m in diameter shall be not less than 6 mm thick in the reference steel or of equivalent thickness in the steel to be used.

6.7.3.4.3   The cylindrical portions, ends (heads) and manhole covers of all shells shall be not less than 4 mm thick regardless of the material of construction.

6.7.3.4.4   The equivalent thickness of a steel other than the thickness prescribed for the reference steel in 6.7.3.4.2 shall be determined using the following formula:

$$e_1 = \frac{21.4 \times e_0}{\sqrt[3]{R_{m1} \times A_1}}$$

where:

e_1 =   required equivalent thickness (in mm) of the steel to be used;

e_0 =   minimum thickness (in mm) of the reference steel specified in 6.7.3.4.2;

R_{m1}=   guaranteed minimum tensile strength (in N/mm^2) of the steel to be used (see 6.7.3.3.3);

A_1 =   guaranteed minimum elongation at fracture (in %) of the steel to be used according to national or international standards.

6.7.3.4.5   In no case shall the wall thickness be less than that prescribed in 6.7.3.4.1 to 6.7.3.4.3. All parts of the shell shall have a minimum thickness as determined by 6.7.3.4.1 to 6.7.3.4.3. This thickness shall be exclusive of any corrosion allowance.

6.7.3.4.6   When mild steel is used (see 6.7.3.1), calculation using the equation in 6.7.3.4.4 is not required.

6.7.3.4.7   There shall be no sudden change of plate thickness at the attachment of the ends (heads) to the cylindrical portion of the shell.

6.7.3.5   Service equipment

6.7.3.5.1   Service equipment shall be so arranged as to be protected against the risk of being wrenched off or damaged during handling and transport. When the connection between the frame and the shell allows relative movement between the sub-assemblies, the equipment shall be so fastened as to permit such movement without risk of damage to working parts. The external discharge fittings (pipe sockets, shut-off devices), the internal stop-valve and its seating shall be protected against the danger of being wrenched off by external forces (for example, by using shear sections). The filling and discharge devices (including flanges or threaded plugs) and any protective caps shall be capable of being secured against unintended opening.

6.7.3.5.1.1 For offshore tank-containers, where positioning of service equipment and the design and strength of protection for such equipment is concerned, the increased danger of impact damage when handling such tanks in open seas shall be taken into account.

6.7.3.5.2 All openings with a diameter of more than 1.5 mm in shells of portable tanks, except openings for pressure relief devices, inspection openings and closed bleed holes, shall be fitted with at least three mutually independent shut-off devices in series, the first being an internal stop-valve, excess flow valve or equivalent device, the second being an external stop-valve and the third being a blank flange or equivalent device.

6.7.3.5.2.1 When a portable tank is fitted with an excess flow valve, the excess flow valve shall be so fitted that its seating is inside the shell or inside a welded flange or, when fitted externally, its mountings shall be designed so that, in the event of impact, its effectiveness shall be maintained. The excess flow valves shall be selected and fitted so as to close automatically when the rated flow specified by the manufacturer is reached. Connections and accessories leading to or from such a valve shall have a capacity for a flow more than the rated flow of the excess flow valve.

6.7.3.5.3 For filling and discharge openings, the first shut-off device shall be an internal stop-valve and the second shall be a stop-valve placed in an accessible position on each discharge and filling pipe.

6.7.3.5.4 For filling and discharge bottom openings of portable tanks intended for the transport of flammable and/or toxic non-refrigerated liquefied gases, the internal stop-valve shall be a quick-closing safety device which closes automatically in the event of unintended movement of the portable tank during filling or discharge or fire engulfment. Except for portable tanks having a capacity of not more than 1000 ℓ, it shall be possible to operate this device by remote control.

6.7.3.5.5 In addition to filling, discharge and gas pressure equalizing orifices, shells may have openings in which gauges, thermometers and manometers can be fitted. Connections for such instruments shall be made by suitable welded nozzles or pockets and not be screwed connections through the shell.

6.7.3.5.6 All portable tanks shall be fitted with manholes or other inspection openings of suitable size to allow for internal inspection and adequate access for maintenance and repair of the interior.

6.7.3.5.7 External fittings shall be grouped together so far as reasonably practicable.

6.7.3.5.8 Each connection on a portable tank shall be clearly marked to indicate its function.

6.7.3.5.9 Each stop-valve or other means of closure shall be designed and constructed to a rated pressure not less than the MAWP of the shell, taking into account the temperatures expected during transport. All stop-valves with a screwed spindle shall close by a clockwise motion of the handwheel. For other stop-valves, the position (open and closed) and direction of closure shall be clearly indicated. All stop-valves shall be designed to prevent unintentional opening.

6.7.3.5.10 Piping shall be designed, constructed and installed so as to avoid the risk of damage due to thermal expansion and contraction, mechanical shock and vibration. All piping shall be of suitable metallic material. Welded pipe joints shall be used wherever possible.

6.7.3.5.11 Joints in copper tubing shall be brazed or have an equally strong metal union. The melting point of brazing materials shall be no lower than 525°C. The joints shall not decrease the strength of tubing, as may happen when cutting threads.

6.7.3.5.12 The burst pressure of all piping and pipe fittings shall be not less than the highest of four times the MAWP of the shell or four times the pressure to which it may be subjected in service by the action of a pump or other device (except pressure relief devices).

6.7.3.5.13 Ductile metals shall be used in the construction of valves and accessories.

6.7.3.6 Bottom openings

6.7.3.6.1 Certain non-refrigerated liquefied gases shall not be transported in portable tanks with bottom openings when portable tank instruction T50 in 4.2.5.2.6 indicates that bottom openings are not allowed. There shall be no openings below the liquid level of the shell when it is filled to its maximum permissible filling limit.

6.7.3.7 Pressure relief devices

6.7.3.7.1 Portable tanks shall be provided with one or more spring-loaded pressure relief devices. The pressure relief devices shall open automatically at a pressure not less than the MAWP and be fully open at a pressure equal to 110% of the MAWP. These devices shall, after discharge, close at a pressure not lower than 10% below the pressure at which discharge starts and shall remain closed at all lower pressures. The pressure relief devices shall be of a type that will resist dynamic forces, including liquid surge. Frangible discs not in series with a spring-loaded pressure relief device are not permitted.

6.7.3.7.2 Pressure relief devices shall be designed to prevent the entry of foreign matter, the leakage of gas and the development of any dangerous excess pressure.

6.7.3.7.3 Portable tanks intended for the transport of certain non-refrigerated liquefied gases identified in portable tank instruction T50 in 4.2.5.2.6 shall have a pressure relief device approved by the competent authority. Unless a portable tank in dedicated service is fitted with an approved relief device constructed of materials compatible with the load, such device shall comprise a frangible disc preceding a spring-loaded device. The space between the frangible disc and the device shall be provided with a pressure gauge or a suitable tell-tale indicator. This arrangement permits the detection of disc rupture, pinholing or leakage which could cause a malfunction of the pressure relief device. The frangible discs shall rupture at a nominal pressure 10% above the start-to-discharge pressure of the relief device.

6.7.3.7.4 In the case of multi-purpose portable tanks, the pressure relief devices shall open at a pressure indicated in 6.7.3.7.1 for the gas having the highest maximum allowable pressure of the gases allowed to be transported in the portable tank.

6.7.3.8 **Capacity of relief devices**

6.7.3.8.1 The combined delivery capacity of the relief devices shall be sufficient that, in the event of total fire engulfment, the pressure (including accumulation) inside the shell does not exceed 120% of the MAWP. Spring-loaded relief devices shall be used to achieve the full relief capacity prescribed. In the case of multi-purpose tanks, the combined delivery capacity of the pressure relief devices shall be taken for the gas which requires the highest delivery capacity of the gases allowed to be transported in portable tanks.

6.7.3.8.1.1 To determine the total required capacity of the relief devices, which shall be regarded as being the sum of the individual capacities of the several devices, the following formula* shall be used:

$$Q = 12.4 \frac{FA^{0.82}}{LC} \sqrt{\frac{ZT}{M}}$$

where:

- Q = minimum required rate of discharge in cubic metres of air per second (m^3/s) at standard conditions: 1 bar and 0°C (273 K);

- F = a coefficient with the following value:
    for uninsulated shells, $F = 1$
    for insulated shells, $F = U(649 - t)/13.6$ but in no case is less than 0.25

    where:
    U = thermal conductance of the insulation, in kW/m·K, at 38°C;
    t = actual temperature of the non-refrigerated liquefied gas during filling (in °C) (when this temperature is unknown, let $t = 15$°C);

    The value of F given above for insulated shells may be taken provided that the Insulation is in conformance with 6.7.3.8.1.2;

- A = total external surface area of shell in square metres;

- Z = the gas compressibility factor in the accumulating condition (when this factor is unknown, let Z equal 1.0);

- T = absolute temperature in kelvin (°C + 273) above the pressure relief devices in the accumulating condition;

- L = the latent heat of vaporization of the liquid, in kJ/kg, in the accumulating condition;

- M = molecular mass of the discharged gas;

- C = a constant which is derived from one of the following formulae as a function of the ratio k of specific heats:

$$k = \frac{C_p}{C_v}$$

---

* This formula applies only to non-refrigerated liquefied gases which have critical temperatures well above the temperature at the accumulating condition. For gases which have critical temperatures near or below the temperature at the accumulating condition, the calculation of the pressure-relief device delivery capacity shall consider further thermodynamic properties of the gas (see, for example, CGA S-1.2-2003 ''Pressure Relief Device Standards – Part 2 – Cargo and Portable Tanks for Compressed Gases'').

where:

C_p = specific heat at constant pressure; and

C_v = specific heat at constant volume.

When $k > 1$:

$$C = \sqrt{k \left(\frac{2}{k+1} \right)^{\frac{k+1}{k-1}}}$$

When $k = 1$ or k is unknown:

$$C = \frac{1}{\sqrt{e}} = 0.607$$

where e is the mathematical constant 2.7183.

C may also be taken from the following table:

k	C	k	C	k	C
1.00	0.607	1.26	0.660	1.52	0.704
1.02	0.611	1.28	0.664	1.54	0.707
1.04	0.615	1.30	0.667	1.56	0.710
1.06	0.620	1.32	0.671	1.58	0.713
1.08	0.624	1.34	0.674	1.60	0.716
1.10	0.628	1.36	0.678	1.62	0.719
1.12	0.633	1.38	0.681	1.64	0.722
1.14	0.637	1.40	0.685	1.66	0.725
1.16	0.641	1.42	0.688	1.68	0.728
1.18	0.645	1.44	0.691	1.70	0.731
1.20	0.649	1.46	0.695	2.00	0.770
1.22	0.652	1.48	0.698	2.20	0.793
1.24	0.656	1.50	0.701		

6.7.3.8.1.2 Insulation systems, used for the purpose of reducing the venting capacity, shall be approved by the competent authority or its authorized body. In all cases, insulation systems approved for this purpose shall:

.1 remain effective at all temperatures up to 649°C; and

.2 be jacketed with a material having a melting point of 700°C or greater.

6.7.3.9 Marking of pressure relief devices

6.7.3.9.1 Every pressure relief device shall be clearly and permanently marked with the following:

.1 the pressure (in bar or kPa) at which it is set to discharge;

.2 the allowable tolerance at the discharge pressure, for spring-loaded devices;

.3 the reference temperature corresponding to the rated pressure, for frangible discs; and

.4 the rated flow capacity of the device in standard cubic metres of air per second (m^3/s).

When practicable, the following information shall also be shown:

.5 the manufacturer's name and relevant catalogue number.

6.7.3.9.2 The rated flow capacity marked on the pressure relief devices shall be determined according to ISO 4126-1:1996.

6.7.3.10 Connections to pressure relief devices

6.7.3.10.1 Connections to pressure relief devices shall be of sufficient size to enable the required discharge to pass unrestricted to the safety device. No stop-valve shall be installed between the shell and the pressure relief devices except when duplicate devices are provided for maintenance or other reasons and the stop-valves serving the devices actually in use are locked open or the stop-valves are interlocked so that at least one of the duplicate devices is always operable and capable of meeting the provisions of 6.7.3.8. There shall be no obstruction in an opening leading to a vent or pressure relief device which might restrict or cut off the flow from the shell to that device. Vents from the pressure relief devices, when used, shall deliver the relieved vapour or liquid to the atmosphere in conditions of minimum back-pressure on the relieving device.

6.7.3.11 **Siting of pressure relief devices**

6.7.3.11.1 Each pressure relief device inlet shall be situated on top of the shell in a position as near the longitudinal and transverse centre of the shell as reasonably practicable. All pressure relief device inlets shall, under maximum filling conditions, be situated in the vapour space of the shell and the devices shall be so arranged as to ensure that the escaping vapour is discharged unrestrictedly. For flammable non-refrigerated liquefied gases, the escaping vapour shall be directed away from the shell in such a manner that it cannot impinge upon the shell. Protective devices which deflect the flow of vapour are permissible provided the required relief-device capacity is not reduced.

6.7.3.11.2 Arrangements shall be made to prevent access to the pressure relief devices by unauthorized persons and to protect the devices from damage caused by the portable tank overturning.

6.7.3.12 **Gauging devices**

6.7.3.12.1 Unless a portable tank is intended to be filled by mass, it shall be equipped with one or more gauging devices. Glass level-gauges and gauges made of other fragile material, which are in direct communication with the contents of the shell, shall not be used.

6.7.3.13 **Portable tank supports, frameworks, lifting and tie-down attachments**

6.7.3.13.1 Portable tanks shall be designed and constructed with a support structure to provide a secure base during transport. The forces specified in 6.7.3.2.9 and the safety factor specified in 6.7.3.2.10 shall be considered in this aspect of the design. Skids, frameworks, cradles or other similar structures are acceptable.

6.7.3.13.2 The combined stresses caused by portable tank mountings (such as cradles, frameworks, etc.) and portable tank lifting and tie-down attachments shall not cause excessive stress in any portion of the shell. Permanent lifting and tie-down attachments shall be fitted to all portable tanks. Preferably they shall be fitted to the portable tank supports but may be secured to reinforcing plates located on the shell at the points of support.

6.7.3.13.3 In the design of supports and frameworks, the effects of environmental corrosion shall be taken into account.

6.7.3.13.4 Forklift pockets shall be capable of being closed off. The means of closing forklift pockets shall be a permanent part of the framework or permanently attached to the framework. Single-compartment portable tanks with a length less than 3.65 m need not have closed-off forklift pockets provided that:

.1 the shell and all the fittings are well protected from being hit by the forklift blades; and

.2 the distance between the centres of the forklift pockets is at least half of the maximum length of the portable tank.

6.7.3.13.5 When portable tanks are not protected during transport, according to 4.2.2.3, the shells and service equipment shall be protected against damage to the shell and service equipment resulting from lateral or longitudinal impact or overturning. External fittings shall be protected so as to preclude the release of the shell contents upon impact or overturning of the portable tank on its fittings. Examples of protection include:

.1 protection against lateral impact, which may consist of longitudinal bars protecting the shell on both sides at the level of the median line;

.2 protection of the portable tank against overturning, which may consist of reinforcement rings or bars fixed across the frame;

.3 protection against rear impact, which may consist of a bumper or frame;

.4 protection of the shell against damage from impact or overturning by use of an ISO frame in accordance with ISO 1496-3:1995.

6.7.3.14 **Design approval**

6.7.3.14.1 The competent authority or its authorized body shall issue a design approval certificate for any new design of a portable tank. This certificate shall attest that the portable tank has been surveyed by that authority, is suitable for its intended purpose and meets the provisions of this chapter and, when appropriate, the provisions for gases provided in portable tank instruction T50 in 4.2.5.2.6. When a series of portable tanks are manufactured without change in the design, the certificate shall be valid for the entire series. The certificate shall refer to the prototype test report, the gases allowed to be transported, the materials of construction of the shell and an approval number. The approval number shall consist of the distinguishing sign or mark of the State in whose territory the approval was granted, i.e. the distinguishing sign for use in international traffic, as prescribed by the Convention on Road Traffic, Vienna, 1968, and a registration number. Any alternative arrangements according to 6.7.1.2 shall be indicated on the certificate. A design approval may serve for the approval of smaller portable tanks made of materials of the same kind and thickness, by the same fabrication techniques and with identical supports, equivalent closures and other appurtenances.

6.7.3.14.2 The prototype test report for the design approval shall include at least the following:

.1 the results of the applicable framework test specified in ISO 1496-3:1995;

.2 the results of the initial inspection and test in 6.7.3.15.3; and

.3 the results of the impact test in 6.7.3.15.1, when applicable.

6.7.3.15 **Inspection and testing**

6.7.3.15.1 Portable tanks meeting the definition of *container* in the International Convention for Safe Containers (CSC), 1972, as amended, shall not be used unless they are successfully qualified by subjecting a representative prototype of each design to the Dynamic, Longitudinal Impact Test prescribed in the United Nations *Manual of Tests and Criteria*, part IV, section 41. This provision only applies to portable tanks which are constructed according to a design approval certificate which has been issued on or after 1 January 2008.

6.7.3.15.2 The shell and items of equipment of each portable tank shall be inspected and tested before being put into service for the first time (initial inspection and test) and thereafter at not more than five-year intervals (5-year periodic inspection and test) with an intermediate periodic inspection and test (2.5-year periodic inspection and test) midway between the 5-year periodic inspections and tests. The 2.5-year periodic inspection and test may be performed within 3 months of the specified date. An exceptional inspection and test shall be performed regardless of the last periodic inspection and test when necessary according to 6.7.3.15.7.

6.7.3.15.3 The initial inspection and test of a portable tank shall include a check of the design characteristics, an internal and external examination of the portable tank and its fittings with due regard to the non-refrigerated liquefied gases to be transported, and a pressure test referring to the test pressures according to 6.7.3.3.2. The pressure test may be performed as a hydraulic test or by using another liquid or gas with the agreement of the competent authority or its authorized body. Before the portable tank is placed into service, a leakproofness test and a test of the satisfactory operation of all service equipment shall also be performed. When the shell and its fittings have been pressure-tested separately, they shall be subjected together after assembly to a leakproofness test. All welds subject to full stress level in the shell shall be inspected during the initial test by radiographic, ultrasonic, or another suitable non-destructive test method. This does not apply to the jacket.

6.7.3.15.4 The 5-year periodic inspection and test shall include an internal and external examination and, as a general rule, a hydraulic pressure test. Sheathing, thermal insulation and the like shall be removed only to the extent required for reliable appraisal of the condition of the portable tank. When the shell and equipment have been pressure-tested separately, they shall be subjected together after assembly to a leakproofness test.

6.7.3.15.5 The intermediate 2.5-year periodic inspection and test shall at least include an internal and external examination of the portable tank and its fittings with due regard to the non-refrigerated liquefied gases intended to be transported, a leakproofness test and a test of the satisfactory operation of all service equipment. Sheathing, thermal insulation and the like shall be removed only to the extent required for reliable appraisal of the condition of the portable tank. For portable tanks intended for the transport of a single non-refrigerated liquefied gas, the 2.5-year internal examination may be waived or substituted by other test methods or inspection procedures specified by the competent authority or its authorized body.

6.7.3.15.6 A portable tank may not be filled and offered for transport after the date of expiry of the last 5-year or 2.5-year periodic inspection and test as required by 6.7.3.15.2. However, a portable tank filled prior to the date of expiry of the last periodic inspection and test may be transported for a period not to exceed three months beyond the date of expiry of the last periodic test or inspection. In addition, a portable tank may be transported after the date of expiry of the last periodic test and inspection:

.1 after emptying but before cleaning, for purposes of performing the next required test or inspection prior to refilling; and

.2 unless otherwise approved by the competent authority, for a period not to exceed six months beyond the date of expiry of the last periodic test or inspection, in order to allow the return of dangerous goods for proper disposal or recycling. Reference to this exemption shall be mentioned in the transport document.

6.7.3.15.7 The exceptional inspection and test is necessary when the portable tank shows evidence of damaged or corroded areas, or leakage, or other conditions that indicate a deficiency that could affect the integrity of the portable tank. The extent of the exceptional inspection and test shall depend on the amount of damage or deterioration of the portable tank. It shall include at least the 2.5-year inspection and test according to 6.7.3.15.5.

6.7.3.15.8 The internal and external examinations shall ensure that:

.1   the shell is inspected for pitting, corrosion, or abrasions, dents, distortions, defects in welds or any other conditions, including leakage, that might render the portable tank unsafe for transport;

.2   the piping, valves, and gaskets are inspected for corroded areas, defects, or any other conditions, including leakage, that might render the portable tank unsafe for filling, discharge or transport;

.3   devices for tightening manhole covers are operative and there is no leakage at manhole covers or gaskets;

.4   missing or loose bolts or nuts on any flanged connection or blank flange are replaced or tightened;

.5   all emergency devices and valves are free from corrosion, distortion and any damage or defect that could prevent their normal operation. Remote closure devices and self-closing stop-valves shall be operated to demonstrate proper operation;

.6   required markings on the portable tank are legible and in accordance with the applicable provisions; and

.7   the framework, the supports and the arrangements for lifting the portable tank are in satisfactory condition.

6.7.3.15.9 The inspections and tests in 6.7.3.15.1, 6.7.3.15.3, 6.7.3.15.4, 6.7.3.15.5 and 6.7.3.15.7 shall be performed or witnessed by an expert approved by the competent authority or its authorized body. When the pressure test is a part of the inspection and test, the test pressure shall be the one indicated on the data plate of the portable tank. While under pressure, the portable tank shall be inspected for any leaks in the shell, piping or equipment.

6.7.3.15.10 In all cases when cutting, burning or welding operations on the shell have been effected, that work shall be to the approval of the competent authority or its authorized body, taking into account the pressure-vessel code used for the construction of the shell. A pressure test to the original test pressure shall be performed after the work is completed.

6.7.3.15.11 When evidence of any unsafe condition is discovered, the portable tank shall not be returned to service until it has been corrected and the pressure test is repeated and passed.

6.7.3.16   **Marking**

6.7.3.16.1 Every portable tank shall be fitted with a corrosion-resistant metal plate permanently attached to the portable tank in a conspicuous place readily accessible for inspection. When, for reasons of portable tank arrangements, the plate cannot be permanently attached to the shell, the shell shall be marked with at least the information required by the pressure-vessel code. As a minimum, at least the following information shall be marked on the plate by stamping or by any other similar method:

Country of manufacture:

| U | Approval | Approval | For alternative arrangements (see 6.7.1.2): |
| N | country | number | "AA" |

Manufacturer's name or mark

Manufacturer's serial number

Authorized body for the design approval

Owner's registration number

Year of manufacture

Pressure-vessel code to which the shell is designed

Test pressure . . . . bar/kPa gauge*

MAWP . . . . bar/kPa gauge*

External design pressure[†] . . . . bar/kPa gauge*

Design temperature range . . . . °C to . . . . °C

Design reference temperature . . . . °C

Water capacity at 20°C . . . . litres

Initial pressure test date and witness identification

Shell material(s) and material standard reference(s)

Equivalent thickness in reference steel . . . . mm

Date and type of most recent periodic test(s):

Month . . . . Year . . . . Test pressure . . . . bar/kPa gauge*

Stamp of expert who performed or witnessed the most recent test.

* The unit used shall be marked.
[†] See 6.7.3.2.8.

6.7.3.16.2 The following information shall be durably marked either on the portable tank itself or on a metal plate firmly secured to the portable tank:

> Name of the operator
>
> Name of non-refrigerated liquefied gas(es) permitted for transport
>
> Maximum permissible load mass for each non-refrigerated liquefied gas permitted . . . . kg
>
> Maximum permissible gross mass (MPGM) . . . . kg
>
> Unladen (tare) mass . . . . kg

6.7.3.16.3 If a portable tank is designed and approved for handling in open seas, the words "OFFSHORE PORTABLE TANK" shall be marked on the identification plate.

6.7.4 Provisions for the design, construction, inspection and testing of portable tanks intended for the transport of refrigerated liquefied gases of class 2

6.7.4.1 Definitions

For the purposes of this section:

Holding time means the time that will elapse from the establishment of the initial filling condition until the pressure has risen due to heat influx to the lowest set pressure of the pressure-limiting device(s);

Jacket means the outer insulation cover or cladding, which may be part of the insulation system;

Leakproofness test means a test, using gas, subjecting the shell and its service equipment to an effective internal pressure not less than 90% of the MAWP;

Maximum allowable working pressure (MAWP) means the maximum effective gauge pressure permissible at the top of the shell of a filled portable tank in its operating position, including the highest effective pressure during filling and discharge;

Maximum permissible gross mass (MPGM) means the sum of the tare mass of the portable tank and the heaviest load authorized for transport;

Minimum design temperature means the temperature which is used for the design and construction of the shell, not higher than the lowest (coldest) temperature (service temperature) of the contents during normal conditions of filling, discharge and transport;

Portable tank means a thermally insulated multimodal tank having a capacity of more than 450 ℓ fitted with service equipment and structural equipment necessary for the transport of refrigerated liquefied gases. The portable tank shall be capable of being filled and discharged without the removal of its structural equipment. It shall possess stabilizing members external to the tank, and shall be capable of being lifted when full. It shall be designed primarily to be loaded onto a transport vehicle or ship and shall be equipped with skids, mountings or accessories to facilitate mechanical handling. Road tank-vehicles, rail tank-wagons, non-metallic tanks, intermediate bulk containers (IBCs), gas cylinders and large receptacles are not considered to fall within the definition for portable tanks;

Reference steel means a steel with a tensile strength of 370 N/mm^2 and an elongation at fracture of 27%;

Service equipment means measuring instruments and filling, discharge, venting, safety, pressurizing, cooling and thermal insulation devices;

Shell means the part of the portable tank which retains the refrigerated liquefied gas intended for transport, including openings and their closures, but does not include service equipment or external structural equipment;

Structural equipment means the reinforcing, fastening, protective and stabilizing members external to the shell;

Tank means a construction which normally consists of either:

(a) a jacket and one or more inner shells where the space between the shell(s) and the jacket is exhausted of air (vacuum insulation) and may incorporate a thermal insulation system; or

(b) a jacket and an inner shell with an intermediate layer of solid thermally insulating material (such as solid foam);

Test pressure means the maximum gauge pressure at the top of the shell during the pressure test.

6.7.4.2   **General design and construction provisions**

6.7.4.2.1   Shells shall be designed and constructed in accordance with the provisions of a pressure-vessel code recognized by the competent authority. Shells and jackets shall be made of metallic materials suitable for forming. Jackets shall be made of steel. Non-metallic materials may be used for the attachments and supports between the shell and jacket, provided their material properties at the minimum design temperature are proven to be sufficient. The materials shall, in principle, conform to national or international material standards. For welded shells and jackets, only materials whose weldability has been fully demonstrated shall be used. Welds shall be skilfully made and afford complete safety. When the manufacturing process or the materials make it necessary, the shell shall be suitably heat-treated to guarantee adequate toughness in the weld and in the heat-affected zones. In choosing the material, the minimum design temperature shall be taken into account with respect to risk of brittle fracture, to hydrogen embrittlement, to stress corrosion cracking and to resistance to impact. When fine-grain steel is used, the guaranteed value of the yield strength shall be not more than 460 N/mm^2 and the guaranteed value of the upper limit of the tensile strength shall be not more than 725 N/mm^2, in accordance with the material specifications. Portable tank materials shall be suitable for the external environment in which they may be transported.

6.7.4.2.2   Any part of a portable tank, including fittings, gaskets and pipe-work, which can be expected normally to come into contact with the refrigerated liquefied gas transported shall be compatible with that refrigerated liquefied gas.

6.7.4.2.3   Contact between dissimilar metals which could result in damage by galvanic action shall be avoided.

6.7.4.2.4   The thermal insulation system shall include a complete covering of the shell(s) with effective insulating materials. External insulation shall be protected by a jacket so as to prevent the ingress of moisture and other damage under normal transport conditions.

6.7.4.2.5   When a jacket is so closed as to be gas-tight, a device shall be provided to prevent any dangerous pressure from developing in the insulation space.

6.7.4.2.6   Portable tanks intended for the transport of refrigerated liquefied gases having a boiling point below -182°C at atmospheric pressure shall not include materials which may react with oxygen or oxygen-enriched atmospheres in a dangerous manner when located in parts of the thermal insulation when there is a risk of contact with oxygen or with oxygen-enriched fluid.

6.7.4.2.7   Insulating materials shall not deteriorate unduly in service.

6.7.4.2.8   A reference holding time shall be determined for each refrigerated liquefied gas intended for transport in a portable tank.

6.7.4.2.8.1 The reference holding time shall be determined by a method recognized by the competent authority on the basis of the following:

    .1   the effectiveness of the insulation system, determined in accordance with 6.7.4.2.8.2;

    .2   the lowest set pressure of the pressure-limiting device(s);

    .3   the initial filling conditions;

    .4   an assumed ambient temperature of 30°C;

    .5   the physical properties of the individual refrigerated liquefied gas intended to be transported.

6.7.4.2.8.2 The effectiveness of the insulation system (heat influx in watts) shall be determined by type testing the portable tank in accordance with a procedure recognized by the competent authority. This test shall consist of either:

    .1   a constant-pressure test (for example at atmospheric pressure), when the loss of refrigerated liquefied gas is measured over a period of time; or

    .2   a closed-system test, when the rise in pressure in the shell is measured over a period of time.

When performing the constant-pressure test, variations in atmospheric pressure shall be taken into account. When performing either test, corrections shall be made for any variation of the ambient temperature from the assumed ambient temperature reference value of 30°C.

Note: For the determination of the actual holding time before each journey, see 4.2.3.7.

6.7.4.2.9   The jacket of a vacuum-insulated double-wall tank shall have either an external design pressure not less than 100 kPa (1 bar) gauge pressure calculated in accordance with a recognized technical code or a calculated critical collapsing pressure of not less than 200 kPa (2 bar) gauge pressure. Internal and external reinforcements may be included in calculating the ability of the jacket to resist the external pressure.

6.7.4.2.10   Portable tanks shall be designed and constructed with supports to provide a secure base during transport and with suitable lifting and tie-down attachments.

6.7.4.2.11 Portable tanks shall be designed to withstand, without loss of contents, at least the internal pressure due to the contents and the static, dynamic and thermal loads during normal conditions of handling and transport. The design shall demonstrate that the effects of fatigue, caused by repeated application of these loads through the expected life of the portable tank, have been taken into account.

6.7.4.2.11.1 For tanks that are intended for use as offshore tank-containers, the dynamic stresses imposed by handling in open seas shall be taken into account.

6.7.4.2.12 Portable tanks and their fastenings under the maximum permissible load shall be capable of absorbing the following separately applied static forces:

   .1   in the direction of travel: twice the MPGM multiplied by the acceleration due to gravity (*g*);*

   .2   horizontally at right angles to the direction of travel: the MPGM (when the direction of travel is not clearly determined, the forces shall be equal to twice the MPGM) multiplied by the acceleration due to gravity (*g*);*

   .3   vertically upwards: the MPGM multiplied by the acceleration due to gravity (*g*);* and

   .4   vertically downwards: twice the MPGM (total loading including the effect of gravity) multiplied by the acceleration due to gravity (*g*).*

6.7.4.2.13 Under each of the forces in 6.7.4.2.12, the safety factor to be observed shall be as follows:

   .1   for materials having a clearly defined yield point, a safety factor of 1.5 in relation to the guaranteed yield strength; or

   .2   for materials with no clearly defined yield point, a safety factor of 1.5 in relation to the guaranteed 0.2% proof strength or, for austenitic steels, the 1% proof strength.

6.7.4.2.14 The values of yield strength or proof strength shall be the values according to national or international material standards. When austenitic steels are used, the specified minimum values according to the material standards may be increased by up to 15% when these greater values are attested in the material inspection certificate. When no material standard exists for the metal in question, or when non-metallic materials are used, the values of yield strength or proof strength shall be approved by the competent authority.

6.7.4.2.15 Portable tanks intended for the transport of flammable refrigerated liquefied gases shall be capable of being electrically earthed.

6.7.4.3   **Design criteria**

6.7.4.3.1 Shells shall be of a circular cross-section.

6.7.4.3.2 Shells shall be designed and constructed to withstand a test pressure not less than 1.3 times the MAWP. For shells with vacuum insulation, the test pressure shall not be less than 1.3 times the sum of the MAWP and 100 kPa (1 bar). In no case shall the test pressure be less than 300 kPa (3 bar) gauge pressure. Attention is drawn to the minimum shell thickness provisions, specified in 6.7.4.4.2 to 6.7.4.4.7.

6.7.4.3.3 For metals exhibiting a clearly defined yield point or characterized by a guaranteed proof strength (0.2% proof strength, generally, or 1% proof strength for austenitic steels), the primary membrane stress σ (sigma) in the shell shall not exceed $0.75R_e$ or $0.50R_m$, whichever is lower, at the test pressure, where:

   R_e = yield strength in N/mm^2, or 0.2% proof strength or, for austenitic steels, 1% proof strength;

   R_m = minimum tensile strength in N/mm^2.

6.7.4.3.3.1 The values of R_e and R_m to be used shall be the specified minimum values according to national or international material standards. When austenitic steels are used, the specified minimum values for R_e and R_m according to the material standards may be increased by up to 15% when greater values are attested in the material inspection certificate. When no material standard exists for the metal in question, the values of R_e and R_m used shall be approved by the competent authority or its authorized body.

6.7.4.3.3.2 Steels which have a R_e/R_m ratio of more than 0.85 are not allowed for the construction of welded shells. The values of R_e and R_m to be used in determining this ratio shall be the values specified in the material inspection certificate.

---

* For calculation purposes, $g = 9.81$ m/s^2.

6.7.4.3.3.3 Steels used in the construction of shells shall have an elongation at fracture, in %, of not less than $10,000/R_m$ with an absolute minimum of 16% for fine-grain steels and 20% for other steels. Aluminium and aluminium alloys used in the construction of shells shall have an elongation at fracture, in %, of not less than $10,000/6R_m$ with an absolute minimum of 12%.

6.7.4.3.3.4 For the purpose of determining actual values for materials, it shall be noted that for sheet metal, the axis of the tensile test specimen shall be at right angles (transversely) to the direction of rolling. The permanent elongation at fracture shall be measured on test specimens of rectangular cross-section in accordance with ISO 6892:1984 using a 50 mm gauge length.

6.7.4.4 **Minimum shell thickness**

6.7.4.4.1 The minimum shell thickness shall be the greater thickness based on:

  .1 the minimum thickness determined in accordance with the provisions in 6.7.4.4.2 to 6.7.4.4.7; and

  .2 the minimum thickness determined in accordance with the recognized pressure-vessel code, including the provisions in 6.7.4.3.

6.7.4.4.2 Shells of not more than 1.80 m in diameter shall be not less than 5 mm thick in the reference steel or of equivalent thickness in the metal to be used. Shells of more than 1.80 m in diameter shall be not less than 6 mm thick in the reference steel or of equivalent thickness in the metal to be used.

6.7.4.4.3 Shells of vacuum-insulated tanks of not more than 1.80 m in diameter shall be not less than 3 mm thick in the reference steel or of equivalent thickness in the metal to be used. Such shells of more than 1.80 m in diameter shall be not less than 4 mm thick in the reference steel or of equivalent thickness in the metal to be used.

6.7.4.4.4 For vacuum-insulated tanks, the aggregate thickness of the jacket and the shell shall correspond to the minimum thickness prescribed in 6.7.4.4.2, the thickness of the shell itself being not less than the minimum thickness prescribed in 6.7.4.4.3.

6.7.4.4.5 Shells shall be not less than 3 mm thick regardless of the material of construction.

6.7.4.4.6 The equivalent thickness of a metal other than the thickness prescribed for the reference steel in 6.7.4.4.2 and 6.7.4.4.3 shall be determined using the following equation:

$$e_1 = \frac{21.4 \times e_0}{\sqrt[3]{R_{m1} \times A_1}}$$

where:

  e_1 = required equivalent thickness (in mm) of the steel to be used;

  e_0 = minimum thickness (in mm) of the reference steel specified in 6.7.4.4.2 and 6.7.4.4.3;

  R_{m1} = guaranteed minimum tensile strength (in N/mm^2) of the metal to be used (see 6.7.4.3.3);

  A_1 = guaranteed minimum elongation at fracture (in %) of the metal to be used according to national or international standards.

6.7.4.4.7 In no case shall the wall thickness be less than that prescribed in 6.7.4.4.1 to 6.7.4.4.5. All parts of the shell shall have a minimum thickness as determined by 6.7.4.4.1 to 6.7.4.4.6. This thickness shall be exclusive of any corrosion allowance.

6.7.4.4.8 There shall be no sudden change of plate thickness at the attachment of the ends (heads) to the cylindrical portion of the shell.

6.7.4.5 **Service equipment**

6.7.4.5.1 Service equipment shall be so arranged as to be protected against the risk of being wrenched off or damaged during handling and transport. When the connection between the frame and the tank or the jacket and the shell allows relative movement, the equipment shall be so fastened as to permit such movement without risk of damage to working parts. The external discharge fittings (pipe sockets, shut-off devices), the stop-valve and its seating shall be protected against the danger of being wrenched off by external forces (for example, by using shear sections). The filling and discharge devices (including flanges or threaded plugs) and any protective caps shall be capable of being secured against unintended opening.

6.7.4.5.1.1 For offshore tank-containers, where positioning of service equipment and the design and strength of protection for such equipment is concerned, the increased danger of impact damage when handling such tanks in open seas shall be taken into account.

6.7.4.5.2 Each filling and discharge opening in portable tanks used for the transport of flammable refrigerated liquefied gases shall be fitted with at least three mutually independent shut-off devices in series, the first being a stop-valve situated as close as reasonably practicable to the jacket, the second being a stop-valve and the third being a blank flange or equivalent device. The shut-off device closest to the jacket shall be a quick-closing device, which closes automatically in the event of unintended movement of the portable tank during filling or discharge or fire engulfment. This device shall also be possible to operate by remote control.

6.7.4.5.3 Each filling and discharge opening in portable tanks used for the transport of non-flammable refrigerated liquefied gases shall be fitted with at least two mutually independent shut-off devices in series, the first being a stop-valve situated as close as reasonably practicable to the jacket, the second a blank flange or equivalent device.

6.7.4.5.4 For sections of piping which can be closed at both ends and where liquid product can be trapped, a method of automatic pressure relief shall be provided to prevent excess pressure build-up within the piping.

6.7.4.5.5 Vacuum-insulated tanks need not have an opening for inspection.

6.7.4.5.6 External fittings shall be grouped together so far as reasonably practicable.

6.7.4.5.7 Each connection on a portable tank shall be clearly marked to indicate its function.

6.7.4.5.8 Each stop-valve or other means of closure shall be designed and constructed to a rated pressure not less than the MAWP of the shell, taking into account the temperature expected during transport. All stop-valves with a screwed spindle shall be closed by a clockwise motion of the handwheel. In the case of other stop-valves, the position (open and closed) and direction of closure shall be clearly indicated. All stop-valves shall be designed to prevent unintentional opening.

6.7.4.5.9 When pressure-building units are used, the liquid and vapour connections to that unit shall be provided with a valve as close to the jacket as reasonably practicable to prevent the loss of contents in case of damage to the pressure-building unit.

6.7.4.5.10 Piping shall be designed, constructed and installed so as to avoid the risk of damage due to thermal expansion and contraction, mechanical shock and vibration. All piping shall be of a suitable material. To prevent leakage due to fire, only steel piping and welded joints shall be used between the jacket and the connection to the first closure of any outlet. The method of attaching the closure to this connection shall be to the satisfaction of the competent authority or its authorized body. Elsewhere, pipe joints shall be welded when necessary.

6.7.4.5.11 Joints in copper tubing shall be brazed or have an equally strong metal union. The melting point of brazing materials shall be no lower than 525°C. The joints shall not decrease the strength of the tubing, as may happen by cutting of threads.

6.7.4.5.12 The materials of construction of valves and accessories shall have satisfactory properties at the lowest operating temperature of the portable tank.

6.7.4.5.13 The burst pressure of all piping and pipe fittings shall be not less than the highest of four times the MAWP of the shell or four times the pressure to which it may be subjected in service by the action of a pump or other device (except pressure relief devices).

6.7.4.6   Pressure relief devices

6.7.4.6.1 Every shell shall be provided with not less than two independent spring-loaded pressure relief devices. The pressure relief devices shall open automatically at a pressure not less than the MAWP and be fully open at a pressure equal to 110% of the MAWP. These devices shall, after discharge, close at a pressure not lower than 10% below the pressure at which discharge starts and shall remain closed at all lower pressures. The pressure relief devices shall be of the type that will resist dynamic forces, including surge.

6.7.4.6.2 Shells for non-flammable refrigerated liquefied gases and hydrogen may in addition have frangible discs in parallel with the spring-loaded devices as specified in 6.7.4.7.2 and 6.7.4.7.3.

6.7.4.6.3 Pressure relief devices shall be designed to prevent the entry of foreign matter, the leakage of gas and the development of any dangerous excess pressure.

6.7.4.6.4 Pressure relief devices shall be approved by the competent authority or its authorized body.

6.7.4.7   **Capacity and setting of pressure relief devices**

6.7.4.7.1   In the case of the loss of vacuum in a vacuum-insulated tank or of loss of 20% of the insulation of a tank insulated with solid materials, the combined capacity of all pressure relief devices installed shall be sufficient so that the pressure (including accumulation) inside the shell does not exceed 120% of the MAWP.

6.7.4.7.2   For non-flammable refrigerated liquefied gases (except oxygen) and hydrogen, this capacity may be achieved by the use of frangible discs in parallel with the required safety relief devices. Frangible discs shall rupture at nominal pressure equal to the test pressure of the shell.

6.7.4.7.3   Under the circumstances described in 6.7.4.7.1 and 6.7.4.7.2 together with complete fire engulfment, the combined capacity of all pressure relief devices installed shall be sufficient to limit the pressure in the shell to the test pressure.

6.7.4.7.4   The required capacity of the relief devices shall be calculated in accordance with a well-established technical code recognized by the competent authority.*

6.7.4.8   **Marking of pressure relief devices**

6.7.4.8.1   Every pressure relief device shall be plainly and permanently marked with the following:

     .1   the pressure (in bar or kPa) at which it is set to discharge;

     .2   the allowable tolerance at the discharge pressure, for spring-loaded devices;

     .3   the reference temperature corresponding to the rated pressure, for frangible discs; and

     .4   the rated flow capacity of the device in standard cubic metres of air per second (m^3/s).

     When practicable, the following information shall also be shown:

     .5   the manufacturer's name and relevant catalogue number.

6.7.4.8.2   The rated flow capacity marked on the pressure relief devices shall be determined according to ISO 4126-1:1996.

6.7.4.9   **Connections to pressure relief devices**

6.7.4.9.1   Connections to pressure relief devices shall be of sufficient size to enable the required discharge to pass unrestricted to the safety device. No stop-valve shall be installed between the shell and the pressure relief devices except when duplicate devices are provided for maintenance or other reasons and the stop-valves serving the devices actually in use are locked open or the stop-valves are interlocked so that the provisions of 6.7.4.7 are always fulfilled. There shall be no obstruction in an opening leading to a vent or pressure relief device which might restrict or cut off the flow from the shell to that device. Pipework to vent the vapour or liquid from the outlet of the pressure relief devices, when used, shall deliver the relieved vapour or liquid to the atmosphere in conditions of minimum back-pressure on the relieving device

6.7.4.10   **Siting of pressure relief devices**

6.7.4.10.1   Each pressure relief device inlet shall be situated on top of the shell in a position as near the longitudinal and transverse centre of the shell as reasonably practicable. All pressure relief device inlets shall, under maximum filling conditions, be situated in the vapour space of the shell and the devices shall be so arranged as to ensure that the escaping vapour is discharged unrestrictedly. For refrigerated liquefied gases, the escaping vapour shall be directed away from the tank and in such a manner that it cannot impinge upon the tank. Protective devices which deflect the flow of vapour are permissible provided the required relief-device capacity is not reduced.

6.7.4.10.2   Arrangements shall be made to prevent access to the devices by unauthorized persons and to protect the devices from damage caused by the portable tank overturning.

6.7.4.11   **Gauging devices**

6.7.4.11.1   Unless a portable tank is intended to be filled by mass, it shall be equipped with one or more gauging devices. Glass level-gauges and gauges made of other fragile material, which are in direct communication with the contents of the shell, shall not be used.

---

* See for example CGA Pamphet S-1.2-2003 "Pressure Relief Device Standards – Part 2 – Cargo and Portable Tanks for Compressed Gases."

6.7.4.11.2   A connection for a vacuum gauge shall be provided in the jacket of a vacuum-insulated portable tank.

6.7.4.12   **Portable tank supports, frameworks, lifting and tie-down attachments**

6.7.4.12.1   Portable tanks shall be designed and constructed with a support structure to provide a secure base during transport. The forces specified in 6.7.4.2.12 and the safety factor specified in 6.7.4.2.13 shall be considered in this aspect of the design. Skids, frameworks, cradles or other similar structures are acceptable.

6.7.4.12.2   The combined stresses caused by portable tank mountings (such as cradles, frameworks, etc.) and portable tank lifting and tie-down attachments shall not cause excessive stress in any portion of the tank. Permanent lifting and tie-down attachments shall be fitted to all portable tanks. Preferably they shall be fitted to the portable tank supports but may be secured to reinforcing plates located on the tank at the points of support.

6.7.4.12.3   In the design of supports and frameworks, the effects of environmental corrosion shall be taken into account.

6.7.4.12.4   Forklift pockets shall be capable of being closed off. The means of closing forklift pockets shall be a permanent part of the framework or permanently attached to the framework. Single-compartment portable tanks with a length less than 3.65 m need not have closed-off forklift pockets provided that:

.1   the tank and all the fittings are well protected from being hit by the forklift blades; and

.2   the distance between the centres of the forklift pockets is at least half of the maximum length of the portable tank.

6.7.4.12.5   When portable tanks are not protected during transport, according to 4.2.3.3, the shells and service equipment shall be protected against damage to the shell and service equipment resulting from lateral or longitudinal impact or overturning. External fittings shall be protected so as to preclude the release of the shell contents upon impact or overturning of the portable tank on its fittings. Examples of protection include:

.1   protection against lateral impact, which may consist of longitudinal bars protecting the shell on both sides at the level of the median line;

.2   protection of the portable tank against overturning, which may consist of reinforcement rings or bars fixed across the frame;

.3   protection against rear impact, which may consist of a bumper or frame;

.4   protection of the shell against damage from impact or overturning by use of an ISO frame in accordance with ISO 1496-3:1995;

.5   protection of the portable tank from impact or overturning by a vacuum insulation jacket.

6.7.4.13   **Design approval**

6.7.4.13.1   The competent authority or its authorized body shall issue a design approval certificate for any new design of a portable tank. This certificate shall attest that a portable tank has been surveyed by that authority, is suitable for its intended purpose and meets the provisions of this chapter. When a series of portable tanks are manufactured without change in the design, the certificate shall be valid for the entire series. The certificate shall refer to the prototype test report, the refrigerated liquefied gases allowed to be transported, the materials of construction of the shell and jacket and an approval number. The approval number shall consist of the distinguishing sign or mark of the State in whose territory the approval was granted, i.e. the distinguishing sign for use in international traffic, as prescribed by the Convention on Road Traffic, Vienna, 1968, and a registration number. Any alternative arrangements according to 6.7.1.2 shall be indicated on the certificate. A design approval may serve for the approval of smaller portable tanks made of materials of the same kind and thickness, by the same fabrication techniques and with identical supports, equivalent closures and other appurtenances.

6.7.4.13.2   The prototype test report for the design approval shall include at least the following:

.1   the results of the applicable framework test specified in ISO 1496-3:1995;

.2   the results of the initial inspection and test in 6.7.4.14.3; and

.3   the results of the impact test in 6.7.4.14.1, when applicable.

6.7.4.14   **Inspection and testing**

6.7.4.14.1   Portable tanks meeting the definition of *container* in the International Convention for Safe Containers (CSC), 1972, as amended, shall not be used unless they are successfully qualified by subjecting a representative prototype of each design to the Dynamic, Longitudinal Impact Test prescribed in the United Nations *Manual of Tests and Criteria*, part IV, section 41. This provision only applies to portable tanks which are constructed according to a design approval certificate which has been issued on or after 1 January 2008.

6

6.7.4.14.2 The tank and items of equipment of each portable tank shall be inspected and tested before being put into service for the first time (initial inspection and test) and thereafter at not more than five-year intervals (5-year periodic inspection and test) with an intermediate periodic inspection and test (2.5-year periodic inspection and test) midway between the 5-year periodic inspections and tests. The 2.5-year periodic inspection and test may be performed within 3 months of the specified date. An exceptional inspection and test shall be performed regardless of the last periodic inspection and test when necessary according to 6.7.4.14.7.

6.7.4.14.3 The initial inspection and test of a portable tank shall include a check of the design characteristics, an internal and external examination of the portable tank shell and its fittings with due regard to the refrigerated liquefied gases to be transported, and a pressure test referring to the test pressures according to 6.7.4.3.2. The pressure test may be performed as a hydraulic test or by using another liquid or gas, with the agreement of the competent authority or its authorized body. Before the portable tank is placed into service, a leakproofness test and a test of the satisfactory operation of all service equipment shall also be performed. When the shell and its fittings have been pressure-tested separately, they shall be subjected together after assembly to a leakproofness test. All welds subject to full stress level shall be inspected during the initial test by radiographic, ultrasonic, or another suitable non-destructive test method. This does not apply to the jacket.

6.7.4.14.4 The 5-year and 2.5-year periodic inspections and tests shall include an external examination of the portable tank and its fittings with due regard to the refrigerated liquefied gases transported, a leakproofness test, a test of the satisfactory operation of all service equipment and a vacuum reading, when applicable. In the case of non-vacuum-insulated tanks, the jacket and insulation shall be removed during a 2.5-year and a 5-year periodic inspection and test, but only to the extent necessary for a reliable appraisal.

6.7.4.14.5 [Reserved]

6.7.4.14.6 A portable tank may not be filled and offered for transport after the date of expiry of the last 5-year or 2.5-year periodic inspection and test as required by 6.7.4.14.2. However, a portable tank filled prior to the date of expiry of the last periodic inspection and test may be transported for a period not to exceed three months beyond the date of expiry of the last periodic test or inspection. In addition, a portable tank may be transported after the date of expiry of the last periodic test and inspection:

.1 after emptying but before cleaning, for purposes of performing the next required test or inspection prior to refilling; and

.2 unless otherwise approved by the competent authority, for a period not to exceed six months beyond the date of expiry of the last periodic test or inspection, in order to allow the return of dangerous goods for proper disposal or recycling. Reference to this exemption shall be mentioned in the transport document.

6.7.4.14.7 The exceptional inspection and test is necessary when the portable tank shows evidence of damaged or corroded areas, leakage, or any other conditions that indicate a deficiency that could affect the integrity of the portable tank. The extent of the exceptional inspection and test shall depend on the amount of damage or deterioration of the portable tank. It shall include at least the 2.5-year periodic inspection and test according to 6.7.4.14.4.

6.7.4.14.8 The internal examination during the initial inspection and test shall ensure that the shell is inspected for pitting, corrosion, or abrasions, dents, distortions, defects in welds or any other conditions that might render the portable tank unsafe for transport.

6.7.4.14.9 The external examination shall ensure that:

.1 the external piping, valves, pressurizing/cooling systems when applicable, and gaskets are inspected for corroded areas, defects, or any other conditions, including leakage, that might render the portable tank unsafe for filling, discharge or transport;

.2 there is no leakage at any manhole covers or gaskets;

.3 missing or loose bolts or nuts on any flanged connection or blank flange are replaced or tightened;

.4 all emergency devices and valves are free from corrosion, distortion and any damage or defect that could prevent their normal operation. Remote closure devices and self-closing stop-valves shall be operated to demonstrate proper operation;

.5 required markings on the portable tank are legible and in accordance with the applicable provisions; and

.6 the framework, the supports and the arrangements for lifting the portable tank are in satisfactory condition.

6.7.4.14.10 The inspections and tests in 6.7.4.14.1, 6.7.4.14.3, 6.7.4.14.4 and 6.7.4.14.7 shall be performed or witnessed by an expert approved by the competent authority or its authorized body. When the pressure test is a part of the inspection and test, the test pressure shall be the one indicated on the data plate of the portable tank. While under pressure, the portable tank shall be inspected for any leaks in the shell, piping or equipment.

6.7.4.14.11 In all cases when cutting, burning or welding operations on the shell of a portable tank have been effected, that work shall be to the approval of the competent authority or its authorized body, taking into account the pressure-vessel code used for the construction of the shell. A pressure test to the original test pressure shall be performed after the work is completed.

6.7.4.14.12 When evidence of any unsafe condition is discovered, the portable tank shall not be returned to service until it has been corrected and the test is repeated and passed.

6.7.4.15   Marking

6.7.4.15.1 Every portable tank shall be fitted with a corrosion-resistant metal plate permanently attached to the portable tank in a conspicuous place readily accessible for inspection. When, for reasons of portable tank arrangements, the plate cannot be permanently attached to the shell, the shell shall be marked with at least the information required by the pressure-vessel code. As a minimum, at least the following information shall be marked on the plate by stamping or by any other similar method:

Country of manufacture:

U	Approval	Approval	For alternative arrangements (see 6.7.1.2):
N	country	number	"AA"

Manufacturer's name or mark

Manufacturer's serial number

Authorized body for the design approval

Owner's registration number

Year of manufacture

Pressure-vessel code to which the tank is designed

Test pressure . . . . bar/kPa gauge*

MAWP . . . . bar/kPa gauge*

Minimum design temperature . . . . °C

Water capacity at 20°C . . . . litres

Initial pressure test date and witness identification

Shell material(s) and material standard reference(s)

Equivalent thickness in reference steel . . . . mm

Date and type of most recent periodic test(s):

Month . . . . Year . . . . Test pressure . . . . bar/kPa gauge*

Stamp of expert who performed or witnessed the most recent test

The name(s), in full, of the gas(es) for whose transport the portable tank is approved

Either "thermally insulated" or "vacuum insulated"

Effectiveness of the insulation system (heat influx) . . . . watts (W)

Reference holding time . . . . days or hours and

initial pressure . . . . bar/kPa gauge* and

degree of filling . . . . kg

for each refrigerated liquefied gas permitted for transport.

    * The unit used shall be marked.

6.7.4.15.2 The following information shall be durably marked either on the portable tank itself or on a metal plate firmly secured to the portable tank:

Name of the owner and the operator

Name of the refrigerated liquefied gas being transported (and minimum mean bulk temperature)

Maximum permissible gross mass (MPGM) . . . . kg

Unladen (tare) mass . . . . kg

Actual holding time for gas being transported . . . . days (or hours)

6.7.4.15.3 If a portable tank is designed and approved for handling in open seas, the words "OFFSHORE PORTABLE TANK" shall be marked on the identification plate.

6

6.7.5 Provisions for the design, construction, inspection and testing of multiple-element gas containers (MEGCs) intended for the transport of non-refrigerated gases

6.7.5.1 Definitions

For the purposes of this section:

Elements are cylinders, tubes or bundles of cylinders;

Leakproofness test means a test, using gas, subjecting the elements and the service equipment of the MEGC to an effective internal pressure of not less than 20% of the test pressure;

Manifold means an assembly of piping and valves connecting the filling and/or discharge openings of the elements;

Maximum permissible gross mass (MPGM) means the sum of the tare mass of the MEGC and the heaviest load authorized for transport;

Service equipment means measuring instruments and filling, discharge, venting and safety devices;

Structural equipment means the reinforcing, fastening, protective and stabilizing members external to the elements.

6.7.5.2 General design and construction provisions

6.7.5.2.1 The MEGC shall be capable of being filled and discharged without the removal of its structural equipment. It shall possess stabilizing members external to the elements to provide structural integrity for handling and transport. MEGCs shall be designed and constructed with supports to provide a secure base during transport and with lifting and tie-down attachments which are adequate for lifting the MEGC, including when loaded to its maximum permissible gross mass. The MEGC shall be designed to be loaded onto or into a cargo transport unit or ship and shall be equipped with skids, mountings or accessories to facilitate mechanical handling.

6.7.5.2.2 MEGCs shall be designed, manufactured and equipped in such a way as to withstand all conditions to which they will be subjected during normal conditions of handling and transport. The design shall take into account the effects of dynamic loading and fatigue.

6.7.5.2.3 Elements of a MEGC shall be made of seamless steel and be constructed and tested according to chapter 6.2. All of the elements in a MEGC shall be of the same design type.

6.7.5.2.4 Elements of MEGCs, fittings and pipework shall be:

.1 compatible with the substances intended to be transported (for gases, see ISO 11114-1:1997 and ISO 1114-2:2000); or

.2 properly passivated or neutralized by chemical reaction.

6.7.5.2.5 Contact between dissimilar metals which could result in damage by galvanic action shall be avoided.

6.7.5.2.6 The materials of the MEGC, including any devices, gaskets, and accessories, shall not adversely affect the gases intended for transport in the MEGC.

6.7.5.2.7 MEGCs shall be designed to withstand, without loss of contents, at least the internal pressure due to the contents, and the static, dynamic and thermal loads during normal conditions of handling and transport. The design shall demonstrate that the effects of fatigue, caused by repeated application of these loads through the expected life of the multiple-element gas container, have been taken into account.

6.7.5.2.8 MEGCs and their fastenings shall, under the maximum permissible load, be capable of withstanding the following separately applied static forces:

.1 in the direction of travel: twice the MPGM multiplied by the acceleration due to gravity (*g*);*

.2 horizontally at right angles to the direction of travel: the MPGM (when the direction of travel is not clearly determined, the forces shall be equal to twice the MPGM) multiplied by the acceleration due to gravity (*g*);*

.3 vertically upwards: the MPGM multiplied by the acceleration due to gravity (*g*);* and

.4 vertically downwards: twice the MPGM (total loading including the effect of gravity) multiplied by the acceleration due to gravity (*g*).*

---

* For calculation purposes, g = 9.81 m/s^2.

6.7.5.2.9 Under the forces defined above, the stress at the most severely stressed point of the elements shall not exceed the values given in either the relevant standards of 6.2.2.1 or, if the elements are not designed, constructed and tested according to those standards, in the technical code or standard recognized or approved by the competent authority of the country of use (see 6.2.3.1).

6.7.5.2.10 Under each of the forces in 6.7.5.2.8, the safety factor for the framework and fastenings to be observed shall be as follows:

.1 for steels having a clearly defined yield point, a safety factor of 1.5 in relation to the guaranteed yield strength; or

.2 for steels with no clearly defined yield point, a safety factor of 1.5 in relation to the guaranteed 0.2% proof strength and, for austenitic steels, the 1% proof strength.

6.7.5.2.11 MEGCs intended for the transport of flammable gases shall be capable of being electrically earthed.

6.7.5.2.12 The elements shall be secured in a manner that prevents undesired movement in relation to the structure and the concentration of harmful localized stresses.

6.7.5.3 Service equipment

6.7.5.3.1 Service equipment shall be configured or designed to prevent damage that could result in the release of the pressure receptacle contents during normal conditions of handling and transport. When the connection between the frame and the elements allows relative movement between the sub-assemblies, the equipment shall be so fastened as to permit such movement without damage to working parts. The manifolds, the discharge fittings (pipe sockets, shut-off devices), and the stop-valves shall be protected from being wrenched off by external forces. Manifold piping leading to shut-off valves shall be sufficiently flexible to protect the valves and the piping from shearing, or releasing the pressure receptacle contents. The filling and discharge devices (including flanges or threaded plugs) and any protective caps shall be capable of being secured against unintended opening.

6.7.5.3.2 Each element intended for the transport of gases of class 2.3 shall be fitted with a valve. The manifold for liquefied gases of class 2.3 shall be so designed that the elements can be filled separately and be kept isolated by a valve capable of being sealed. For the transport of gases of class 2.1, the elements shall be divided into groups of not more than 3000 litres each isolated by a valve.

6.7.5.3.3 For filling and discharge openings of the MEGC, two valves in series shall be placed in an accessible position on each discharge and filling pipe. One of the valves may be a non-return valve. The filling and discharge devices may be fitted to a manifold. For sections of piping which can be closed at both ends and where a liquid product can be trapped, a pressure relief valve shall be provided to prevent excessive pressure build-up. The main isolation valves on an MEGC shall be clearly marked to indicate their directions of closure. Each stop-valve or other means of closure shall be designed and constructed to withstand a pressure equal to or greater than 1.5 times the test pressure of the MEGC. All stop-valves with screwed spindles shall close by a clockwise motion of the handwheel. For other stop-valves, the positions (open and closed) and direction of closure shall be clearly indicated. All stop-valves shall be designed and positioned to prevent unintentional opening. Ductile metals shall be used in the construction of valves or accessories.

6.7.5.3.4 Piping shall be designed, constructed and installed so as to avoid damage due to expansion and contraction, mechanical shock and vibration. Joints in tubing shall be brazed or have an equally strong metal union. The melting point of brazing materials shall be no lower than 525°C. The rated pressure of the service equipment and of the manifold shall be not less than two thirds of the test pressure of the elements.

6.7.5.4 Pressure relief devices

6.7.5.4.1 The elements of MEGCs used for the transport of UN 1013 carbon dioxide and UN 1070 nitrous oxide shall be divided into groups of not more than 3000 litres each isolated by a valve. Each group shall be fitted with one or more pressure relief devices. MEGCs for other gases shall be fitted with pressure relief devices as specified by the competent authority for the country of use.

6.7.5.4.2 When pressure relief devices are fitted, every element or group of elements of an MEGC that can be isolated shall then be fitted with one or more pressure relief devices. Pressure relief devices shall be of a type that will resist dynamic forces, including liquid surge, and shall be designed to prevent the entry of foreign matter, the leakage of gas and the development of any dangerous excess pressure.

6.7.5.4.3 MEGCs used for the transport of certain non-refrigerated gases identified in instruction T50 in 4.2.5.2.6 may have a pressure relief device as required by the competent authority of the country of use. Unless an MEGC in dedicated service is fitted with an approved pressure relief device constructed of materials compatible with the load, such a device shall comprise a frangible disc preceding a spring-loaded device. The space between the frangible disc and the spring-loaded device may be equipped with a pressure gauge or a suitable tell-tale indicator. This arrangement permits the detection of disc rupture, pinholing or leakage which could cause a malfunction of the pressure relief device. The frangible disc shall rupture at a nominal pressure 10% above the start-to-discharge pressure of the spring-loaded device.

6.7.5.4.4   In the case of multi-purpose MEGCs used for the transport of low-pressure liquefied gases, the pressure relief devices shall open at a pressure as specified in 6.7.3.7.1 for the gas having the highest maximum allowable working pressure of the gases allowed to be transported in the MEGC.

6.7.5.5   Capacity of pressure relief devices

6.7.5.5.1   The combined delivery capacity of the pressure relief devices when fitted shall be sufficient that, in the event of complete fire engulfment of the MEGC, the pressure (including accumulation) inside the elements does not exceed 120% of the set pressure of the pressure relief device. The formula provided in CGA S-1.2-2003 "Pressure Relief Device Standards, Part 2, Cargo and Portable Tanks for Compressed Gases" shall be used to determine the minimum total flow capacity for the system of pressure relief devices. CGA S-1.1-2003 "Pressure Relief Device Standards, Part 1, Cylinders for Compressed Gases" may be used to determine the relief capacity of individual elements. Spring-loaded pressure relief devices may be used to achieve the full relief capacity prescribed in the case of low-pressure liquefied gases. In the case of multi-purpose MEGCs, the combined delivery capacity of the pressure relief devices shall be taken for the gas which requires the highest delivery capacity of the gases allowed to be transported in the MEGC.

6.7.5.5.2   To determine the total required capacity of the pressure relief devices installed on the elements for the transport of liquefied gases, the thermodynamic properties of the gas shall be considered (see, for example, CGA S-1.2-2003 "Pressure Relief Device Standards, Part 2, Cargo and Portable Tanks for Compressed Gases" for low-pressure liquefied gases and CGA S-1.1-2003 "Pressure Relief Device Standards, Part 1, Cylinders for Compressed Gases" for high-pressure liquefied gases).

6.7.5.6   Marking of pressure relief devices

6.7.5.6.1   Pressure relief devices shall be clearly and permanently marked with the following:

(a) the manufacturer's name and relevant catalogue number;

(b) the set pressure and/or the set temperature;

(c) the date of the last test.

6.7.5.6.2   The rated flow capacity marked on spring-loaded pressure relief devices for low-pressure liquefied gases shall be determined according to ISO 4126-1:1991.

6.7.5.7   Connections to pressure relief devices

6.7.5.7.1   Connections to pressure relief devices shall be of sufficient size to enable the required discharge to pass unrestricted to the pressure relief device. No stop-valve shall be installed between the element and the pressure relief devices, except when duplicate devices are provided for maintenance or other reasons, and the stop-valves serving the devices actually in use are locked open, or the stop-valves are interlocked so that at least one of the duplicate devices is always operable and capable of meeting the requirements of 6.7.5.5. There shall be no obstruction in an opening leading to or leaving from a vent or pressure relief device which might restrict or cut off the flow from the element to that device. The opening through all piping and fittings shall have at least the same flow area as the inlet of the pressure relief device to which it is connected. The nominal size of the discharge piping shall be at least as large as that of the pressure relief device outlet. Vents from the pressure relief devices, when used, shall deliver the relieved vapour or liquid to the atmosphere in conditions of minimum back-pressure on the relieving device.

6.7.5.8   Siting of pressure relief devices

6.7.5.8.1   Each pressure relief device shall, under maximum filling conditions, be in communication with the vapour space of the elements for the transport of liquefied gases. The devices, when fitted, shall be so arranged as to ensure that the escaping vapour is discharged upwards and unrestrictedly so as to prevent any impingement of escaping gas or liquid upon the MEGC, its elements or personnel. For flammable and pyrophoric and oxidizing gases, the escaping gas shall be directed away from the element in such a manner that it cannot impinge upon the other elements. Heat-resistant protective devices which deflect the flow of gas are permissible provided the required pressure relief device capacity is not reduced.

6.7.5.8.2   Arrangements shall be made to prevent access to the pressure relief devices by unauthorized persons and to protect the devices from damage caused by the MEGC overturning.

6.7.5.9   Gauging devices

6.7.5.9.1   When a MEGC is intended to be filled by mass, it shall be equipped with one or more gauging devices. Level-gauges made of glass or other fragile material shall not be used.

6.7.5.10   MEGC supports, frameworks, lifting and tie-down attachments

6.7.5.10.1   MEGCs shall be designed and constructed with a support structure to provide a secure base during transport. The forces specified in 6.7.5.2.8 and the safety factor specified in 6.7.5.2.10 shall be considered in this aspect of the design. Skids, frameworks, cradles or other similar structures are acceptable.

6.7.5.10.2   The combined stresses caused by element mountings (e.g., cradles, frameworks, etc.) and MEGC lifting and tie-down attachments shall not cause excessive stress in any element. Permanent lifting and tie-down attachments shall be fitted to all MEGCs. In no case shall mountings or attachments be welded onto the elements.

6.7.5.10.3   In the design of supports and frameworks, the effects of environmental corrosion shall be taken into account.

6.7.5.10.4   When MEGCs are not protected during transport, according to 4.2.4.3, the elements and service equipment shall be protected against damage resulting from lateral or longitudinal impact or overturning. External fittings shall be protected so as to preclude the release of the elements' contents upon impact or overturning of the MEGC on its fittings. Particular attention shall be paid to the protection of the manifold. Examples of protection include:

    .1   protection against lateral impact, which may consist of longitudinal bars;

    .2   protection against overturning, which may consist of reinforcement rings or bars fixed across the frame;

    .3   protection against rear impact, which may consist of a bumper or frame;

    .4   protection of the elements and service equipment against damage from impact or overturning by use of an ISO frame in accordance with the relevant provisions of ISO 1496-3:1995.

6.7.5.11   Design approval

6.7.5.11.1   The competent authority or its authorized body shall issue a design approval certificate for any new design of a MEGC. This certificate shall attest that the MEGC has been surveyed by that authority, is suitable for its intended purpose and meets the requirements of this chapter, the applicable provisions for gases of chapter 4.1 and of packing instruction P200. When a series of MEGCs are manufactured without change in the design, the certificate shall be valid for the entire series. The certificate shall refer to the prototype test report, the materials of construction of the manifold, the standards to which the elements are made and an approval number. The approval number shall consist of the distinguishing sign or mark of the country granting the approval, i.e. the distinguishing sign for use in international traffic, as prescribed by the Convention on Road Traffic, Vienna, 1968, and a registration number. Any alternative arrangements according to 6.7.1.2 shall be indicated on the certificate. A design approval may serve for the approval of smaller MEGCs made of materials of the same type and thickness, by the same fabrication techniques and with identical supports, equivalent closures and other appurtenances.

6.7.5.11.2   The prototype test report for the design approval shall include at least the following:

    .1   the results of the applicable framework test specified in ISO 1496-3:1995;

    .2   the results of the initial inspection and test specified in 6.7.5.12.3;

    .3   the results of the impact test specified in 6.7.5.12.1; and

    .4   certification documents verifying that the cylinders and tubes comply with the applicable standards.

6.7.5.12   Inspection and testing

6.7.5.12.1   MEGCs meeting the definition of *container* in the International Convention for Safe Containers (CSC), 1972, as amended, shall not be used unless they are successfully qualified by subjecting a representative prototype of each design to the Dynamic, Longitudinal Impact Test prescribed in the United Nations *Manual of Tests and Criteria*, part IV, section 41. This provision only applies to MEGCs which are constructed according to a design approval certificate which has been issued on or after 1 January 2008.

6.7.5.12.2   The elements and items of equipment of each MEGC shall be inspected and tested before being put into service for the first time (initial inspection and test). Thereafter, MEGCs shall be inspected at no more than five-year intervals (5-year periodic inspection). An exceptional inspection and test shall be performed, regardless of the last periodic inspection and test, when necessary according to 6.7.5.12.5.

6.7.5.12.3 The initial inspection and test of an MEGC shall include a check of the design characteristics, an external examination of the MEGC and its fittings with due regard to the gases to be transported, and a pressure test performed at the test pressures according to packing instruction P200. The pressure test of the manifold may be performed as a hydraulic test or by using another liquid or gas with the agreement of the competent authority or its authorized body. Before the MEGC is placed into service, a leakproofness test and a test of the satisfactory operation of all service equipment shall also be performed. When the elements and their fittings have been pressure-tested separately, they shall be subjected together after assembly to a leakproofness test.

6.7.5.12.4 The 5-year periodic inspection and test shall include an external examination of the structure, the elements and the service equipment in accordance with 6.7.5.12.6. The elements and the piping shall be tested at the periodicity specified in packing instruction P200 and in accordance with the provisions described in 6.2.1.6. When the elements and equipment have been pressure-tested separately, they shall be subjected together after assembly to a leakproofness test.

6.7.5.12.5 An exceptional inspection and test is necessary when the MEGC shows evidence of damaged or corroded areas, leakage, or other conditions that indicate a deficiency that could affect the integrity of the MEGC. The extent of the exceptional inspection and test shall depend on the amount of damage or deterioration of the MEGC. It shall include at least the examinations required under 6.7.5.12.6.

6.7.5.12.6 The examinations shall ensure that:

.1 the elements are inspected externally for pitting, corrosion, abrasions, dents, distortions, defects in welds or any other conditions, including leakage, that might render the MEGC unsafe for transport;

.2 the piping, valves, and gaskets are inspected for corroded areas, defects, and other conditions, including leakage, that might render the MEGC unsafe for filling, discharge or transport;

.3 missing or loose bolts or nuts on any flanged connection or blank flange are replaced or tightened;

.4 all emergency devices and valves are free from corrosion, distortion and any damage or defect that could prevent their normal operation. Remote closure devices and self-closing stop-valves shall be operated to demonstrate proper operation;

.5 required markings on the MEGC are legible and in accordance with the applicable requirements; and

.6 the framework, the supports and the arrangements for lifting the MEGC are in satisfactory condition.

6.7.5.12.7 The inspections and tests in 6.7.5.12.1, 6.7.5.12.3, 6.7.5.12.4 and 6.7.5.12.5 shall be performed or witnessed by a body authorized by the competent authority. When the pressure test is a part of the inspection and test, the test pressure shall be the one indicated on the data plate of the MEGC. While under pressure, the MEGC shall be inspected for any leaks in the elements, piping or equipment.

6.7.5.12.8 When evidence of any unsafe condition is discovered, the MEGC shall not be returned to service until it has been corrected and the applicable tests and verifications are passed.

6.7.5.13 Marking

6.7.5.13.1 Every MEGC shall be fitted with a corrosion-resistant metal plate permanently attached to the MEGC in a conspicuous place readily accessible for inspection. The elements shall be marked in accordance with chapter 6.2. At least the following information shall be marked on the plate by stamping or by any other similar method:

Country of manufacture:

U	Approval	Approval	For alternative arrangements (see 6.7.1.2):
N	country	number	''AA''

Manufacturer's name or mark

Manufacturer's serial number

Authorized body for the design approval

Year of manufacture

Test pressure: . . . . bar/kPa gauge*

Design temperature range . . . . °C to . . . . °C

Number of elements . . . .

Total water capacity . . . . litres

Initial pressure test date and identification of the authorized body

Date and type of most recent periodic test(s):

Year . . . . Month . . . .

Stamp of the authorized body who performed or witnessed the most recent test.

* The unit used shall be marked.

Note: No metal plate may be fixed to the elements.

6.7.5.13.2 The following information shall be marked on a metal plate firmly secured to the MEGC:

Name of the operator

Maximum permissible load mass . . . . kg

Working pressure at 15°C: . . . . bar gauge

Maximum permissible gross mass (MPGM) . . . . kg

Unladen (tare) mass . . . . kg

6

Chapter 6.8

Provisions for road tank vehicles

6.8.1 General

6.8.1.1 Tank support frameworks, fitting and tie-down attachments*

6.8.1.1.1 Road tank vehicles shall be designed and manufactured with supports to provide a secure base during transport and with suitable tie-down attachments. The tie-down attachments shall be located on the tank support or vehicle structure in such a manner that the suspension system is not left in free play.

6.8.1.1.2 Tanks shall be carried only on vehicles whose fastenings are capable, in conditions of maximum permissible loading of the tanks, of absorbing the forces specified in 6.7.2.2.12, 6.7.3.2.9 and 6.7.4.2.12.

6.8.2 Road tank vehicles for long international voyages for substances of classes 3 to 9

6.8.2.1 Design and construction

6.8.2.1.1 A road tank vehicle for long international voyages shall be fitted with a tank complying with the provisions of chapters 4.2 and 6.7 and shall comply with the relevant provisions for tank supports, frameworks, lifting and tie-down attachments,* except for the provisions for forklift pockets, and in addition comply with the provisions of 6.8.1.1.1.

6.8.2.2 Approval, testing and marking

6.8.2.2.1 For approval, testing and marking of the tank, see 6.7.2.

6.8.2.2.2 The tank supports and tie-down attachments* of vehicles for long international voyages shall be included in the visual external inspection provided for in 6.7.2.19.

6.8.2.2.3 The vehicle of a road tank vehicle shall be tested and inspected in accordance with the road transport provisions of the competent authority of the country in which the vehicle is operated.

6.8.3 Road tank vehicles for short international voyages

6.8.3.1 Road tank vehicles for substances of classes 3 to 9 (IMO type 4)

6.8.3.1.1 *General provisions*

6.8.3.1.1.1 An IMO type 4 tank shall comply with either:

.1 the provisions of 6.8.2; or

.2 the provisions of 6.8.3.1.2 and 6.8.3.1.3.

6.8.3.1.2 *Design and construction*

6.8.3.1.2.1 An IMO type 4 tank shall comply with the provisions of 6.7.2, with the exception of:

.1 6.7.2.3.2; however, they shall have been subjected to a test pressure not less than that specified according to the appropriate tank instruction assigned to the substance;

---

* See also IMO Assembly resolution A.581(14) of 20 November 1985, Guidelines for securing arrangements for the transport of road vehicles on ro–ro ships.

> .2 6.7.2.4; however, the thickness of cylindrical portions and ends in reference steel shall be:
>
> > .1 not more than 2 mm thinner than the thickness specified according to the appropriate tank instruction assigned to the substance;
> >
> > .2 subject to an absolute minimum thickness of 4 mm of reference steel; and
> >
> > .3 for other materials, subject to an absolute minimum thickness of 3 mm;
>
> .3 6.7.2.2.13; however, the safety factor shall be not less than 1.3;
>
> .4 6.7.2.2.1 to 6.7.2.2.7; however, the materials of construction shall comply with the provisions of the competent authority for road transport;
>
> .5 6.7.2.5.1; however, the protection of valves and accessories shall comply with the provisions of the competent authority for road transport;
>
> .6 6.7.2.5.3; however, IMO type 4 tanks shall be provided with manholes or other openings in the tank which comply with the provisions of the competent authority for road transport;
>
> .7 6.7.2.5.2 and 6.7.2.5.4; however, tank nozzles and external fittings shall comply with the provisions of the competent authority for road transport;
>
> .8 6.7.2.6; however, IMO type 4 tanks with bottom openings shall not be used for substances for which bottom openings are not permitted in the appropriate tank instruction assigned to the substance. In addition, existing openings and hand inspection holes shall be either closed by bolted flanges mounted both internally and externally, fitted with product-compatible gaskets, or by welding as specified in 6.7.2.6.1. The closing of openings and hand inspection holes shall be approved by the competent authority for sea transport;
>
> .9 6.7.2.7 to 6.7.2.15; however, IMO type 4 tanks shall be fitted with pressure relief devices of the type required according to the appropriate tank instruction assigned to the substance. The devices shall be acceptable to the competent authority for the road transport for the substances to be transported. The start-to-discharge pressure of the spring-loaded pressure relief devices shall in no case be less than the maximum allowable working pressure, nor greater than 25% above that pressure; and
>
> .10 6.7.2.17; however, tank supports on permanently attached IMO type 4 tanks shall comply with the provisions of the competent authority for road transport.

6.8.3.1.2.2 For IMO type 4 tanks, the maximum effective gauge pressure developed by the substances to be transported shall not exceed the maximum allowable working pressure of the tank.

6.8.3.1.3 *Approval, testing and marking*

6.8.3.1.3.1 IMO type 4 tanks shall be approved for road transport by the competent authority.

6.8.3.1.3.2 The competent authority for sea transport shall issue additionally, in respect of an IMO type 4 tank, a certificate attesting compliance with the relevant design, construction and equipment provisions of this subsection and the special provisions for certain substances, as applicable.

6.8.3.1.3.3 IMO type 4 tanks shall be periodically tested and inspected in accordance with the provisions of the competent authority for road transport.

6.8.3.1.3.4 An IMO type 4 tank shall be marked in accordance with 6.7.2.20. However, where the marking required by the competent authority for road transport is substantially in agreement with that of 6.7.2.20, it will be sufficient to endorse the metal plate attached to the IMO type 4 tank with "IMO 4".

6.8.3.1.3.5 IMO type 4 tanks which are not permanently attached to the chassis shall be marked "IMO type 4" in letters at least 32 mm high.

6.8.3.2 **Road tank vehicles for non-refrigerated liquefied gases of class 2 (IMO type 6)**

6.8.3.2.1 *General provisions*

6.8.3.2.1.1 An IMO type 6 tank shall comply with either:

> .1 the provisions of 6.7.3; or
>
> .2 the provisions of 6.8.3.2.2 and 6.8.3.2.3.

6.8.3.2.1.2 For an IMO type 6 tank, the design temperature range is defined in 6.7.3.1. The temperature to be taken is to be agreed by the competent authority for road transport.

6.8.3.2.2 *Design and construction*

6.8.3.2.2.1 An IMO type 6 tank shall comply with the provisions of 6.7.3, with the exception of:

> .1 the safety factor of 1.5 in 6.7.3.2.10; however, the safety factor shall not be less than 1.3;

.2   6.7.3.5.7;

.3   6.7.3.6.1, if bottom openings are approved by the competent authority for sea transport;

.4   6.7.3.7.1; however, the devices shall open at a pressure not less than the MAWP and be fully open at a pressure not exceeding the test pressure of the tank;

.5   6.7.3.8, if the delivery capacity of the pressure relief devices is approved by the competent authorities for sea and road transport;

.6   the location of the pressure relief device inlets in 6.7.3.11.1, which need not be in the longitudinal centre of the shell;

.7   the provisions for forklift pockets; and

.8   6.7.3.13.5.

6.8.3.2.2.2   If the landing legs of an IMO type 6 tank vehicle are to be used as support structure, the loads specified in 6.7.3.2.9 shall be taken into account in their design and method of attachment. Any bending stress induced in the shell as a result of this manner of support shall also be included in the design calculations.

6.8.3.2.2.3   Securing arrangements (tie-down attachments) shall be fitted to the tank support structure and the towing vehicle of an IMO type 6 tank. Semi-trailers unaccompanied by a towing vehicle shall be accepted for shipment only if the trailer supports and the securing arrangements and the position of stowage are agreed by the competent authority for sea transport, unless the approved Cargo Securing Manual includes this arrangement.

6.8.3.2.3   *Approval, testing and marking*

6.8.3.2.3.1   IMO type 6 tanks shall be approved for road transport by the competent authority for road transport.

6.8.3.2.3.2   The competent authority for sea transport shall issue additionally, in respect of an IMO type 6 tank, a certificate attesting compliance with the relevant design, construction and equipment provisions of this chapter and, where appropriate, the special provisions for the gases listed in the Dangerous Goods List. The certificate shall list the gases allowed to be transported.

6.8.3.2.3.3   An IMO type 6 tank shall be periodically tested and inspected in accordance with the provisions of the competent authority for road transport.

6.8.3.2.3.4   An IMO type 6 tank shall be marked in accordance with 6.7.3.16. However, where the marking required by the competent authority for road transport is substantially in agreement with that of 6.7.3.16.1, it will be sufficient to endorse the metal plate attached to the road tank vehicle with "IMO 6".

6.8.3.3   **Road tank vehicles for refrigerated liquefied gases of class 2 (IMO type 8)**

6.8.3.3.1   *General provisions*

6.8.3.3.1.1   An IMO type 8 tank shall comply with either:

.1   the provisions of 6.7.4; or

.2   the provisions of 6.8.3.3.2 and 6.8.3.3.3.

6.8.3.3.1.2   An IMO type 8 tank shall not be offered for transport by sea in a condition that would lead to venting during the voyage under normal conditions of transport.

6.8.3.3.2   *Design and construction*

6.8.3.3.2.1   An IMO type 8 tank shall comply with the provisions of 6.7.4, with the exception:

.1   that aluminium jackets may be used, with the approval of the competent authority for sea transport;

.2   that IMO type 8 tanks may have a shell thickness less than 3 mm, subject to the approval of the competent authority for sea transport;

.3   that for IMO type 8 tanks used for non-flammable refrigerated gases, one of the valves may be replaced by a frangible disc. The frangible disc shall rupture at a nominal pressure equal to the test pressure;

.4   of the provisions of 6.7.4.7.3 for the combined capacity of all pressure relief devices under complete fire-engulfment conditions;

.5   of the safety factor of 1.5 in 6.7.4.2.13; however, the safety factor shall not be less than 1.3;

.6   of 6.7.4.8; and

.7   of the provisions for forklift pockets.

6.8.3.3.2.2   If the landing legs of an IMO type 8 tank are to be used as support structure, the loads agreed as in 6.7.4.2.12 shall be taken into account in their design and method of attachment. Bending stress induced in the shell as a result of this manner of support shall be included in design calculations.

6.8.3.3.2.3   Securing arrangements (tie-down attachments) shall be fitted to the tank support structure and the towing vehicle of an IMO type 8 tank. Semi-trailers unaccompanied by a towing vehicle shall be accepted for shipment only if the trailer supports and the securing arrangements and the position of stowage are agreed by the competent authority for sea transport, unless the approved Cargo Securing Manual includes this arrangement.

6.8.3.3.3   *Approval, testing and marking*

6.8.3.3.3.1   IMO type 8 tanks shall be approved for road transport by the competent authority for road transport.

6.8.3.3.3.2   The competent authority for sea transport shall issue additionally, in respect of an IMO type 8 tank, a certificate attesting compliance with the relevant design, construction and equipment provisions of this subsection and, where appropriate, the special tank type provisions for the gases in the Dangerous Goods List. The certificate shall list the gases allowed to be transported.

6.8.3.3.3.3   IMO type 8 tanks shall be periodically tested and inspected in accordance with the provisions of the competent authority for road transport.

6.8.3.3.3.4   IMO type 8 tanks shall be marked in accordance with 6.7.4.15. However, where the marking required by the competent authority for road transport is substantially in agreement with that of 6.7.4.15.1, it will be sufficient to endorse the metal plate attached to the road tank vehicle with "IMO type 8"; the reference to holding time may be omitted.

6

Chapter 6.9

Provisions for the design, construction, inspection and testing of bulk containers

Note: Sheeted bulk containers shall not be used for sea transport.

6.9.1   Definitions

For the purposes of this section:

Closed bulk containers are totally closed bulk containers having a rigid roof, sidewalls, end walls and floor (including hopper-type bottoms), including bulk containers with an opening roof, or side or end wall that can be closed during transport. Closed bulk containers may be equipped with openings to allow for the exchange of vapours and gases with air and which prevent, under normal conditions of transport, the release of solid contents as well as the penetration of rain and splash water.

Sheeted bulk containers are open-top bulk containers with rigid bottom (including hopper-type bottom), side and end walls and a non-rigid covering.

6.9.2   Application and general provisions

6.9.2.1   Bulk containers and their service and structural equipment shall be designed and constructed to withstand, without loss of contents, the internal pressure of the contents and the stresses of normal handling and transport.

6.9.2.2   Where a discharge valve is fitted, it shall be capable of being made secure in the closed position and the whole discharge system shall be suitably protected from damage. Valves having lever closures shall be able to be secured against unintended opening and the open or closed position shall be readily apparent.

6.9.2.3   **Code for designating types of bulk container**

The following table indicates the codes to be used for designating types of bulk containers:

Types of bulk container	Code
Sheeted bulk container (Not allowed for sea transport)	BK1
Closed bulk container	BK2

6.9.2.4   In order to take account of progress in science and technology, the use of alternative arrangements which offer at least equivalent safety as provided by the provisions of this chapter may be considered by the competent authority.

6.9.3   Provisions for the design, construction, inspection and testing of freight containers used as bulk containers

6.9.3.1   **Design and construction provisions**

6.9.3.1.1   The general design and construction provisions in this section are deemed to be met if the bulk container complies with the requirements of ISO 1496-4:1991 "Series 1 freight containers – Specification and testing – Part 4: Non-pressurized containers for dry bulk" and the container is siftproof.

6.9.3.1.2   Freight containers designed and tested in accordance with ISO 1496-1:1990 "Series 1 freight containers – Specification and testing – Part 1: General cargo containers for general purposes" shall be equipped with operational equipment which is, including its connection to the freight container, designed to strengthen the end walls and to improve the longitudinal restraint as necessary to comply with the test requirements of ISO 1496-4:1991, as relevant.

6.9.3.1.3 Bulk containers shall be siftproof. Where a liner is used to make the container siftproof, it shall be made of a suitable material. The strength of the material used for, and the construction of, the liner shall be appropriate to the capacity of the container and its intended use. Joins and closures of the liner shall withstand pressures and impacts liable to occur under normal conditions of handling and transport. For ventilated bulk containers, any liner shall not impair the operation of ventilating devices.

6.9.3.1.4 The operational equipment of bulk containers designed to be emptied by tilting shall be capable of withstanding the total filling mass in the tilted orientation.

6.9.3.1.5 Any movable roof or side or end wall or roof section shall be fitted with locking devices with securing devices designed to show the locked state to an observer at ground level.

6.9.3.2 Service equipment

6.9.3.2.1 Filling and discharge devices shall be so constructed and arranged as to be protected against the risk of being wrenched off or damaged during transport and handling. The filling and discharge devices shall be capable of being secured against unintended opening. The open and closed position and direction of closure shall be clearly indicated.

6.9.3.2.2 Seals of openings shall be so arranged as to avoid any damage by the operation, filling and emptying of the bulk container.

6.9.3.2.3 Where ventilation is required, bulk containers shall be equipped with means of air exchange, either by natural convection, e.g., by openings, or active elements, e.g., fans. The ventilation shall be designed to prevent negative pressures in the container at all times. Ventilating elements of bulk containers for the transport of flammable substances or substances emitting flammable gases or vapours shall be designed so as not to be a source of ignition.

6.9.3.3 Inspection and testing

6.9.3.3.1 Freight containers used, maintained and qualified as bulk containers in accordance with the requirements of this section shall be tested and approved in accordance with the International Convention for Safe Containers (CSC), 1972, as amended.

6.9.3.3.2 Freight containers used and qualified as bulk containers shall be inspected periodically according to that Convention.

6.9.3.4 Marking

6.9.3.4.1 Freight containers used as bulk containers shall be marked with a Safety Approval Plate in accordance with the International Convention for Safe Containers.

6.9.4 Provisions for the design, construction and approval of bulk containers other than freight containers

6.9.4.1 Bulk containers covered in this section include skips, offshore bulk containers, bulk bins, swap bodies, trough-shaped containers, roller containers, and load compartments of vehicles.

6.9.4.2 These bulk containers shall be designed and constructed so as to be strong enough to withstand the shocks and loadings normally encountered during transport, including, as applicable, transhipment between modes of transport.

6.9.4.3 Load compartments of vehicles shall comply with the requirements of, and be acceptable to, the competent authority responsible for land transport of the dangerous goods to be transported in bulk.

6.9.4.4 These bulk containers shall be approved by the competent authority and the approval shall include the code for designating types of bulk containers in accordance with 6.9.2.3 and the provisions for inspection and testing, as appropriate.

6.9.4.5 Where it is necessary to use a liner in order to retain the dangerous goods, it shall meet the provisions of 6.9.3.1.3.

6.9.4.6 The following statement shall be shown on the transport document:

"Bulk container BK2 approved by the competent authority of . . . .".

6

PART 7

PROVISIONS CONCERNING
TRANSPORT OPERATIONS

Chapter 7.1

Stowage

7.1.1 General provisions

7.1.1.1   Except in class 1 – *Explosives* (see 7.1.7), ships are divided into two groupings for the purpose of making appropriate stowage recommendations:

   .1   cargo ships or passenger ships carrying a number of passengers limited to not more than 25 or to 1 passenger per 3 metres of overall length, whichever is the greater number;

   .2   other passenger ships in which the limiting number of passengers transported is exceeded.

7.1.1.2   Stowage categories

   Substances, materials and articles shall be stowed as indicated in the Dangerous Goods List in accordance with one of the categories specified below (see also appendix B).

7.1.1.2.1   *Stowage category A*

   Cargo ships or passenger ships carrying a number of passengers limited to not more than 25 or to 1 passenger per 3 metres of overall length, whichever is the greater number   } ON DECK OR UNDER DECK

   Other passenger ships in which the limiting number of passengers transported is exceeded   } ON DECK OR UNDER DECK

7.1.1.2.2   *Stowage category B*

   Cargo ships or passenger ships carrying a number of passengers limited to not more than 25 or to 1 passenger per 3 metres of overall length, whichever is the greater number   } ON DECK OR UNDER DECK

   Other passenger ships in which the limiting number of passengers transported is exceeded   } ON DECK ONLY

7.1.1.2.3   *Stowage category C*

   Cargo ships or passenger ships carrying a number of passengers limited to not more than 25 or to 1 passenger per 3 metres of overall length, whichever is the greater number   } ON DECK ONLY

   Other passenger ships in which the limiting number of passengers transported is exceeded   } ON DECK ONLY

7.1.1.2.4   *Stowage category D*

   Cargo ships or passenger ships carrying a number of passengers limited to not more than 25 or to 1 passenger per 3 metres of overall length, whichever is the greater number   } ON DECK ONLY

   Other passenger ships in which the limiting number of passengers transported is exceeded   } **PROHIBITED**

7.1.1.2.5   *Stowage category E*

   Cargo ships or passenger ships carrying a number of passengers limited to not more than 25 or to 1 passenger per 3 metres of overall length, whichever is the greater number   } ON DECK OR UNDER DECK

   Other passenger ships in which the limiting number of passengers transported is exceeded   } **PROHIBITED**

7.1.1.3   Because of the rapidity with which an accident involving dangerous goods may affect the whole ship, the transport of some particularly dangerous substances, materials or articles is not permitted aboard "other passenger ships" where large numbers of people may need to be evacuated at short notice. This is indicated in the Dangerous Goods List.

7

7.1.1.4   If spillages or leakages of dangerous goods occur in an *under-deck* cargo space, precautions shall be taken to prevent the inadvertent pumping of such spillages or leakages through the machinery space bilge piping and pumps.

7.1.1.5   The minimum stacking height for testing packagings intended to contain dangerous cargoes in accordance with chapter 6.1 is 3 metres, for IBCs and large packagings the stacking test load shall be determined in accordance with 6.5.4.6.4 and 6.6.5.3.3.4 respectively. At the discretion of the master, stowing to a greater height is allowed, taking into account the conditions of stowage and the degree of support and reinforcement provided.

7.1.1.5.1   Drums containing dangerous goods shall always be stowed in an upright position unless otherwise authorized by the competent authority.

7.1.1.6   Where *on or under deck* stowage is permitted, stowage *under deck* is recommended wherever possible, except that, for certain articles of class 1 whose principal hazard is the production of smoke or toxic fumes, stowage *on deck* is recommended (see also 7.1.7.1.7.2).

7.1.1.7   Fibreboard boxes and other packages susceptible to water damage shall be stowed *under deck* or, if they are stowed *on deck*, they shall be so protected that at no time they are exposed to weather or to seawater.

7.1.1.8   Stowage *on deck only* has been prescribed in cases where:

.1   constant supervision is required; or

.2   accessibility is particularly required; or

.3   there is a substantial risk of formation of explosive gas mixtures, development of highly toxic vapours, or unobserved corrosion of the ship.

7.1.1.9   When dangerous goods are stowed *on deck*, hydrants, sounding pipes and the like and access thereto shall be kept free and clear of such deck cargo.

7.1.1.10   At all times the stowage of dangerous goods shall be so arranged:

.1   as to ensure clear walkways and access to all the facilities necessary for the safe working of the ship; and

.2   that, for goods possessing a particular hazard, the special provisions regarding stowage, which are included in the Dangerous Goods List or in this chapter, are observed.

7.1.1.11   Notwithstanding the stowage provisions given in the Dangerous Goods List, empty uncleaned receptacles which shall be stowed *on deck only* when full may be stowed *on deck or under deck* in a mechanically ventilated cargo space. Empty uncleaned cylinders which carry a label of class 2.3 shall be stowed *on deck only* (see also 4.1.1.11).

7.1.1.12   For stowage of dangerous goods in limited quantities, see 3.4.3.

7.1.1.13   Where it is necessary to prevent pressure build-up, decomposition or polymerization of a substance, the packages shall be stowed *shaded from radiant heat*, which includes protection from strong sunlight.

7.1.1.14   When it is indicated in the Dangerous Goods List that the substances shall be stowed *shaded from radiant heat*, stowage *under deck* shall be "away from" sources of heat.

7.1.1.15   Where, for certain dangerous goods, protection from sources of heat is required, this shall be taken to include sparks, flames, steam pipes, heating coils, top or side walls of heated fuel and cargo tanks, and bulkheads of machinery spaces (see regulation II-1/2.8 of SOLAS, 1974 (as amended)); alternatively, for the latter, such bulkheads shall be insulated to A-60 standards or equivalent, except that in the case of explosives, in addition to an A-60 bulkhead, "away from" stowage shall be maintained.

7.1.1.16   Portable tanks shall not be overstowed by other cargo transport units unless they are designed for that purpose and transported in specially designed ships, or unless they are specially protected to the satisfaction of the competent authority.

7.1.2   Stowage in relation to living quarters

7.1.2.1   Where stowage *clear of living quarters* is required, in deciding the stowage, consideration shall be given to the possibility that leaking vapours may penetrate the accommodation, machinery spaces and other work areas via entrances or other openings in bulkheads or through ventilation ducts.

7.1.2.2   The criteria for identifying the substances, materials and articles for which such stowage is required are:

.1   volatile toxic substances;

.2   volatile corrosive substances;

    .3 substances which, in moist air, produce toxic or corrosive vapours;

    .4 substances which evolve strongly narcotic vapours;

    .5 flammable, toxic or corrosive gases of class 2.

7.1.2.3   For those substances which shall be stowed *clear of living quarters*, this is indicated in column 16 of the Dangerous Goods List.

7.1.2.4   All infectious substances shall be stowed "separated by a complete compartment or hold from" living quarters.

7.1.3   Stowage in relation to undeveloped films and plates, and mailbags

Undeveloped photographic films and plates, and mailbags (which shall be assumed to contain them), shall be segregated from class 7 materials in accordance with 7.2.9.8.

7.1.4   Stowage of marine pollutants

7.1.4.1   Taking into account the severe hazards to the marine environment to which incidents involving marine pollutants may lead, it is necessary that these substances are properly stowed and secured so as to minimize these hazards without impairing the safety of the ship and the persons on board.

7.1.4.2   Where stowage is permitted *on deck or under deck*, *under deck* stowage is preferred except when a weather deck provides equivalent protection.

7.1.4.3   Where stowage *on deck only* is required, preference shall be given to stowage on well-protected decks or to stowage inboard in sheltered areas of exposed decks.

7.1.5   Stowage in relation to foodstuffs

7.1.5.1   Substances and articles for which toxicity is indicated by a label of class 6.1, packing groups I and II, or a label of class 2.3 shall be stowed "separated from" foodstuffs except when the substances and the foodstuffs are in different closed cargo transport units. In such cases, no segregation is required between units.

7.1.5.2   All infectious substances shall be stowed "separated by a complete compartment or hold from" all foodstuffs.

7.1.5.3   Material for which radioactivity is indicated by a label of class 7 shall be stowed "separated from" foodstuffs.

7.1.5.4   Substances and articles for which corrosivity is indicated by a label of class 8 and substances for which toxicity is indicated by a label of class 6.1, packing group III shall be stowed "away from" foodstuffs.

7.1.5.5   For the definitions of "separated by a complete compartment or hold from", "separated from" and "away from", see chapter 7.2.

7.1.6   Stowage of solutions and mixtures

7.1.6.1   Solutions or mixtures containing one or more non-dangerous substances and a dangerous substance identified by name in this Code shipped under a generic or N.O.S. entry shall be stowed in accordance with the stowage category assigned to this generic or N.O.S. entry.

7.1.7   Stowage and handling of goods of class 1

7.1.7.1   **Definitions for stowage of class 1**

For the purpose of this section, the following types of stowage are referred to in column 16 of the Dangerous Goods List.

7.1.7.1.1   *Closed cargo transport unit* means a unit which fully encloses the contents by permanent structures and can be secured to the ship's structure, and includes a magazine. Cargo transport units with fabric sides or tops are not closed cargo transport units. Where this stowage is specified, stowage in small compartments such as deck-houses and mast lockers are acceptable alternatives. The floor of any closed cargo transport unit or compartment shall either be constructed of wood, close-boarded or arranged that goods are stowed on sparred gratings, wooden pallets or dunnage. Provided that the necessary additional specifications are met, a closed cargo transport unit may be used for type "A" or "C" class 1 stowage or as a magazine.

7.1.7.1.2   *Magazine* means a closed cargo transport unit or a *compartment* in the ship designed to protect certain goods of class 1 from damage by other cargo during loading and unloading, and adverse weather conditions when in transit, and to prevent unauthorized access. Magazines may also be a fixed compartment in a ship. Magazines may be positioned in any part of the ship conforming with the general stowage conditions for goods of class 1 (see 7.1.7.4) but magazines which are fixed structures shall be sited so that their doors, where fitted, are easily accessible.

7.1.7.1.3   *Secured to the ship's structure* in the context of on-deck stowage of goods of class 1 means any closed cargo transport unit or large unpackaged article (see 4.1.5.15), which shall be securely stowed and lashed to prevent the shifting of the goods.

7.1.7.1.4   *Magazine stowage types "A", "C"* and *special stowage*. The stowage of class 1 substances and certain articles is subject to varying levels of containment (except for compatibility group S substances) when stowed below deck. The levels are dependent on hazard presented by the nature of the particular goods involved. The different levels of containment are defined below as *"A", "C"* and *"special"*; Magazine stowage type *"A"* is given to those substances which shall be kept clear of steelwork. All other substances except SUBSTANCES, EXPLOSIVE, N.O.S. in compatibility groups G or L and those in compatibility group A are given *closed cargo transport unit* stowage. Substances in compatibility group A are given *magazine stowage type "C"*. SUBSTANCES, EXPLOSIVE, N.O.S. in compatibility groups G and L and some articles in compatibility groups G, H, L and K which are particularly hazardous are given *special stowage*. Column 16 of the Dangerous Goods List specifies the type of stowage applicable to each substance or article.

7.1.7.1.5   *Magazine stowage type "A"* means that the inner sides and floors of cargo transport units and compartments on the ship shall be close-boarded with wood. The roof or deckhead shall be clean and free of rust or scale. It need not be battened. The top of the stow shall be at least 300 mm from the roof or deckhead. This form of stowage guards against friction between any spilled contents from packages and the sides of magazines or the ship's sides and bulkheads. When utilized as part of the structure of the space, the ship's sides and bulkheads shall be clean and free from rust or scale and shall be protected by battening or sweatboards spaced not more than 150 mm apart. All stanchions and other unprotected ironwork shall be similarly clean and battened. When other goods of class 1 are stowed in the unit or space with goods requiring *magazine stowage type "A"*, it is essential to ensure that their packagings have no exposed external parts made of ferrous metal or aluminium alloy. When in the square of a cargo space, loading shall not take place from the top unless special precautions are taken.

7.1.7.1.6   *Magazine stowage type "C"* means a closed cargo transport unit positioned as near as practicable to the centreline of the ship; it shall not be positioned closer to the ship's side than a distance equal to one eighth of the beam or 2.4 m, whichever is the lesser.

7.1.7.1.7   *Special stowage*

   .1   Goods of class 1 allocated to this category shall be stowed as far away as practicable from living quarters and from work areas, and shall not be overstowed. Closed cargo transport units used for goods of this category shall not be positioned closer to the ship's side than a distance equal to one eighth of the beam or 2.4 m, whichever is the lesser.

   .2   This stowage is allocated to certain articles of which the principal hazard is that of fire and leakage of the contents, accompanied by dense smoke or tear-producing or toxic fumes (compatibility group G, H, or K), and also to substances and articles which present a special risk (compatibility group L). Where on-deck stowage is recommended but not possible, the goods shall always be subject to special stowage.

   .3   Goods in compatibility groups G or H may be transported in steel magazines. A steel cargo transport unit which prevents leakage of contents may also be used for this purpose. Alternative arrangements may also be agreed by the competent authority concerned.

   .4   Goods of only one compatibility group shall be stowed in any one compartment. When separate compartments are not available, the competent authority may allow goods in compatibility groups G and H to be stowed in the same compartment not less than 3 m apart, provided they are placed in separate steel magazines.

   .5   Goods in compatibility group K or L shall be transported in steel magazines.

7.1.7.2   Stowage categories

   For the purpose of column 16 in the Dangerous Goods List, class 1 goods (see 7.1.7.1) shall be stowed as indicated in column 16 of the Dangerous Goods List in accordance with one of the categories specified below. Where categories indicate that goods of class 1 may be transported in a passenger ship, the maximum net explosive mass that may be transported on any passenger ship shall be determined in accordance with 7.1.7.5.

Stowage category 01	Cargo ship (up to 12 passengers)	ON DECK OR UNDER DECK
	Passenger ship	ON DECK OR UNDER DECK

Stowage category 02	Cargo ship (up to 12 passengers)	ON DECK OR UNDER DECK
	Passenger ship	ON DECK IN CLOSED CARGO TRANSPORT UNITS OR UNDER DECK IN CLOSED CARGO TRANSPORT UNITS
Stowage category 03	Cargo ship (up to 12 passengers)	ON DECK OR UNDER DECK
	Passenger ship	ON DECK ONLY IN CLOSED CARGO TRANSPORT UNITS
Stowage category 04	Cargo ship (up to 12 passengers)	ON DECK OR UNDER DECK
	Passenger ship	**PROHIBITED**
Stowage category 05	Cargo ship (up to 12 passengers)	ON DECK IN CLOSED CARGO TRANSPORT UNITS OR UNDER DECK
	Passenger ship	ON DECK IN CLOSED CARGO TRANSPORT UNITS OR UNDER DECK
Stowage category 06	Cargo ship (up to 12 passengers)	ON DECK IN CLOSED CARGO TRANSPORT UNITS OR UNDER DECK
	Passenger ship	ON DECK IN CLOSED CARGO TRANSPORT UNITS OR UNDER DECK IN CLOSED CARGO TRANSPORT UNITS
Stowage category 07	Cargo ship (up to 12 passengers)	ON DECK IN CLOSED CARGO TRANSPORT UNITS OR UNDER DECK
	Passenger ship	ON DECK ONLY IN CLOSED CARGO TRANSPORT UNITS
Stowage category 08	Cargo ship (up to 12 passengers)	ON DECK IN CLOSED CARGO TRANSPORT UNITS OR UNDER DECK
	Passenger ship	**PROHIBITED**
Stowage category 09	Cargo ship (up to 12 passengers)	ON DECK IN CLOSED CARGO TRANSPORT UNITS OR UNDER DECK IN CLOSED CARGO TRANSPORT UNITS
	Passenger ship	ON DECK IN CLOSED CARGO TRANSPORT UNITS OR UNDER DECK IN CLOSED CARGO TRANSPORT UNITS
Stowage category 10	Cargo ship (up to 12 passengers)	ON DECK IN CLOSED CARGO TRANSPORT UNITS OR UNDER DECK IN CLOSED CARGO TRANSPORT UNITS
	Passenger ship	ON DECK ONLY IN CLOSED CARGO TRANSPORT UNITS
Stowage category 11	Cargo ship (up to 12 passengers)	ON DECK IN CLOSED CARGO TRANSPORT UNITS OR UNDER DECK IN MAGAZINE STOWAGE TYPE "C"
	Passenger ship	ON DECK ONLY IN CLOSED CARGO TRANSPORT UNITS
Stowage category 12	Cargo ship (up to 12 passengers)	ON DECK IN CLOSED CARGO TRANSPORT UNITS OR UNDER DECK IN MAGAZINE STOWAGE TYPE "C"
	Passenger ship	**PROHIBITED**
Stowage category 13	Cargo ship (up to 12 passengers)	ON DECK IN CLOSED CARGO TRANSPORT UNITS OR UNDER DECK IN MAGAZINE STOWAGE TYPE "A"
	Passenger ship	ON DECK ONLY IN CLOSED CARGO TRANSPORT UNITS
Stowage category 14	Cargo ship (up to 12 passengers)	ON DECK ONLY IN CLOSED CARGO TRANSPORT UNITS
	Passenger ship	**PROHIBITED**
Stowage category 15	Cargo ship (up to 12 passengers)	ON DECK IN CLOSED CARGO TRANSPORT UNITS OR UNDER DECK IN CLOSED CARGO TRANSPORT UNITS
	Passenger ship	**PROHIBITED**

7.1.7.3   Application of stowage provisions for class 1

Goods of class 1 requiring *under deck* and *on deck* stowage shall be stowed in accordance with 7.1.7.4. However, the provisions of 7.1.7.4.4, 7.1.7.4.5 and 7.1.7.4.6 need not be applied to goods of division 1.4, compatibility group S. Such goods may be stowed together with all other goods of class 1 except those in compatibility group A or L (see 7.2.7.2.1.4).

7.1.7.4   Stowage provisions for goods of class 1

7.1.7.4.1   *General*

7.1.7.4.1.1   For under-deck stowage of class 1 goods in stowage categories 09 and 10:

.1   avoid stowage of other cargo in the same compartment or hold if it is readily combustible (such as items packaged in straw);

.2   maintain direct access to hatchways by not overstowing goods with goods other than class 1; and

.3 in all cases, all goods, including goods of class 1 stowed in cargo transport units, within the compartment or hold shall be so secured as to eliminate the possibility of significant movement. Where an entire deck is utilized as a magazine, the stowage shall be so arranged that the goods stowed therein will be removed from the ship before working any cargo in any decks above or below that deck in the same hold.

7.1.7.4.1.2 Goods of class 1, with the exception of goods in division 1.4, shall not be stowed in the outermost row.

7.1.7.4.2 *Sources of heat*

.1 Goods of class 1 shall be stowed in a cool part of the ship and shall be kept as cool as practicable while on board. Stowage shall be "away from" (see 7.2.2.2.1) all sources of heat (see 7.1.1.15).

.2 The compartments shall be clean. In order to reduce the risk of ignition, the space shall be free of dust from other cargoes, such as grain or coal dust.

7.1.7.4.3 *Wetness*

Compartments where goods of class 1 are to be stowed *under deck* shall be dry. In the event of the contents of packages being affected by water whilst on board, immediate advice shall be sought from the shipper; pending this advice, handling of the packages shall be avoided.

7.1.7.4.4 *Securement*

Goods of class 1 shall be properly secured to prevent significant movement during the voyage. Cargo transport units which contain goods of class 1 or large unpackaged articles shall be securely stowed and lashed to prevent the shifting of the goods. Goods within a compartment, within a hold or within a cargo transport unit which also contains goods of class 1 shall be secured so as to eliminate the possibility of significant movement. Where necessary, precautions shall be taken to prevent cargo sliding down between the frames at the ship's side.

7.1.7.4.5 *Stowage of rockets and rocket motors*

.1 Rockets or rocket motors of small or medium size – i.e. those normally transported in the assembled condition – which are fitted with their complete ignition system (self-propulsive) may be transported, whether in palletized unit loads or not, without restriction on the stowage configuration, provided that they are EFFECTIVELY mechanically restrained from significant flight by strapping or other means embodied in the packing design, or that they embody one or more of the following safeguards:

.1 Electro-explosive devices incorporated in the ignition system shall be effectively protected against stray currents from any source and the venturi shall be effectively protected to prevent accidental ignition.

.2 In the case of percussion ignition systems, the percussion device shall be effectively protected.

.3 The firing route from igniter to propellant charge shall be interrupted by a mechanical shutter or by displacement of part of the explosives train and the venturi shall be effectively capped to prevent accidental ignition.

.4 The rockets or rocket motors shall be fitted with aerodynamic "spoilers" – or, better still, flight spoilers – of an approved design.

.2 Rockets or rocket motors of large size – i.e. those normally transported in an unassembled condition – shall always be moved under the following stowage restrictions when in the self-propulsive state:

.1 the OUTER packaging shall be marked to indicate the head end of the rocket or rocket motor, and

.2 the rockets or rocket motors shall be stowed with heads towards and not more than 30 cm away from a bulkhead, deck, deckhead or the ship's side.

.3 Rockets or rocket motors of ANY size which do not meet the requirements stated in paragraph .1.1 to .1.4 above shall be moved under the stowage restrictions detailed in paragraph .2.

7.1.7.4.6 *Separation from living quarters and machinery spaces*

.1 Goods of class 1 shall be stowed as far away as possible from living quarters and machinery spaces and shall not be stowed directly above or below such spaces. Where the provisions of this subsection are less stringent than those of SOLAS 1974, as amended, the Convention's provisions shall be satisfied for ships to which they are applicable.

.2 There shall be a permanent steel "A" class bulkhead between living quarters and a compartment containing goods of class 1. Goods in division 1.1, 1.2, 1.3, or 1.5 shall not be stowed within 3 m of this bulkhead; in the decks immediately above or below, they shall be stowed at least 3 m from the line of this bulkhead projected vertically.

.3 There shall be a permanent steel "A" class bulkhead between a compartment containing goods of class 1 and a machinery space. Goods of class 1 (except those in division 1.4, compatibility group S) shall not be stowed within 3 m of this bulkhead; in the decks above or below, they shall be stowed at least 3 m from the line of this bulkhead projected vertically. Unless the separation bulkhead between the machinery space of category "A" and a compartment containing goods of class 1 is insulated to class "A-60" standard, additional measures, as indicated in appendix 2 of this chapter, shall be taken for goods other than in division 1.4, compatibility group S; see also 7.1.7.4.6.5.

.4 Where goods of class 1 are stowed "away from" bulkheads bounding living quarters or machinery spaces, the intervening space may be filled with cargo that is not readily combustible.

.5 In ships the keels of which were laid before 1 September 1984 and where these requirements may prove impracticable, alternative arrangements as detailed in appendix 2 of this chapter may be approved by the competent authority of the flag State.

.6 Goods of class 1 shall not be stowed within a horizontal distance of 6 m from any open fire, machinery exhausts, galley uptakes, lockers used for combustible stores, or other potential sources of ignition. They shall always be so stowed as to ensure clear walkways and be "away from" all other facilities necessary for the safe working of a ship and be clear of fire hydrants, steam pipes and means of access, and be not less than a horizontal distance of 8 m from the bridge, living quarters and life-saving appliances.

7.1.7.4.7 *Electrical equipment and cables*

.1 Electrical equipment and cables shall not generally be installed in compartments in which goods of class 1 are to be transported. Where they are installed but do not need to be energized during the voyage or where they do not meet the required standard (see appendix 3), they shall be isolated from the supply so that no part of the circuit within the compartment is energized. The method of isolation may be by opening switches or circuit breakers, or by disconnection from busbars or by the removal of links in the system. In any case, the means, or access to the means, of disconnection and of reconnection shall be padlocked off and under the control of a responsible person.

.2 When electrical equipment and cables in a compartment in which goods of class 1 are to be transported need to be energized during the voyage for the safe operation of the ship, they shall meet the recognized standards (see appendix 3 of this chapter). All electrical equipment and cables shall be tested by a skilled person to ensure that they are safe, and to determine satisfactory insulation resistance and continuity of the cable cores and continuity and earthing of metal sheathing or armouring, and shall be so certified by that person.

.3 All goods of class 1 shall be stowed in a safe position relative to electrical equipment and cables. Additional physical protection shall be provided, where necessary, to minimize possible damage to the electrical equipment and cables, especially during loading and unloading.

.4 Cable joints in compartments shall be avoided where possible. When joints are unavoidable, they shall be enclosed in metal-clad junction boxes of the recognized standard (see appendix 3 of this chapter).

.5 All lighting shall be of the fixed type and shall meet the relevant inspection, test and installation standards of this section.

.6 Standards required for electrical equipment and cables in compartments, including permanently fixed magazines, where explosives dust may be encountered or where articles containing a flammable liquid may be stowed are specified in appendix 3 of this chapter. In all other cases, equipment and cables appropriate to the compartment may be used only if tested in accordance with 7.1.7.4.7.2.

7.1.7.4.8 *Lightning protection*

A lightning conductor, earthed to the sea, shall be provided on any mast or structure, unless effective electrical bonding is provided between the sea and the mast or structure from its extremity and throughout to the main body of the hull structure. Steel masts in ships of all-welded construction may be considered to comply with this requirement.

7.1.7.4.9 *Security*

All compartments, magazines and cargo transport units shall be locked or suitably secured in order to prevent unauthorized access. The means of locking and securing shall be such that, in the case of emergency, access can be gained without delay.

7.1.7.4.10 *Loading and unloading operations*

In the event that a package containing goods of class 1 is found to be suffering from breakage or leakage, expert advice should be obtained for its safe handling and disposal (see 7.3.1.3). Loading and unloading procedures and equipment used should be of such a nature that sparks are not produced, in particular where the floors of the cargo compartment are not constructed of close-boarded wood. All cargo handlers should be briefed by the shipper or receiver of the possible risks and necessary precautions, prior to commencing the handling of explosives.

7.1.7.5 **Transport of goods of class 1 on passenger ships**

7.1.7.5.1 For the purpose of stowage in this class, the terms "passenger ship" and "cargo ship" are used as in SOLAS 1974, as amended.

7.1.7.5.2 Explosives in division 1.4, compatibility group S, may be transported in any amount on passenger ships. No other explosives may be transported on passenger ships except any one of the following:

.1 explosive articles for life-saving purposes listed in the Dangerous Goods List, if the total net explosives mass of such articles does not exceed 50 kg per ship; or

.2 goods in compatibility groups C, D and E, if the net explosives mass does not exceed 10 kg per ship; or

.3 articles in compatibility group G other than those requiring special stowage, if the total net explosives mass does not exceed 10 kg per ship; or

.4 articles in compatibility group B, if the total net explosives mass does not exceed 10 kg per ship.

7.1.7.5.3 Notwithstanding the provisions of 7.1.7.5.2, additional quantities or types of goods of class 1 may be transported in passenger ships in which there are special safety measures approved by the competent authority.

7.1.7.5.4 Articles in compatibility group N shall only be allowed in passenger ships if the total net explosives mass does not exceed 50 kg per ship and no other explosives apart from division 1.4, compatibility group S, are transported.

7.1.7.5.5 Goods of class 1 which may be transported in passenger ships are identified in the Dangerous Goods List. They shall be stowed in accordance with the following table:

Division	Samples, explosive	Compatibility group												
		A	B	C	D	E	F	G	H	J	K	L	N	S
1.1	d	c	e	e	e	e	c	e	–	c	–	c	–	–
1.2	d	–	e	e	e	e	c	e	c	c	c	c	–	–
1.3	d	–	–	e	–	–	c	e	c	c	c	c	–	–
1.4	d	–	b	b	b	b	c	b	–	–	–	–	–	a
1.5	d	–	–	–	e	–	–	–	–	–	–	–	–	–
1.6	d	–	–	–	–	–	–	–	–	–	–	–	e	–

a = As for cargo ships, *on deck or under deck*.

b = As for cargo ships, *on deck or under deck*, in magazines only.

c = Prohibited; this provision overrides all others.

d = As specified by the competent authority of the country concerned, with regard to the provisions of 7.1.7.

e = In containers or the like, *on deck only*.

7.1.8 Stowage of goods of class 2

7.1.8.1 General stowage precautions for goods of class 2

7.1.8.1.1 Receptacles shall be kept as cool as reasonably practicable during transport and should be stowed "away from" all sources of heat.

7.1.8.1.2 Receptacles shall be stowed in the following manner:

.1 Receptacles shall be dunnaged to prevent their resting directly on a steel deck. They shall be stowed and chocked as necessary to prevent movement unless mounted in a frame as a unit. Receptacles for liquefied gases shall be stowed such that the liquid phase is not in contact with any pressure relief device.

.2 When receptacles are stowed in a vertical position they shall be stowed in a block, cribbed or boxed in with suitable sound lumber and the box or crib dunnaged to provide clearance from a steel deck. Receptacles in a box or crib shall be braced to prevent any movement. The box or crib shall be securely chocked and lashed to prevent movement in any direction.

.3 When stowed *on deck*, receptacles shall be protected from radiant heat, which includes protection from strong sunlight.

.4 Receptacles stowed *under deck* shall be stowed in mechanically ventilated cargo spaces.

7.1.8.1.3 Adequate measures shall be taken to prevent the penetration of leaking gases into any other part of the ship. Gases may not necessarily be lighter than air and may sink to the lower levels of a cargo space, where they may be accidentally ignited and "flashback" may occur. Attention shall also be paid in this respect when toxic or suffocating gas is transported.

7.1.8.1.4 Whenever gases are transported, stowage shall be such that leaking vapours are unlikely to penetrate the accommodation, machinery spaces and other work areas via entrances or other openings in bulkheads or through ventilation ducts.

7.1.8.1.5 Where gases are loaded in a closed cargo transport unit, special attention shall be paid to the provisions of 7.4.2.5.2.

7.1.8.2 **General stowage precautions for flammable or toxic gases**

.1 Adequate precautions shall be taken to protect flammable gases from heat. Mechanical ventilation shall be provided which shall effectively remove flammable vapours from enclosed cargo spaces.

.2 On ships carrying passengers, these gases shall be stowed well "away from" any deck or spaces provided for the use of passengers. When such gases are transported on board roll-on/roll-off ships, special attention shall be given to the relevant provisions of chapter 7.4.

7.1.9 Stowage of goods of class 3

7.1.9.1 The vapours from all substances of class 3 have a narcotic effect, and prolonged inhalation may result in unconsciousness. Deep or prolonged narcosis may lead to death.

7.1.9.2 Class 3 substances shall be stowed as indicated in the Dangerous Goods List. However, substances with a flashpoint of less than 23°C c.c. packaged in jerricans, plastics (3H1, 3H2), drums, plastics (1H1, 1H2) and plastics receptacles in a plastic drum (6HH1, 6HH2) shall be stowed *on deck only* unless packed in a closed cargo transport unit.

7.1.9.3 The substances of this class shall be kept as cool as reasonably practicable during transit. They should be stowed "away from" all possible sources of heat.

7.1.9.4 Adequate precautions shall be taken to protect the flammable liquids from heat emanating from bulkheads or other sources. Ventilation shall be provided which shall effectively remove flammable vapours from the cargo space.

7.1.9.5 Adequate measures shall be taken to prevent the penetration of leaking liquid or vapour into any other part of the ship. Vapours may not necessarily be lighter than air and may sink to the lower levels of a cargo space, where they may be accidentally ignited and a "flashback" to the flammable liquids may occur.

7.1.9.6 Whenever flammable liquids with a flashpoint of less than 23°C c.c. are transported in portable tanks, the stowage shall be such that leaking vapours are unlikely to penetrate the accommodation, machinery spaces and other work areas via entrances or other openings in bulkheads or through ventilation ducts.

7.1.9.7 Where it is deemed necessary for a substance of this class to be stowed "clear of living quarters", it is included in the Dangerous Goods List.

7.1.9.8 On ships carrying passengers, substances in this class shall be stowed well away from any deck or spaces provided for the use of passengers. When such substances are transported on board roll-on/roll-off ships, see chapter 7.4.

7.1.10 Stowage of goods of classes 4.1, 4.2 and 4.3

7.1.10.1 **General stowage precautions for goods of classes 4.1, 4.2 and 4.3**

7.1.10.1.1 The substances of these classes shall be kept as cool as reasonably practicable during transit. They should be stowed "away from" all sources of heat.

7.1.10.1.2 Provision shall be made, where a substance is liable to give off vapours or dust which can form an explosive mixture with air, for stowage to be in a well-ventilated space.

7.1.10.1.3 It may be necessary during the voyage to jettison a package or packages of a consignment of a substance in these classes if there is danger of involvement in a fire. This shall be borne in mind when stowage is permitted *under deck*.

7.1.10.1.4 On ships carrying passengers, substances of these classes shall be stowed well away from any deck or spaces provided for the use of passengers. When such substances are transported on board roll-on/roll-off ships, see chapter 7.4.

7.1.10.2 **Additional stowage precautions for self-reactive substances, UN 2956, UN 3241, UN 3242, UN 3251 and solid desensitized explosives**

7.1.10.2.1 During transport, packages containing self-reactive substances, UN 2956, UN 3241, UN 3242, UN 3251 or solid desensitized explosives shall be shaded from radiant heat, which includes protection from direct sunlight.

7

7.1.10.3 **Stowage precautions for FISHMEAL, UNSTABILIZED (UN 1374, packing group III) and FISHMEAL, STABILIZED (UN 2216, class 9)**

7.1.10.3.1 For loose packagings:

.1 Temperature readings shall be taken 3 times a day during the voyage and recorded.

.2 If the temperature of the cargo exceeds 55°C and continues to increase, ventilation to the hold shall be restricted. If self-heating continues, then carbon dioxide or inert gas shall be introduced. The ship shall be equipped with facilities for introducing carbon dioxide or inert gas into the holds.

.3 The cargo shall be stowed well clear of pipes and bulkheads which are liable to become heated (such as engine-room bulkheads).

.4 For UN 1374, where *loose bags* are being carried, double strip stowage is recommended, provided there is good surface and through ventilation. The diagram in 7.1.10.3.3 shows how this can be achieved. For UN 2216, where *loose bags* are being carried, no special ventilation is required for block stowage of bagged cargo.

7.1.10.3.2 For containers:

.1 After packing, the doors and other openings shall be sealed to prevent the penetration of air into the unit.

.2 Temperature readings in the hold shall be taken once a day early in the morning during the voyage and recorded.

.3 If the temperature of the hold rises excessively above ambient and continues to increase, the possible need to apply copious quantities of water in an emergency and the consequent risk to the stability of the ship shall be considered.

.4 The cargo shall be stowed well clear of pipes and bulkheads which are liable to become heated (such as engine-room bulkheads).

7.1.10.3.3

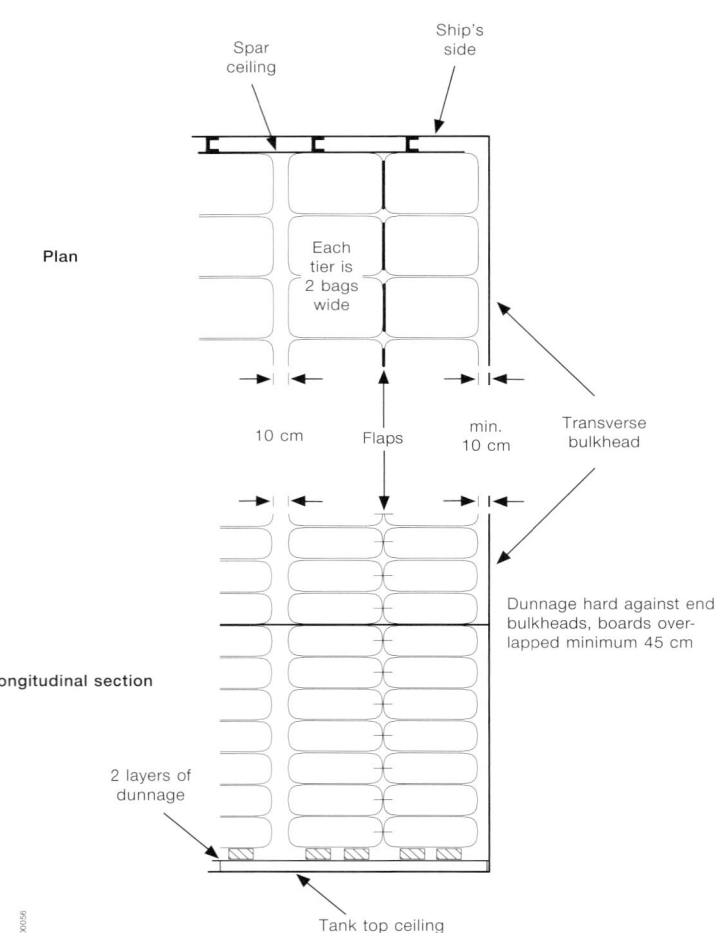

Double strip stowage

7.1.10.4   **Stowage precautions for SEED CAKE (UN 1386)**

7.1.10.4.1   Stowage precautions for SEED CAKE, containing vegetable oil (a) mechanically expelled seeds, containing more than 10% oil or more than 20% oil and moisture combined:

.1   through and surface ventilation is required;

.2   if the voyage exceeds 5 days, the ship shall be equipped with facilities for introducing carbon dioxide or inert gas into the cargo spaces;

.3   bags shall always be stowed in double strip, as shown in 7.1.10.3.3 of this Code for fishmeal, unstabilized; and

.4   regular temperature readings shall be taken at varying depths in the cargo space and recorded. If the temperature of the cargo exceeds 55°C and continues to increase, ventilation to the cargo spaces shall be restricted. If self-heating continues, then carbon dioxide or inert gas shall be introduced.

7.1.10.4.2   Stowage precautions for SEED CAKE, containing vegetable oil (b) solvent extractions and expelled seeds containing not more than 10% of oil and, when the amount of moisture is higher than 10%, not more than 20% of oil and moisture combined:

.1   surface ventilation is required to assist in removing any residual solvent vapour;

.2   if bags are stowed without provision for ventilation to circulate throughout the stow and the voyage exceeds 5 days, regular temperature readings shall be taken at varying depths in the hold and recorded; and

.3   if the voyage exceeds 5 days, the vessel shall be equipped with facilities for introducing carbon dioxide or inert gas into the cargo spaces.

7.1.11   Stowage of goods of class 5.1

7.1.11.1   Except for cargo spaces for the stowage of cargo transport units, cargo spaces shall be cleaned before oxidizing substances are loaded into them. Attention shall be paid to the removal of all combustible materials which are not necessary for the stowage of such cargoes.

7.1.11.2   As far as reasonably practicable, non-combustible securing and protecting materials and only a minimum of clean dry wooden dunnage shall be used.

7.1.11.3   Precautions shall be taken to avoid the penetration of oxidizing substances into other cargo spaces, bilges, etc., which may contain combustible material.

7.1.11.4   After discharge, cargo spaces used for the transport of oxidizing substances shall be inspected for contamination. A space that has been contaminated shall be properly cleaned and examined before being used for other cargoes, especially foodstuffs.

7.1.11.5   **Stowage precautions for AMMONIUM NITRATE, UN 1942 and**
AMMONIUM NITRATE BASED FERTILIZER, UN 2067

7.1.11.5.1   AMMONIUM NITRATE, UN 1942 and AMMONIUM NITRATE BASED FERTILIZER, UN 2067 may be stowed under deck in a clean cargo space capable of being opened up in an emergency. The possible need to open hatches in case of fire to provide maximum ventilation and to apply water in an emergency and the consequent risk to the stability of the ship through flooding of cargo space shall be considered before loading.

7.1.11.5.2   The compatibility of non-hazardous ammonium nitrate mixtures with other materials which may be stowed in the same cargo space shall be considered before loading.

7.1.12   Stowage of goods of class 5.2

7.1.12.1   Organic peroxides shall be stowed in accordance with stowage category D, as specified in 7.1.1.2.

7.1.12.2   When organic peroxides are transported on roll-on/roll-off ships, see the relevant provisions of chapter 7.4.

7.1.12.3   Organic peroxides shall be stowed ''away from'' living quarters or access to them.

7.1.12.4 Organic peroxides shall be stowed "away from" all sources of heat. Packages containing organic peroxides shall be protected from direct sunshine and stowed in a cool, well-ventilated place.

7.1.12.5 When stowage arrangements are made, it shall be borne in mind that it may become necessary to take the appropriate emergency action, such as jettisoning.

7.1.13 Stowage of goods of class 6.1

7.1.13.1 General stowage precautions for goods of class 6.1

7.1.13.1.1 After discharge, spaces used for the transport of substances of this class shall be inspected for contamination. A space which has been contaminated shall be properly cleaned and examined before being used for other cargoes, especially foodstuffs.

7.1.13.2 Additional stowage precautions for toxic substances which are also flammable liquids

.1 On ships carrying passengers, these substances shall be stowed "away from" any deck or spaces provided for the use of passengers. When such substances are transported on board roll-on/roll-off ships, see the relevant provisions of chapter 7.4.

.2 These substances shall be stowed in a mechanically ventilated space and be kept as cool as reasonably practicable during transit. They should be stowed "away from" all sources of heat.

7.1.14 Stowage of goods of class 7

7.1.14.1 Radioactive material shall be stowed as indicated in the Dangerous Goods List for class 7 in 3.2, in accordance with the appropriate stowage category specified in 7.1.1.2.

7.1.14.2 The total activity in a single cargo space of an inland water craft, or in another conveyance, for transport of LSA material or SCO in Type IP-1, Type IP-2, Type IP-3 packaging or unpackaged shall not exceed the limits shown in the table hereunder.

Conveyance activity limits for LSA material and SCO in industrial packages or unpackaged

Nature of material	Activity limit for conveyances other than by inland waterway	Activity limit for a cargo space of an inland water craft
LSA-I	No limit	No limit
LSA-II and LSA-III non-combustible solids	No limit	$100A_2$
LSA-II and LSA-III combustible solids, and all liquids and gases	$100A_2$	$10A_2$
SCO	$100A_2$	$10A_2$

7.1.14.3 Consignments shall be securely stowed.

7.1.14.4 Provided that its average surface heat flux does not exceed 15 W/m^2 and that the immediately surrounding cargo is not in sacks or bags, a package or overpack may be transported or stored among packaged general cargo without any special stowage provisions except as may be specifically required by the competent authority in an applicable approval certificate.

7.1.14.5 Loading of freight containers and accumulation of packages, overpacks and freight containers shall be controlled as follows:

.1 Except under the condition of exclusive use, the total number of packages, overpacks and freight containers aboard a single conveyance shall be so limited that the total sum of the transport indexes aboard the conveyance does not exceed the values shown in the table hereunder. For consignments of LSA-I material there shall be no limit on the sum of the transport indexes.

TI limits for freight containers and conveyances not under exclusive use

Type of freight container or conveyance	Limit on total sum of transport indexes in a freight container or aboard a conveyance
Freight container – small	50
Freight container – large	50
Vehicle	50

Aircraft
    Passenger
    Cargo
	
	50
	200

Inland waterway vessel	50

Seagoing vessel[a]
1 *Hold, compartment or defined area:*
    Packages, overpacks, small freight containers
    Large freight containers
 	
2 *Total vessel:*
    Packages, overpacks, small freight containers
    Large freight containers
	
	50
	200
	
	200
	No limit

[a] Packages or overpacks transported in or on a vehicle which are in accordance with the provisions of 7.1.14.7 may be transported by vessels provided that they are not removed from the vehicle at any time while on board the ship.

.2   Where a consignment is transported under exclusive use, there shall be no limit on the sum of the transport indexes aboard a single conveyance.

.3   The radiation level under routine conditions of transport shall not exceed 2 mSv/h at any point on, and 0.1 mSv/h at 2 m from, the external surface of the conveyance, except for consignments transported under exclusive use by road or rail, for which the radiation limits around the vehicle are specified in 7.1.14.7.2 and 7.1.14.7.3.

.4   The total sum of the criticality safety indexes in a freight container and aboard a conveyance shall not exceed the values shown in the table hereunder.

CSI limits for freight containers and conveyances containing fissile material

Type of freight container or conveyance	Limit on total sum of criticality safety indexes in a freight container or aboard a conveyance	
	Not under exclusive use	Under exclusive use
Freight container – small	50	n.a.
Freight container – large	50	100
Vehicle	50	100

Aircraft
    Passenger
    Cargo
	
	50
	50
		
		n.a.
		100

Inland waterway vessel	50	100

Seagoing vessel[a]
1 *Cargo space or defined deck area:*
    Packages, overpacks, small freight containers
    Large freight containers
 	
2 *Total vessel:*
    Packages, overpacks, small freight containers
    Large freight containers
	
	50
	50
	
	200[b]
	No limit[b]
		
		100
		100
		
		200[c]
		No limit[c]

[a] Packages or overpacks transported in or on a vehicle which are in accordance with the provisions of 7.1.14.7 may be transported by ships provided that they are not removed from the vehicle at any time while on board the ship. In that case, the entries under the heading "Under exclusive use" apply.

[b] The consignment shall be so handled and stowed that the total sum of CSI's in any group does not exceed 50, and that each group is handled and stowed so that the groups are separated from each other by at least 6 m.

[c] The consignment shall be so handled and stowed that the total sum of CSI's in any group does not exceed 100, and that each group is handled and stowed so that the groups are separated from each other by at least 6 m. The intervening space between groups may be occupied by other cargo.

7.1.14.6   Any package or overpack having a transport index greater than 10, or any consignment having a criticality safety index greater than 50, shall be transported only under exclusive use.

7

7.1.14.7   For consignments under exclusive use, the radiation level shall not exceed:

.1   10 mSv/h at any point on the external surface of any package or overpack, and may only exceed 2 mSv/h provided that:

.1   the vehicle is equipped with an enclosure which, during routine conditions of transport, prevents the access of unauthorized persons to the interior of the enclosure, and

.2   provisions are made to secure the package or overpack so that its position within the vehicle enclosure remains fixed during routine conditions of transport, and

.3   there is no loading or unloading during the shipment;

.2   2 mSv/h at any point on the outer surfaces of the vehicle, including the upper and lower surfaces, or, in the case of an open vehicle, at any point on the vertical planes projected from the outer edges of the vehicle, on the upper surface of the load, and on the lower external surface of the vehicle; and

.3   0.1 mSv/h at any point 2 m from the vertical planes represented by the outer lateral surfaces of the vehicle, or, if the load is transported in an open vehicle, at any point 2 m from the vertical planes projected from the outer edges of the vehicle.

7.1.14.8   In the case of road vehicles, no persons other than the driver and assistants shall be permitted in vehicles carrying packages, overpacks or freight containers bearing category II – YELLOW or III – YELLOW labels.

7.1.14.9   Packages or overpacks having a surface radiation level greater than 2 mSv/h, unless being transported in or on a vehicle under exclusive use in accordance with the table under 7.1.14.5, footnote (a), shall not be transported by ship except under special arrangement.

7.1.14.10   The transport of consignments by means of a special use ship which, by virtue of its design or by reason of its being chartered, is dedicated to the purpose of carrying radioactive material shall be excepted from the provisions specified in 7.1.14.5 provided that the following conditions are met:

.1   a radiation protection programme for the shipment shall be approved by the competent authority of the flag State of the ship and, when requested, by the competent authority at each port of call;

.2   stowage arrangements shall be predetermined for the whole voyage, including any consignments to be loaded at ports of call *en route*; and

.3   the loading, transport and unloading of the consignments shall be supervised by persons qualified in the transport of radioactive material.

7.1.14.11   Any conveyance and equipment used regularly for the transport of radioactive material shall be periodically checked to determine the level of contamination. The frequency of such checks shall be related to the likelihood of contamination and the extent to which radioactive material is transported.

7.1.14.12   Except as provided in 7.1.14.13, any conveyance, or equipment or part thereof, which has become contaminated above the limits specified in 4.1.9.1.2 in the course of the transport of radioactive material or which shows a radiation level in excess of 5 μSv/h at the surface, shall be decontaminated as soon as possible by a qualified person and shall not be re-used unless the non-fixed contamination does not exceed the limits specified in 4.1.9.1.2, and the radiation level resulting from the fixed contamination on surfaces after decontamination is less than 5 μSv/h at the surface.

7.1.14.13   A freight container, tank, IBC or conveyance dedicated to the transport of unpackaged radioactive material under exclusive use shall be excepted from the provisions of 4.1.9.1.4 and 7.1.14.12 solely with regard to its internal surfaces and only for as long as it remains under that specific exclusive use.

7.1.14.14   Where a consignment is undeliverable, the consignment shall be placed in a safe location and the appropriate competent authority shall be informed as soon as possible and a request made for instructions on further action.

7.1.15   Stowage of goods of class 8

7.1.15.1   General stowage precautions for goods of class 8

7.1.15.1.1   The substances of this class shall be kept as dry as reasonably practicable, since in the presence of moisture they may be corrosive to most metals and some also react violently with water.

7.1.15.1.2   All substances of this class for which an unprotected plastics packaging is permitted shall be kept as cool as reasonably practicable as the resistance of most plastics decreases at higher temperatures.

7.1.15.2   Additional stowage precautions for corrosive substances which are also flammable liquids

7.1.15.2.1   On ships carrying passengers, these substances shall be stowed well "away from" any deck or spaces provided for the use of passengers. When such substances are transported on board roll-on/roll-off ships, special attention shall be given to the relevant provisions of chapter 7.4.

7.1.15.2.2   These substances shall be stowed in a mechanically ventilated space and be kept as cool as reasonably practicable during transit. They should be stowed "away from" all sources of heat.

7.1.16   Stowage of goods of class 9

7.1.16.1   Stowage precautions for AMMONIUM NITRATE BASED FERTILIZER, UN 2071

7.1.16.1.1   AMMONIUM NITRATE BASED FERTILIZER, UN 2071 shall be stowed in a clean cargo space capable of being opened up in an emergency. In the case of bagged fertilizer or fertilizer in containers, it is sufficient if, in the case of an emergency, the cargo is accessible through free approaches (hatch entries), and mechanical ventilation enables the master to exhaust any gases or fumes resulting from decomposition. The possible need to open hatches in case of fire to provide maximum ventilation and to apply water in an emergency, and the consequent risk to the stability of the ship through flooding of the cargo space, shall be considered before loading.

7.1.16.1.2   If suppression of decomposition should prove impracticable (such as in bad weather), there would not necessarily be immediate danger to the structure of the ship. However, the residue left after decomposition may have only half the mass of the original cargo; this loss of mass may also affect the stability of the ship and shall be considered before loading.

7.1.16.1.3   AMMONIUM NITRATE BASED FERTILIZER, UN 2071 shall be stowed out of direct contact with a metal engine-room bulkhead. In the case of bagged material, this may be done, for example, by using wooden boards to provide an air space between the bulkhead and the cargo. This requirement need not apply to short international voyages.

7.1.16.1.4   In the case of ships not fitted with smoke-detecting or other suitable devices, arrangements shall be made during the voyage to inspect cargo spaces containing these fertilizers at intervals not exceeding 4 hours (such as to sniff at the ventilators serving them) to ensure early detection of decomposition should that occur.

7.1.16.2   Stowage precautions for FISHMEAL, STABILIZED (UN 2216, class 9)

For stowage precautions for FISHMEAL, STABILIZED (UN 2216, class 9), see 7.1.10.3.

Appendix 1

Deck stowage

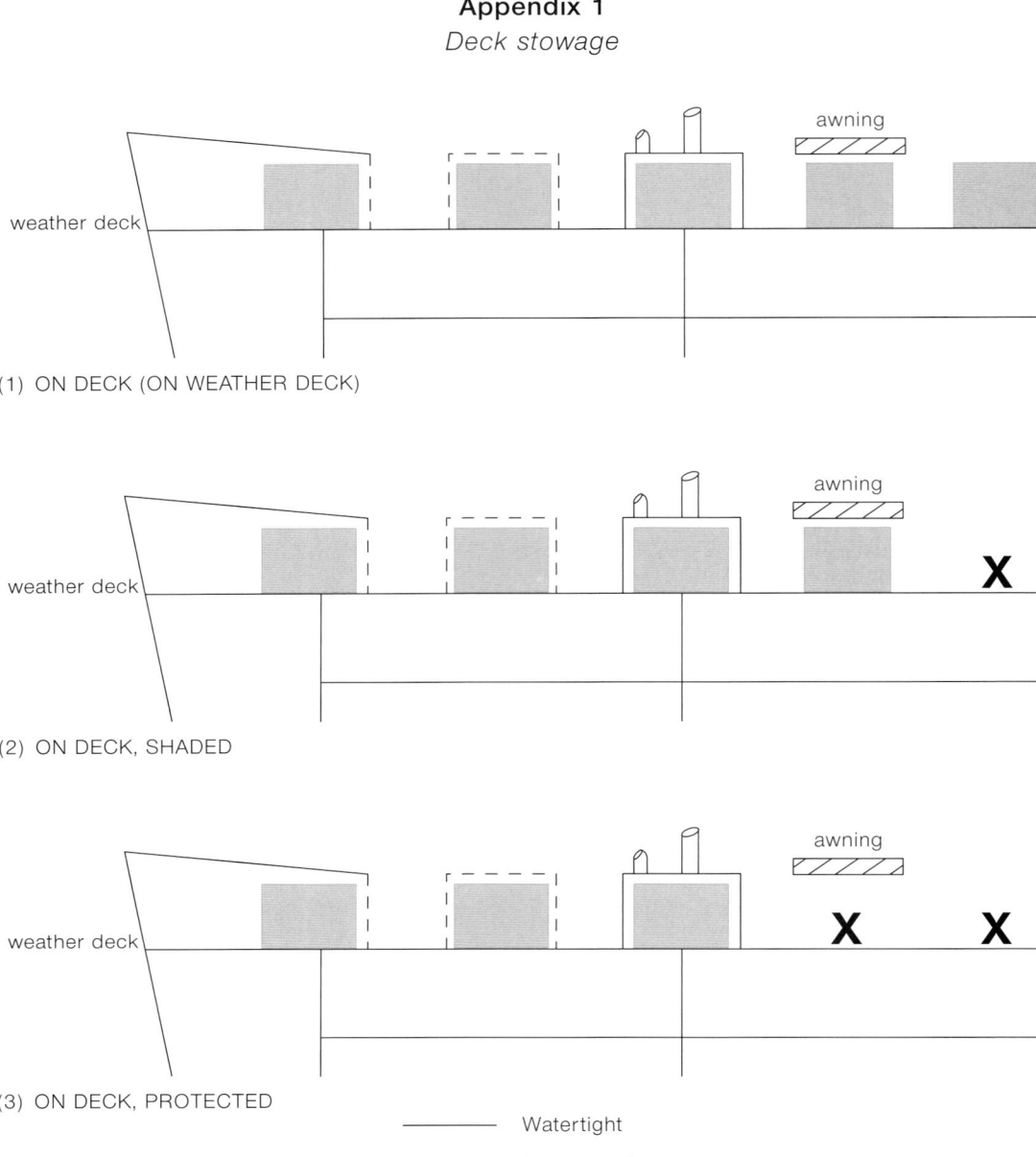

(1) ON DECK (ON WEATHER DECK)

(2) ON DECK, SHADED

(3) ON DECK, PROTECTED

——————— Watertight

– – – – Spray-proof

X Not permitted

Appendix 2

Separation from machinery spaces

1   Paragraph 7.1.7.4.6.3 prescribes the degree of separation between goods of class 1 (other than those in division 1.4, compatibility group S) and a category "A" machinery space. The separation required is an "A-60" bulkhead and in addition a distance of at least 3 m from the bulkhead.

2   In a ship the keel of which was laid before 1 September 1984 and which is not provided with a separation bulkhead of class "A-60" standard, the following alternatives are acceptable:

     .1   stowage at least 9 m away from an "A-0" bulkhead; or

     .2   stowage at least 3 m away from one of the alternative constructional provisions specified in 3 below, combined with the additional safety measures given in 4.

3   **Construction provisions**

     .1   two bulkheads of steel not less than 0.6 m apart forming a floodable cofferdam; or

     .2   one watertight bulkhead of steel and one temporary bulkhead positioned not less than 0.6 m away from the former, suitably constructed of timber and faced on the engine-room side with an approved fire insulation material of the type and thickness which would be applied to a division of "A-30" standard.

4   **Additional safety measures**

     .1   a fixed fire-detection and fire-alarm system and a fixed fire-extinguishing installation meeting the standards of SOLAS 1974, as amended, shall be fitted to the main machinery space, but a temporary system of at least equivalent capacity may be accepted;

     .2   a power-operated pump which, together with its source of power and permanent sea connections, shall be located outside the machinery space; and

     .3   at least two sets of self-contained breathing apparatus are available for fire fighting.

Appendix 3

Electrical standards

(See paragraph 7.1.7.4.7 of this chapter)

	Risk involved	Requirement for electrical equipment, including junction boxes and vent fans*
1	Explosive dust only	Equipment to have enclosure IP6X and temperature class T5.
2	Flammable vapour only	Equipment to be Ex i(b) IIAT5 or Ex d IIAT5: luminaries only may be Ex e IIT5.
3	Explosive dust and flammable vapour	Equipment to be Ex i(b) IIAT5 with IP6X enclosures or Ex d IIAT5 with IP6X enclosures. Luminaries may only be Ex e IIT5 with IP6X enclosures.

In all the above cases, cables shall be:

     .1   enclosed in heavy gauge, solid-drawn or continuously butt-welded and galvanized conduit; or

     .2   protected by electrically continuous metal sheathing or metallic wire armour, braid or tape; or

     .3   of the mineral-insulated metal-covered type.

---

* Reference is made to the Recommendations published by the International Electrotechnical Commission (IEC) and, in particular, to publication 529 – *Classification of degrees of protection provided by enclosures.*

Chapter 7.2

Segregation

7.2.1 General

7.2.1.1 The provisions of this chapter shall apply to all cargo spaces *on deck or under deck* of all types of ships and to cargo transport units.

7.2.1.2 Incompatible goods shall be segregated from one another.

7.2.1.3 For the implementation of this requirement, two substances or articles are considered mutually incompatible when their stowage together may result in undue hazards in case of leakage or spillage, or any other accident.

7.2.1.4 The extent of the hazard arising from possible reactions between incompatible dangerous goods may vary and so the segregation arrangements required may also vary as appropriate. Such segregation is obtained by maintaining certain distances between incompatible dangerous goods or by requiring the presence of one or more steel bulkheads or decks between them, or a combination thereof. Intervening spaces between such dangerous goods may be filled with other cargo compatible with the dangerous substances or articles in question.

7.2.1.5 The following segregation terms are used throughout this Code:

.1 ''Away from'';

.2 ''Separated from'';

.3 ''Separated by a complete compartment or hold from'';

.4 ''Separated longitudinally by an intervening complete compartment or hold from''.

These terms are defined in 7.2.2 and their use in regard to the different modes of sea transport is explained further in the other subsections of this chapter.

7.2.1.6 The general provisions for segregation between the various classes of dangerous goods are shown in the ''segregation table'' of 7.2.1.16. In addition to the general provisions, there may be a need to segregate a particular substance, material or article from other goods which could contribute to its hazard. Particular provisions for segregation are indicated in the Dangerous Goods List and, in the case of conflicting provisions, always take precedence over the general provisions.

For example:

In the Dangerous Goods List entry for ACETYLENE, DISSOLVED, class 2.1, UN 1001, the following particular segregation requirement is specified:
''separated from'' chlorine.

In the Dangerous Goods List entry for BARIUM CYANIDE, class 6.1, UN 1565, the following particular segregation is specified:
''separated from'' acids.

7.2.1.6.1 Where the Code indicates a single secondary hazard (one subsidiary risk label), the segregation provisions applicable to that hazard shall take precedence where they are more stringent than those of the primary hazard.

7.2.1.6.2 Except for class 1, the segregation provisions for substances, materials or articles having more than two hazards (2 or more subsidiary risk labels) are given in the Dangerous Goods List.

For example:

In the Dangerous Goods List entry for BROMINE CHLORIDE, class 2.3, UN 2901, subsidiary risks 5.1 and 8, the following particular segregation is specified:
segregation as for class 5.1 but ''separated from'' class 7.

7.2.1.7 Segregation groups

7.2.1.7.1 For the purpose of segregation, dangerous goods having certain similar chemical properties have been grouped together in segregation groups as listed in 7.2.1.7.2. The entries allocated to these segregation groups are listed in 3.1.4.4. Where in the Dangerous Goods List entry in column 16 (stowage and segregation) a particular segregation requirement refers to a group of substances, such as ''acids'', the particular segregation requirement applies to the goods allocated to the respective segregation group.

7

7.2.1.7.2 Segregation groups referred to in the Dangerous Goods List:

.1 acids

.2 ammonium compounds

.3 bromates

.4 chlorates

.5 chlorites

.6 cyanides

.7 heavy metals and their salts (including their organometallic compcunds)

.8 hypochlorites

.9 lead and its compounds

.10 liquid halogenated hydrocarbons

.11 mercury and mercury compounds

.12 nitrites and their mixtures

.13 perchlorates

.14 permanganates

.15 powdered metals

.16 peroxides

.17 azides

.18 alkalis

7.2.1.7.3 It is recognized that not all substances falling within a segregation group are listed in this Code by name. These substances are shipped under N.O.S. entries. Although these N.O.S. entries are not listed themselves in the above groups, the shipper shall decide whether allocation under the segregation group is appropriate. Mixtures, solutions or preparations containing substances falling within a segregation group and shipped under an N.O.S. entry are also considered to fall within that segregation group.

7.2.1.7.4 The segregation groups in this Code do not cover substances which fall outside the classification criteria of this Code. It is recognized that some non-hazardous substances have similar chemical properties as substances listed in the segregation groups. A shipper or the person responsible for packing the goods into a cargo transport unit who does have knowledge of the chemical properties of such non-dangerous goods may decide to implement the segregation requirements of a related segregation group on a voluntary basis.

7.2.1.8 In the case of segregation from combustible material, this shall be understood not to include packaging materials or dunnage.

7.2.1.9 Whenever dangerous goods are stowed together, whether or not in a cargo transport unit, the segregation of such dangerous goods from others shall always be in accordance with the most stringent provisions for any of the dangerous goods concerned.

7.2.1.10 For the purposes of 7.2.1.6.1, the segregation provisions corresponding to a subsidiary risk label of class 1 are those for class 1, division 1.3.

7.2.1.11 Notwithstanding 7.2.1.6.1, 7.2.1.6.2 and 7.2.1.13, substances of the same class may be stowed together without regard to segregation required by secondary hazards (subsidiary risk label(s)), provided the substances do not react dangerously with each other and cause:

.1 combustion and/or evolution of considerable heat;

.2 evolution of flammable, toxic or asphyxiant gases;

.3 the formation of corrosive substances; or

.4 the formation of unstable substances.

7.2.1.12 Where the Dangerous Goods List specifies that "segregation as for class . . ." applies, the segregation provisions applicable to that class in 7.2.1.16 shall be applied. However, for the purposes of interpreting 7.2.1.11, which permits substances of the same class to be stowed together provided they do not react dangerously with each other, the segregation provisions of the class as represented by the primary hazard class in the Dangerous Goods List shall be applied.

For example:

UN 2965 – BORON TRIFLUORIDE DIMETHYL ETHERATE, class 4.3

The Dangerous Goods List entry specifies "segregation as for class 3, but "away from" classes 3, 4.1 and 8".

For the purposes of establishing the segregation provisions applicable in 7.2.1.16, the class 3 column shall be consulted.

This substance may be stowed together with other class 4.3 substances where they do not react dangerously with each other; see 7.2.1.11.

7.2.1.13 **Special provisions for segregation**

7.2.1.13.1 No segregation needs to be applied:

.1 between dangerous goods of different classes which comprise the same substance but vary only in their water content, such as sodium sulphide in classes 4.2 and 8 or for class 7 if the difference is due to quantity only;

.2 between dangerous goods which belong to a group of substances of different classes but for which scientific evidence exists that they do not react dangerously when in contact with each other. Substances within the same table shown below are compatible with one another.

Table 1

UN Number	Proper Shipping Name	Class	Subsidiary risk(s)	Packing group
2014	HYDROGEN PEROXIDE, AQUEOUS SOLUTION with not less than 20% but not more than 60% hydrogen peroxide (stabilized as necessary)	5.1	8	II
2984	HYDROGEN PEROXIDE, AQUEOUS SOLUTION with not less than 8% but less than 20% hydrogen peroxide (stabilized as necessary)	5.1		III
3105	ORGANIC PEROXIDE TYPE D, LIQUID (peroxyacetic acid, type D, stabilized)	5.2	8	
3107	ORGANIC PEROXIDE TYPE E, LIQUID (peroxyacetic acid, type E, stabilized)	5.2	8	
3109	ORGANIC PEROXIDE TYPE F, LIQUID (peroxyacetic acid, type F, stabilized)	5.2	8	
3149	HYDROGEN PEROXIDE AND PEROXYACETIC ACID MIXTURE with acid(s), water and not more than 5% peroxyacetic acid, STABILIZED	5.1	8	II

Table 2

UN Number	Proper Shipping Name	Class	Subsidiary risk(s)	Packing group
1295	TRICHLOROSILANE	4.3	3/8	I
1818	SILICON TETRACHLORIDE	8	–	II
2189	DICHLOROSILANE	2.3	2.1/8	–

7.2.1.13.2 Notwithstanding the provisions of 7.2.1.7.1 to 7.2.1.7.4, substances of class 8, packing group II or III, that would otherwise be required to be segregated from one another due to the provisions pertaining to segregation groups as identified by an entry in column (16) of the Dangerous Goods List indicating ''away from'' or ''separated from'' ''acids'' or ''away from'' or ''separated from'' ''alkalis'', may be transported in the same cargo transport unit, whether in the same packaging or not, provided:

.1 the substances comply with the provisions of 7.2.1.11;

.2 the package does not contain more than 30 litres for liquids or 30 kg for solids;

.3 the transport document includes the statement required by 5.4.1.5.11.3; and

.4 a copy of the test report that verifies that the substances do not react dangerously with each other shall be provided if requested by the competent authority.

7.2.1.14 Where, for the purposes of segregation, terms such as ''away from class ...'' are used in the Dangerous Goods List, ''class ...'' is deemed to include:

.1 all substances within ''class ...''; and

.2 all substances for which a subsidiary risk label of ''class ...'' is required.

7.2.1.15 Stowage in a shelter-'tween-deck cargo space is not considered to be *on deck* stowage.

7.2.1.16 **Segregation table**

The following table shows the general provisions for segregation between the various classes of dangerous goods.

SINCE THE PROPERTIES OF SUBSTANCES, MATERIALS OR ARTICLES WITHIN EACH CLASS MAY VARY GREATLY, THE DANGEROUS GOODS LIST SHALL ALWAYS BE CONSULTED FOR PARTICULAR PROVISIONS FOR SEGREGATION AS, IN THE CASE OF CONFLICTING PROVISIONS, THESE TAKE PRECEDENCE OVER THE GENERAL PROVISIONS.

SEGREGATION SHALL ALSO TAKE ACCOUNT OF A SINGLE SUBSIDIARY RISK LABEL.

CLASS		1.1 1.2 1.5	1.3 1.6	1.4	2.1	2.2	2.3	3	4.1	4.2	4.3	5.1	5.2	6.1	6.2	7	8	9
Explosives	1.1, 1.2, 1.5	*	*	*	4	2	2	4	4	4	4	4	4	2	4	2	4	X
Explosives	1.3, 1.6	*	*	*	4	2	2	4	3	3	4	4	4	2	4	2	2	X
Explosives	1.4	*	*	*	2	1	1	2	2	2	2	2	2	X	4	2	2	X
Flammable gases	2.1	4	4	2	X	X	X	2	1	2	X	2	2	X	4	2	1	X
Non-toxic, non-flammable gases	2.2	2	2	1	X	X	X	1	X	1	X	X	1	X	2	1	X	X
Toxic gases	2.3	2	2	1	X	X	X	2	X	2	X	X	2	X	2	1	X	X
Flammable liquids	3	4	4	2	2	1	2	X	X	2	1	2	2	X	3	2	X	X
Flammable solids (including self-reactive substances and solid desensitized explosives)	4.1	4	3	2	1	X	X	X	X	1	X	1	2	X	3	2	1	X
Substances liable to spontaneous combustion	4.2	4	3	2	2	1	2	2	1	X	1	2	2	1	3	2	1	X
Substances which, in contact with water, emit flammable gases	4.3	4	4	2	X	X	X	1	X	1	X	2	2	X	2	2	1	X
Oxidizing substances (agents)	5.1	4	4	2	2	X	X	2	1	2	2	X	2	1	3	1	2	X
Organic peroxides	5.2	4	4	2	2	1	2	2	2	2	2	2	X	1	3	2	2	X
Toxic substances	6.1	2	2	X	X	X	X	X	X	1	X	1	1	X	1	X	X	X
Infectious substances	6.2	4	4	4	4	2	2	3	3	3	2	3	3	1	X	3	3	X
Radioactive material	7	2	2	2	2	1	1	2	2	2	2	1	2	X	3	X	2	X
Corrosive substances	8	4	2	2	1	X	X	X	1	1	1	2	2	X	3	2	X	X
Miscellaneous dangerous substances and articles	9	X	X	X	X	X	X	X	X	X	X	X	X	X	X	X	X	X

Numbers and symbols relate to the following terms as defined in this chapter:
1 – "Away from"
2 – "Separated from"
3 – "Separated by a complete compartment or hold from"
4 – "Separated longitudinally by an intervening complete compartment or hold from"
X – The segregation, if any, is shown in the Dangerous Goods List
* – See 7.2.7.2 of this chapter

7.2.1.17 For the purposes of the segregation provisions for the various means of transport by sea, this chapter has been subdivided as follows:

.1 segregation of packages: 7.2.2;

.2 segregation of cargo transport units on board container ships: 7.2.3;

.3 segregation of cargo transport units on board roll-on/roll-off ships: 7.2.4;

.4 segregation in shipborne barges and on board barge-carrying ships: 7.2.5;

.5 segregation between bulk materials possessing chemical hazards and dangerous goods in packaged form: 7.2.6.

7.2.2 Segregation of packages

7.2.2.1 **Applicability**

7.2.2.1.1 The provisions of this subsection apply to the segregation of:

.1 packages containing dangerous goods and stowed in the conventional way;

.2   dangerous goods within cargo transport units; and

.3   dangerous goods stowed in the conventional way from those packed in such cargo transport units.

7.2.2.2   Segregation of packages containing dangerous goods and stowed in the conventional way

7.2.2.2.1   *Definitions of the segregation terms*

.1   *Away from:*
Effectively segregated so that the incompatible goods cannot interact dangerously in the event of an accident but may be transported in the same compartment or hold or *on deck*, provided a minimum horizontal separation of **3 metres, projected vertically,** is obtained.

.2   *Separated from:*
In different compartments or holds when stowed *under deck*. Provided the intervening deck is resistant to fire and liquid, a vertical separation, i.e. in different compartments, may be accepted as equivalent to this segregation. For *on deck* stowage, this segregation means a separation by a distance of **at least 6 metres horizontally**.

.3   *Separated by a complete compartment or hold from:*
Either a vertical or a horizontal separation. If the intervening decks are not resistant to fire and liquid, then only a longitudinal separation, i.e. by an intervening complete compartment or hold, is acceptable. For *on deck* stowage, this segregation means a separation by a distance of **at least 12 metres horizontally**. The same distance has to be applied if one package is stowed *on deck* and the other one in an upper compartment.

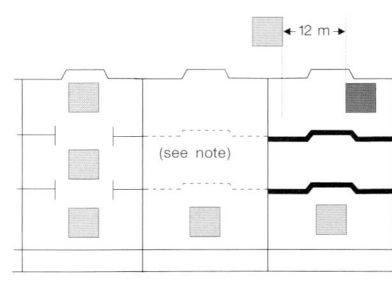

Note: One of the two decks must be resistant to fire and to liquid.

.4   *Separated longitudinally by an intervening complete compartment or hold from:*
Vertical separation alone does not meet this requirement. Between a package *under deck* and one *on deck*, a minimum distance of 24 metres, including a complete compartment, must be maintained longitudinally. For *on deck* stowage, this segregation means a separation by a distance of **at least 24 metres longitudinally**.

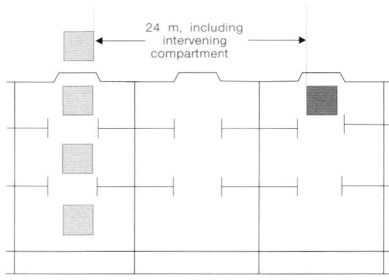

Legend

(1)   Reference *package* . . . .

(2)   *Package* containing incompatible goods . . . .

(3)   Deck resistant to fire and liquid . . . .

NOTE: Full vertical lines represent transverse bulkheads between cargo spaces (compartments or holds) resistant to fire and liquid.

7.2.2.3   Segregation in cargo transport units

Dangerous goods which have to be segregated from each other shall not be transported in the same cargo transport unit with the exception of dangerous goods which shall be segregated "away from" each other which may be transported in the same cargo transport unit with the approval of the competent authority. In such cases an equivalent standard of safety shall be maintained.

7.2.2.4 Segregation of dangerous goods stowed in the conventional way from those transported in cargo transport units

7.2.2.4.1 Dangerous goods stowed in the conventional way shall be segregated from goods transported in open cargo transport units in accordance with 7.2.2.2.

7.2.2.4.2 Dangerous goods stowed in the conventional way shall be segregated from goods transported in closed cargo transport units in accordance with 7.2.2.2 except that:

.1 where "away from" is required, no segregation between the packages and the closed cargo transport units is required; and

.2 where "separated from" is required, the segregation between the packages and the closed cargo transport units may be as for "away from" as defined in 7.2.2.2.1.1.

7.2.3 Segregation of cargo transport units on board container ships

7.2.3.1 Applicability and definitions

7.2.3.1.1 The provisions of this subsection apply to the segregation of cargo transport units which are transported on board full container ships or on decks, or in holds and compartments of other types of ships provided that these cargo spaces are properly fitted to give a permanent stowage of the containers during transport (see 7.2.3.2). For the open holds of hatchless container ships, see table 7.2.3.3.

7.2.3.1.2 *Container space* means a distance of not less than 6 m fore and aft or not less than 2.4 m athwartships.

7.2.3.1.3 For ships which incorporate conventional cargo spaces or any other method of stowage, the appropriate subsection of this chapter shall apply to the relevant cargo space.

7.2.3.2 Table of segregation of freight containers on board container ships

SEGREGATION REQUIREMENT	VERTICAL CLOSED VERSUS CLOSED	VERTICAL CLOSED VERSUS OPEN	VERTICAL OPEN VERSUS OPEN		HORIZONTAL CLOSED VERSUS CLOSED ON DECK	HORIZONTAL CLOSED VERSUS CLOSED UNDER DECK	HORIZONTAL CLOSED VERSUS OPEN ON DECK	HORIZONTAL CLOSED VERSUS OPEN UNDER DECK	HORIZONTAL OPEN VERSUS OPEN ON DECK	HORIZONTAL OPEN VERSUS OPEN UNDER DECK
"AWAY FROM" .1	ONE ON TOP OF THE OTHER PERMITTED	OPEN ON TOP OF CLOSED PERMITTED OTHERWISE AS FOR "OPEN VERSUS OPEN"	*NOT* IN THE SAME VERTICAL LINE *UNLESS* SEGREGATED BY A DECK	FORE AND AFT	NO RESTRICTION	NO RESTRICTION	NO RESTRICTION	NO RESTRICTION	*ONE* CONTAINER SPACE	*ONE* CONTAINER SPACE OR *ONE* BULKHEAD
				ATHWARTSHIPS	NO RESTRICTION	NO RESTRICTION	NO RESTRICTION	NO RESTRICTION	*ONE* CONTAINER SPACE	*ONE* CONTAINER SPACE
"SEPARATED FROM" .2	*NOT* IN THE SAME VERTICAL LINE *UNLESS* SEGREGATED BY A DECK	AS FOR "OPEN VERSUS OPEN"		FORE AND AFT	*ONE* CONTAINER SPACE	*ONE* CONTAINER SPACE OR *ONE* BULKHEAD	*ONE* CONTAINER SPACE	*ONE* CONTAINER SPACE OR *ONE* BULKHEAD	*ONE* CONTAINER SPACE	*ONE* BULKHEAD
				ATHWARTSHIPS	*ONE* CONTAINER SPACE	*ONE* CONTAINER SPACE	*ONE* CONTAINER SPACE	*TWO* CONTAINER SPACES	*TWO* CONTAINER SPACES	*ONE* BULKHEAD
"SEPARATED BY A COMPLETE COMPARTMENT OR HOLD FROM" .3				FORE AND AFT	*ONE* CONTAINER SPACE	*ONE* BULKHEAD	*ONE* CONTAINER SPACE	*ONE* BULKHEAD	*TWO* CONTAINER SPACES	*TWO* BULKHEADS
				ATHWARTSHIPS	*TWO* CONTAINER SPACES	*ONE* BULKHEAD	*TWO* CONTAINER SPACES	*ONE* BULKHEAD	*THREE* CONTAINER SPACES	*TWO* BULKHEADS
"SEPARATED LONGITUDINALLY BY AN INTERVENING COMPLETE COMPARTMENT OR HOLD FROM" .4	PROHIBITED			FORE AND AFT	MINIMUM HORIZONTAL DISTANCE OF 24 METRES	ONE BULKHEAD AND MINIMUM HORIZONTAL DISTANCE OF 24 METRES*	MINIMUM HORIZONTAL DISTANCE OF 24 METRES	TWO BULKHEADS	MINIMUM HORIZONTAL DISTANCE OF 24 METRES	TWO BULKHEADS
				ATHWARTSHIPS	PROHIBITED	PROHIBITED	PROHIBITED	PROHIBITED	PROHIBITED	PROHIBITED

* CONTAINERS NOT LESS THAN 6 METRES FROM INTERVENING BULKHEAD.

NOTE: ALL BULKHEADS AND DECKS SHALL BE RESISTANT TO FIRE AND LIQUID.

7.2.3.2.1   *Illustrations of segregation of cargo transport units on board container ships*

7.2.3.2.1.1   The illustrations of this subsection apply to the segregation of cargo transport units which are transported on board full container ships or on decks, or in holds and compartments of other type of ships provided that these cargo spaces are properly fitted to give permanent stowage of the cargo transport units during transport.*

7.2.3.2.1.2   To determine locations in which cargo transport units are not permitted to contain dangerous goods that are incompatible with those in a reference cargo transport unit, the following method shall be used: container spaces (such as one container space, two container spaces) are identified in accordance with the applicable segregation provisions in the direct fore-and-aft and athwartships directions from the reference cargo transport unit. Lines are projected between the outermost corners of the cargo transport units occupying these spaces as shown in the figure. Cargo transport units located partially or completely between these lines and the reference cargo transport unit shall not contain dangerous goods that are incompatible with those in the reference cargo transport unit.

7.2.3.2.1.3   The deck/hold layout used for the illustrations is:
  - two 20′ containers stowed in a 40′ container space
  - distance between two 40′ container spaces is 2 ft/60 cm

7.2.3.2.1.4   *Definitions of the segregation terms*

(1)   Reference cargo transport unit (CTU) . . . .

(2)   CTU containing incompatible goods NOT permitted . . . .

(3)   CTU containing incompatible goods permitted . . . .

(4)   Distance atwarthships (a) one container space . . . .

(b) two container spaces . . . .

(c) three container spaces . . . .

(5)   Distance fore and aft   (a) one container space . . . .

(b) two container spaces . . . .

Note: All bulkheads and decks shall be resistant to fire and liquids.

---

* For container ships with partly hatchless container cargo spaces, the illustrations of 7.2.3.3.1 apply to such spaces.

Situation fore & aft + athwartships: 1 container space

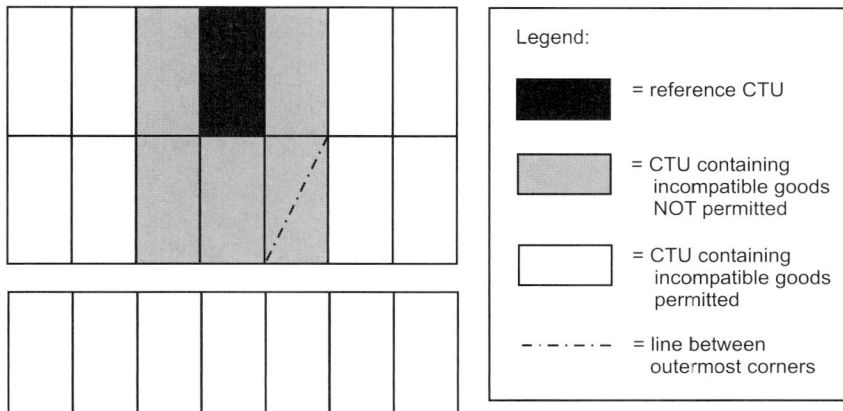

Situation fore & aft: 1 container space
& athwartships: 2 container spaces

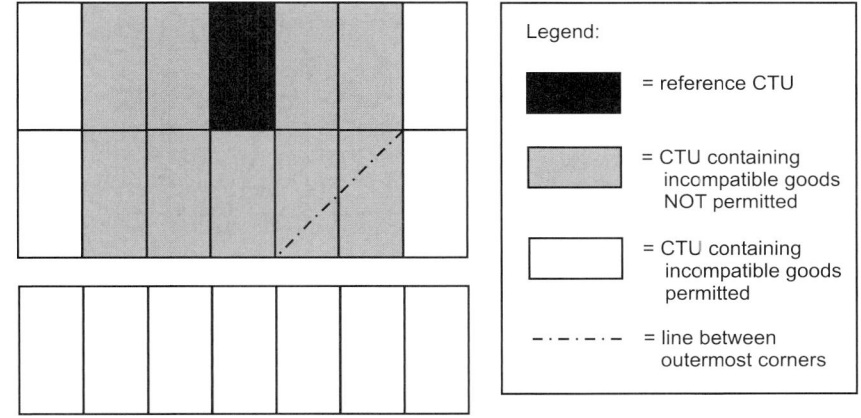

Situation fore & aft: 2 container spaces
& athwartships: 3 container spaces

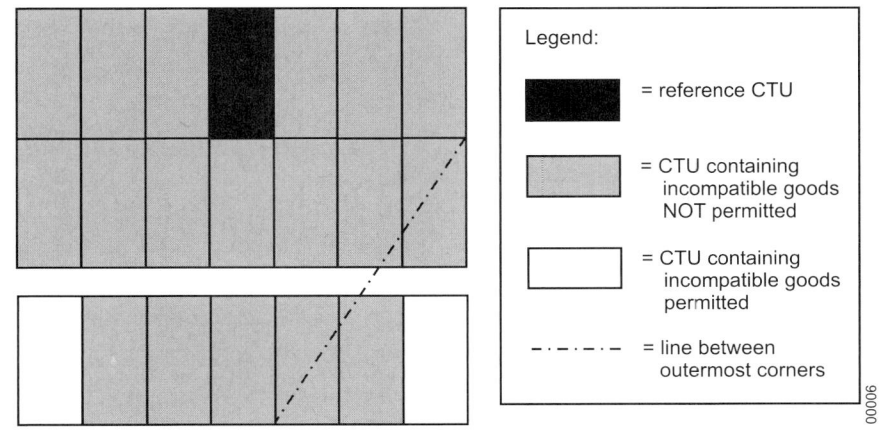

Note: All bulkheads and decks shall be resistant to fire and liquids.

"AWAY FROM" .1			
CLOSED VERSUS CLOSED	**HORIZONTAL**		**VERTICAL**
	ON DECK	**UNDER DECK**	
FORE AND AFT	No restriction	No restriction	One on top of the other permitted
ATHWARTSHIPS	No restriction	No restriction	

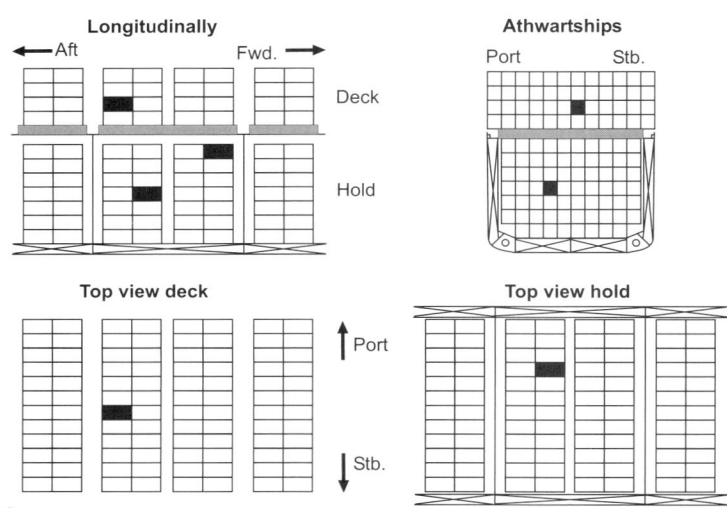

1 – Situation *closed* versus *closed*

Note: All bulkheads and decks shall be resistant to fire and liquids.

"AWAY FROM" .1			
CLOSED VERSUS OPEN	**HORIZONTAL**		**VERTICAL**
	ON DECK	**UNDER DECK**	
FORE AND AFT	No restriction	No restriction	Open on top of closed permitted
ATHWARTSHIPS	No restriction	No restriction	otherwise NOT in the same vertical line unless segregated by a deck

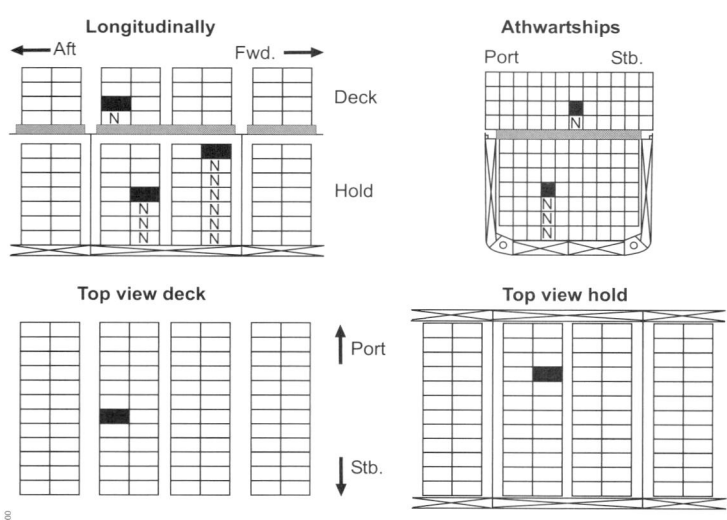

1 – Situation *closed* versus *open*

Note: All bulkheads and decks shall be resistant to fire and liquids.

"AWAY FROM" .1			
OPEN VERSUS OPEN	**HORIZONTAL**		**VERTICAL**
	ON DECK	UNDER DECK	
FORE AND AFT	One container space	One container space or one bulkhead	NOT in the same vertical line unless segregated by a deck
ATHWARTSHIPS	One container space	One container space	

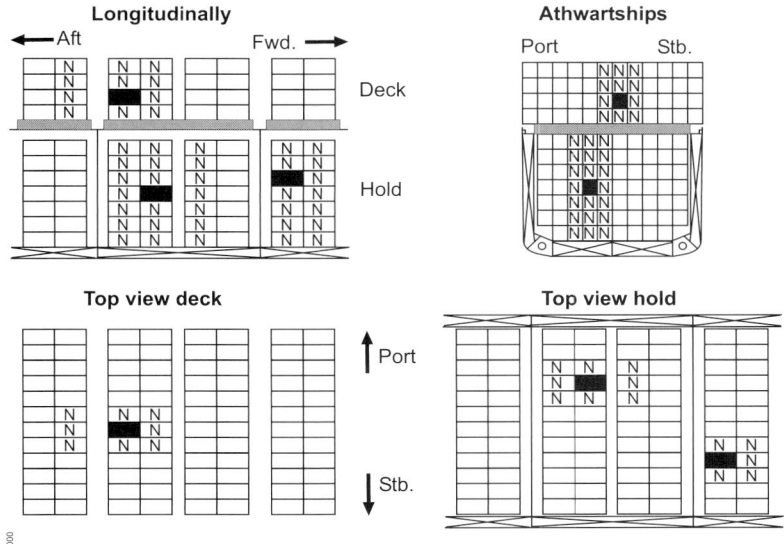

1 – Situation *open* versus *open*

Note: All bulkheads and decks shall be resistant to fire and liquids.

"SEPARATED FROM" .2			
CLOSED VERSUS CLOSED	**HORIZONTAL**		**VERTICAL**
	ON DECK	UNDER DECK	
FORE AND AFT	One container space	One container space or one bulkhead	NOT in the same vertical line unless segregated by a deck
ATHWARTSHIPS	One container space	One container space	

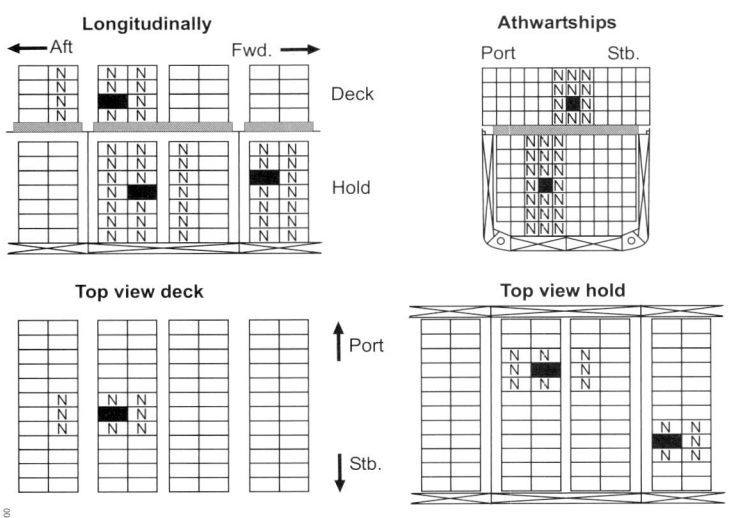

2 – Situation *closed* versus *closed*

Note: All bulkheads and decks shall be resistant to fire and liquids.

"SEPARATED FROM" .2			
CLOSED VERSUS OPEN	HORIZONTAL		VERTICAL
	ON DECK	UNDER DECK	
FORE AND AFT	One container space	One container space or one bulkhead	NOT in the same vertical line unless segregated by a deck
ATHWARTSHIPS	One container space	Two container spaces	

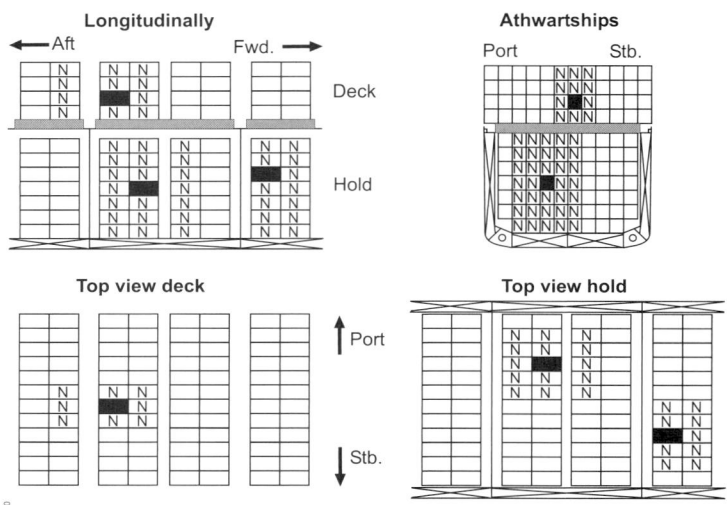

2 – Situation *closed* versus *open*

Note: All bulkheads and decks shall be resistant to fire and liquids.

"SEPARATED FROM" .2			
OPEN VERSUS OPEN	HORIZONTAL		VERTICAL
	ON DECK	UNDER DECK	
FORE AND AFT	One container space	One bulkhead	NOT in the same vertical line unless segregated by a deck
ATHWARTSHIPS	Two container spaces	One bulkhead	

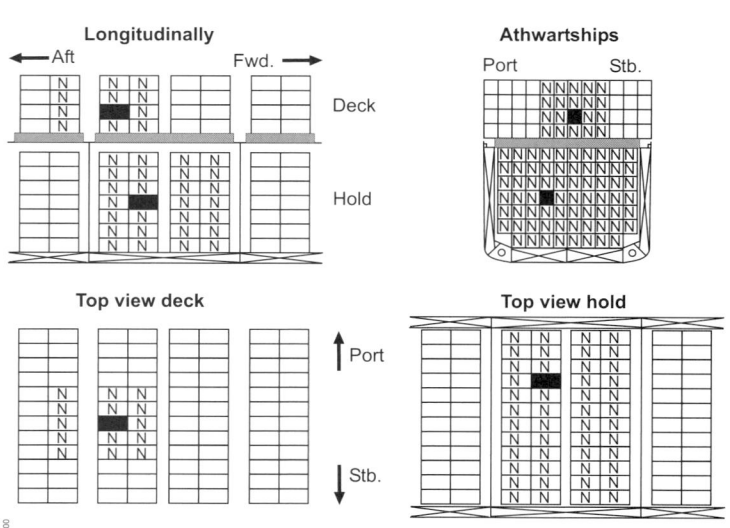

2 – Situation *open* versus *open*

Note: All bulkheads and decks shall be resistant to fire and liquids.

"SEPARATED BY A COMPLETE COMPARTMENT OR HOLD FROM" .3			
CLOSED VERSUS CLOSED OR CLOSED VERSUS OPEN	HORIZONTAL		VERTICAL
	ON DECK	UNDER DECK	
FORE AND AFT	One container space	One bulkhead	NOT in the same vertical line unless segregated by a deck
ATHWARTSHIPS	Two container spaces	One bulkhead	

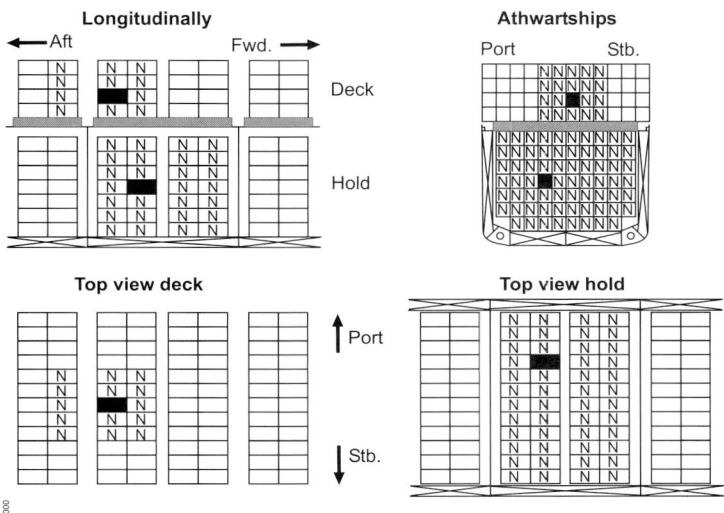

3 – Situations *closed* **versus** *closed* **and** *closed* **versus** *open*

Note: All bulkheads and decks shall be resistant to fire and liquids.

"SEPARATED BY A COMPLETE COMPARTMENT OR HOLD FROM" .3			
OPEN VERSUS OPEN	HORIZONTAL		VERTICAL
	ON DECK	UNDER DECK	
FORE AND AFT	Two container spaces	Two bulkheads	NOT in the same vertical line unless segregated by a deck
ATHWARTSHIPS	Three container spaces	Two bulkheads	

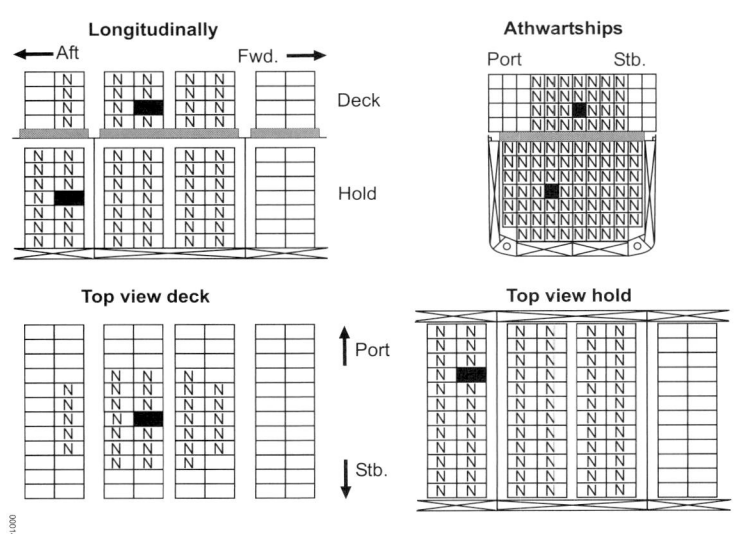

3 – Situation *open* **versus** *open*

Note: All bulkheads and decks shall be resistant to fire and liquids.

"SEPARATED LONGITUDINALLY BY AN INTERVENING COMPLETE COMPARTMENT OR HOLD FROM" .4			
CLOSED VERSUS CLOSED	HORIZONTAL		VERTICAL
	ON DECK	UNDER DECK	
FORE AND AFT	Minimum horizontal distance of 24 metres	One bulkhead and minimum horizontal distance of 24 metres*	Prohibited
ATHWARTSHIPS	Prohibited	Prohibited	

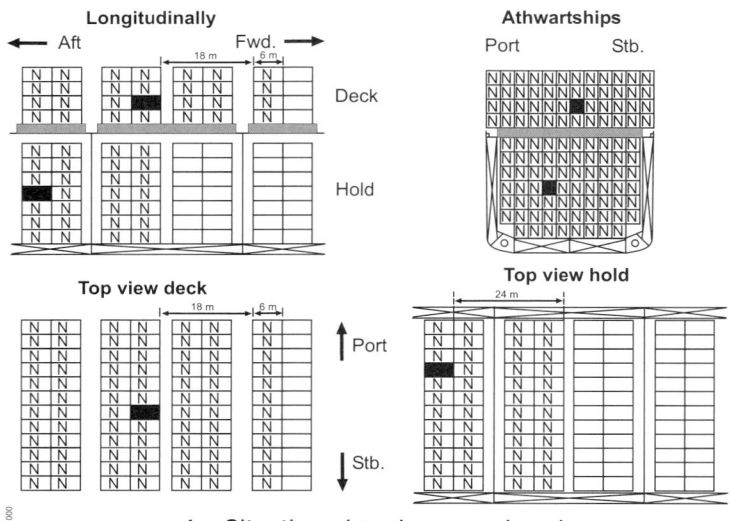

4 – Situation *closed* versus *closed*

Note: All bulkheads and decks shall be resistant to fire and liquids.

* Containers not less than 6 m from intervening bulkhead.

"SEPARATED LONGITUDINALLY BY AN INTERVENING COMPLETE COMPARTMENT OR HOLD FROM" .4			
CLOSED VERSUS OPEN	HORIZONTAL		VERTICAL
	ON DECK	UNDER DECK	
FORE AND AFT	Minimum horizontal distance of 24 metres	Two bulkheads	Prohibited
ATHWARTSHIPS	Prohibited	Prohibited	

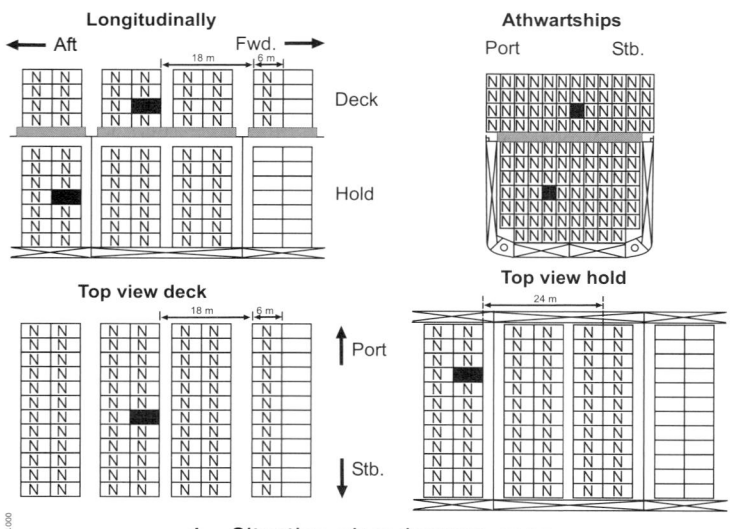

4 – Situation *closed* versus *open*

Note: All bulkheads and decks shall be resistant to fire and liquids.

"SEPARATED LONGITUDINALLY BY AN INTERVENING COMPLETE COMPARTMENT OR HOLD FROM" .4			
OPEN VERSUS OPEN	HORIZONTAL		VERTICAL
	ON DECK	UNDER DECK	
FORE AND AFT	Minimum horizontal distance of 24 metres	Two bulkheads	Prohibited
ATHWARTSHIPS	Prohibited	Prohibited	

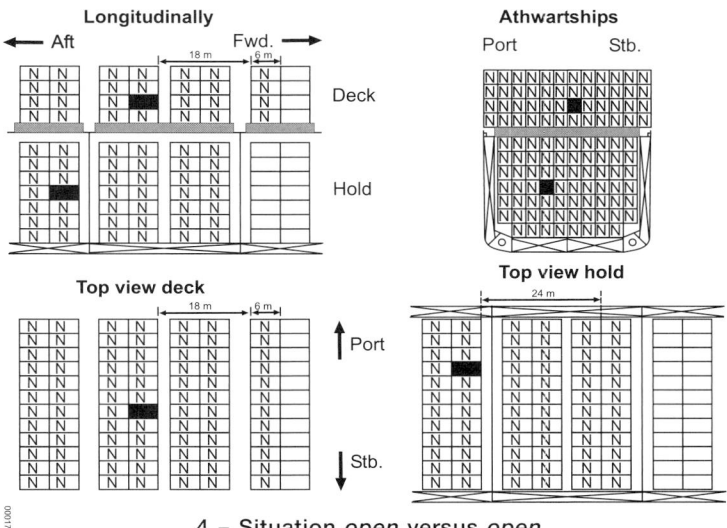

4 – Situation *open* versus *open*

Note: All bulkheads and decks shall be resistant to fire and liquids.

7.2.3.3   Table of segregation of cargo transport units on board hatchless container ships

SEGREGATION REQUIREMENT	VERTICAL				HORIZONTAL					
	CLOSED VERSUS CLOSED	CLOSED VERSUS OPEN	OPEN VERSUS OPEN		CLOSED VERSUS CLOSED		CLOSED VERSUS OPEN		OPEN VERSUS OPEN	
					ON DECK	UNDER DECK	ON DECK	UNDER DECK	ON DECK	UNDER DECK
"AWAY FROM" .1	ONE ON TOP OF THE OTHER PERMITTED	OPEN ON TOP OF CLOSED PERMITTED OTHERWISE AS FOR "OPEN VERSUS OPEN"	*NOT* IN THE SAME VERTICAL LINE	FORE AND AFT	NO RESTRICTION	NO RESTRICTION	NO RESTRICTION	NO RESTRICTION	*ONE* CONTAINER SPACE	*ONE* CONTAINER SPACE *OR ONE* BULKHEAD
				ATHWART-SHIPS	NO RESTRICTION	NO RESTRICTION	NO RESTRICTION	NO RESTRICTION	*ONE* CONTAINER SPACE	*ONE* CONTAINER SPACE
"SEPARATED FROM" .2	*NOT* IN THE SAME VERTICAL LINE	AS FOR "OPEN VERSUS OPEN"		FORE AND AFT	*ONE* CONTAINER SPACE	*ONE* CONTAINER SPACE OR *ONE* BULKHEAD	*ONE* CONTAINER SPACE	*ONE* CONTAINER SPACE OR *ONE* BULKHEAD	*ONE* CONTAINER SPACE AND NOT IN OR ABOVE SAME HOLD	*ONE* BULKHEAD
				ATHWART-SHIPS	*ONE* CONTAINER SPACE	*ONE* CONTAINER SPACE	*TWO* CONTAINER SPACES	*TWO* CONTAINER SPACES	*TWO* CONTAINER SPACES AND NOT IN OR ABOVE SAME HOLD	*ONE* BULKHEAD
"SEPARATED BY A COMPLETE COMPARTMENT OR HOLD FROM" .3				FORE AND AFT	*ONE* CONTAINER SPACE AND NOT IN OR ABOVE SAME HOLD	*ONE* BULKHEAD	*ONE* CONTAINER SPACE AND NOT IN OR ABOVE SAME HOLD	*ONE* BULKHEAD	*TWO* CONTAINER SPACES AND NOT IN OR ABOVE SAME HOLD	*TWO* BULKHEADS
				ATHWART-SHIPS	*TWO* CONTAINER SPACES AND NOT IN OR ABOVE SAME HOLD	*ONE* BULKHEAD	*TWO* CONTAINER SPACES AND NOT IN OR ABOVE SAME HOLD	*ONE* BULKHEAD	*THREE* CONTAINER SPACES AND NOT IN OR ABOVE SAME HOLD	*TWO* BULKHEADS
"SEPARATED LONGITUDINALLY BY AN INTERVENING COMPLETE COMPARTMENT OR HOLD FROM" .4	PROHIBITED			FORE AND AFT	MINIMUM HORIZONTAL DISTANCE OF 24 METRES AND NOT IN OR ABOVE SAME HOLD	*ONE* BULKHEAD *AND* MINIMUM HORIZONTAL DISTANCE OF 24 METRES*	MINIMUM HORIZONTAL DISTANCE OF 24 METRES AND NOT IN OR ABOVE SAME HOLD	*TWO* BULKHEADS	MINIMUM HORIZONTAL DISTANCE OF 24 METRES AND NOT IN OR ABOVE SAME HOLD	*TWO* BULKHEADS
				ATHWART-SHIPS	PROHIBITED	PROHIBITED	PROHIBITED	PROHIBITED	PROHIBITED	PROHIBITED

* CONTAINERS NOT LESS THAN 6 METRES FROM INTERVENING BULKHEAD.
NOTE: ALL BULKHEADS AND DECKS SHALL BE RESISTANT TO FIRE AND LIQUID.

7.2.3.3.1   *Illustrations of segregation of cargo transport units on board hatchless container ships*

7.2.3.3.1.1   The illustrations of this subsection apply to the segregation of cargo transport units which are transported on board hatchless container ships provided that the cargo spaces are properly fitted to give permanent stowage of the cargo transport units during transport.*

7.2.3.3.1.2   To determine locations in which cargo transport units are not permitted to contain dangerous goods that are incompatible with those in a reference cargo transport unit, the following method shall be used: container spaces (such as one container space, two container spaces) are identified in accordance with the applicable segregation provisions in the direct fore-and-aft and athwartship directions from the reference cargo transport unit. Lines are projected between the outermost corners of the cargo transport units occupying these spaces as shown in the figure. Cargo transport units located partially or completely between these lines and the reference cargo transport unit shall not contain dangerous goods that are incompatible with those in the reference cargo transport unit.

7.2.3.3.1.3   The deck/hold layout used for the illustrations is:

– two 20′ containers stowed in a 40′ container space

– distance between two 40′ container spaces is 2 ft/60 cm

---

* For partly hatchless container ships with conventional container cargo spaces, the illustrations of 7.2.3.2.1 apply to such spaces.

7.2.3.3.1.4 *Definitions of the segregation terms*

(1)   Reference cargo transport unit (CTU) . . . . ■ ■

(2)   CTU containing incompatible goods NOT permitted . . . . [N] [N]

(3)   CTU containing incompatible goods permitted . . . . [ ] [ ]

(4)   Distance athwartships (a) one container space . . . . [N] ■ [N]

(b) two container spaces . . . . [N|N] ■ [N|N]

(c) three container spaces . . . . [N|N|N] ■ [N|N|N]

(5)   Distance fore and aft (a) one container space . . . . [N] ■ [N]

(b) two container spaces . . . . [N | N] ■ [N | N]

Note: All bulkheads and decks shall be resistant to fire and liquids.

Situation fore & aft + athwartships: 1 container space

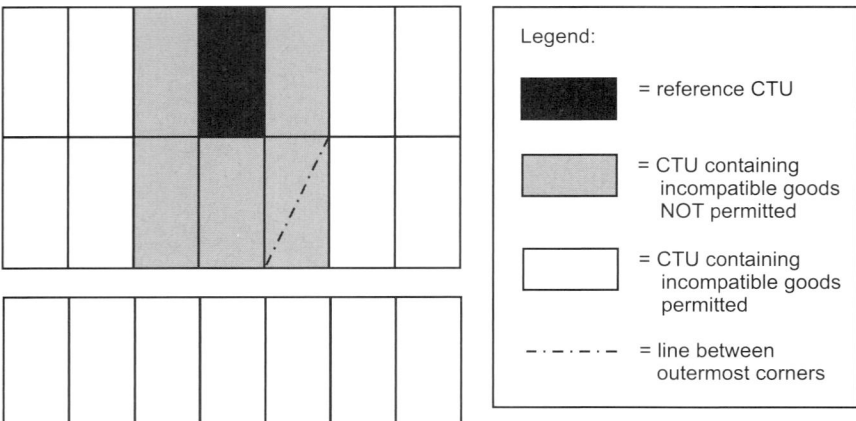

**Situation fore & aft: 1 container space
& athwartships: 2 container spaces**

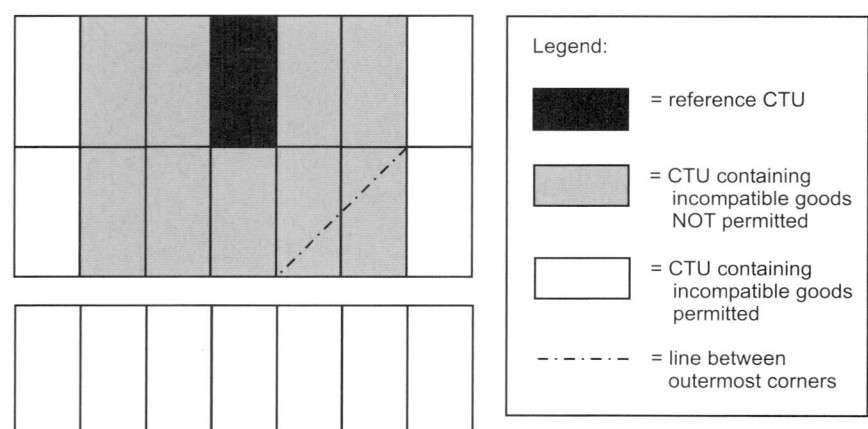

**Situation fore & aft: 2 container spaces
& athwartships: 3 container spaces**

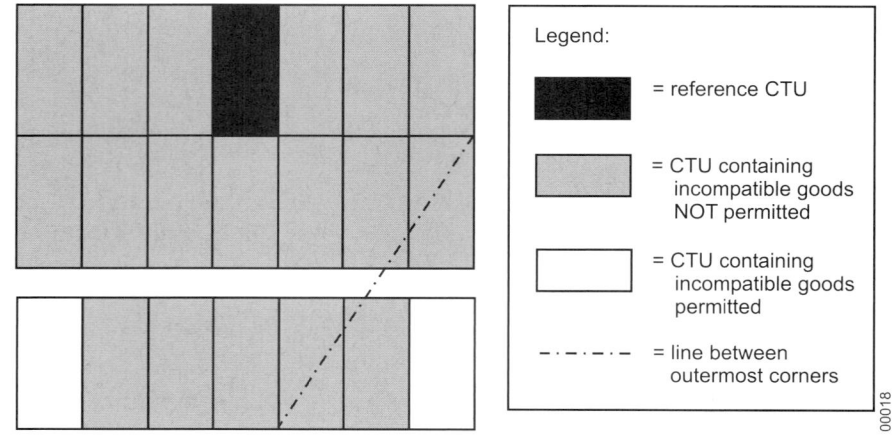

Note: All bulkheads and decks shall be resistant to fire and liquids.

"AWAY FROM" .1		
CLOSED VERSUS CLOSED	HORIZONTAL	VERTICAL
	ON DECK	
FORE AND AFT	No restriction	One on top of the other permitted
ATHWARTSHIPS	No restriction	

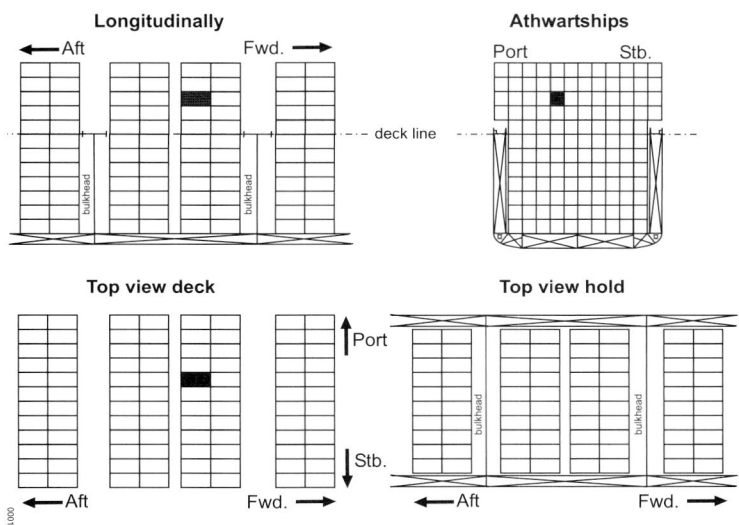

1 – Situation *closed* versus *closed* – ON DECK

Note: All bulkheads and decks shall be resistant to fire and liquids.

"AWAY FROM" .1		
CLOSED VERSUS CLOSED	HORIZONTAL	VERTICAL
	UNDER DECK	
FORE AND AFT	No restriction	One on top of the other permitted
ATHWARTSHIPS	No restriction	

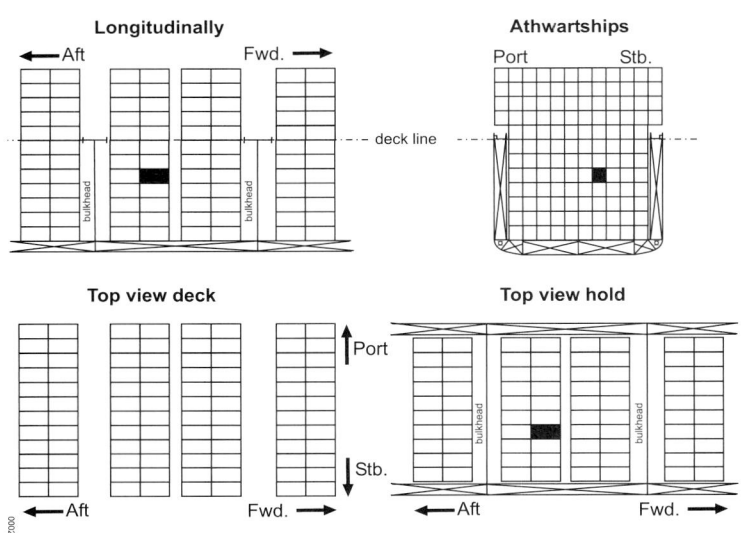

1 – Situation *closed* versus *closed* – UNDER DECK

Note: All bulkheads and decks shall be resistant to fire and liquids.

"AWAY FROM" .1		
CLOSED VERSUS OPEN	HORIZONTAL	VERTICAL
	ON DECK	
FORE AND AFT	No restriction	Open on top of closed permitted
ATHWARTSHIPS	No restriction	otherwise NOT in the same vertical line

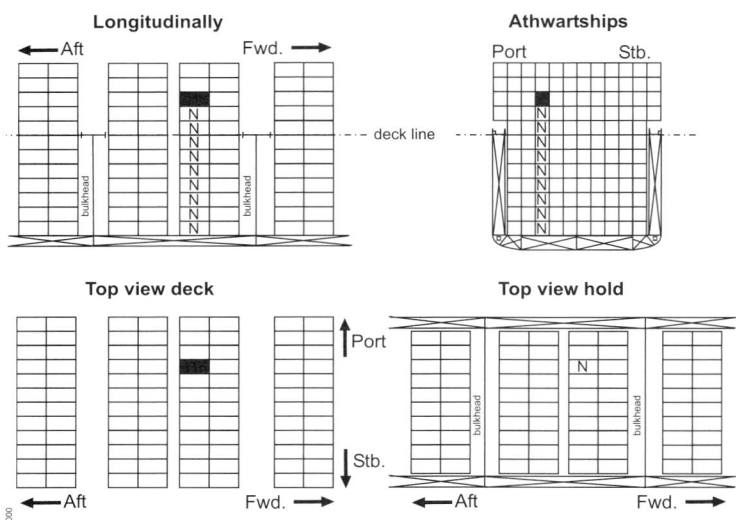

1 – Situation *closed* versus *open* – ON DECK

Note: All bulkheads and decks shall be resistant to fire and liquids.

"AWAY FROM" .1		
CLOSED VERSUS OPEN	HORIZONTAL	VERTICAL
	UNDER DECK	
FORE AND AFT	No restriction	Open on top of closed permitted
ATHWARTSHIPS	No restriction	otherwise NOT in the same vertical line

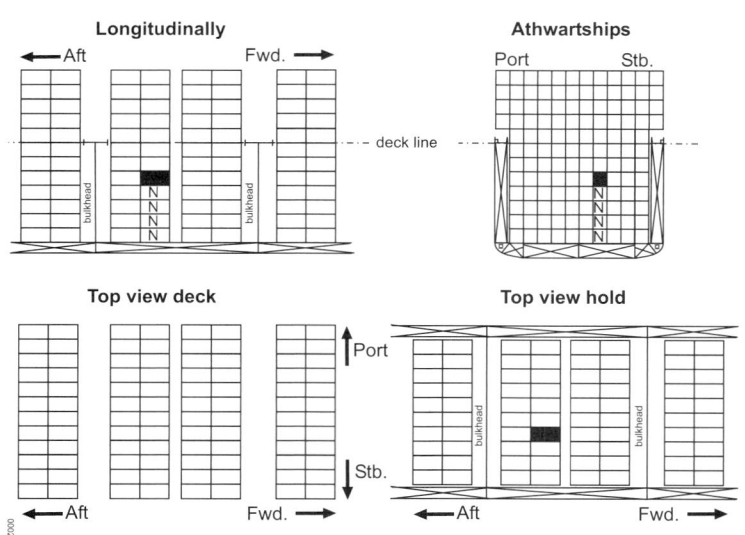

1 – Situation *closed* versus *open* – UNDER DECK

Note: All bulkheads and decks shall be resistant to fire and liquids.

"AWAY FROM" .1		
OPEN VERSUS OPEN	HORIZONTAL	VERTICAL
	ON DECK	
FORE AND AFT	One container space	NOT in the same vertical line
ATHWARTSHIPS	One container space	

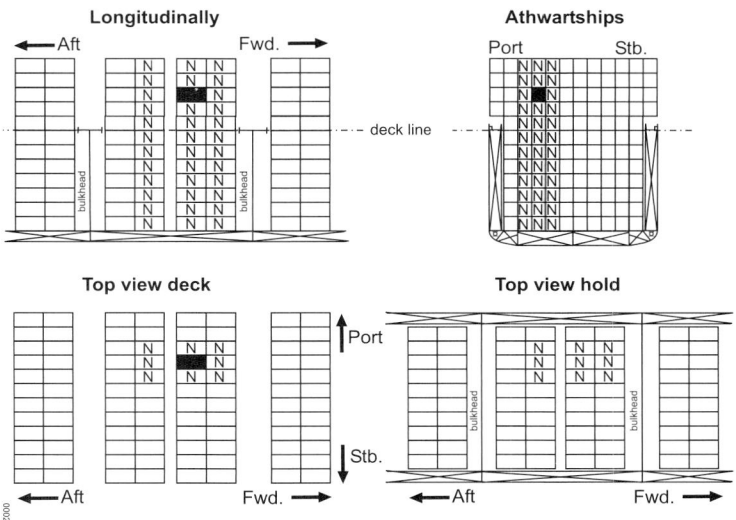

1 – Situation *open* versus *open* – ON DECK

Note: All bulkheads and decks shall be resistant to fire and liquids.

"AWAY FROM" .1		
OPEN VERSUS OPEN	HORIZONTAL	VERTICAL
	UNDER DECK	
FORE AND AFT	One container space or one bulkhead	NOT in the same vertical line
ATHWARTSHIPS	One container space	

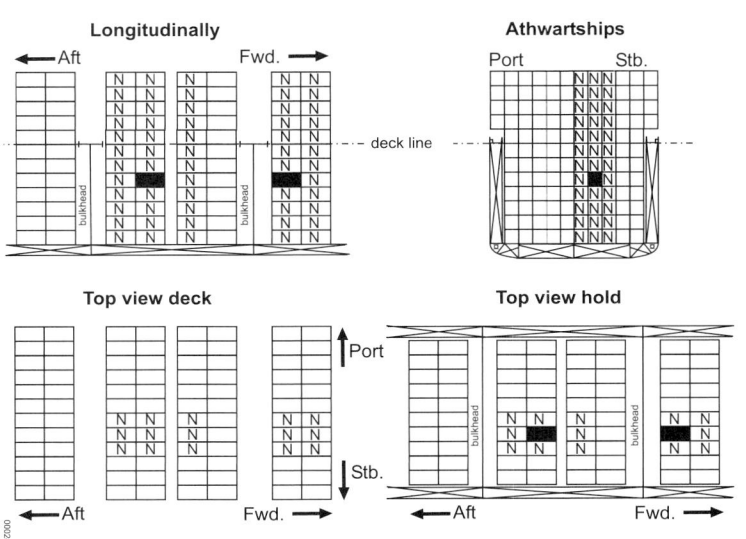

1 – Situation *open* versus *open* – UNDER DECK

Note: All bulkheads and decks shall be resistant to fire and liquids.

"SEPARATED FROM" .2		
CLOSED VERSUS CLOSED	HORIZONTAL	VERTICAL
	ON DECK	
FORE AND AFT	One container space	NOT in the same vertical line
ATHWARTSHIPS	One container space	

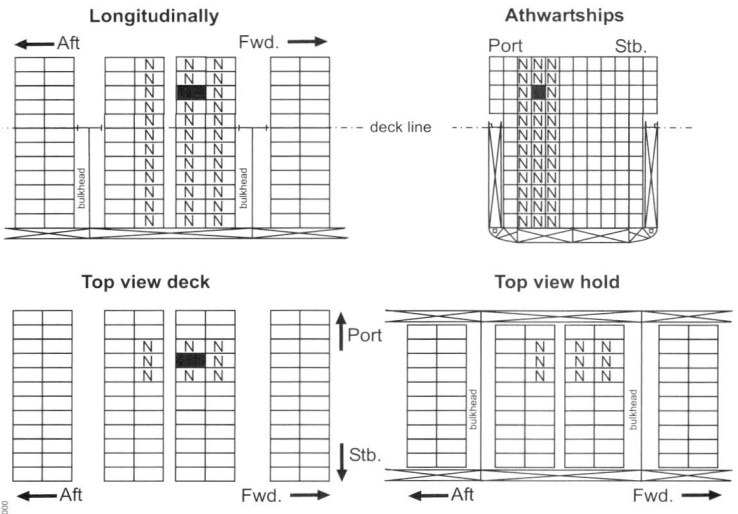

2 – Situation *closed* versus *closed* – ON DECK
Note: All bulkheads and decks shall be resistant to fire and liquids.

"SEPARATED FROM" .2		
CLOSED VERSUS CLOSED	HORIZONTAL	VERTICAL
	UNDER DECK	
FORE AND AFT	One container space or one bulkhead	NOT in the same vertical line
ATHWARTSHIPS	One container space	

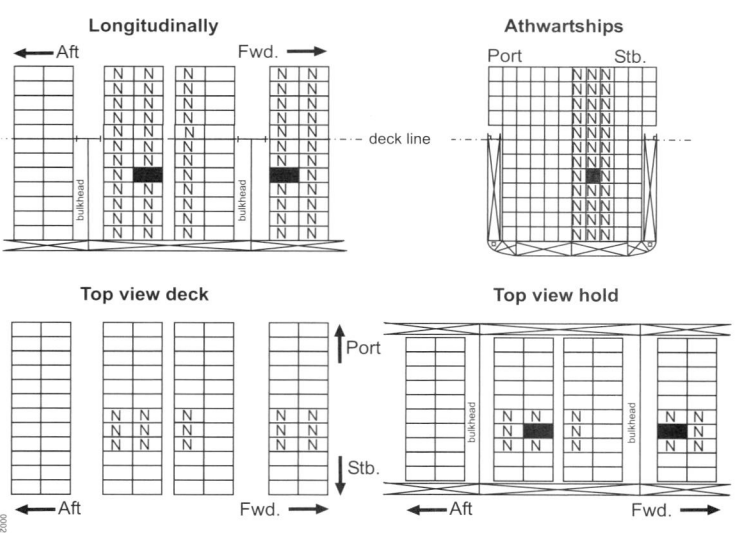

2 – Situation *closed* versus *closed* – UNDER DECK
Note: All bulkheads and decks shall be resistant to fire and liquids.

"SEPARATED FROM" .2		
CLOSED VERSUS OPEN	HORIZONTAL	VERTICAL
	ON DECK	
FORE AND AFT	One container space	NOT in the same vertical line
ATHWARTSHIPS	Two container spaces	

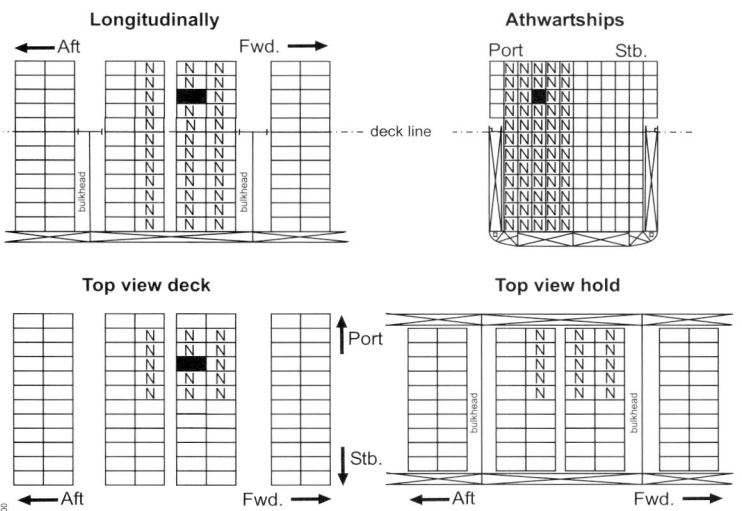

2 – Situation *closed* versus *open* – ON DECK

Note: All bulkheads and decks shall be resistant to fire and liquids.

"SEPARATED FROM" .2		
CLOSED VERSUS OPEN	HORIZONTAL	VERTICAL
	UNDER DECK	
FORE AND AFT	One container space or one bulkhead	NOT in the same vertical line
ATHWARTSHIPS	Two container spaces	

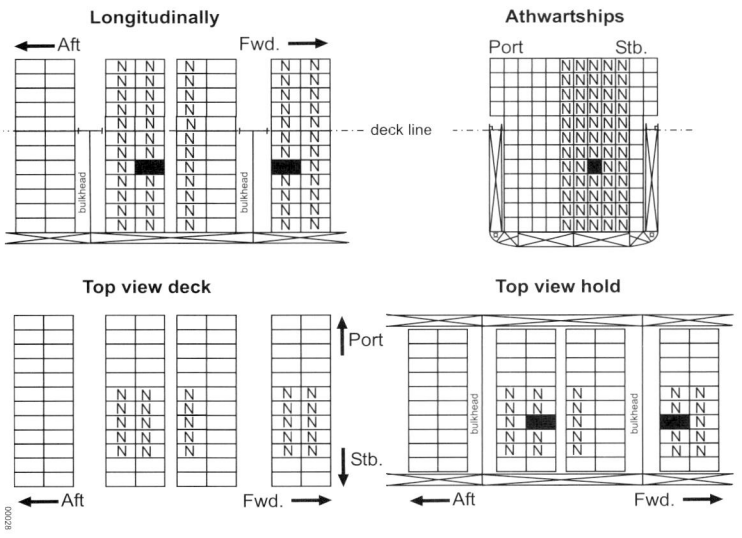

2 – Situation *closed* versus *open* – UNDER DECK

Note: All bulkheads and decks shall be resistant to fire and liquids.

"SEPARATED FROM" .2		
OPEN VERSUS OPEN	HORIZONTAL	VERTICAL
	ON DECK	
FORE AND AFT	One container space and not in or above same hold	NOT in the same vertical line
ATHWARTSHIPS	Two container spaces and not in or above same hold	

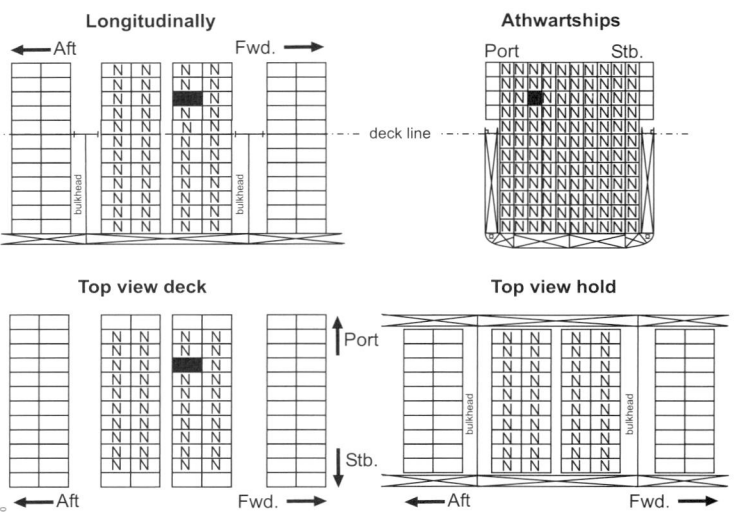

2 – Situation *open* versus *open* – ON DECK
Note: All bulkheads and decks shall be resistant to fire and liquids.

"SEPARATED FROM" .2		
OPEN VERSUS OPEN	HORIZONTAL	VERTICAL
	UNDER DECK	
FORE AND AFT	One bulkhead	NOT in the same vertical line
ATHWARTSHIPS	One bulkhead	

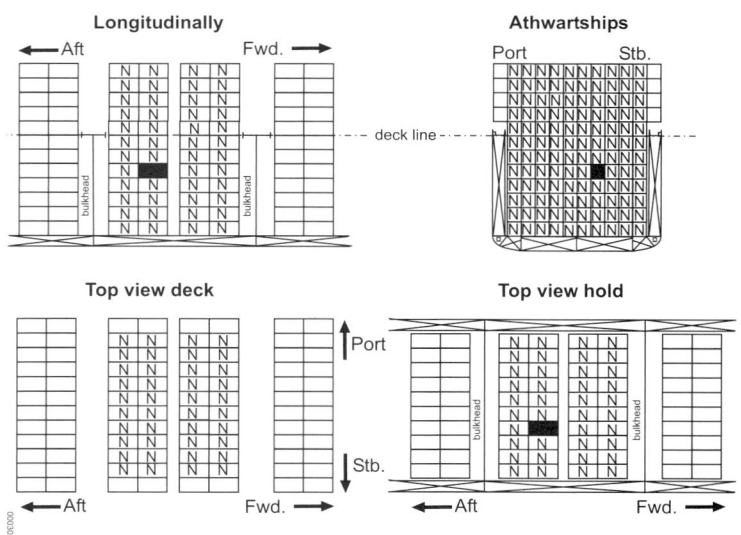

2 – Situation *open* versus *open* – UNDER DECK
Note: All bulkheads and decks shall be resistant to fire and liquids.

"SEPARATED BY A COMPLETE COMPARTMENT OR HOLD FROM" .3		
CLOSED VERSUS CLOSED	HORIZONTAL	VERTICAL
	ON DECK	
FORE AND AFT	One container space and not in or above same hold	NOT in the same vertical line
ATHWARTSHIPS	Two container spaces and not above same hold	

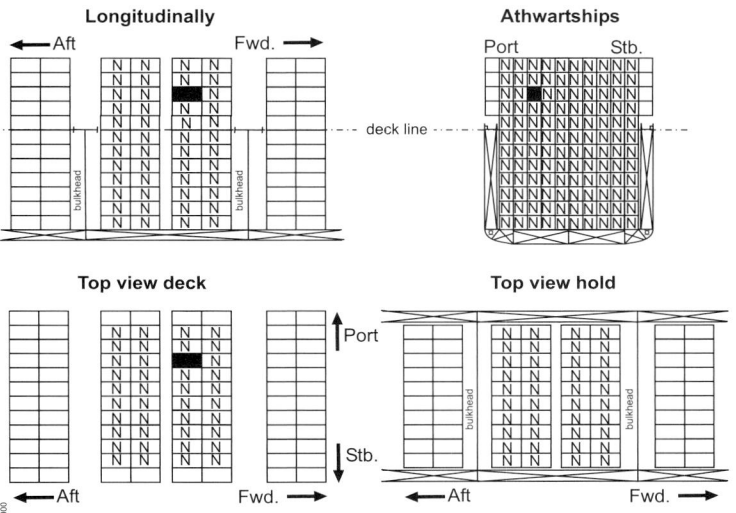

3 – Situation *closed* versus *closed* – ON DECK

Note: All bulkheads and decks shall be resistant to fire and liquids.

"SEPARATED BY A COMPLETE COMPARTMENT OR HOLD FROM" .3		
CLOSED VERSUS CLOSED	HORIZONTAL	VERTICAL
	UNDER DECK	
FORE AND AFT	One bulkhead	NOT in the same vertical line
ATHWARTSHIPS	One bulkhead	

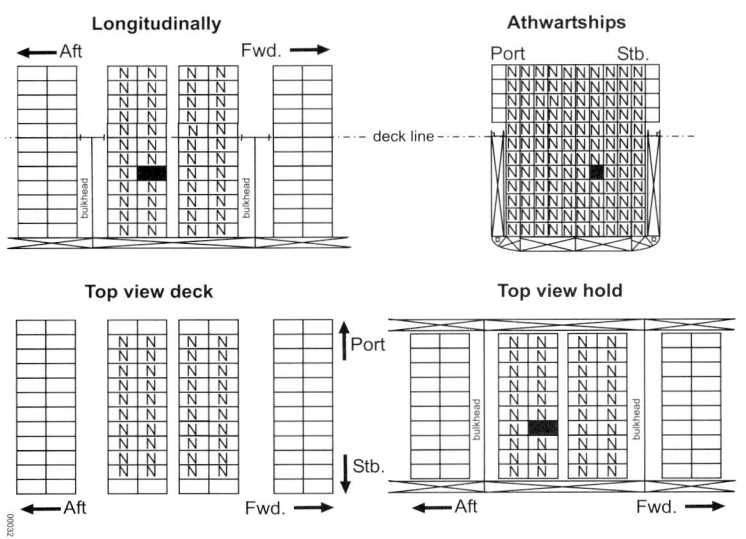

3 – Situation *closed* versus *closed* – UNDER DECK

Note: All bulkheads and decks shall be resistant to fire and liquids.

"SEPARATED BY A COMPLETE COMPARTMENT OR HOLD FROM" .3		
CLOSED VERSUS OPEN	HORIZONTAL	VERTICAL
	ON DECK	
FORE AND AFT	One container space and not in or above same hold	NOT in the same vertical line
ATHWARTSHIPS	Two container spaces and not above same hold	

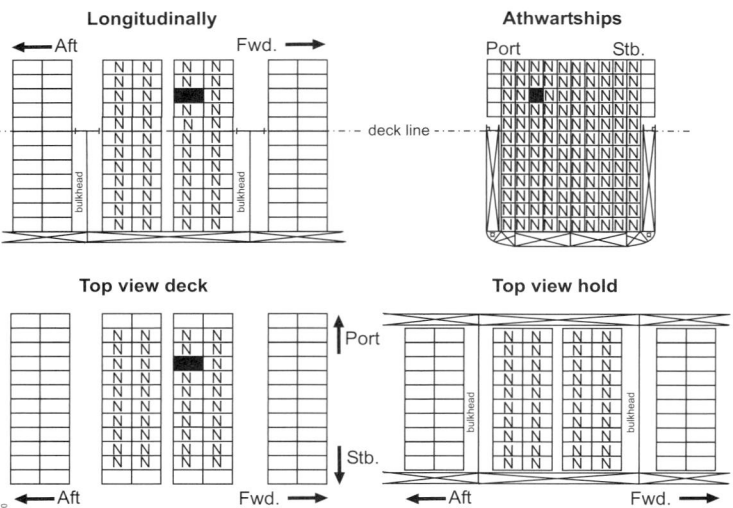

3 – Situation *closed* **versus** *open* **– ON DECK**

Note: All bulkheads and decks shall be resistant to fire and liquids.

"SEPARATED BY A COMPLETE COMPARTMENT OR HOLD FROM" .3		
CLOSED VERSUS OPEN	HORIZONTAL	VERTICAL
	UNDER DECK	
FORE AND AFT	One bulkhead	NOT in the same vertical line
ATHWARTSHIPS	One bulkhead	

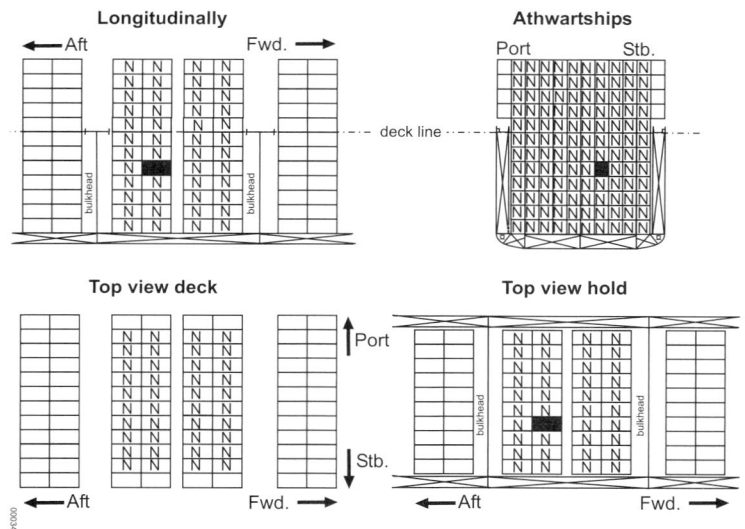

3 – Situation *closed* **versus** *open* **– UNDER DECK**

Note: All bulkheads and decks shall be resistant to fire and liquids.

"SEPARATED BY A COMPLETE COMPARTMENT OR HOLD FROM" .3		
OPEN VERSUS OPEN	HORIZONTAL	VERTICAL
	ON DECK	
FORE AND AFT	Two container spaces and not in or above same hold	NOT in the same vertical line
ATHWARTSHIPS	Three container spaces and not above same hold	

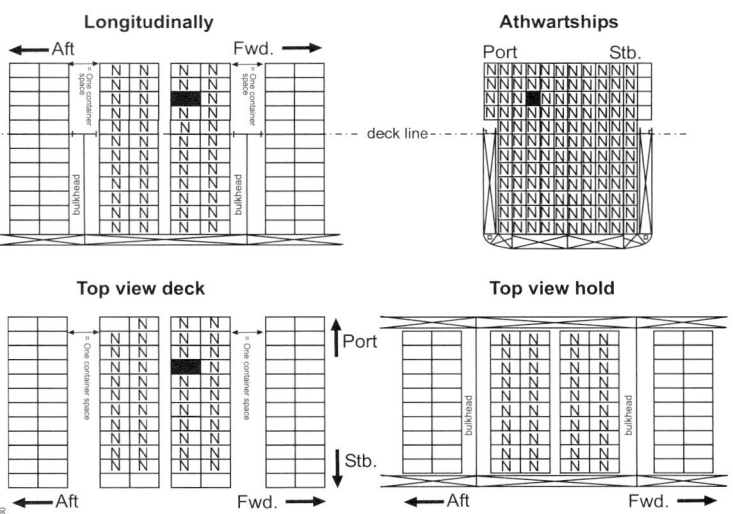

3 – Situation *open* versus *open* – ON DECK

Note: All bulkheads and decks shall be resistant to fire and liquids.

"SEPARATED BY A COMPLETE COMPARTMENT OR HOLD FROM" .3		
OPEN VERSUS OPEN	HORIZONTAL	VERTICAL
	UNDER DECK	
FORE AND AFT	Two bulkheads	NOT in the same vertical line
ATHWARTSHIPS	Two bulkheads	

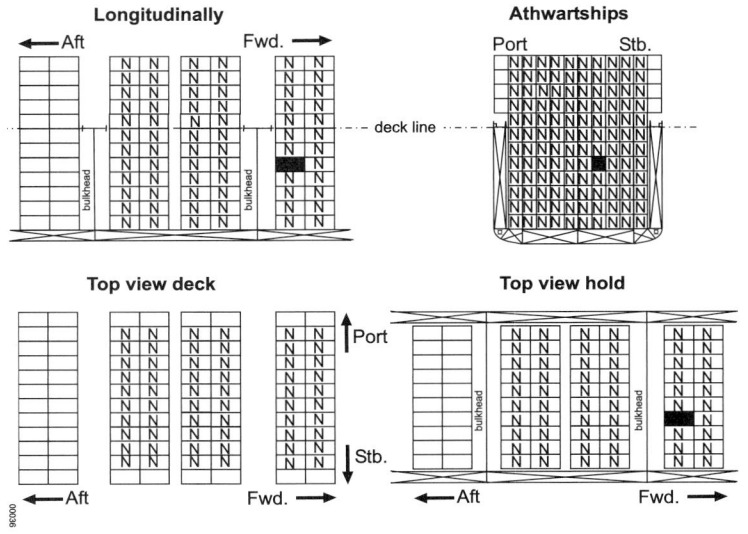

3 – Situation *open* versus *open* – UNDER DECK

Note: All bulkheads and decks shall be resistant to fire and liquids.

"SEPARATED LONGITUDINALLY BY AN INTERVENING COMPLETE COMPARTMENT OR HOLD FROM" .4		
CLOSED VERSUS CLOSED	HORIZONTAL	VERTICAL
	ON DECK	
FORE AND AFT	Minimum horizontal distance of 24 metres and not above same hold	Prohibited
ATHWARTSHIPS	Prohibited	

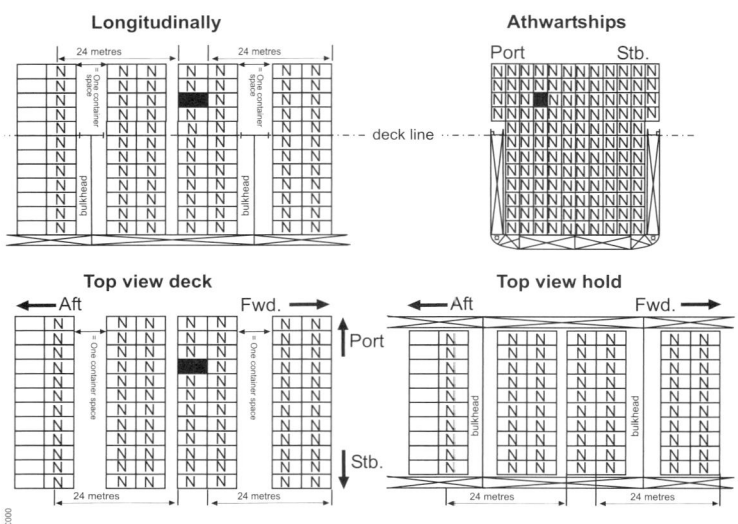

4 – Situation *closed* versus *closed* – ON DECK

Note: All bulkheads and decks shall be resistant to fire and liquids.

"SEPARATED LONGITUDINALLY BY AN INTERVENING COMPLETE COMPARTMENT OR HOLD FROM" .4		
CLOSED VERSUS CLOSED	HORIZONTAL	VERTICAL
	UNDER DECK	
FORE AND AFT	One bulkhead and minimum horizontal distance of 24 metres*	Prohibited
ATHWARTSHIPS	Prohibited	

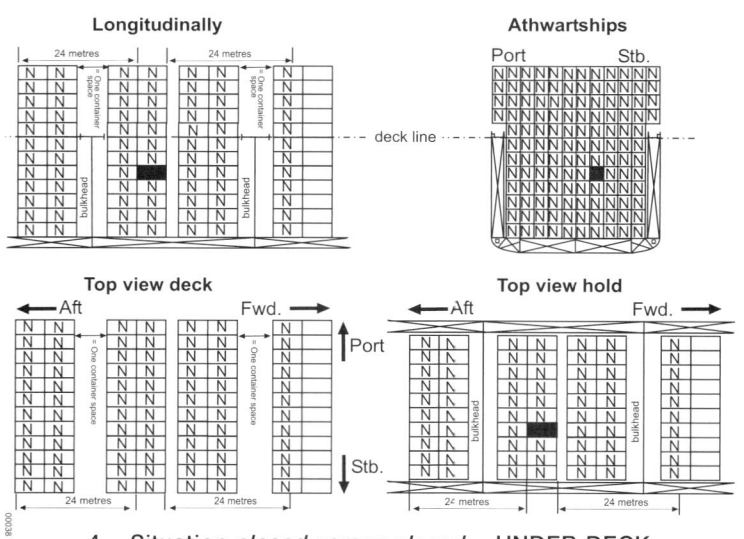

4 – Situation *closed versus closed* – UNDER DECK

Note: All bulkheads and decks shall be resistant to fire and liquids.

* Containers not less than 6 m from intervening bulkhead.

"SEPARATED LONGITUDINALLY BY AN INTERVENING COMPLETE COMPARTMENT OR HOLD FROM" .4		
CLOSED VERSUS OPEN OR OPEN VERSUS OPEN	HORIZONTAL	VERTICAL
	ON DECK	
FORE AND AFT	Minimum horizontal distance of 24 metres and not above same hold	Prohibited
ATHWARTSHIPS	Prohibited	

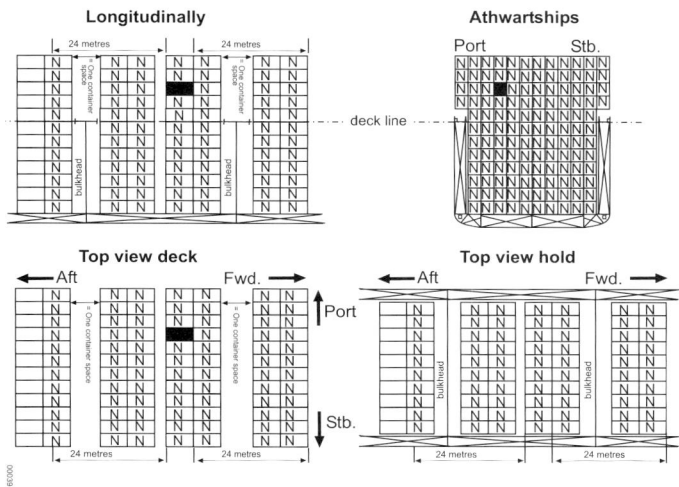

4 – Situations *closed* versus *open* and *open* versus *open* – ON DECK

Note: All bulkheads and decks shall be resistant to fire and liquids.

"SEPARATED LONGITUDINALLY BY AN INTERVENING COMPLETE COMPARTMENT OR HOLD FROM" .4		
CLOSED VERSUS OPEN OR OPEN VERSUS OPEN	HORIZONTAL	VERTICAL
	UNDER DECK	
FORE AND AFT	Two bulkheads	Prohibited
ATHWARTSHIPS	Prohibited	

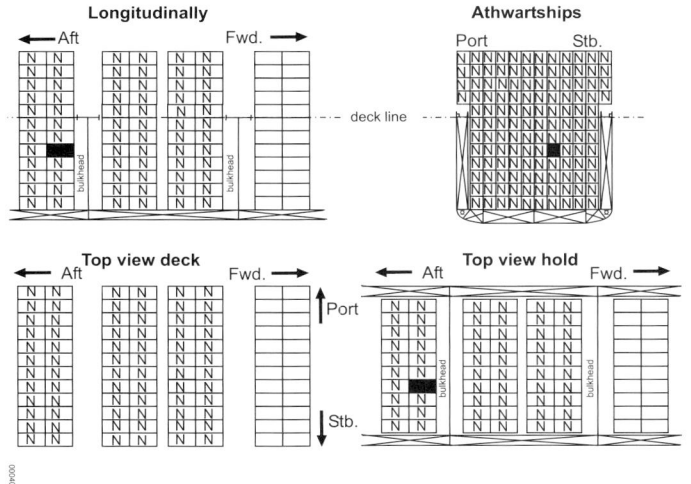

4 – Situations *closed* versus *open* and *open* versus *open* – UNDER DECK

Note: All bulkheads and decks shall be resistant to fire and liquids.

7.2.4   Segregation of cargo transport units on board roll-on/roll-off ships

7.2.4.1   Applicability

7.2.4.1.1   These provisions apply to the segregation of cargo transport units which are transported on board roll-on/roll-off ships or in roll-on/roll-off cargo spaces.

7.2.4.1.2   For roll-on/roll-off ships which carry cargo transport units on decks or in holds, and when these cargo spaces are properly arranged for the permanent stowage of such cargo transport units during transport, the provisions of 7.2.3 shall apply to such spaces.

7.2.4.1.3   For roll-on/roll-off ships which incorporate conventional cargo spaces or any other method of stowage, the appropriate paragraph of this chapter shall apply to the relevant cargo space.

7.2.4.2   Table of segregation of cargo transport units on board ro–ro ships

SEGREGATION REQUIREMENT		HORIZONTAL					
		CLOSED VERSUS CLOSED		CLOSED VERSUS OPEN		OPEN VERSUS OPEN	
		ON DECK	UNDER DECK	ON DECK	UNDER DECK	ON DECK	UNDER DECK
"AWAY FROM"	FORE AND AFT	NO RESTRICTION	NO RESTRICTION	NO RESTRICTION	NO RESTRICTION	AT LEAST 3 METRES	AT LEAST 3 METRES
.1	ATHWART-SHIPS	NO RESTRICTION	NO RESTRICTION	NO RESTRICTION	NO RESTRICTION	AT LEAST 3 METRES	AT LEAST 3 METRES
"SEPARATED FROM"	FORE AND AFT	AT LEAST 6 METRES	AT LEAST 6 METRES OR *ONE* BULKHEAD	AT LEAST 6 METRES	AT LEAST 6 METRES OR *ONE* BULKHEAD	AT LEAST 6 METRES	AT LEAST 12 METRES OR *ONE* BULKHEAD
.2	ATHWART-SHIPS	AT LEAST 3 METRES	AT LEAST 3 METRES OR *ONE* BULKHEAD	AT LEAST 3 METRES	AT LEAST 6 METRES OR *ONE* BULKHEAD	AT LEAST 6 METRES	AT LEAST 12 METRES or *ONE* BULKHEAD
"SEPARATED BY A COMPLETE COMPARTMENT OR HOLD FROM"	FORE AND AFT	AT LEAST 12 METRES	AT LEAST 24 METRES + DECK	AT LEAST 24 METRES	AT LEAST 24 METRES + DECK	AT LEAST 36 METRES	*TWO* DECKS OR *TWO* BULKHEADS
.3	ATHWART-SHIPS	AT LEAST 12 METRES	AT LEAST 24 METRES + DECK	AT LEAST 24 METRES	AT LEAST 24 METRES + DECK	PROHIBITED	PROHIBITED
"SEPARATED LONGITUDINALLY BY AN INTERVENING COMPLETE COMPARTMENT OR HOLD FROM"	FORE AND AFT	AT LEAST 36 METRES	*TWO* BULKHEADS OR AT LEAST 36 METRES + *TWO* DECKS	AT LEAST 36 METRES	AT LEAST 48 METRES INCLUDING *TWO* BULKHEADS	AT LEAST 48 METRES	PROHIBITED
.4	ATHWART-SHIPS	PROHIBITED	PROHIBITED	PROHIBITED	PROHIBITED	PROHIBITED	PROHIBITED

NOTE: ALL BULKHEADS AND DECKS SHALL BE RESISTANT TO FIRE AND LIQUID.

7.2.4.2.1   *Illustrations of segregation of cargo transport units on board ro–ro ships*

7.2.4.2.1.1   The illustrations of this subsection apply to the segregation of cargo transport units which are transported on board roll-on/roll-off ships or in roll-on/roll-off cargo spaces.*

7.2.4.2.1.2   To determine locations in which cargo transport units are not permitted to contain dangerous goods that are incompatible with those in a reference cargo transport unit, the following method shall be used: locations where incompatible dangerous goods are not permitted with respect to the reference cargo transport unit are first determined in the direct fore-and-aft and athwartships directions. Lines are projected between the outermost corners of the cargo transport units occupying these spaces as shown in the figure. Cargo transport units located partially or completely between these lines and the reference cargo transport unit shall not contain dangerous goods that are incompatible with those in the reference cargo transport unit.

7.2.4.2.1.3   The standard dimension of a cargo transport unit used for the illustrations is:

– length: 12 m
– width: 2.50 m

---

* For ro–ro ships which carry cargo transport units on decks or in holds, the illustrations of 7.2.3.2.1 apply to such spaces.

7.2.4.2.1.4   *Definitions of the segregation terms*

(1) Reference cargo transport unit ....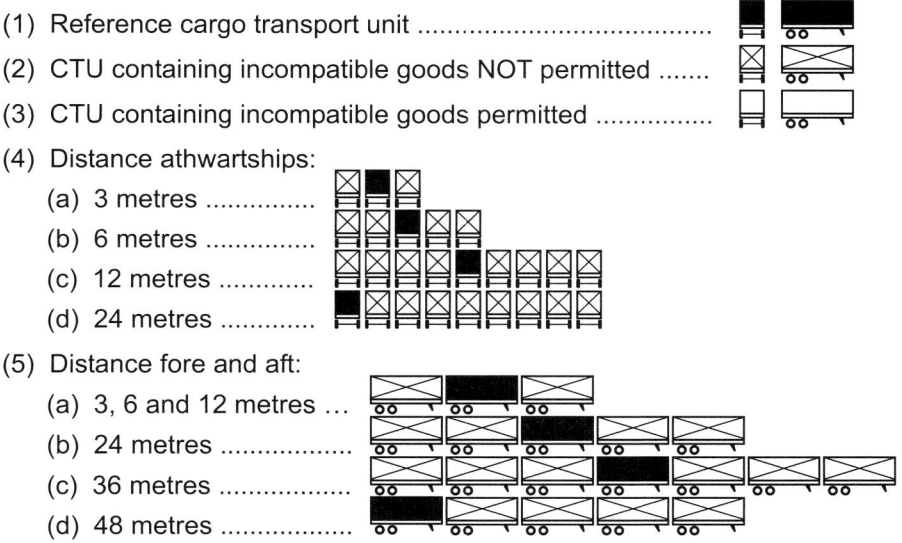

(2) CTU containing incompatible goods NOT permitted ....

(3) CTU containing incompatible goods permitted ....

(4) Distance athwartships:

    (a) 3 metres ....

    (b) 6 metres ....

    (c) 12 metres ....

    (d) 24 metres ....

(5) Distance fore and aft:

    (a) 3, 6 and 12 metres ...

    (b) 24 metres ....

    (c) 36 metres ....

    (d) 48 metres ....

Situation fore and aft: 3 + 6 metres athwartships: 3 metres

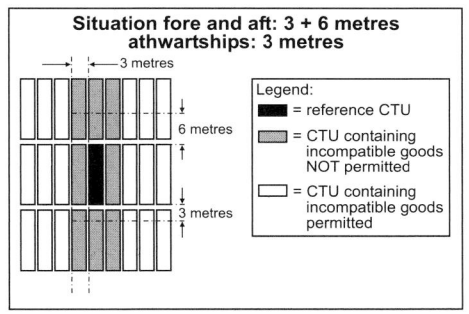

Legend:
- = reference CTU
- = CTU containing incompatible goods NOT permitted
- = CTU containing incompatible goods permitted

Situation fore and aft: 6 metres athwartships: 6 metres

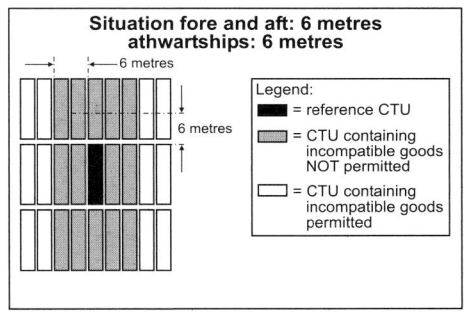

Legend:
- = reference CTU
- = CTU containing incompatible goods NOT permitted
- = CTU containing incompatible goods permitted

Situation fore and aft: 12 metres athwartships: 12 metres

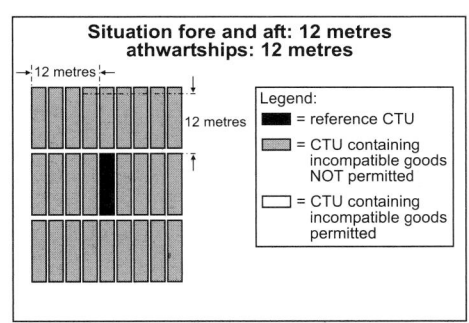

Legend:
- = reference CTU
- = CTU containing incompatible goods NOT permitted
- = CTU containing incompatible goods permitted

Situation fore and aft: 24 metres athwartships: 24 metres

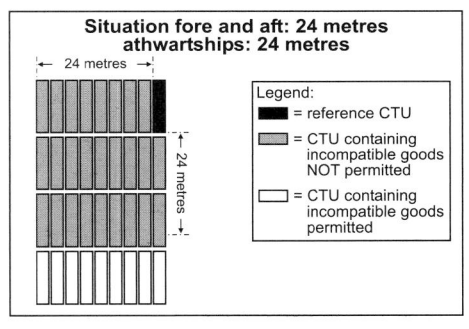

Legend:
- = reference CTU
- = CTU containing incompatible goods NOT permitted
- = CTU containing incompatible goods permitted

Situation fore and aft: 36 metres athwartships: 36 metres

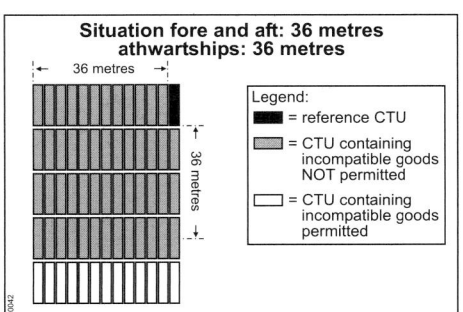

Legend:
- = reference CTU
- = CTU containing incompatible goods NOT permitted
- = CTU containing incompatible goods permitted

Situation fore and aft: 36 + 48 metres athwartships: prohibited

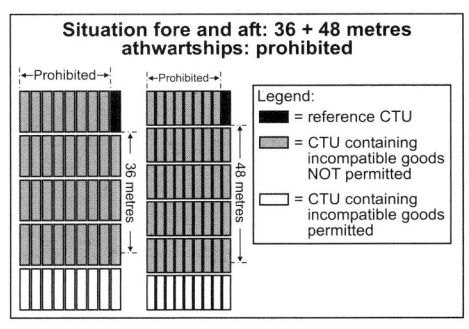

Legend:
- = reference CTU
- = CTU containing incompatible goods NOT permitted
- = CTU containing incompatible goods permitted

Note: All bulkheads and decks shall be resistant to fire and liquids.

"AWAY FROM" .1		
CLOSED VERSUS CLOSED OR CLOSED VERSUS OPEN	ON DECK	UNDER DECK
FORE AND AFT	No restriction	No restriction
ATHWARTSHIPS	No restriction	No restriction

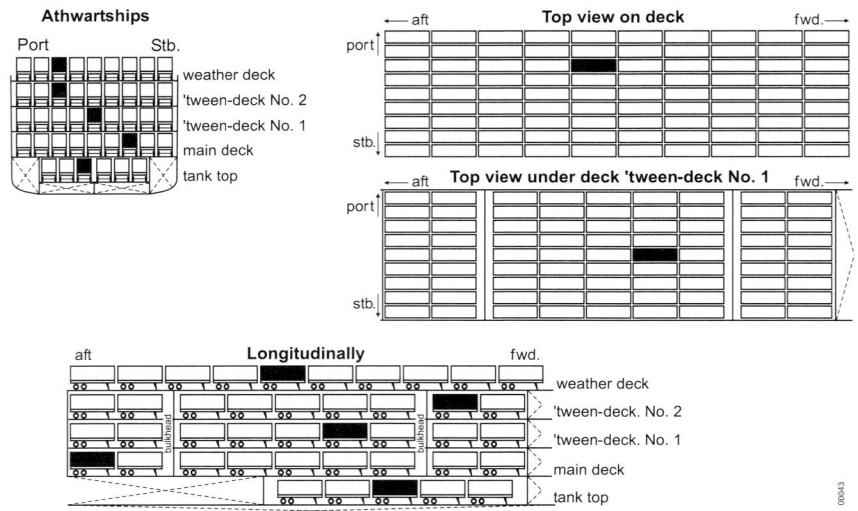

1 – Situation *closed* versus *closed* and *closed* versus *open*

Note: All bulkheads and decks shall be resistant to fire and liquids.

"AWAY FROM" .1		
OPEN VERSUS OPEN	ON DECK	UNDER DECK
FORE AND AFT	At least 3 metres	At least 3 metres
ATHWARTSHIPS	At least 3 metres	At least 3 metres

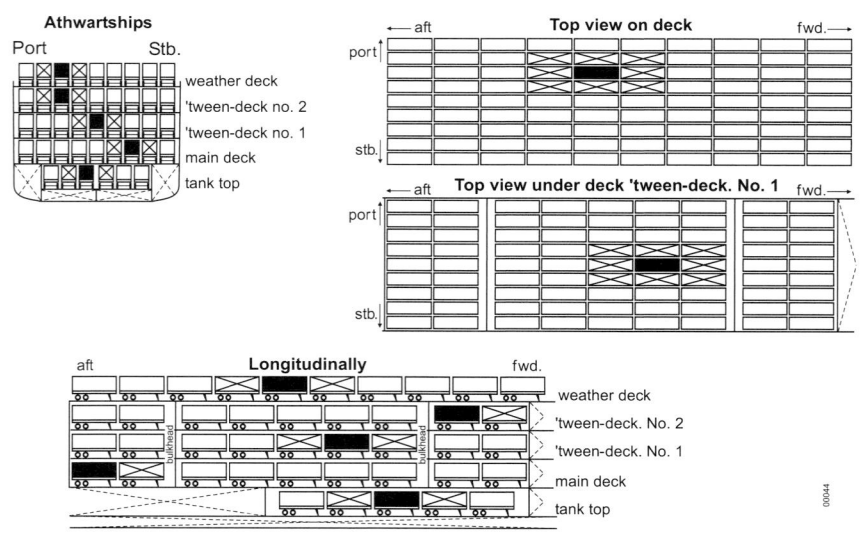

1 – Situation *open* versus *open*

Note: All bulkheads and decks shall be resistant to fire and liquids.

"SEPARATED FROM" .2		
CLOSED VERSUS CLOSED	ON DECK	UNDER DECK
FORE AND AFT	At least 6 metres	At least 6 metres or ONE bulkhead
ATHWARTSHIPS	At least 3 metres	At least 3 metres or ONE bulkhead

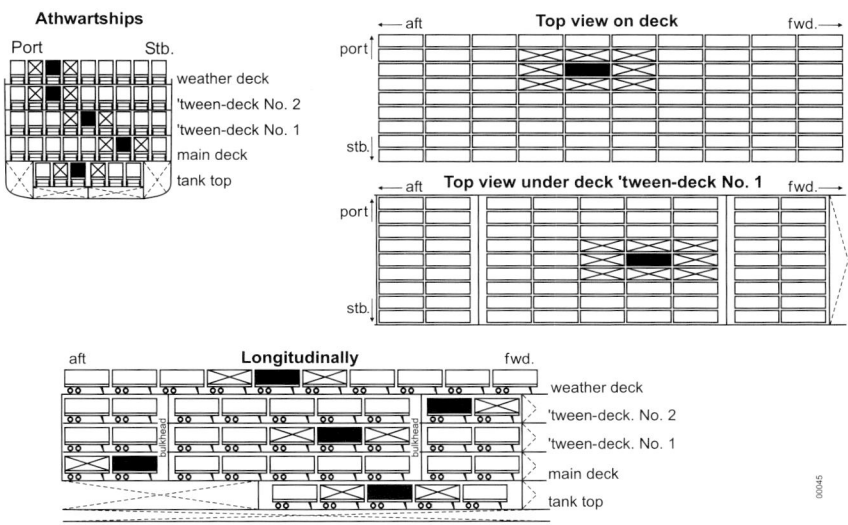

2 – Situation *closed* versus *closed*

Note: All bulkheads and decks shall be resistant to fire and liquids.

"SEPARATED FROM" .2		
CLOSED VERSUS OPEN	ON DECK	UNDER DECK
FORE AND AFT	At least 6 metres	At least 6 metres or ONE bulkhead
ATHWARTSHIPS	At least 3 metres	At least 6 metres or ONE bulkhead

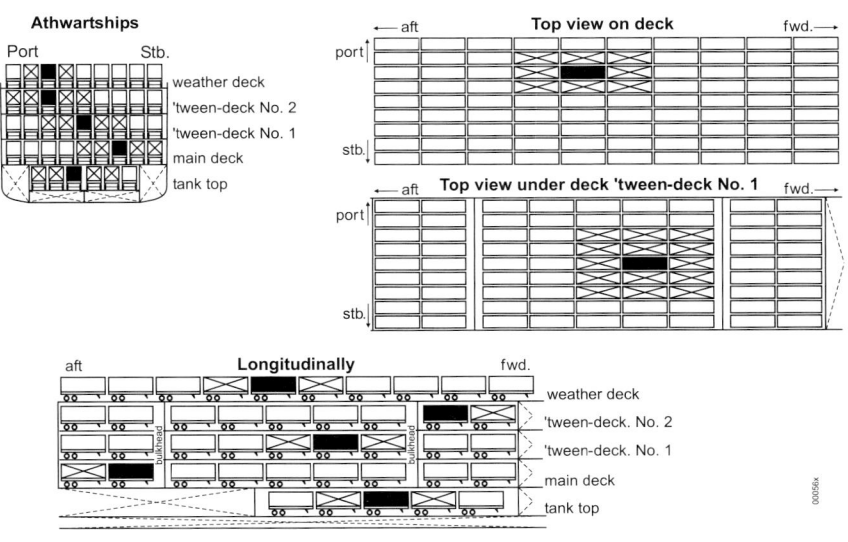

2 – Situation *closed* versus *open*

Note: All bulkheads and decks shall be resistant to fire and liquids.

7

"SEPARATED FROM" .2		
OPEN VERSUS OPEN	**ON DECK**	**UNDER DECK**
FORE AND AFT	At least 6 metres	At least 12 metres or ONE bulkhead
ATHWARTSHIPS	At least 6 metres	At least 12 metres or ONE bulkhead

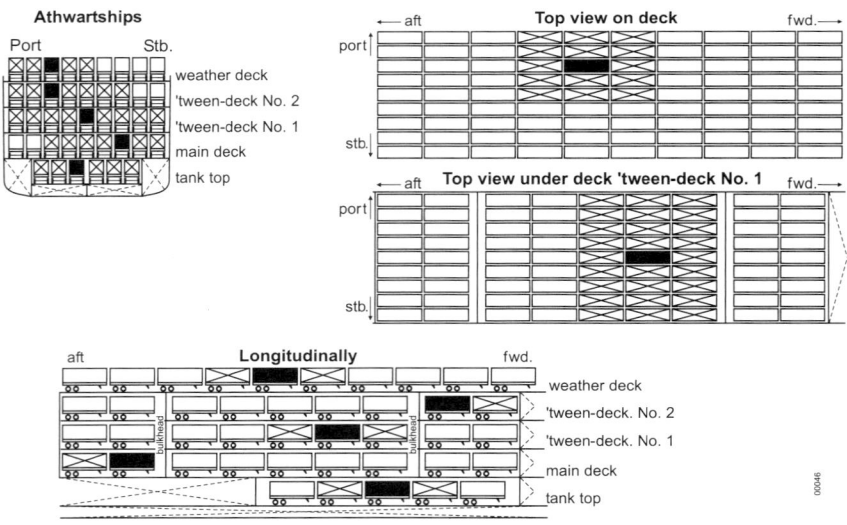

2 – Situation *open* **versus** *open*

Note: All bulkheads and decks shall be resistant to fire and liquids.

"SEPARATED BY A COMPLETE COMPARTMENT OR HOLD FROM" .3		
CLOSED VERSUS CLOSED	**ON DECK**	**UNDER DECK**
FORE AND AFT	At least 12 metres	At least 24 metres + deck
ATHWARTSHIPS	At least 12 metres	At least 24 metres + deck

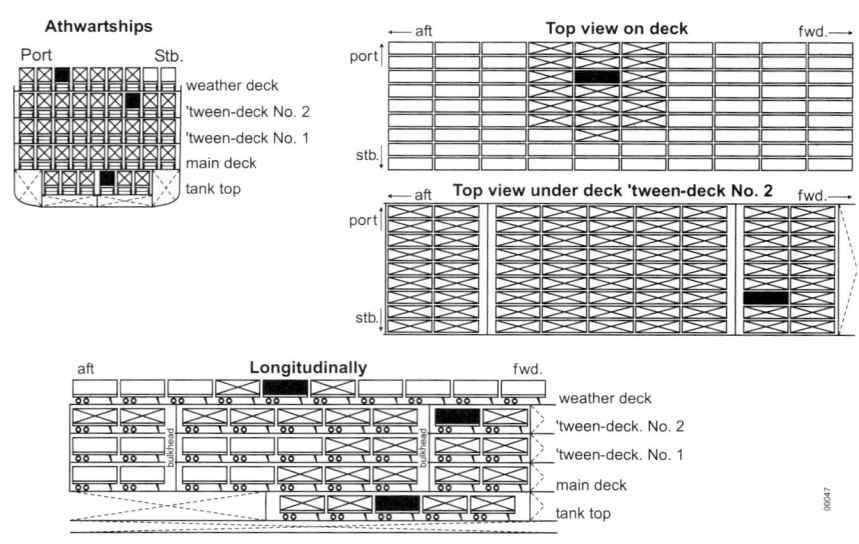

3 – Situation *closed* **versus** *closed*

Note: All bulkheads and decks shall be resistant to fire and liquids.

"SEPARATED BY A COMPLETE COMPARTMENT OR HOLD FROM" .3		
CLOSED VERSUS OPEN	ON DECK	UNDER DECK
FORE AND AFT	At least 24 metres	At least 24 metres + deck
ATHWARTSHIPS	At least 24 metres	At least 24 metres + deck

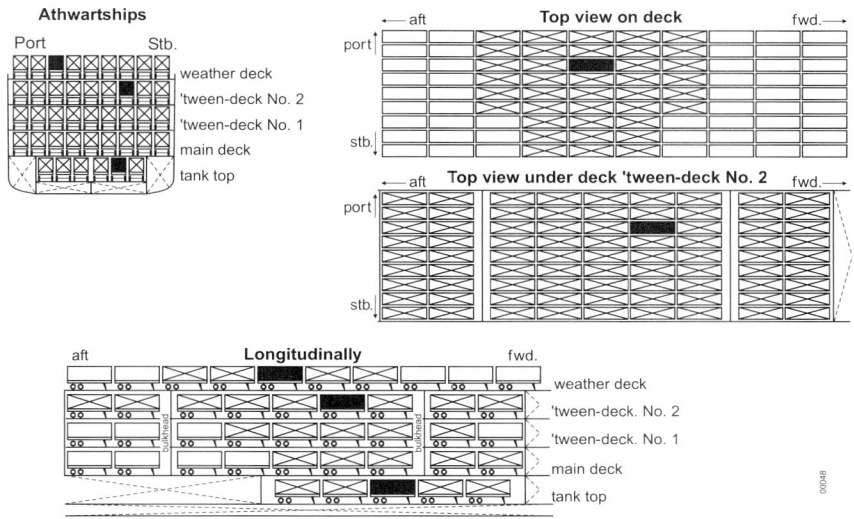

3 – Situation *closed* versus *open*

Note: All bulkheads and decks shall be resistant to fire and liquids.

"SEPARATED BY A COMPLETE COMPARTMENT OR HOLD FROM" .3		
OPEN VERSUS OPEN	ON DECK	UNDER DECK
FORE AND AFT	At least 36 metres	Two decks or TWO bulkheads
ATHWARTSHIPS	Prohibited	Prohibited

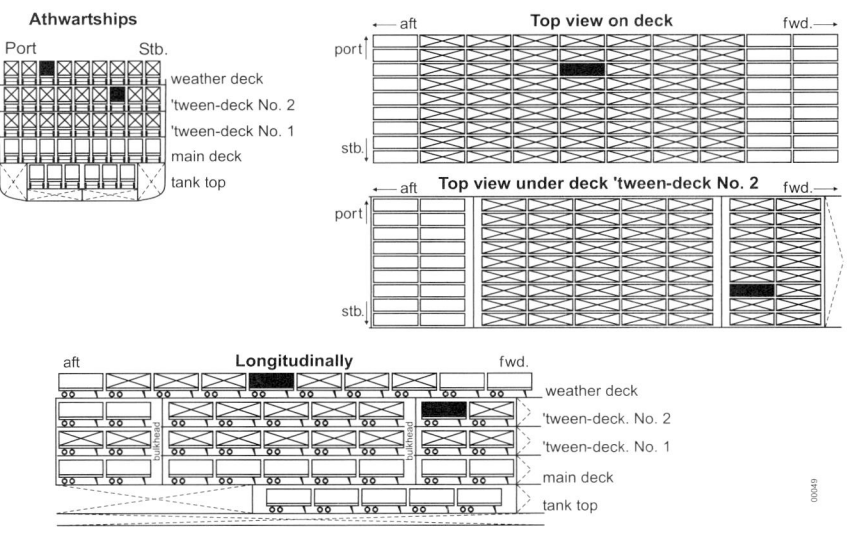

3 – Situation *open* versus *open*

Note: All bulkheads and decks shall be resistant to fire and liquids

7

"SEPARATED LONGITUDINALLY BY AN INTERVENING COMPLETE COMPARTMENT OR HOLD FROM" .4		
CLOSED VERSUS CLOSED	ON DECK	UNDER DECK
FORE AND AFT	At least 36 metres	Two bulkheads or at least 36 metres + two decks
ATHWARTSHIPS	Prohibited	Prohibited

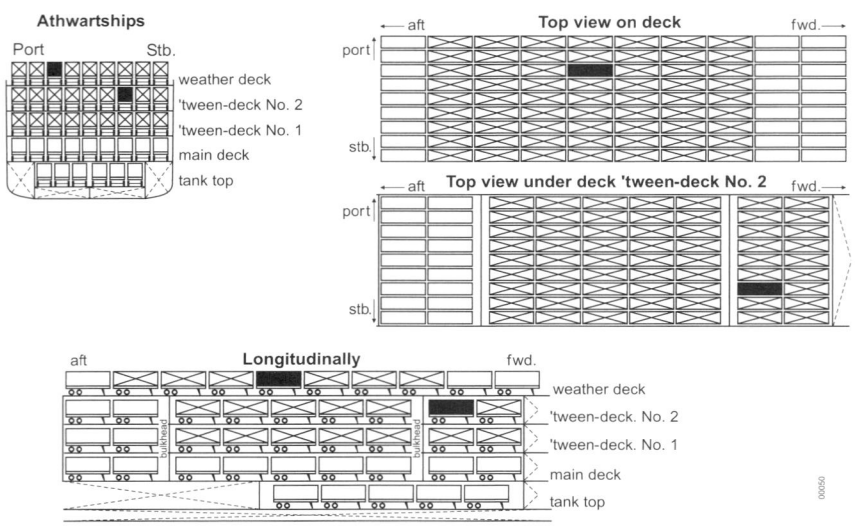

4 – Situation *closed* versus *closed*
Note: All bulkheads and decks shall be resistant to fire and liquids

"SEPARATED LONGITUDINALLY BY AN INTERVENING COMPLETE COMPARTMENT OR HOLD FROM" .4		
CLOSED VERSUS OPEN	ON DECK	UNDER DECK
FORE AND AFT	At least 36 metres	At least 48 metres including TWO bulkheads
ATHWARTSHIPS	Prohibited	Prohibited

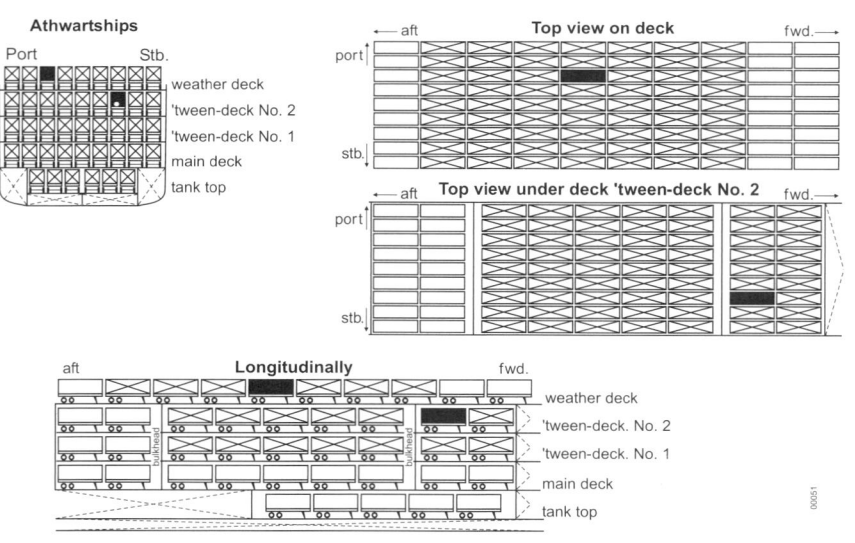

4 – Situation *closed* versus *open*
Note: All bulkheads and decks shall be resistant to fire and liquids

"SEPARATED LONGITUDINALLY BY AN INTERVENING COMPLETE COMPARTMENT OR HOLD FROM" .4		
OPEN VERSUS OPEN	ON DECK	UNDER DECK
FORE AND AFT	At least 48 metres	Prohibited
ATHWARTSHIPS	Prohibited	Prohibited

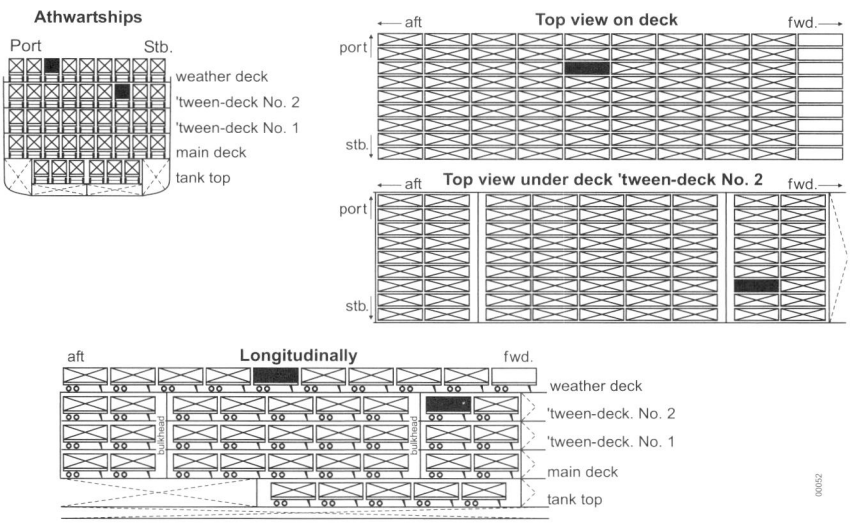

4 – Situation *open* versus *open*

Note: All bulkheads and decks shall be resistant to fire and liquids

7.2.5 Segregation in shipborne barges and on board barge-carrying ships

7.2.5.1 Applicability

7.2.5.1.1 The provisions of this subsection apply to the segregation in shipborne barges as well as the segregation between shipborne barges transported on board ships specially designed and equipped to carry such barges, see also chapter 7.6.

7.2.5.1.2 For barge-carrying ships which incorporate other cargo spaces or any other method of stowage, the appropriate subsection of this chapter shall apply to the relevant cargo space.

7.2.5.2 Segregation in shipborne barges

For segregation in shipborne barges, the appropriate subsections of this chapter shall apply.

7.2.5.3 Segregation between shipborne barges on barge-carrying ships

7.2.5.3.1 When a shipborne barge is loaded with two or more substances with different provisions for segregation, the most stringent segregation applicable shall be applied.

7.2.5.3.2 "Away from" and "separated from" require no segregation between shipborne barges.

7.2.5.3.3 "Separated by a complete compartment or hold from" means, for barge-carrying ships with vertical holds, that separate holds are required. On barge-carrying ships having horizontal barge levels, separate barge levels are required and the barges shall not be in the same vertical line.

7.2.5.3.4 "Separated longitudinally by an intervening complete compartment or hold from" means, for barge-carrying ships with vertical holds, that separation by an intervening hold or engine-room is required. On barge-carrying ships having horizontal barge levels, separate barge levels and a longitudinal separation by at least two intervening barge spaces is required.

7.2.6 Segregation between bulk materials possessing chemical hazards and dangerous goods in packaged form

7.2.6.1 Applicability

7.2.6.1.1 Unless otherwise required in this chapter or in the Dangerous Goods List, segregation between bulk materials possessing chemical hazards and dangerous goods in packaged form shall be in accordance with the following table.

7.2.6.1.2 *Segregation table*

Bulk materials (classified as dangerous goods)	CLASS	Dangerous goods in packaged form															
		1.1 1.2 1.5	1.3 1.6	1.4	2.1	2.2 2.3	3	4.1	4.2	4.3	5.1	5.2	6.1	6.2	7	8	9
Flammable solids	**4.1**	4	3	2	2	2	2	X	1	X	1	2	X	3	2	1	X
Substances liable to spontaneous combustion	**4.2**	4	3	2	2	2	2	1	X	1	2	2	1	3	2	1	X
Substances which, in contact with water, emit flammable gases	**4.3**	4	4	2	1	X	2	X	1	X	2	2	X	2	2	1	X
Oxidizing substances (agents)	**5.1**	4	4	2	2	X	2	1	2	2	X	2	1	3	1	2	X
Toxic substances	**6.1**	2	2	X	X	X	X	X	1	X	1	1	X	1	X	X	X
Radioactive material	**7**	2	2	2	2	2	2	2	2	2	1	2	X	3	X	2	X
Corrosive substances	**8**	4	2	2	1	X	1	1	1	1	2	2	X	3	2	X	X
Miscellaneous dangerous substances and articles	**9**	X	X	X	X	X	X	X	X	X	X	X	X	X	X	X	X
Materials hazardous only in bulk (MHB)		X	X	X	X	X	X	X	X	X	X	X	X	3	X	X	X

Numbers and symbols relate to the following terms as defined in this chapter:
1 – "Away from"
2 – "Separated from"
3 – "Separated by a complete compartment or hold from"
4 – "Separated longitudinally by an intervening complete compartment or hold from"
X – The segregation, if any, is shown in the Dangerous Goods List or the individual entries in the Code of Safe Practice for Solid Bulk Cargoes

7.2.6.1.3 *Definitions of the segregation terms*

7.2.6.1.3.1 *Away from:*
Effectively segregated so that incompatible materials cannot interact dangerously in the event of an accident but may be transported in the same compartment or hold or *on deck* provided a minimum horizontal separation of 3 metres, projected vertically, is provided.

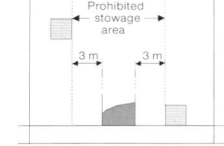

7.2.6.1.3.2 *Separated from:*
In different holds when stowed *under deck*. Provided an intervening deck is resistant to fire and liquid, a vertical separation, i.e. in different compartments, may be accepted as equivalent to this segregation.

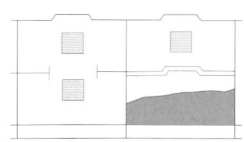

7.2.6.1.3.3 *Separated by a complete compartment or hold from:*
Either a vertical or a horizontal separation. If the decks are not resistant to fire and liquid, then only a longitudinal separation, i.e. by an intervening complete compartment, is acceptable.

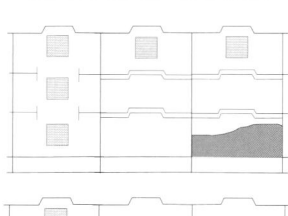

7.2.6.1.3.4 *Separated longitudinally by an intervening complete compartment or hold from:*
Vertical separation alone does not meet this requirement.

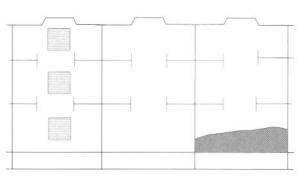

Legend

(1) Reference *bulk material* . . . .

(2) *Package* containing incompatible goods . . . .

(3) Deck resistant to fire and liquid . . . .

Note: Vertical lines represent transverse watertight bulkheads between cargo spaces.

7.2.7   Segregation of goods of class 1

7.2.7.1   Segregation from dangerous goods of other classes

7.2.7.1.1   Notwithstanding the segregation provisions of this chapter, AMMONIUM NITRATE (UN 1942), AMMONIUM NITRATE FERTILIZERS (UN 2067), alkali metal nitrates (e.g., UN 1486) and alkaline earth metal nitrates (e.g., UN 1454) may be stowed together with blasting explosives (except EXPLOSIVE, BLASTING, TYPE C, UN 0083) provided the aggregate is treated as blasting explosives under class 1.

7.2.7.1.2   For the segregation of goods of class 1 from solid bulk materials possessing chemical hazards, see 7.2.6.

7.2.7.1.3   *Dangerous goods of extreme flammability*

7.2.7.1.3.1   Certain dangerous substances, because of their extreme flammability, may not be transported in a ship carrying goods of class 1. This restriction is indicated in the following Dangerous Goods List entries:

Proper Shipping Name	UN No.	Class
CARBON DISULPHIDE	1131	3
NICKEL CARBONYL	1259	6.1
PYROPHORIC LIQUID, ORGANIC, N.O.S.	2845	4.2
PYROPHORIC LIQUID, INORGANIC, N.O.S.	3194	4.2
ORGANOMETALLIC SUBSTANCE, LIQUID, PYROPHORIC	3392	4.2
ORGANOMETALLIC SUBSTANCE, LIQUID, PYROPHORIC, WATER-REACTIVE	3394	4.2

7.2.7.1.3.2   The restriction of 7.2.7.1.3.1 does not apply in the case of:

.1   goods in division 1.4, compatibility group S; or

.2   explosive articles for life-saving purposes as identified in the individual schedules, if the total net explosives mass of such articles does not exceed 50 kg per ship; or

.3   goods in compatibility groups C, D and E, if the total net explosives mass does not exceed 10 kg per ship; or

.4   articles in compatibility group G other than fireworks and those requiring special stowage, if the total net explosives mass does not exceed 10 kg per ship.

7.2.7.1.3.3   Notwithstanding the provisions of 7.2.7.1.3.1, additional quantities or types of goods of class 1 in excess of those mentioned in 7.2.7.1.3.2 may be transported together with dangerous goods of extreme flammability only with the approval of the competent authority.

7.2.7.1.3.4   Where the ship is carrying goods of class 1 and dangerous goods of extreme flammability, they are to be segregated in accordance with this chapter and care shall be taken that they are stowed in parts of the ship as remote as possible from each other.

7.2.7.2   Segregation within class 1

7.2.7.2.1   *General*

7.2.7.2.1.1   Goods of class 1 may be stowed within the same compartment, magazine, or cargo transport unit as indicated in 7.2.7.2.1.4. In other cases, they shall be stowed in separate containers except as provided in 7.2.7.2.2 and 7.2.7.2.1.5.

7.2.7.2.1.2   When goods requiring different stowage arrangements are permitted by 7.2.7.2.1.4 to be transported in the same compartment, magazine, cargo transport unit or vehicle, the appropriate stowage arrangement shall conform to the most stringent provisions for the entire load.

7.2.7.2.1.3   Where a mixed load of different divisions is transported within the same compartment, magazine, or cargo transport unit, the entire load shall be treated as if belonging to the hazard division in the order 1.1 (most dangerous), 1.5, 1.2, 1.3, 1.6 and 1.4 (least dangerous) and the stowage arrangement shall conform to the most stringent provisions for the entire load.

7

7.2.7.2.1.4   *Permitted mixed stowage for goods of class 1*

Compatibility group	A	B	C	D	E	F	G	H	J	K	L	N	S
A	X												
B		X											X
C			X	X^6	X^6		X^1					X^4	X
D			X^6	X	X^6		X^1					X^4	X
E			X^6	X^6	X		X^1					X^4	X
F						X							X
G			X^1	X^1	X^1		X						X
H								X					X
J									X				X
K										X			X
L											X^2		
N			X^4	X^4	X^4							X^3	X^5
S		X	X	X	X	X	X	X	X	X		X^5	X

"X" indicates that goods of the corresponding compatibility groups may be stowed in the same compartment, magazine, cargo transport unit or vehicle

Notes:

[1] Explosive articles in compatibility group G (other than fireworks and those requiring special stowage) may be stowed with explosive articles of compatibility groups C, D and E provided no explosive substances are transported in the same compartment, magazine, cargo transport unit or vehicle.

[2] A consignment of one type in compatibility group L shall only be stowed with a consignment of the same type within compatibility group L.

[3] Different types of articles of division 1.6, compatibility group N, may only be transported together when it is proven that there is no additional risk of sympathetic detonation between the articles. Otherwise they shall be treated as division 1.1.

[4] When articles of compatibility group N are transported with articles or substances of compatibility groups C, D or E, the goods of compatibility group N shall be treated as compatibility group D.

[5] When articles of compatibility group N are transported together with articles or substances of compatibility group S, the entire load shall be treated as compatibility group N.

[6] Any combination of articles in compatibility groups C, D and E shall be treated as compatibility group E. Any combination of substances in compatibility groups C and D shall be treated as the most appropriate compatibility group shown in 2.1.2.3, taking into account the predominant characteristics of the combined load. This overall classification code shall be displayed on any label or placard placed on a unit load or cargo transport unit as prescribed in 5.2.2.2.2.

7.2.7.2.1.5   Cargo transport units carrying different goods of class 1 do not require segregation from each other provided 7.2.7.2.1 and 7.2.7.2.2 authorize the goods to be transported together. Where this is not permitted by 7.2.7.2.1.4, cargo transport units shall be "separated from" one another.

7.2.7.2.2   *Segregation on deck*

When goods in different compatibility groups are transported on deck, they shall be stowed not less than 6 m apart unless their mixed stowage is allowed according to 7.2.7.2.1.4.

7.2.7.2.3   *Segregation in single-hold ships*

In a single-hold ship carrying dangerous goods other than those of class 1, segregation shall be as for larger ships except that:

.1   Goods in division 1.1 or 1.2 of compatibility group B may be stowed in the same hold as substances of compatibility group D provided:

  –   the net explosives mass of goods of compatibility group B does not exceed 50 kg; and

  –   such goods are stowed in a steel magazine which is stowed at least 6 m from the substances of compatibility group D.

.2   Goods in division 1.4 of compatibility group B may be stowed in the same hold as substances of compatibility group D provided they are separated either by a distance of at least 6 m or by a steel division.

7.2.7.3 **Segregation from non-dangerous goods**

7.2.7.3.1 In general, it is not necessary to segregate goods of class 1 from other non-dangerous cargo.

7.2.7.3.2 Mail, baggage, personal effects and household effects, however, shall not be stowed in the same compartment, or in compartments immediately above or below goods of class 1 other than those in compatibility group S.

7.2.7.3.3 Where goods of class 1 are stowed against an intervening bulkhead, any mail on the other side of the bulkhead shall be stowed "away from" it, preferably with the intervening space filled by other non-dangerous cargo.

7.2.8 **Segregation provisions for goods of class 4.1 and class 5.2**

.1 Segregation as for class 1, division 1.3, shall be applied for packages carrying a subsidiary risk label of class 1.

7.2.9 **Segregation for goods of class 7**

7.2.9.1 Radioactive material shall be segregated sufficiently from crew and passengers. The following values for dose shall be used for the purpose of calculating segregation distances or radiation levels:

(a) for crew in regularly occupied working areas, a dose of 5 mSv in a year;

(b) for passengers, in areas where the passengers have regular access, a dose of 1 mSv in a year to the critical group, taking account of the exposures expected to be delivered by all other relevant sources and practices under control.

7.2.9.2 Radioactive material shall be sufficiently segregated from undeveloped photographic film. The basis for determining segregation distances for this purpose shall be that the radiation exposure of undeveloped photographic film due to the transport of radioactive material be limited to 0.1 mSv per consignment of such film.

7.2.9.3 Category II – YELLOW or III – YELLOW packages or overpacks shall not be transported in spaces occupied by passengers, except those exclusively reserved for couriers specially authorized to accompany such packages or overpacks.

7.2.9.4 Any group of packages, overpacks, and freight containers containing fissile material stored in transit in any one storage area shall be so limited that the total sum of the criticality safety indexes in the group does not exceed 50. Each group shall be stored so as to maintain a spacing of at least 6 m from other such groups.

7.2.9.5 Where the total sum of the criticality safety indexes on board a conveyance or in a freight container exceeds 50, as permitted in the table under 7.1.14.5.4, storage shall be such as to maintain a spacing of at least 6 m from other groups of packages, overpacks or freight containers containing fissile material or other conveyances carrying radioactive material.

7.2.9.6 Any departure from the segregation provisions shall be approved by the competent authority of the flag State of the ship and, when requested, by the competent authority at each port of call.

7.2.9.7 The segregation requirements specified in 7.2.9.1. may be established in one of the following two ways:

– By following the segregation tables (I and III hereafter) in respect of living quarters or spaces regularly occupied by persons. Table III includes comprehensive provisions which are of general applicability. Table I provides simplified information which is applicable to certain ship sizes, or

– By demonstration that, for the following indicated exposure times, the direct measurement of the radiation level in regularly occupied spaces and living quarters is less than:

for the crew:
0.0070 mSv/h up to 700 hours in a year, or
0.0018 mSv/h up to 2750 hours in a year; and

for the passengers:
0.0018 mSv/h up to 550 hours in a year,

taking into account any relocation of cargo during the voyage. In all cases, the measurements of radiation level must be made and documented by a suitably qualified person.

7.2.9.8 The radiation exposure of undeveloped photographic film and plates shall be based upon a single-voyage exposure of 0.1 mSv. One of the segregation tables (II and III hereafter) shall be followed. Table III includes comprehensive provisions which are of general applicability. Table II provides simplified information which is applicable to certain ship sizes and voyage durations only.

7.2.9.9 As an alternative to the use of tables II and III, separation distances may be estimated by the use of the nomograph in 7.2.9.10. This nomograph will be particularly useful in cases where stowage factors (cargo density or thickness of cargo) are significantly different from the figures given in tables II and III.

TABLE I
CLASS 7 – Radioactive material
Simplified segregation table for persons

Sum of transport indices (TI)	Segregation distance of radioactive material from passengers and crew			
	General cargo ship[1]		Ferry etc.[2]	Offshore support vessel[3]
	Break-bulk (metres)	Containers (TEUs)[4]		
Up to 10	6	1	Stow at bow or stern furthest from living quarters and regularly occupied work areas	Stow at stern or at platform midpoint
More than 10 but not more than 20	8	1	as above	as above
More than 20 but not more than 50	13	2	as above	not applicable
More than 50 but not more than 100	18	3	as above	not applicable
More than 100 but not more than 200	26	4	as above	not applicable
More than 200 but not more than 400	36	6	as above	not applicable

[1] General cargo, break-bulk or ro–ro container ship of 150 m minimum length.

[2] Ferry or cross-channel, coastal and inter-island ship of 100 m minimum length.

[3] Offshore support vessel of 50 m minimum length. (In this case the practical maximum sum of TIs carried is 20.)

[4] TEU means "20 ft Equivalent Unit" (this is equivalent to a standard freight container of 6 m nominal length).

TABLE II
CLASS 7 – Radioactive material
Simplified segregation table for photographic films and plates

Sum of transport indices (TI)	Duration of voyage in days				
	Not more than 1[1,2]	More than 1 but not more than 4[1,2]	More than 4 but not more than 10[2]	More than 10 but not more than 30[2]	More than 30 but not more than 50[2]
Not more than 10	$\frac{1}{3}$ ship length			$\frac{1}{2}$ ship length	
More than 10 but not more than 20					
More than 20 but not more than 50		$\frac{3}{4}$ ship length	$\frac{1}{3}$ ship length (shielding required)[3]		
More than 50 but not more than 400					

[1] Ferry or cross-channel, coastal and inter-island ship of 100 m minimum length.

[2] General cargo, break-bulk or ro–ro container ship of 150 m minimum length.

[3] Shielding required in the form of intervening cargo, either as a complete layer of filled containers or as a cargo space with 6 m (minimum) carried between the film and class 7 packages.

TABLE III
CLASS 7 – Radioactive material
Segregation table in metres
Safe distances for persons and undeveloped photographic films and plates

Sum of transport indices (Note (7))	Living quarters: Nil	1	2	1-day: Nil	1	2	2-day: Nil	1	2	4-day: Nil	1	2	10-day: Nil	1	2	20-day: Nil	1	2	30-day: Nil	1	2	40-day: Nil	1	2	50-day: Nil	1	2
0.5	2	X	X	2	X	X	3	X	X	4	X	X	6	2	X	8	2	X	10	3	X	11	3	X	12	3	X
1	2	X	X	3	X	X	4	X	X	5	2	X	8	2	X	11	3	X	13	4	X	15	4	X	17	4	X
2	3	X	X	4	X	X	5	2	X	7	2	X	11	3	X	15	4	X	19	5	X	22	5	X	24	6	X
3	4	X	X	5	X	X	6	2	X	9	2	X	13	4	X	19	5	X	23	6	X	27	7	X	30	7	X
5	4	X	X	6	2	X	8	2	X	11	3	X	17	4	X	24	6	X	30	7	X	34	8	X	38	9	3
10	6	2	X	8	2	X	11	3	X	15	4	X	24	6	X	34	8	X	42	10	3	48	12	3	54	13	3
20	8	2	X	11	3	X	15	4	X	22	5	X	34	8	X	48	12	3	59	14	4	68	16	4	76	18	5
30	10	3	X	13	4	X	19	5	X	26	7	X	42	10	3	59	14	4	72	17	4	83	20	5	93	22	6
50	13	3	X	17	4	X	24	6	X	34	8	X	54	13	3	76	18	5	92	23	6	110	26	7	120	29	7
100	18	5	X	24	6	X	34	8	X	48	12	3	76	18	5	110	25	6	130	32	8	150	36	9	170	40	10
150	22	6	X	30	7	X	42	10	3	59	14	4	93	22	6	130	31	8	160	39	10	185	45	11	*	50	12
200	26	6	X	34	8	X	48	12	3	68	16	4	110	26	7	150	36	9	185	43	11	*	51	13	*	58	14
300	32	8	3	42	10	X	59	14	4	83	20	5	130	32	8	185	44	11	*	55	13	*	63	15	*	70	17
400	36	9	3	48	12	3	68	16	4	95	23	6	150	36	9	*	50	13	*	63	15	*	73	18	*	81	20

NOTES:

(1) X indicates that thickness of screening cargo is sufficient without any additional segregation distance.

(2) By using 2 m of intervening unit-density cargo for persons, and 3 m for films and plates, no distance shielding is necessary for any length of voyage specified.

(3) Using 1 steel bulkhead or steel deck – multiply segregation distance by 0.8.
Using 2 steel bulkheads or steel decks – multiply segregation distance by 0.64.

(4) "Cargo of unit density" means cargo stowed at a density of 1 tonne per cubic metre; where the density is less than this, the depth of cargo specified must be increased in proportion.

(5) "Minimum distance" means the least distance in any direction, whether vertical or horizontal, from the outer surface of the nearest package.

(6) The figures in the 300 and 400 rows shall be used in those cases where the appropriate provisions of this class permit the total transport index to exceed 200.

(7) Transport indices of packages, overpacks, freight containers and tanks, as appropriate.

* Not to be carried unless screening by other cargo and bulkheads can be arranged in accordance with the other columns.

7.2.9.10   **Rules for the use of the nomographs**

7.2.9.10.1   When there is no intervening cargo between the radioactive material and the persons or the undeveloped photographic film or plates, calculate the safe distance as follows:

.1   for persons – use the FG scales, read off the safe separation distance in metres (D_p) on the G scale adjacent to the sum of the transport indices (N) on the F scale; and

.2   for film and plates – draw a straight line between the length of the voyage (t), I scale, and the sum of the transport indices (N), F scale; separation distance in metres (D_f) will be the intersection on the H scale.

7.2.9.10.2   When there is intervening cargo between the radioactive material and the persons or undeveloped photographic film or plates, calculate the safe distance as follows:

.1   for persons – draw a straight line through the thickness of cargo (S) in metres, A scale, and the stowage factor (u), B scale, which is the cargo density, intersecting the CD scales. From this intersection draw another straight line through the value of the sum of the transport indices (as at 1 metre from the external surface), E scale, cutting the G scale at the safe separation distance figure (D_p); and

.2   for film and plates – as for persons, but from intersection on FG scales draw a straight line to the I scale; this line will cut the H scale at the separation distance for film and plates in metres (D_f).

Note: For thickness of cargo (S) up to 2.5 m, use the left of A scale and the left (lower) of B scale. For S between 2.5 m and 7.5 m, use the right of A scale and the right (or upper) of B scale. For S in excess of 7.5 m, divide both S and u by 10 and use the corresponding parts of A and B scales. When there is no intervening bulkhead, use the lower lines of B scale, for one bulkhead the middle lines, and for two bulkheads the top lines.

7.2.9.10.3   Other problems, such as estimating the minimum thickness of cargo or determining the stowage factor of intervening cargo when the thickness of the cargo is known, can also be solved by means of the nomographs.

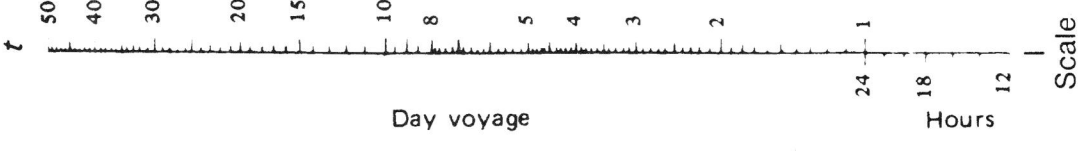

Day voyage   Hours

Minimum distance from undeveloped film or plates, in metres

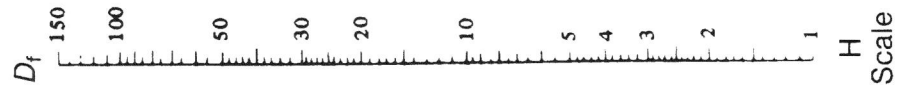

Minimum distance from persons, in metres

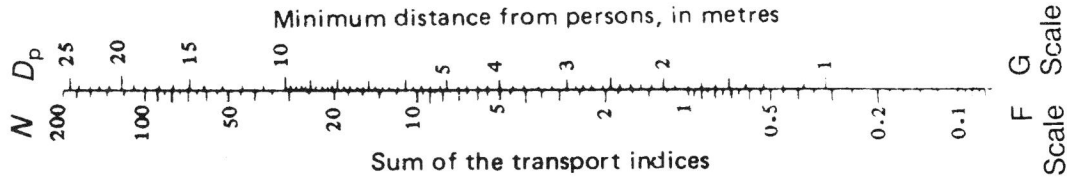

Sum of the transport indices

Sum of the transport indices (as at 1 metre from the external surface)

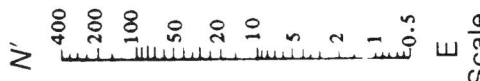

Relaxation values of radiation

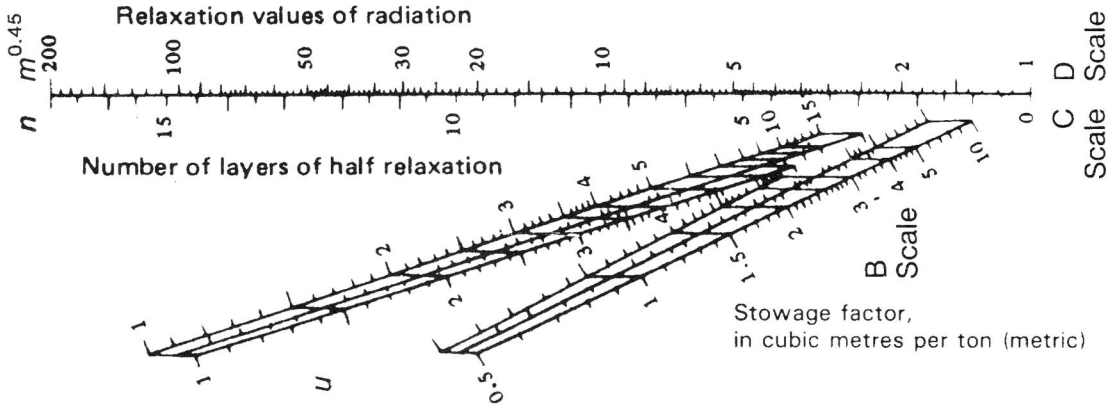

Number of layers of half relaxation

Stowage factor,
in cubic metres per ton (metric)

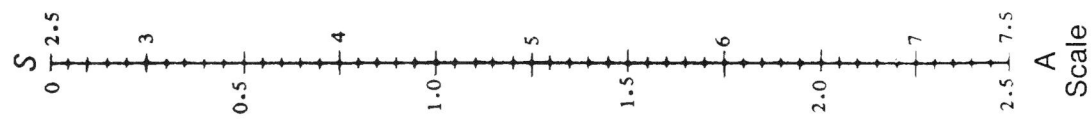

Thickness of cargo, in metres

00055

Chapter 7.3

Special provisions in the event of an incident and fire precautions involving dangerous goods

Note: The provisions of this chapter are not mandatory.

7.3.1   General

7.3.1.1   In the event of an incident involving dangerous goods, detailed recommendations are contained in *The EmS Guide: Emergency Response Procedures for Ships Carrying Dangerous Goods*.

7.3.1.2   In the event of personnel exposure during an incident involving dangerous goods, detailed recommendations are contained in *Medical First Aid Guide for Use in Accidents Involving Dangerous Goods (MFAG)*.

7.3.1.3   In the event that a package containing dangerous goods is found to be suffering from breakage or leakage while the ship is in port, the port authorities should be informed and appropriate procedures should be followed.

7.3.2   General provisions in the event of incidents

7.3.2.1   Recommendations on emergency action may differ depending on whether or not the goods are stowed *on deck* or *under deck* or whether a substance is gaseous, liquid or solid. When dealing with incidents involving flammable gases, or flammable liquids with a flashpoint of 60°C closed-cup (c.c.) or below, all sources of ignition (such as naked lights, unprotected light bulbs, electric handtools) should be avoided.

7.3.2.2   In general, the recommendation is to wash spillages *on deck* overboard with copious quantities of water and, where there is likely to be a dangerous reaction with water, from as far away as practicable. Disposal of spilt dangerous goods overboard is a matter for judgement by the master, bearing in mind that the safety of the crew has priority over pollution of the sea. If it is safe to do so, spillages and leakages of substances, articles and materials identified in this Code as MARINE POLLUTANT should be collected for safe disposal. Inert absorbent material should be used for liquids.

7.3.2.3   Toxic, corrosive and/or flammable vapours in *under deck* cargo spaces should, where possible, be dispersed before undertaking any emergency action. Where a mechanical ventilation system is used, care will be necessary to ensure that flammable vapours are not ignited.

7.3.2.4   If there is any reason to suspect leakage of these substances, entry into a hold or cargo space should not be permitted until the master or responsible officer has taken all safety considerations into account and is satisfied that it is safe to do so.

7.3.2.5   Emergency entry into the hold under other circumstances should only be undertaken by trained crew wearing self-contained breathing apparatus and other protective clothing.

7.3.2.6   A careful inspection for structural damage should be carried out after dealing with spillages of substances corrosive to steel and cryogenic liquids.

7.3.3   Special provisions for incidents involving infectious substances

7.3.3.1   If any person responsible for the transport or opening of packages containing infectious substances becomes aware of damage to or leakage from such packages, he should:

.1   avoid handling the package or keep handling to a minimum;

.2   inspect adjacent packages for contamination and put aside any that have been contaminated;

.3   inform the appropriate public health authority or veterinary authority, and provide information on any other countries of transit where persons may have been exposed to danger; and

.4   notify the consignor and/or the consignee.

7.3.3.2   Decontamination

A cargo transport unit, a bulk container or a cargo space of a ship, which has been used to transport infectious substances, shall be inspected for release of the substance before re-use. If infectious substances were released during transport, the cargo transport unit, the bulk container or the cargo space of a ship shall be decontaminated before it is re-used. Decontamination may be achieved by any means which effectively inactivates the infectious substance released.

7.3.4   Special provisions for incidents involving radioactive material

7.3.4.1   If it is evident that a package is damaged or leaking, or if it is suspected that the package may have leaked or been damaged, access to the package should be restricted and a qualified person should, as soon as possible, assess the extent of contamination and the resultant radiation level of the package. The scope of the assessment should include the package, the conveyance, the adjacent loading and unloading areas, and, if necessary, all other material which has been transported in the conveyance. When necessary, additional steps for the protection of persons, property and the environment, in accordance with provisions established by the relevant competent authority, should be taken to overcome and minimize the consequences of such leakage or damage.

7.3.4.2   Packages damaged or leaking radioactive contents in excess of allowable limits for normal conditions of transport may be removed to an acceptable interim location under supervision, but should not be forwarded until repaired or reconditioned and decontaminated.

7.3.4.3   In the event of accidents or incidents during the transport of radioactive material, emergency provisions, as established by relevant national and/or international organizations, should be observed to protect persons, property and the environment. Appropriate guidelines for such provisions are contained in the International Atomic Energy Agency's document "Planning and Preparing for Emergency Response to Transport Accidents involving Radioactive Material", Safety Standard Series No. TS-G-1.2 (ST-3), IAEA, Vienna (2002).

7.3.4.4   Attention is drawn to the latest versions of both *The EmS Guide: Emergency Response Procedures for Ships Carrying Dangerous Goods* and the *Medical First Aid Guide for Use in Accidents Involving Dangerous Goods (MFAG)*.

7.3.4.5   Emergency response procedures should take into account the formation of other dangerous substances that may result from the reaction between the contents of a consignment and the environment in the event of an accident.

7.3.4.6   In the event of a package containing radioactive material suffering from breakage or leakage while the ship is in port, the port authorities should be informed and advice obtained from them or from the competent authority.* Procedures have been drawn up in many countries for summoning radiological assistance in any such emergency.

7.3.5   General fire precautions

7.3.5.1   The prevention of fire in a cargo of dangerous goods is achieved by practising good seamanship, observing in particular the following precautions:

    .1   keep combustible material away from ignition sources;

    .2   protect a flammable substance by adequate packing;

    .3   reject damaged or leaking packages;

    .4   stow packages protected from accidental damage or heating;

    .5   segregate packages from substances liable to start or spread fire;

    .6   where appropriate and practicable, stow dangerous goods in an accessible position so that packages in the vicinity of a fire may be protected;

    .7   enforce prohibition of smoking in dangerous areas and display clearly recognizable "NO SMOKING" notices or signs; and

    .8   the dangers from short-circuits, earth leakages or sparking will be apparent. Lighting and power cables and fittings should be maintained in good condition. Cables or equipment found to be unsafe should be disconnected. Where a bulkhead is required to be suitable for segregation purposes, cables and conduit penetrations of the decks and bulkheads should be sealed against the passage of gas and vapours. When stowing dangerous goods on deck, the position and design of auxiliary machinery, electrical equipment and cable runs should be considered in order to avoid sources of ignition.

---

* Reference is made to chapter 7.9 and the IAEA list of national competent authorities responsible for approvals and authorizations in respect of the transport of radioactive material. The list is updated annually.

7

7.3.5.2 Fire precautions applying to individual classes, and where necessary to individual substances, are recommended in 7.3.2 and 7.3.6 to 7.3.9 and in the Dangerous Goods List.

7.3.6 Special fire precautions for class 1

7.3.6.1 .1 The greatest risk in the handling and transport of goods of class 1 is that of fire from a source external to the goods, and it is vital that any fire should be detected and extinguished before it can reach such goods. Consequently, it is essential that fire precautions, fire-fighting measures and equipment are of a high standard and ready for immediate application and use.

.2 Compartments containing goods of class 1 and adjacent cargo spaces should be provided with a fire-detection system. If such spaces are not protected by a fixed fire-extinguishing system, they should be accessible for fire-fighting operations.

.3 No repair work should be carried out in a compartment containing goods of class 1. Special care should be exercised in carrying out repairs in any adjacent space. No welding, burning, cutting, or riveting operations involving the use of fire, flame, spark, or arc-producing equipment should be carried out in any space other than machinery spaces and workshops where fire-extinguishing arrangements are available, except in any emergency and, if in port, with prior authorization of the port authority.

7.3.7 Special fire precautions for class 2

7.3.7.1 Effective ventilation should be provided to remove any leakage of gas from within the cargo space or spaces, bearing in mind that some gases are heavier than air and may accumulate in dangerous concentrations in the lower part of the ship.

7.3.7.2 Measures should be taken to prevent leaking gases from penetrating into any other part of the ship.

7.3.7.3 .1 If there is any reason to suspect leakage of a gas, entry into cargo spaces or other enclosed spaces should not be permitted until the master or responsible officer has taken all safety considerations into account and is satisfied that it is safe to do so. Emergency entry under other circumstances should only be undertaken by trained crew wearing self-contained breathing apparatus, and protective clothing when recommended, and always under the supervision of a responsible officer.

.2 Leakage from pressure receptacles containing flammable gases may give rise to explosive mixtures with air. Such mixtures, if ignited, may result in explosion and fire.

7.3.8 Special fire precautions for class 3

7.3.8.1 Flammable liquids give off flammable vapours which, especially in an enclosed space, form explosive mixtures with air. Such vapours, if ignited, may cause a "flashback" to the place in which the substances are stowed. Due regard should be paid to the provision of adequate ventilation to prevent accumulation of vapours.

7.3.9 Special fire precautions and fire fighting for class 7

7.3.9.1 The radioactive contents of Excepted, Industrial, and Type A packages are so restricted that, in the event of an accident and damage to the package, there is a high probability that any material released, or shielding efficiency lost, would not give rise to such radiological hazard as to hamper fire-fighting or rescue operations.

7.3.9.2 Type B(U) packages, Type B(M) packages and Type C packages are designed to be strong enough to withstand severe fire without significant loss of contents or dangerous loss of radiation shielding.

Chapter 7.4

Transport of cargo transport units on board ships

7.4.1 Applicability

7.4.1.1 The provisions of this chapter apply to the transport, loading and unloading of dangerous goods in cargo transport units on board ships.

7.4.2 General provisions for cargo transport units

7.4.2.1 Cargo transport units used for the transport of dangerous goods shall be of adequate strength to resist the possible stress imposed by the conditions of the services in which they are employed. They shall be adequately maintained.

7.4.2.2 Unless otherwise specified, the applicable provisions of the International Convention for Safe Containers (CSC) 1972, as amended, shall be followed for the use of any cargo transport unit which meets the definition of a "container" within the terms of that convention.

7.4.2.3 The International Convention for Safe Containers does not apply to offshore containers that are handled in open seas. The design and testing of offshore containers shall take into account the dynamic lifting and impact forces that may occur when a container is handled in open seas in adverse weather and sea conditions. The requirement for such containers shall be determined by the approving competent authority. Such provisions shall be based on MSC/Circ.860 "Guidelines for the approval of offshore containers handled in open seas". Such containers shall be clearly marked with the words "OFFSHORE CONTAINER" on the safety approval plate.

7.4.2.4 Loading of cargo transport units on board ships

7.4.2.4.1 Before loading, cargo transport units used for the transport of dangerous goods shall be examined for external signs of damage, leakage or sifting of contents. Any cargo transport unit found to be damaged, leaking or sifting shall not be accepted for shipment until repairs have been effected or damaged packages have been removed.

7.4.2.5 Ventilation* and condensation

7.4.2.5.1 Unless otherwise specified in this Code, the provisions concerning ventilation that are set out in various places in this Code shall be taken to refer to the space aboard the ship in which cargo transport units are stowed and shall not be interpreted to require ventilation into the cargo transport unit.

7.4.2.5.2 When, for any reason, it is necessary to open the doors of a cargo transport unit, the nature of the contents and the possibility that leakage may have caused an unsafe concentration of toxic or flammable vapours or have produced an oxygen-enriched or -depleted atmosphere shall be considered. If such a possibility exists, then the interior of the cargo transport unit shall be approached with caution.

7.4.2.5.3 Where class 4.3 substances are to be packed in a cargo transport unit, the possibility that the cargo transport unit could suffer from heavy condensation on the internal surface shall be kept in mind. The degree of such condensation is dependent upon the amount of moisture contained within the closed cargo transport unit, in addition to the temperature differences experienced. The risk is minimized if the moisture content of the packaging and securing materials is kept low.

---

* For cargo transport units under fumigation, see MSC.1/Circ.1265 "Recommendations on the safe use of pesticides in ships applicable to the fumigation of cargo transport units".

7.4.2.6   Heat protection

7.4.2.6.1   Where it is required that dangerous goods shall be kept as cool as practicable, this requirement shall be applied to the cargo transport unit as a whole.

> **Note**: The surface of a cargo transport unit can heat rapidly when in direct sunlight in nearly windless conditions and the cargo may also become heated.

7.4.3   Fumigated units

7.4.3.1   Cargo transport units under fumigation (fumigated units) shall be carried on board ships in accordance with the provisions of this Code relevant to the Proper Shipping Name FUMIGATED UNIT and UN number UN 3359 shown in the Dangerous Goods List in chapter 3.2. Particular transport conditions concerning UN 3359 are set out in special provision 910 in chapter 3.3.

7.4.3.2   A fumigated unit shall not be allowed on board until a sufficient period has elapsed to attain a reasonable uniform gas concentration throughout the cargo in it. Because of variations due to types and amounts of fumigants and commodities and temperature levels, the period between fumigant application and loading of the fumigated unit on board the ship shall be determined by the competent authority. Twenty-four hours is normally sufficient for this purpose. Unless the doors of a fumigated unit have been opened to allow the fumigant gas(es) and residues to be completely ventilated or the unit has been mechanically ventilated, the shipment shall conform to the provisions of this Code concerning UN 3359. Ventilated containers shall be marked with the date of ventilation on the fumigated warning sign(s). When the fumigated goods or materials have been unloaded, the fumigation warning sign(s) shall be removed.

7.4.3.3   The master shall be informed prior to the loading of a fumigated unit.

7.4.4   Stowage of cargo transport units in cargo spaces other than ro–ro cargo spaces

7.4.4.1   The following provisions shall apply to the stowage of cargo transport units on board ships in cargo spaces other than ro–ro cargo spaces:

   .1   A cargo transport unit packed or loaded with flammable gases or liquids having a flashpoint of less than 23°C c.c. shall only be stowed *under deck* together in the same cargo space with refrigerated or heated cargo transport units, the coolant or heating equipment of which could provide a possible source of ignition, if:

   –   the cooling compartment and the cooling or heating equipment of the cargo transport units comply with 7.7.3; and

   –   the design, construction and equipment of the cargo space complies with the provisions of regulation II-2/19 of SOLAS 74, as amended, or regulation II-2/54 of SOLAS 74, as amended by the resolutions indicated in II-2/1.2.1, as applicable;

   otherwise the stowage is restricted to *on deck only*.

   .2   A temperature-controlled cargo transport unit packed or loaded with flammable gases or liquids having a flashpoint of less than 23°C c.c. shall only be stowed *under deck* if the provisions under .1 above are met; otherwise the stowage is restricted to *on deck only*.

   .3   A cargo transport unit packed or loaded with flammable gas or flammable liquid having a flashpoint of less than +23°C c.c. transported *on deck* shall be stowed "away from" (as defined in 7.2.2.2.1) possible sources of ignition. In the case of container ships, a distance equivalent to one container space athwartships away from possible sources of ignition applied in any direction will satisfy this requirement.

7.4.4.2   Additional provisions for hatchless container holds

7.4.4.2.1   Dangerous goods shall only be transported in or vertically above hatchless container holds if:

   .1   the dangerous goods are permitted for *under deck* stowage as specified in the Dangerous Goods List; and

   .2   the hatchless container hold is in full compliance with the provisions of regulation II-2/19 of SOLAS 74, as amended, or regulation II-2/54 of SOLAS 74, as amended by the resolutions indicated in II-2/1.2.1, as applicable.

7.4.5   Stowage of cargo transport units in ro–ro cargo spaces

7.4.5.1   Loading and unloading operations on each vehicle deck shall take place under the supervision of either a working party consisting of officers and other crew members or responsible persons appointed by the master.

7.4.5.2   Passengers and other unauthorized persons shall be excluded from vehicle decks on which dangerous goods have been loaded. All doors leading directly to these decks shall be securely closed during the voyage and notices or signs prohibiting entrance to such decks shall be conspicuously displayed.

7.4.5.3   During the voyage, access to such decks by passengers and other unauthorized persons shall only be permitted when such persons are accompanied by an authorized crew member.

7.4.5.4   The transport of dangerous goods shall be prohibited on any vehicle deck on which the foregoing provisions cannot be met.

7.4.5.5   Closing arrangements for the openings between ro–ro cargo spaces and machinery and accommodation spaces shall be such as to avoid the possibility of dangerous vapours and liquids entering such spaces. Such openings shall normally be kept securely closed when dangerous cargo is on board, except to permit access by authorized persons or for emergency use.

7.4.5.6   Ro–ro ships may carry dangerous goods in cargo transport units or stowed in the conventional way on vehicle decks, in cargo holds or on weather decks. The provisions for such stowage shall be in compliance with the relevant provisions laid down elsewhere in this Code.

7.4.5.7   Dangerous goods required to be carried *on deck only* shall not be carried on closed vehicle decks, but may be carried on open vehicle decks when authorized by the competent authority concerned.

7.4.5.8   Flammable gases or liquids having a flashpoint of less than 23°C c.c. shall not be stowed in a closed ro–ro space or special category space unless:

–   the design, construction and equipment of the space comply with the provisions of regulation II-2/19 of SOLAS 74, as amended, or regulation II-2/54 of SOLAS 74, as amended by the resolutions indicated in II-2/1.2.1, as applicable, and the ventilation system is operated to maintain at least six air changes per hour; or

–   the ventilation system of the space is operated to maintain at least ten air changes per hour and non-certified safe electrical systems in the space are capable of being isolated by means other than removal of fuses in the event of failure of the ventilation system or any other circumstance likely to cause accumulation of flammable vapours.

Otherwise stowage is restricted to *on deck only*.

7.4.5.9   The provisions in this paragraph are without prejudice to relevant ventilation requirements of SOLAS 74, as amended.

In stowage conditions defined in 7.1.1, if continuous ventilation is impracticable in a closed ro–ro cargo space other than a special category space, ventilation fans shall be operated daily for a limited period, as weather permits. In any case, prior to discharge, the fans shall be operated for a reasonable period. The ro–ro cargo space shall be proved gas-free at the end of the period. When the ventilation is not continuous, electrical systems which are not certified safe shall be isolated.

7.4.5.10   Certain dangerous goods are required "to be stowed in a mechanically ventilated space". When such goods are transported in a closed ro–ro cargo space or a special category space, this space shall be mechanically ventilated.

7.4.5.11   Cargo transport units packed or loaded with flammable gases or liquids having a flashpoint of less than 23°C c.c. and transported on deck shall be stowed "away from" (as defined in 7.2.2.2.1.1) possible sources of ignition.

7.4.5.12   Mechanically operated refrigeration or heating equipment fitted to any cargo transport unit shall not be operated during the voyage when stowed in a closed ro–ro cargo space or a special category space.

7.4.5.13   Electrically operated refrigeration or heating equipment fitted to any cargo transport unit stowed in a closed ro–ro cargo space or special category space shall not be operated when flammable gases or liquids having a flashpoint of less than 23°C c.c. may be present in the cargo transport unit or in the same space, unless:

–   the design, construction and equipment of the space comply with the provisions of regulation II-2/19 of SOLAS 1974, as amended, or regulation II-2/54 of SOLAS 74, as amended by the resolutions indicated in II-2/1.2.1, as applicable, and the refrigeration or heating equipment of the cargo transport unit complies with paragraph 7.7.3; or

–   the ventilation system of the space is operated to maintain at least ten air changes per hour and all electrical systems in the space are capable of being isolated by means other than removal of fuses in the event of ventilation failure or other circumstance likely to cause accumulation of flammable vapours.

7

7.4.5.14 Stowage of portable tanks, road tank vehicles and railway tank wagons containing dangerous goods shall be in accordance with the provisions of the Dangerous Goods List and chapter 7.1

7.4.5.15 The master of a ship carrying dangerous goods on vehicle decks shall ensure that, during loading and unloading operations and during the voyage, regular inspections of these decks are made by an authorized crew member or responsible person in order to achieve early detection of any hazard.

7.4.6 Transport of dangerous goods of class 1 in cargo transport units

7.4.6.1 Special structural provisions may be applicable to cargo transport units used for the stowage of class 1 dangerous goods. The special provisions that are applicable are indicated under "stowage" in the Dangerous Goods List.

7.4.6.2 In ships other than specially fitted container ships, cargo transport units shall be stowed in the bottom layer only.

7.4.6.3 Loading and unloading cargo transport units packed with goods of class 1 on to a ship needs special care, and the precautions detailed in the IMO *Recommendations on the Safe Transport of Dangerous Cargoes and Related Activities in Port Areas* should be observed.

7.4.6.4 Structural serviceability of freight containers and vehicles packed with goods of class 1

7.4.6.4.1 Freight containers used for substances requiring magazine stowage type "A" shall be fitted with a close-boarded floor and shall have a non-metallic lining.

7.4.6.4.2 Freight containers and vehicles shall not be offered for the transport of goods of class 1 other than division 1.4 unless the container or the vehicle is structurally serviceable, as witnessed by a current International Convention for Safe Containers (CSC) approval plate (applicable to freight containers only) and a detailed visual examination, as follows:

.1 prior to packing a freight container or vehicle with goods of class 1, it shall be checked to ensure it is free of any residue of previous cargo and is structurally serviceable, and that the interior floor and walls are free from protrusions;

.2 "structurally serviceable" means the freight container or vehicle shall not have major defects in its structural components, e.g. top and bottom side rails, top and bottom end rails, door sill and header, floor cross-members, corner posts, and corner fittings in a freight container. Major defects are: dents or bends in the structural members greater than 19 mm in depth, regardless of length; cracks or breaks in structural members; more than one splice (e.g. a lapped splice) in top or bottom end rails or door headers; more than two splices in any one top or bottom side rail or any splice in a door sill or corner post; door hinges and hardware that are seized, twisted, broken, missing or otherwise inoperative; gaskets and seals that do not seal; or, for freight containers, any distortion of the overall configuration great enough to prevent proper alignment of handling equipment, mounting and securing on chassis or vehicle, or insertion into ship's cells;

.3 in addition, deterioration in any component of the freight container or vehicle, regardless of the material of construction, such as rusted-out metal in sidewalls or disintegrated fibreglass, is unacceptable. Normal wear, however, including oxidation (rust), slight dents and scratches and other damage that does not affect serviceability or the weathertight integrity of the units, is acceptable; and

.4 for free-flowing powdery substances of 1.1C, 1.1D, 1.1G, 1.3C and 1.3G and fireworks of 1.1G, 1.2G and 1.3G, the floor of the freight container shall have a non-metallic surface or covering.

Chapter 7.5

Packing of cargo transport units

7.5.1 General provisions for cargo transport units

7.5.1.1 Cargo transport units used for the transport of dangerous goods shall be of adequate strength to resist the possible stress imposed by the conditions of the services in which they are employed. They shall be adequately maintained.

7.5.1.2 Unless otherwise specified, the applicable provisions of the International Convention for Safe Containers (CSC) 1972, as amended, shall be followed for the use of any cargo transport unit which meets the definition of a "container" within the terms of that Convention.

7.5.1.3 The International Convention for Safe Containers does not apply to offshore containers that are handled in open seas. The design and testing of offshore containers shall take into account the dynamic lifting and impact forces that may occur when a container is handled in open seas in adverse weather and sea conditions. The requirement for such containers shall be determined by the approving competent authority. Such provisions should be based on MSC/Circ.860 "Guidelines for the approval of offshore containers handled in open seas". Such containers shall be clearly marked with the words "OFFSHORE CONTAINER" on the safety approval plate.

7.5.2 Packing of cargo transport units*

7.5.2.1 Packages shall be examined and any found to be damaged, leaking or sifting shall not be packed into a cargo transport unit. Care shall be taken to see that excessive water, snow, ice or foreign matter adhering to packages is removed before packing into a cargo transport unit.

7.5.2.2 Packaged dangerous goods and any other goods within the same cargo transport unit shall be tightly packed and adequately braced and secured for the voyage. The packages shall be packed in such a way that there will be a minimum likelihood of damage to fittings during transport. Such fittings on packages shall be adequately protected.

7.5.2.3 When a dangerous goods consignment forms only part of the load of a cargo transport unit, it shall, preferably, be packed so as to be accessible (such as packing near the doors of the cargo transport unit).

7.5.2.4 If the doors of a cargo transport unit are locked, the means of locking shall be such that, in cases of emergency, the doors can be opened without delay.

7.5.2.5 Before being packed, cargo transport units shall be examined visually for damage, and if there is evidence of material damage the cargo transport unit shall not be packed.

7.5.2.6 Irrelevant markings, labels, placards, orange panels, signs and marine pollutant marks shall be removed or masked before packing a cargo transport unit.

7.5.2.7 Those responsible for the packing of dangerous goods into a cargo transport unit shall provide a "container/vehicle packing certificate": see chapter 5.4. This document is not required for tanks.

7.5.2.8 Cargo transport units shall be loaded so that the cargo is uniformly distributed consistent with the referenced guidelines.*

7.5.3 Empty cargo transport units

7.5.3.1 After a cargo transport unit carrying dangerous goods has been unpacked or unloaded, precautions shall be taken to ensure that there is no contamination likely to make the cargo transport unit dangerous.

7.5.3.2 After unpacking or unloading corrosive substances, particular attention shall be paid to cleaning, as residues may be highly corrosive to the metal structures.

---

* See *IMO/ILO/UNECE Guidelines for Packing of Cargo Transport Units.*

7

Chapter 7.6

Transport of dangerous goods in shipborne barges on barge-carrying ships

7.6.1 Applicability

7.6.1.1 Because of the structural differences of barge-carrying ships from other ships, this chapter contains alternative and special or additional provisions for the transport of packaged dangerous goods or solid bulk materials possessing chemical hazards on these ships.

7.6.1.2 The provisions of this chapter are applicable to shipborne barges which carry packaged dangerous goods or solid bulk materials possessing chemical hazards while aboard barge-carrying ships or barge feeder vessels.

7.6.1.3 The provisions of this chapter are not intended to apply to shipborne barges designed for transport aboard barge-carrying ships while the shipborne barges are operating independently of the barge-carrying ship.

7.6.1.4 Barges used for the shipborne transport of packaged dangerous goods or solid bulk materials possessing chemical hazards shall be of proper design and adequate strength to resist the stresses imposed by the conditions of the services in which they are employed and they shall be adequately maintained. Shipborne barges shall be approved in accordance with provisions for certification of a recognized classification society, or any organization approved by and acting on behalf of the competent authority of the countries concerned.

7.6.1.5 Except as otherwise specified in this chapter, all the provisions laid down for each substance in this Code shall apply to the transport of packaged dangerous goods or solid bulk materials possessing chemical hazards in shipborne barges on barge-carrying ships.

7.6.1.6 The provisions of this chapter apply only to shipborne barges of steel construction. Packaged dangerous goods or solid bulk materials possessing chemical hazards shall only be transported in shipborne barges constructed of other materials, including the hatch covers, under conditions specified by the competent authority concerned.

7.6.2 Definitions

7.6.2.1 *Loading and unloading*, for the purpose of this chapter, means the placement or removal of cargo into or out of a shipborne barge.

7.6.2.2 *Stowage* means, for the purposes of this chapter, the placement of a shipborne barge aboard the barge-carrying ship or barge feeder vessel.

7.6.3 Permitted shipments

7.6.3.1 Packaged dangerous goods or solid bulk materials possessing chemical hazards shall only be transported aboard shipborne barges on barge-carrying ships when they are packaged in accordance with chapter 4 except as provided in 7.6.4.2, 7.6.4.3 and 7.6.4.4 below.

7.6.3.2 Portable tanks (tank containers) containing liquid dangerous goods in bulk aboard shipborne barges shall comply with the applicable provisions of chapter 4.

7.6.3.3 Certain dry dangerous goods in bulk may be transported in shipborne barges; this is indicated in the packing instructions in chapter 4.

7.6.3.4 Because of their particular hazard, certain commodities may not be shipped or may only be shipped in shipborne barges aboard barge-carrying ships under conditions specified in this Code or by the competent authority concerned after taking due account of the circumstances of the intended voyage.

7.6.4 Barge loading

7.6.4.1 Packages shall be examined and any found to be damaged, leaking or sifting shall not be loaded into a shipborne barge. Care shall be taken to ensure that excessive water, snow, ice or foreign matter adhering to packages shall be removed before loading into a shipborne barge.

7.6.4.2 Packages containing dangerous goods, portable tanks, cargo transport units and any other goods within a shipborne barge shall be properly immobilized by stowage and adequately braced and secured for the voyage. Packages shall be loaded in such a way that there will be a minimum likelihood of damage to them and to any fittings during transport. Fittings on packages or portable tanks (tank containers) shall be adequately protected.

7.6.4.3 Where solid bulk materials possessing chemical hazards are transported in shipborne barges, it shall be ensured that at all times the cargo is evenly distributed, properly trimmed and secured.

7.6.4.4 Shipborne barges into which packaged dangerous goods or solid bulk materials possessing chemical hazards are to be loaded shall be examined visually for hull or hatch cover damage which could impair watertight integrity. If there is evidence of such damage, the shipborne barge may not be used for the transport of packaged dangerous goods or solid bulk materials possessing chemical hazards and shall not be loaded.

7.6.4.5 For segregation on shipborne barges and on board barge-carrying ships, see 7.2.5.

7.6.5 Stowage of shipborne barges

7.6.5.1 Stowage of shipborne barges carrying packaged dangerous goods or solid bulk materials possessing chemical hazards aboard barge-carrying ships shall be as required for the substance in the Dangerous Goods List in this Code. When a shipborne barge is loaded with more than one substance, and the stowage locations differ for the substances (i.e. some substances require *on deck* stowage while other substances require *under deck* stowage), the shipborne barge containing these substances shall be stowed on deck.

7.6.6 Ventilation and condensation

7.6.6.1 The provisions concerning ventilation that are specified for various substances or materials in this Code shall be taken to apply to the cargo in the shipborne barge in which that substance or material is loaded.

7.6.6.2 Provision shall be made to ensure that shipborne barges stowed under deck and loaded with cargoes requiring ventilation because of their dangerous nature are ventilated to the extent necessary.

7.6.6.3 Where class 4.3 substances or materials hazardous only in bulk (MHB only)* having similar properties or being subject to the same segregation provisions or substances liable to spontaneous heating are transported in shipborne barges, the possibility that the shipborne barges could suffer from heavy condensation on the internal surface shall be kept in mind. The degree of such condensation is dependent upon the amount of moisture contained within a closed shipborne barge, in addition to the temperature variances experienced. The risk is minimized if the moisture content of packagings and securing materials is kept low.

7.6.6.4 When, for any reason, it is necessary to remove the hatch cover from a shipborne barge, the nature of the contents and the possibility that leakage may have caused an unsafe concentration of toxic or flammable vapour or have produced an oxygen-rich or oxygen-depleted atmosphere shall be considered.

7.6.6.5 Shipborne barges containing a residue of a dangerous cargo or shipborne barges loaded with empty packagings still containing a residue of a dangerous substance shall comply with the same provisions as barges loaded with the substance itself.

7.6.6.6 For barges containing solid goods under fumigation, see 7.4.3.

7.6.7 Fire protection

7.6.7.1 Shipborne barges loaded with significant quantities of packaged dangerous goods or solid bulk materials possessing chemical hazards shall be stowed as far as practicable from accommodation and navigational areas.

7.6.7.2 Where it is recommended that a cargo shall be kept as cool as practicable, this provision shall be applied to the shipborne barge as a whole, unless suitable alternative measures are provided.

---

* Reference is made to the Code of Safe Practice for Solid Bulk Cargoes, 2004, as may be amended.

7.6.7.3 When packaged dangerous goods or solid bulk materials possessing chemical hazards are loaded in shipborne barges aboard barge-carrying ships having the capability of providing fixed fire-fighting systems or fire-detection systems to individual barges, care shall be taken to ensure that these systems are attached to the shipborne barge and operating properly.

7.6.7.4 When packaged dangerous goods or solid bulk materials possessing chemical hazards are loaded in shipborne barges aboard barge-carrying ships having fixed fire-fighting systems or fire-detection systems installed in individual barge holds, care shall be taken to ensure that the ventilation closures on the shipborne barges are open to permit the fire-fighting medium to enter the barges in case of fire.

7.6.7.5 When ventilation ducts are provided to individual shipborne barges, the ventilation fans shall be secured when fire-fighting medium is introduced into the hold to permit the medium to enter the shipborne barges.

7.6.8 Transport of goods of class 1 in shipborne barges

7.6.8.1 General stowage provisions for goods of class 1 are given in 7.1.7.3. Stowage arrangements under deck and on deck are described in 7.1.7.4 and 7.1.7.5 respectively.

7.6.8.2 Fixed magazines may be built within a shipborne barge. Cargo transport units may also be used as magazines within such a barge.

7.6.8.3 Shipborne barges may be used for the transport of all types of goods of class 1. When carrying those requiring special stowage, the following shall apply:

.1 goods in compatibility group G or H shall be in cargo transport units unless other arrangements are approved by the competent authority; and

.2 goods in compatibility group K or L shall be in steel magazines at all times.

7.6.8.4 Goods of different compatibility groups in class 1 may not be stowed within the same shipborne barge unless 7.2.7.2.1 and 7.2.7.2.2 would permit them to be stowed together.

Chapter 7.7

Temperature control provisions

7.7.1 Preamble

7.7.1.1 If the temperature of certain substances (such as organic peroxides and self-reactive substances) exceeds a value which is typical of the substance as packaged for transport, a self-accelerating decomposition, possibly of explosive violence, may result. To prevent such decomposition, it is necessary to control the temperature of such substances during transport. Other substances not requiring temperature control for safety reasons may be transported under controlled temperature conditions for commercial reasons.

7.7.1.2 The provisions for the temperature control of certain specified substances are based on the assumption that the temperature in the immediate surroundings of the cargo does not exceed 55°C during transport and attains this value for a relatively short time only during each period of 24 hours.

7.7.1.3 If a substance which is not normally temperature-controlled is transported under conditions where the temperature may exceed 55°C, it may require temperature control; in such cases, adequate measures shall be taken.

7.7.2 General provisions

7.7.2.1 A self-accelerating decomposition temperature (SADT)* shall be determined in order to decide if a substance shall be subjected to temperature control during transport. The relationship between SADT, the control temperature and the emergency temperature is as follows:

Type of receptacle	SADT*	Control temperature	Emergency temperature
Single packagings and IBCs	20°C or less over 20°C to 35°C over 35°C	20°C below SADT 15°C below SADT 10°C below SADT	10°C below SADT 10°C below SADT 5°C below SADT
Portable tanks	<50°C	10°C below SADT	5°C below SADT

7.7.2.2 The substances for which a control temperature and an emergency temperature are indicated in 2.4.2.3.2.3 or 2.5.3.2.4 shall be transported under conditions of temperature control such that the temperature of the immediate surroundings of the cargo does not exceed the control temperature.

7.7.2.3 The actual transport temperature may be lower than the control temperature but shall be selected so as to avoid dangerous separation of phases.

7.7.2.4 During transport, the temperature (see 7.7.3) shall be monitored at regular intervals (at least once every four to six hours) and the temperature readings shall be logged. If, during transport, the control temperature is exceeded, an alerting procedure shall be initiated involving either repair of the refrigeration machinery or an increase in the cooling capacity (such as by adding liquid or solid refrigerants). If an adequate cooling capacity is not restored, emergency procedures shall be started.

7.7.2.5 The stowage of the cargo shall be such as to ensure that, if disposal is necessary at sea, the packages or closed cargo transport unit can be jettisoned[†] with reasonable safety.

7.7.2.6 The refrigeration system shall be subjected to a thorough inspection and a test prior to the cargo transport unit being packed to ensure that all parts are functioning properly.

---

* The self-accelerating decomposition temperature (SADT) shall be determined in accordance with the latest version of the United Nations *Recommendations on the Transport of Dangerous Goods, Manual of Tests and Criteria*. Test methods for determining flammability are given in Part III, 32.4 of the United Nations *Manual of Tests and Criteria*. Because organic peroxides may react vigorously when heated, it is recommended to determine their flashpoint using small sample sizes such as described in ISO 3679.

[†] See also Assembly resolution A.851(20), General principles for ship reporting systems and ship reporting requirements, including guidelines for reporting incidents involving dangerous goods, harmful substances and/or marine pollutants.

7.7.2.7 When a cargo transport unit is to be filled with packages containing substances having different control temperatures, all packages shall be pre-cooled to avoid exceeding the lowest control temperature.

7.7.2.7.1 In the event that non-temperature-controlled substances are transported in the same cargo transport unit as temperature-controlled substances, the package(s) containing substances that require refrigeration shall be stowed in such a way as to be readily accessible from the door(s) of the cargo transport unit.

7.7.2.7.2 If substances with different control temperatures are loaded in the cargo transport unit, the substances with the lowest control temperature shall be stowed in the most readily accessible position from the doors of the cargo transport unit.

7.7.2.7.3 The door(s) shall be capable of being opened readily in case of emergency so that the package(s) can be removed. The carrier shall be informed about the location of the different substances within the unit. The cargo shall be secured to prevent packages from falling when the door(s) is (are) opened. The packages shall be securely stowed so as to allow for adequate air circulation throughout the cargo.

7.7.2.8 The master shall be provided with operating instructions for the refrigeration system, procedures to be followed in the event of loss of control and instructions for regular monitoring of operating temperatures. Spare parts shall be carried for the systems described in 7.7.3.2.3 and 7.7.3.2.4 so that they are available for emergency use should the refrigeration system malfunction during transport.

7.7.2.9 In cases where it may not be possible to carry specific substances acording to the general provisions, full details of the proposed method of shipment shall be submitted to the competent authority concerned for approval.

7.7.3 Methods of temperature control

7.7.3.1 The suitability of a particular means of temperature control for transport depends on a number of factors. Among those to be considered are:

.1 the control temperature(s) of the substance(s) to be transported;

.2 the difference between the control temperature and the anticipated ambient temperature conditions;

.3 the effectiveness of the thermal insulation of the cargo transport unit. The overall heat-transfer coefficient shall not be more than 0.4 W/(m^2·K) for cargo transport units and 0.6 W/(m^2·K) for tanks; and

.4 the duration of the voyage.

7.7.3.2 Suitable methods for preventing the control temperature being exceeded are, in order of increasing capability:

.1 thermal insulation, provided that the initial temperature of the substance is sufficiently below the control temperature;

.2 thermal insulation with a coolant system, provided that:

– an adequate quantity of non-flammable coolant (such as liquid nitrogen or solid carbon dioxide), allowing a reasonable margin for delay, is carried;

– liquid oxygen or air is not used as a coolant;

– there is uniform cooling effect even when most of the coolant has been consumed; and

– the need to ventilate the cargo transport unit before entering is clearly indicated by a warning on the door(s);

.3 single mechanical refrigeration, provided that the unit is thermally insulated and, for substances with a flashpoint lower than the sum of the emergency temperature plus 5°C, explosion-proof electrical fittings are used within the cooling compartment to prevent ignition of flammable vapours from the substances;

.4 combined mechanical refrigeration system and coolant system, provided that:

– the two systems are independent of one another; and

– the provisions of 7.7.3.2.2 and 7.7.3.2.3 are met;

.5 dual mechanical refrigeration system, provided that:

– apart from the integral power supply unit, the two systems are independent of one another;

– each system alone is capable of maintaining adequate temperature control; and

– for substances with a flashpoint lower than the sum of the emergency temperature plus 5°C, explosion-proof electrical fittings are used within the coolant compartment to prevent ignition of flammable vapours from the substances.

7.7.3.3 The refrigeration equipment and its controls shall be readily and safely accessible and all electrical connections weatherproof. Inside the cargo transport unit, the temperature shall be measured continuously. The measurement shall be taken in the air-space of the unit, using two measuring devices independent of each other. The type and place of the measuring devices shall be selected so that their results are representative of the actual temperature in the cargo. At least one of the two measurements shall be recorded in such a manner that temperature changes are easily detectable.

7.7.3.4 If substances are transported with a control temperature less than +25°C, the cargo transport unit shall be equipped with a visible and audible alarm effectively set at no higher than the control temperature. The alarms shall work independently from the power supply of the refrigeration system.

7.7.3.5 If an electrical supply is necessary for the cargo transport unit to operate the refrigeration or heating equipment, it shall be ensured that the correct connecting plugs are fitted. For under-deck stowage, plugs shall, as a minimum, be of an IP 55 enclosure in accordance with IEC Publication 529,* with the specification for electrical equipment of temperature class T4 and explosion group IIB. However, when stowed on deck, these plugs shall be of an IP 56 enclosure in accordance with IEC Publication 529.*

7.7.4 Special provisions for self-reactive substances (class 4.1) and organic peroxides (class 5.2)

7.7.4.1 For self-reactive substances (class 4.1) identified by UN Nos. 3231 and 3232 and organic peroxides (class 5.2) identified by UN Nos. 3111 and 3112, one of the following methods of temperature control described in 7.7.3.2 shall be used:

.1 the methods referred to under 7.7.3.2.4 or 7.7.3.2.5; or

.2 the method referred to under 7.7.3.2.3 when the maximum ambient temperature to be expected during transport is at least 10°C below the control temperature.

7.7.4.2 For self-reactive substances (class 4.1) identified by UN Nos. 3233 to 3240 and organic peroxides (class 5.2) identified by UN Nos. 3113 to 3120, one of the following methods shall be used:

.1 the methods referred to under 7.7.3.2.4 or 7.7.3.2.5;

.2 the method referred to under 7.7.3.2.3 when the maximum ambient temperature to be expected during transport does not exceed the control temperature by more than 10°C; or

.3 for short international voyages only (see 1.2.1), the methods referred to under 7.7.3.2.1 and 7.7.3.2.2 when the maximum ambient temperature to be expected during transport is at least 10°C below the control temperature.

7.7.5 Special provisions applicable to the transport of substances stabilized by temperature control (other than self-reactive substances and organic peroxides)

7.7.5.1 These provisions apply to the transport of substances:

.1 the Proper Shipping Name of which contains the word "STABILIZED"; and

.2 for which the SADT (see 7.7.2.1) as presented for transport in the package, IBC or tank is 50°C or lower.

When chemical inhibition is not used to stabilize a reactive substance which may generate dangerous amounts of heat and gas, or vapour, under normal transport conditions, these substances shall be transported under temperature control. These provisions do not apply to substances which are stabilized by the addition of chemical inhibitors such that the SADT is greater than 50°C.

7.7.5.2 The provisions in 7.7.2.1 to 7.7.2.3 and 7.7.3 apply to substances meeting criteria .1 and .2 in 7.7.5.1.

7.7.5.3 The actual transport temperature may be lower than the control temperature (see 7.7.2.1) but shall be selected so as to avoid dangerous separation of phases.

7.7.5.4 When these substances are transported in IBCs or portable tanks, the provisions for a SELF-REACTIVE LIQUID TYPE F, TEMPERATURE CONTROLLED shall apply. For transport in IBCs, see the special provisions in 4.1.7.2 and the "additional provisions" in packing instruction IBC520; for transport in portable tanks, see the additional provisions in 4.2.1.13.

---

* Reference is made to the Recommendations published by the International Electrotechnical Commission (IEC) and, in particular, to Publication 529 – *Classification of Degrees of Protection Provided by Enclosures.*

7

7.7.5.5 If a substance the Proper Shipping Name of which contains the word "STABILIZED" and which is not normally required to be transported under temperature control is transported under conditions where the temperature may exceed 55°C, it may require temperature control.

7.7.6 Special provisions for flammable gases or liquids having a flashpoint of less than 23°C c.c. transported under temperature control

7.7.6.1 When flammable gases or liquids having a flashpoint of less than 23°C c.c. are packed or loaded in a cargo transport unit equipped with a refrigerating or heating system, the cooling or heating equipment shall comply with 7.7.3.

7.7.6.2 When flammable liquids having a flashpoint of less than 23°C c.c. and not requiring temperature control for safety reasons are transported under temperature control conditions for commercial reasons, explosion-proof electrical fittings are not required, when the substances are pre-cooled to and transported at a control temperature of at least 10°C below the flashpoint. In case of failure of the refrigerating system, the system shall be disconnected from the power supply.

7.7.7 Special provisions for vehicles transported on ships

7.7.7.1 Insulated, refrigerated and mechanically refrigerated vehicles shall conform to the provisions of 7.7.3 and 7.7.4 or 7.7.5, as appropriate. In addition, the refrigerating appliance of a mechanically refrigerated vehicle shall be capable of operating independently of the engine used to propel the vehicle.

7.7.8 Approval

7.7.8.1 The competent authority may approve that less stringent means of temperature control may be used or that artificial refrigeration may be dispensed with under conditions of transport such as short international voyages or low ambient temperatures.

Chapter 7.8

Transport of wastes

7.8.1 Preamble

Wastes, which are dangerous goods, shall be transported in accordance with the relevant international recommendations and conventions and, in particular, where it concerns transport by sea, with the provisions of this Code.

7.8.2 Applicability

7.8.2.1 The provisions of this chapter are applicable to the transport of wastes by ships and shall be considered in conjunction with all other provisions of this Code.

7.8.2.2 Substances, solutions, mixtures or articles containing or contaminated with radioactive material are subject to the applicable provisions for radioactive material in class 7, and are not to be considered as wastes for the purposes of this chapter.

7.8.3 Transboundary movements under the Basel Convention*

7.8.3.1 Transboundary movement of wastes is permitted to commence only when:

.1 notification has been sent by the competent authority of the country of origin, or by the generator or exporter through the channel of the competent authority of the country of origin, to the country of final destination; and

.2 the competent authority of the country of origin, having received the written consent of the country of final destination stating that the wastes will be safely incinerated or treated by other methods of disposal, has given authorization to the movement.

7.8.3.2 In addition to the transport document required in chapter 5.4, all transboundary movements of wastes shall be accompanied by a waste movement document from the point at which a transboundary movement commences to the point of disposal. This document shall be available at all times to the competent authorities and to all persons involved in the management of waste transport operations.

7.8.3.3 The transport of solid wastes in bulk in cargo transport units and road vehicles is only permitted with the approval of the competent authority of the country of origin.

7.8.3.4 In the event that packages and cargo transport units containing wastes are suffering from leakage or spillage, the competent authorities of the countries of origin and destination shall be immediately informed and advice on the action to be taken obtained from them.

7.8.4 Classification of wastes

7.8.4.1 A waste containing only one constituent which is a dangerous substance subject to the provisions of this Code shall be regarded as being that particular substance. If the concentration of the constituent is such that the waste continues to present a hazard inherent in the constituent itself, it shall be classified according to the criteria of the applicable classes.

7.8.4.2 A waste containing two or more constituents which are dangerous substances subject to the provisions of this Code shall be classified under the applicable class in accordance with their dangerous characteristics and properties as described in 7.8.4.3 and 7.8.4.4.

---

* Basel Convention on the Control of Transboundary Movements of Hazardous Wastes and their Disposal (1989).

7

7.8.4.3 The classification according to the dangerous characteristics and properties shall be carried out as follows:

.1 determination of the physical and chemical characteristics and physiological properties by measurement or calculation followed by classification according to the criteria of the applicable class(es); or

.2 if the determination is not practicable, the waste shall be classified according to the constituent presenting the predominant hazard.

7.8.4.4 In determining the predominant hazard, the following criteria shall be taken into account:

.1 if one or more constituents fall within a certain class and the waste presents a hazard inherent in these constituents, the waste shall be included in that class; or

.2 if there are constituents falling under two or more classes, the classification of the waste shall take into account the order of precedence applicable to dangerous substances with multiple hazards set out in 2.0.3.

7.8.4.5 Wastes harmful to the marine environment only shall be transported under the class 9 entries for ENVIRONMENTALLY HAZARDOUS SUBSTANCE, LIQUID, N.O.S., UN 3082, or ENVIRONMENTALLY HAZARDOUS SUBSTANCE, SOLID, N.O.S., UN 3077, with the addition of the word "WASTE". However, this is not applicable to substances which are covered by individual entries in this Code.

7.8.4.6 Wastes not otherwise subject to the provisions of this Code but covered under the Basel Convention may be transported under the class 9 entries for ENVIRONMENTALLY HAZARDOUS SUBSTANCE, LIQUID, N.O.S., UN 3082 or ENVIRONMENTALLY HAZARDOUS SUBSTANCE, SOLID, N.O.S., UN 3077.

Chapter 7.9

Exemptions, approvals and certificates

7.9.1   Exemptions

Note 1   The provisions of this section do not apply to exemptions mentioned in chapters 1 to 7.8 of this Code (e.g. exemptions for limited quantities in 3.4.7) and to approvals (including permits, authorizations or agreements) and certificates which are referred to in chapters 1 to 7.8 of this Code. For the said approvals and certificates, see 7.9.2.

Note 2   The provisions of this section do not apply to class 7. For consignments of radioactive material for which conformity with any provision of this Code applicable to class 7 is impracticable, refer to 1.5.4.

7.9.1.1   Where this Code requires that a particular provision for the transport of dangerous goods shall be complied with, a competent authority or competent authorities (port State of departure, port State of arrival or flag State) may authorize any other provision by exemption if satisfied that such provision is at least as effective and safe as that required by this Code. Acceptance of an exemption authorized under this section by a competent authority not party to it is subject to the discretion of that competent authority. Accordingly, prior to any shipment covered by the exemption, the recipient of the exemption shall notify other competent authorities concerned.

7.9.1.2   Competent authority or competent authorities which have taken the initiative with respect to the exemption:

    .1   shall send a copy of such exemption to the International Maritime Organization which shall bring it to the attention of the Contracting Parties to SOLAS and/or MARPOL, as appropriate, and

    .2   if appropriate, take action to amend the IMDG Code to include the provisions covered by the exemption.

7.9.1.3   The period of validity of the exemption shall be not more than five years from the date of authorization. An exemption that is not covered under 7.9.1.2.2 may be renewed in accordance with the provisions of this section.

7.9.1.4   A copy of the exemption shall accompany each consignment when offered to the carrier for transport under the terms of the exemption. A copy of the exemption or an electronic copy thereof shall be maintained on board each ship transporting dangerous goods in accordance with the exemption, as appropriate.

7.9.2   Approvals (including permits, authorizations or agreements) and certificates

7.9.2.1   Approvals, including permits, authorizations or agreements, and certificates referred to in chapters 1 to 7.8 of this Code and issued by the competent authority (authorities when the Code requires a multilateral approval) or a body authorized by that competent authority (e.g. approvals for alternative packaging in 4.1.3.7, approval for segregation as in 7.2.2.3 or certificates for portable tanks in 6.7.2.18.1) shall be recognized, as appropriate:

    .1   by other contracting parties to SOLAS if they comply with the requirements of the International Convention for the Safety of Life at Sea (SOLAS), 1974, as amended; and/or

    .2   by other contracting parties to MARPOL if they comply with the requirements of the International Convention for the Prevention of Pollution from Ships, 1973, as modified by the Protocol of 1978 relating thereto (MARPOL 73/78, Annex III), as amended.

7.9.3   Contact information for the main designated national competent authorities

Contact information for the main designated national competent authorities concerned is given in this paragraph.[*] Corrections to these addresses should be sent to the Organization.[†]

---

[*] Reference is made to MSC.1/Circ.1254#, as may be amended, which provides a more comprehensive listing of contact information for competent authorities and bodies.

[†] International Maritime Organization
  4 Albert Embankment
  London SE1 7SR
  United Kingdom
  Email: info@imo.org
  Fax: +44 207587 3120

7

LIST OF CONTACT INFORMATION FOR THE MAIN
DESIGNATED NATIONAL COMPETENT AUTHORITIES

Country	Contact information for the main designated national competent authority
ALGERIA	Ministère des Transports/Direction de la Marine Marchande 119 Rue Didouche Mourad Alger ALGÉRIE Telephone: +213 26061 46 Telex: 66063 DGAF DZ
AMERICAN SAMOA	Silila Patane Harbour Master Port Administration Pagopago American Samoa AMERICAN SAMOA 96799
ANGOLA	National Director Marine Safety, Shipping and Ports National Directorate of Merchant Marine and Ports Rua Rainha Ginga 74, 4 Andar Luanda ANGOLA Telephone: +244 239 0034/397 984 Fax: +244 231 0375 Mobile: +244 924 393 36 Email: ispscode_angola@snet.co.ao
ARGENTINA	Prefectura Naval Argentina (Argentine Coast Guard) Dirección de protección ambiental Departamento de protección ambiental y mercancías peligrosas Division mercancías y residuos peligrosos Avda. Eduardo Madero 235 4° piso, Oficina 4.36 y 4.37 Buenos Aires (C1106ACC) REPÚBLICA ARGENTINA Telephone: +54 11 4318 7669 Fax: +54 11 4318 7474 Email: dpma-mp@prefecturanaval.gov.ar
AUSTRALIA	Manager, Ship Inspection Maritime Operations Australian Maritime Safety Authority GPO Box 2181 Canberra ACT 2601 AUSTRALIA Telephone: +61 2 6279 5048 Fax: +61 2 6279 5058 Email: psc@amsa.gov.au Website: www.amsa.gov.au
BAHAMAS	The Director Bahamas Maritime Authority 120 Old Broad Street London, EC2N 1AR UNITED KINGDOM Telephone: +44 (0)20 7562 1300 Fax: +44 (0)20 7614 0650 Website: www.bahamasmaritime.com

Country	Contact information for the main designated national competent authority
BARBADOS	Director of Maritime Affairs Ministry of Tourism and International Transport 2nd Floor, Carlisle House Hincks Street Bridgetown St. Michael BARBADOS Telephone: +1 246 426 2710/3342 Fax: +1 246 426 7882 Email: ctech@sunbeach.net
BELGIUM	Federal Public Service Mobility and Transport Directorate-General Maritime Transport Rue de Progrès 56 B-1210 Brussels BELGIUM Telephone: +32 2 277 3500 Fax: +32 2 277 4051 Email: dg.mar@mobilit.fgov.be Website: www.mobilit.fgov.be
BELIZE	Ports Commissioner Belize Port Authority P.O. Box 633 Belize City BELIZE, C.A. Telephone: +501 227 2540/0981 Fax: +501 227 2500
BRAZIL	Diretoria de Portos e Costas (DPC-20) Rua Teófilo Otoni No. 04 Centro Rio de Janeiro CEP 20090-070 BRAZIL Telephone: +55 21 2104 5203 Fax: +55 21 2104 5202 Email: secom@dpc.mar.mil.br
BULGARIA	*Head office:* Captain Petar Petrov, Director Directorate "Quality Management" Bulgarian Maritime Administration 9 Dyakon Ignatii Str. Sofia 1000 REPUBLIC OF BULGARIA Telephone: +359 2 93 00 910 / 93 00 912 Fax: +359 2 93 00 920 Email: bma@marad.bg petrov@marad.bg *Regional offices:* Harbour-Master Directorate "Maritime Administration" – Bourgas 3 Kniaz Alexander Batemberg Str. Bourgas 8000 REPUBLIC OF BULGARIA Telephone: +359 56 875 775 Fax: +359 56 840 064 Email: hm_bs@marad.bg

7

Country	Contact information for the main designated national competent authority
BULGARIA *(continued)*	Harbour-Master Directorate "Maritime Administration" – Varna 5 Primorski Bvd Varna 9000 REPUBLIC OF BULGARIA Telephone: +359 52 684 922 Fax: +359 52 602 378 Email: hm_vn@marad.bg
BURUNDI	Minister Ministère des Transports, Postes et Télécommunications B.P. 2000 Bujumbura BURUNDI Telephone: +257 219 324 Fax: +257 217 773

CANADA	The Chairman
	Board of Steamship Inspection
	Transport Canada – Marine Safety
	Tower C, Place de Ville
	330 Sparks Street, 10th Floor
	Ottawa, Ontario
	K1A ON5
	CANADA
	
	Telephone: +1 613 991 3132
	    +1 613 991 3143
	    +1 613 991 3139/40
	Fax: +1 613 993 8196

CAPE VERDE	The Director General Ministry of Infrastructure and Transport S. Vincente CAPE VERDE Telephone: +238 2 328 199/238 2 585 4643 Email: dgmp@cvtelecom.cv

CHILE	Dirección General del Territorio Marítimo y de Marina Mercante
	Dirección de Seguridad y Operaciones Marítimas
	Servicio de Inspecciones Marítimas
	Divisón Prevención de Riesgos y Cargas Peligrosas
	Subida Cementerio No. 300
	Playa Ancha
	Valparaiso
	CHILE
	
	Telephone: +56 32 2208699
	    +56 32 2208654
	    +56 32 2208692
	Email: cargaspeligrosas@directemar.cl

CHINA	Maritime Safety Administration People's Republic of China 11 Jianguomen Nei Avenue Beijing 100736 CHINA Telephone: +86 10 6529 2588 +86 10 6529 2218 Fax: +86 10 6529 2245 Telex: 222258 CMSAR CN

Country	Contact information for the main designated national competent authority
COMOROS	Ministre d'État Ministère du développement des infrastructures des postes et des télécommunications et des transports internationaux Moroni UNION DES COMORES Telephone: +269 744 287/735 794 Fax: +269 734 241/834 241 Mobile: +269 340 248 Email: houmedms@yahoo.fr
CROATIA	Ministry of Maritime Affairs, Transport and Communication Marine Safety Division Prisavlje 14 1000 Zagreb REPUBLIC OF CROATIA Telephone: +385 1 611 5966 Fax: +385 1 611 5968 Email: pomorski-promet@zg.tel.hr
CUBA	Ministerio del Transporte Dirección de Seguridad e Inspección Marítima Boyeros y Tulipán Plaza Ciudad de la Habana CUBA Telephone: +53 7 881 6607 +53 7 881 9498 Fax: +53 7 881 1514 Email: dsim@mitrans.transnet.cu
CYPRUS	Department of Merchant Shipping Ministry of Communications and Works Kylinis Street Mesa Geitonia CY-4007 Lemesos P.O. Box 56193 CY-3305 Lemesos CYPRUS Telephone: +357 5 848 100 Fax: +357 5 848 200 Telex: 2004 MERSHIP CY Email: dms@cytanet.com.cy
CZECH REPUBLIC	Ministry of Transport of the Czech Republic Navigation and Waterways Division Nábr. L. Svobody 12 110 15 Praha 1 CZECH REPUBLIC Telephone: +42 (0)2 230 312 25 Fax: +42 (0)2 248 105 96 Telex: +42 (0)2 12 10 96 Domi C
DENMARK	Danish Maritime Authority P.O. Box 2605 Vermundsgade 38C 2100 Copenhagen Ø DENMARK Telephone: +45 39 17 44 00 Fax: +45 39 17 44 01 Email: SFS@dma.dk

Country	Contact information for the main designated national competent authority
DJIBOUTI	Director of Maritime Affairs Ministère de l'equipement et des transports P.O. Box 59 Djibouti DJIBOUTI Telephone: +253 357 913 Fax: +253 351 538/253 931/355 879
ECUADOR	Dirección General de la Marine Mercante y del Litoral P.O. Box 7412 Guayaquil ECUADOR Telephone: +593 4 526 760 Fax: +593 4 324 246 Telex: 04 3325 DIGMER ED
EQUATORIAL GUINEA	The Director General (Maritime Affairs) Ministerio de Transportes, Tecnologia, Correos y Telecomunicaciones Malabo REPUBLICA DE GUINEA ECUATORIAL Telephone: +240 275 406 Fax: +240 092 618
ERITREA	Director General Department of Maritime Transport Ministry of Transport and Communications ERITREA Telephone: +291 1 121 317/189 156/185 251 Fax: +291 1 184 690 / 186 541 Email: motcrez@eol.com.er
ESTONIA	Estonian Maritime Administration Maritime Safety Division Valge 4 EST-11413 Tallinn ESTONIA Telephone: +372 6205 700/715 Fax: +372 6205 706 Email: mot@vta.ee
ETHIOPIA	Director General Ministry of Transport and Communications P.O. Box 2504 Addis Ababa ETHIOPIA Telephone: +251 11 551 02 44 Fax: +251 11 551 07 15
FINLAND	Finnish Maritime Administration P.O. Box 171 FI-00181 Helsinki FINLAND Telephone: +358 20 448 1 Fax: +358 20 448 4500 +358 20 448 4336 Email: kirjaamo@fmo.fi

Country	Contact information for the main designated national competent authority
FRANCE	MTETM/DGMT/MMD Arche sud 92055 La Défense cedex FRANCE Telephone: +33 (0)1 40 81 86 49 Fax: +33 (0)1 40 81 10 65 Email: olga.lefevre@equipement.gouv.fr
GAMBIA	The Director General Gambia Port Authority P.O. Box 617 Banjul THE GAMBIA Telephone: +220 4 227 270/4 227 260/4 227 266 Fax: +220 4 227 268
GERMANY	Federal Ministry of Transport, Building and Urban Affairs Division A 33 – Transport of Dangerous Goods P.O. Box 20 01 00 D 53170 Bonn GERMANY Telephone: +49 228 3000 or 300-extension +49 228 300 2643 Fax: +49 228 300 3428 Email: Ref-A33@bmvbs.bund.de
GHANA	The Director General Ghana Maritime Authority P.M.B. 34, Ministries Post Office Ministries – Accra GHANA Telephone: +233 21 662 122/684 392 Fax: +233 21 677 702 Email: info@ghanamaritime.org
GREECE	Ministry of Mercantile Marine Safety of Navigation Division International Relations Department 150 Gr. Lambraki Av. 185 18 Piraeus GREECE Telephone: +301 4191188 Fax: +301 4128150 Telex: +212022, 212239 YEN GR Email: dan@yen.gr
GUINEA BISSAU	The Minister Ministry of Transport & Communication Av. 3 de Agosto, Bissau GUINEA BISSAU Telephone: +245 212 583/245 211 308

7

Country	Contact information for the main designated national competent authority

GUYANA	Guyana Maritime Authority/Administration
	Ministry of Public Works and Communications Building
	Top Floor
	Fort Street
	Kingston
	Georgetown
	REPUBLIC OF GUYANA
	
	Telephone: +592 226 3356
	                   +592 225 7330
	                   +592 226 7842
	Fax: +592 226 9581
	Email: MARAD@networksgy.com

ICELAND	Iceland Maritime Administration Verturvör 2 IS-202 Kópavogur ICELAND Telephone: +354 560 0000 Fax: +354 560 0060 E-mail: skrifstofa@vh.is
INDIA	The Directorate General of Shipping Jahz Bhawan Walchand Hirachand Marg Bombay 400 001 INDIA Telephone: +91 22 263651 Telex: +DEGESHIP 2813-BOMBAY
INDONESIA	Director of Marine Safety Directorate-General of Sea Communication (Department Perhubungan) Jl. Medan Merdeka Barat No. 8 Jakarta Pusat INDONESIA Telephone: +62 381 3269 Fax: +62 384 0788
IRAN (ISLAMIC REPUBLIC OF)	Ports and Shipping Organization PSO Building, South Didar Ave. Shahid Haghani Highway, Vanak Square Tehran IRAN Telephone: +98 21 8493 2201 Fax: +98 21 8493 2227
IRELAND	The Chief Surveyor Marine Survey Office Department of Transport Leeson Lane Dublin 2 IRELAND Telephone: +353 1 604 14 20 Fax: +353 1 604 14 C8 E-mail: mso@transport.ie

Country	Contact information for the main designated national competent authority
ISRAEL	Shipping and Ports Inspectorate Itzhak Rabin Government Complex Building 2 Pal-Yam 15a Haifa 31999 ISRAEL Telephone: +972 4 8632080 Fax: +972 4 8632118 Email: techni@mot.gov.il

ITALY	Italian Coast Guard Headquarters
	Ponte dei Mille
	16100 Genoa
	ITALY
	
	Telephone: +39 010 25 18 154 + 102
	   +39 010 25 18 154 + 111
	Fax: +39 010 24 78 245
	Email: 001@sicnavge.it
	   005@sicnavge.it

JAMAICA	The Maritime Authority of Jamaica 4th Floor, Dyoll Building 40 Knutsford Boulevard Kingston 5 JAMAICA, W.I. Telephone: +1 876 929 2201 +1 876 754 7260/5 Telex: +1 876 7256 Email: maj@jamaicaships.com Website: www.jamaicaships.com
JAPAN	Inspection and Measurement Division Maritime Bureau Ministry of Land, Infrastructure and Transport 2-1-3 Kasumigaseki, Chiyoda-ku Tokyo JAPAN Telephone: +81 3 5253 8639 Fax: +81 3 5253 1644 Email: MRB_KSK@mlit.go.jp

KENYA	Director General
	Kenya Maritime Authority
	P.O. Box 95076 (80104)
	Mombasa
	KENYA
	
	Telephone: +254 041 2318398/9
	Fax: +254 041 2318397
	Email: nkarigithu@yahoo.co.uk
	   info@maritimeauthority.co.ke
	   karigithu@ikenya.com
	
	Ministry of Transport & Communications
	P.O. Box 52692
	Nairobi
	KENYA
	
	Telephone: +254 020 2729200
	Fax: +254 020 2724553
	Email: motc@insightkenya.com
	   peterthuo_2004@yahoo.com

Country	Contact information for the main designated national competent authority

LATVIA	Maritime Administration of Latvia
	5 Trijadibas iela
	LV-1048 Riga
	LATVIA
	
	Telephone: +371 70 62 171
	                 +371 70 62 120
	                 +371 70 62 117
	Fax: +371 78 60 082

LIBERIA	Commissioner/Administration Bureau of Maritime Affairs P.O. Box 10-9042 1000 Monrovia 10 Monrovia LIBERIA Telephone: +231 227 744/37747/510 201 Fax: +231 226 069 Email: maritime@liberia.net
MADAGASCAR	Director Agence Portuaire Maritime et Fluviale (APMF) P.O. Box 581 Antananarivo-101 MADAGASCAR Telephone: +261 20 242 5701 Telephone/Fax: +261 20 22 258 60 Mobile: +261 320 229 259 Email: spapmf.dt@mttpat.gov.mg

MALAWI	Director of Marine Services
	Marine Department
	Ministry of Transport & Civil Aviation
	Private Bag A81
	Capital City Lilongwe
	MALAWI
	
	Telephone: +265 1 755 546/752 666
	                 Direct line: 753 531
	Fax: +265 1 750 157/758 894
	Email: marinedepartment@malawi.net
	                 marinesafety@africa-online.net

MALAYSIA	Director Marine Department Peninsular Malaysia P.O. Box 12 42007 Port Kelang Selangor MALAYSIA Telex: MA 39748
MARSHALL ISLANDS	Office of the Maritime Administrator Maritime Operations Department Republic of the Marshall Islands 11495 Commerce Park Drive Reston, Virginia 20191-1507 USA Telephone: +1 703 620 4880 Fax: +1 703 476 8522 Telex: 248403 IRI UR Email: maritime@register-iri.com

Country	Contact information for the main designated national competent authority
MAURITIUS	Director of Shipping Ministry of Land Transport, Shipping and Public Safety New Government Centre, 4 Floor Port Louis MAURITIUS Telephone: +230 201 2115
MEXICO	Coordinación General de Puertos y Marina Mercante Secretaria de Comunicaciones y Transportes Nuevo León 210, Piso 3, Colonia Hipódromo Col. Santa Cruz Atoyac D.F., C.P. 06100 MEXICO Telephone: +52 55 526 53220 Fax: +52 55 557 43902 Email: jtlozano@sct.gob.mx
MONTENEGRO	Ministry of Interior and Public Administration of the Republic of Montenegro Department for Contingency Plans and Civil Security REPUBLIC OF MONTENEGRO Telephone: +382 81 241 590 Fax: +382 81 246 779 Email: mup.emergency@cg.yu

MOROCCO	Direction de la Marine Marchande et des Pêches Maritimes
	Boulevard El Hansali
	Casablanca
	MOROCCO
	
	Telephone: +1 212 227 8092
	    +1 212 222 1931
	Telex: 24613 MARIMAR M
	    22824

MOZAMBIQUE	General Director National Maritime Authority (INAMAR) Av. Marquês do Pombal No. 297 P.O. Box 4317 Maputo MOZAMBIQUE Telephone: +258 21 320 552 Fax: +258 21 324 007 Mobile: +258 82 153 0280 Email: inamar@tvcabo.co.mz *Testing and certification of packaging, intermediate bulk containers and large packaging:* Instituto Nacional de Normalização e Qualidade (INNOQ) Av. 25 de Setembro No. 1179, 2nd Floor Maputo MOZAMBIQUE Telephone: +258 21 303 822/3 Fax: +258 21 304 206 Mobile: +258 823 228 840 Email: innoq@emilmoz.com

7

Country	Contact information for the main designated national competent authority
NAMIBIA	Director of Maritime Affairs Ministry of Works, Transport and Communications Private Bag 13341 6719 Bell Street Snyman Circle, Windhoek NAMIBIA Telephone: +264 61 208 8025/6 Direct line: 208 8111 Fax: +264 61 240 024/224 060 Mobile: +264 811 220 599 Email: mmnangolo@mwtc.gov.na
NETHERLANDS	Ministry of Transport, Public Works and Water Management Directorate-General for Civil Aviation and Freight Transport P.O. Box 20904 2500 EX The Hague THE NETHERLANDS Telephone: +31 70 351 6171 Fax: +31 70 351 1479 Ministry of Transport, Public Works and Water Management Transport Information Centre P.O. Box 90653 2509 LR The Hague THE NETHERLANDS Telephone: +31 70 456 2444 Fax: +31 70 456 2424 Email: vervoerinfo@ivw.nl
NEW ZEALAND	Maritime New Zealand Level 10, Optimation House 1 Grey Street Wellington 6011 NEW ZEALAND Telephone: +64 4 494 1273 Fax: +64 4 494 8901 Email: enquiries@maritimenz.govt.nz Website: www.maritimenz.govt.nz PO Box 27-006 Wellington 6141 NEW ZEALAND Telephone: +64 4 473 0111 Fax: +64 4 494 1263
NIGERIA	Nigerian Maritime Administration and Safety Agency (NIMASA) Maritime House 4 Burma Road, Apapa PMB 12861, GPO Marina Lagos NIGERIA Telephone: +234 587 2214 / 580 4800–9 Fax: +234 587 1329 Telex: 23891 NAMARING Website: www.nimasa.gov.ng
NORWAY	Norwegian Maritime Directorate Smedasundet 50A N-5528 Haugesund NORWAY Telephone: +47 5274 5000 Fax: +47 5274 5001 Email: postmottak@sjofartsdir.no

Country	Contact information for the main designated national competent authority

PAKISTAN	Mercantile Marine Department
	70/4 Timber Hard
	N.M. Reclamation
	Keamari, Post Box No. 4534
	Karachi 75620
	PAKISTAN
	
	Telephone: +92 21 2851306
	               +92 21 2851307
	Fax: +92 21 4547472 (24 hours)
	               +92 21 4547897
	Telex: 29822 DGPS PK (24 hours)

PANAMA	Autoridad Marítima de Panamá Edificio 5534 Diablo Heights P.O. Box 8062 Panama 7 REPUBLIC OF PANAMA Telephone: +507 232 5100/5295 Fax: +507 232 5527 Email: ampadmon@amp.gob.pa Website: www.amp.gob.pa
PAPUA NEW GUINEA	First Assistant Secretary Department of Transport Division of Marine P.O. Box 457 Konedobu PAPUA NEW GUINEA (PNG) Telephone: +675 211866 Telex: 22203
PERU	Dirección General de Capitanías y Guardacostas Autoridad Maritima del Peru Direccion de Medio Ambiente Jr. Independencia No. 150 Callao PERU Fax: +51 1 613 6857 Email: dicapi.medioambiente@dicapi.mil.peru
PHILIPPINES	Philippines Ports Authority Port of Manila Safety Staff P.O. Box 193 Port Area Manila 2803 PHILIPPINES Telephone: +63 2473441 to 49
POLAND	Ministry of Maritime Economy Department of Maritime Safety 00-928 Warsaw ul. Chałubińskiego 4/6 POLAND Telephone: +48 22 630 15 40 Fax: +48 22 830 09 47

Country	Contact information for the main designated national competent authority
PORTUGAL	Direcçao-Geral de Navegaçao e dos Transportes Maritimos Praça Luis de Camoes, 22 – 2°Dto 1200 Lisboa PORTUGAL Telephone: +351 1 373821 Fax: +351 1 373826 Telex: 16753 SEMM PO
REPUBLIC OF KOREA	Maritime Technology Team Maritime Safety Bureau Ministry of Maritime Affairs and Fisheries 140-2 Gye-dong, Jongno-gu, Seoul, 110-793 REPUBLIC OF KOREA Telephone: +82 2 3674 6323 Fax: +82 2 3674 6327
RUSSIAN FEDERATION*	Department of State Policy for Maritime and River Transport Ministry of Transport of the Russian Federation Rozhdestvenka Street, 1, bldg. 1 Moscow 109012 RUSSIAN FEDERATION Telephone: +7 495 926 14 74
SAINT KITTS AND NEVIS	Department of Maritime Affairs Director of Maritime Affairs Ministry of Transport P.O. Box 186 Needsmust ST. KITTS, W.I. Telephone: +869 466-7032/4846 Fax: +869 465-0604/9475 E-mail: Maritimeaffairs@yahoo.com
SAO TOME & PRINCIPE	The Minister Ministry of Public Works, Infrastructure & Land Planning C.P. 171 SAO TOME & PRINCIPE Telephone: +239 223 203/239 226 368 Fax: +239 222 824
SAUDI ARABIA	Port Authority Saudi Arabia Civil Defence Riyadh SAUDI ARABIA Telephone: +966 1 464 9477
SEYCHELLES	Director General Seychelles Maritime Safety Administration P.O. Box 912 Victoria, Mahe SEYCHELLES Telephone: +248 224 866 Fax: +248 224 829 Email: dg@msa.sc

* Except for governmental explosives.

Country	Contact information for the main designated national competent authority
SIERRA LEONE	The Executive Director Sierra Leone Maritime Administration Maritime House Government Wharf Ferry Terminal P.O. Box 313 Freetown SIERRA LEONE Telephone: +232 22 221 211 Fax: +232 22 221 215 Email: slma@sierratel.sl slmaoffice@yahoo.com
SINGAPORE	Maritime and Port Authority of Singapore Shipping Division 21st Storey PSA Building 460 Alexandra Road SINGAPORE 119963 Telephone: +65 375 1931/6223/1600 Fax: +65 375 6231 Email: shipping@mpa.gov.sg
SLOVENIA	Uprava Republike Slovenije za pomorstvo Ukmarjev trg 2 66 000 Koper SLOVENIA Telephone: +386 66 271 216 Fax: +386 66 271 447 Telex: +34 235 UP POM SI
SOUTH AFRICA	South African Maritime Safety Authority P.O. Box 13186 Hatfield 0028 Pretoria SOUTH AFRICA Telephone: +27 12 342 3049 Fax: +27 12 342 3160 South African Maritime Safety Authority Hatfield Gardens, Block E (Ground Floor) Corner Arcadia and Grosvenor Street Hatfield 0083 Pretoria SOUTH AFRICA
SPAIN	Dirección General de la Marina Mercante Subdirección General de Seguridad Marítima y Contaminación c/Ruiz de Alarcón, 1 28071 Madrid SPAIN Telephone: +34 91 597 92 69/70 Fax: +34 91 597 92 87 Email: mercancias.peligrosas@fomento.es pmreal@fomento.es

Country	Contact information for the main designated national competent authority
SUDAN	Director Ministry of Transport, Roads and Bridges Maritime Administration Directorate Port Sudan P.O. Box 531 SUDAN Telephone: +249 311 825 660 Fax: +249 311 831 276 Mobile: +249 912 51 105/310 997 Telephone/Fax: +249 1 837 742 15 Email: smaco22@yahoo.com
SWEDEN	Swedish Maritime Administration Maritime Safety Inspectorate Ship Technical Division SE-601 78 Norrköping SWEDEN Telephone: +46 11 191000 Fax: +46 11 239934 E-mail: inspektion@sjofartsverket.se
SWITZERLAND	Office suisse de la navigation maritime Nauenstrasse 49 P.O. Box CH-4002 Basel SWITZERLAND Telephone: +41 61 270 91 20 Fax: +41 61 270 91 29 Email: dv-ssa@eda.admin.ch
TANZANIA	Director General Surface & Marine Transport Regulatory Authority (SUMATRA) P.O. Box 3093 Dar es Salaam TANZANIA Telephone: +255 22 213 5081 Mobile: +255 744 781 865 Fax: +255 22 211 6697 Email: dg@sumatra.or.tz Ministry of Infrastructure Development P.O. Box 9144 Dar es Salaam TANZANIA Telephone: +255 22 212 2268 Fax: +255 22 211 2751/212 2079 Mobile: +254 748 7404/748 5404 Email: brufunjo@yahoo.com
THAILAND	Ministry of Transport and Communications Ratchadamnoen-Nok Avenue Bangkok 10100 THAILAND Telephone: +66 2 2813422 Fax: +66 2 2801714 Telex: 70000 MINOCOM TH

Country	Contact information for the main designated national competent authority
TUNISIA	Ministère du Transport Direction Générale de la Marine Marchande Avenue 7 novembre (près de l'aéroport) 2035 Tunis B.P. 179 Tunis cedex TUNISIA Telephone: +216 71 806 362 Fax: +216 71 806 413
UNITED ARAB EMIRATES	National Authority of Communications Marine Affairs Department P.O. Box 900 Abu Dhabi UNITED ARAB EMIRATES Telephone: +9712 4182 124 Fax: +9712 4491 500 Email: marine@naoc.gov.ae
UNITED KINGDOM	Maritime and Coastguard Agency Bay 2/21 Spring Place 105 Commercial Road Southampton, SO15 1EG UNITED KINGDOM Telephone: +44 23 8032 9100 Fax: +44 23 8032 9204 Email: dangerous.goods@mcga.gov.uk
UNITED STATES	US Department of Transportation Pipeline and Hazardous Materials Safety Administration Office of International Standards East building / PHH-70 1200 New Jersey Ave, S.E. Washington, D.C. 20590 USA Telephone: +1 202 366 0656 Fax: +1 202 366 5713 Email: infocntr@dot.gov Website: hazmat.dot.gov United States Coast Guard Hazardous Materials Standards Division (CG-3PSO-3) 2100 Second Street, S.W. Washington, D.C. 20593-0001 USA Telephone: +1 202 372 1420 +1 202 372 1426 Fax: +1 202 372 1926
URUGUAY	Prefectura del Puerto de Montevideo Rambla 25 de Agosto de 1825 S/N Montevideo URUGUAY Telephone: +598 2 960123 +598 2 960022 Telex: 23929 COMAPRE-UY

Country	Contact information for the main designated national competent authority
VANUATU	Commissioner of Maritime Affairs Vanuatu Maritime Authority P.O Box 320 Port Vila VANUATU Telephone: +678 23128 Fax: +678 22949 Email: vma@vanuatu.com.vu
VIETNAM	Dr. Tran Dac Suu Director General Vietnam Inland Waterway Administration 80 Tran Hung Dao Hanoi VIETNAM Telephone: +84 4 9421 887 Fax: +84 4 9420 788 Email: cuctruong.viwa@mt.gov.vn For further information, contact Ms. Yen International Relation Department Telephone: +84 4 9424 750 Mobile: +84 913 599 801 Email: yenton197@gmail.com viwa.inter.re@hn.vnn.vn
YEMEN	Executive Chairman Maritime Affairs Authority P.O. Box 19395 Sana'a REPUBLIC OF YEMEN Telephone: +967 1 414 412 / 419 914 / 423 005 Fax: +967 1 414 645 E-mail: MAA-Headoffice@y.net.ye Website: www.MAA.gov.ye
ZAMBIA	Department of Maritime & Inland Waterways Ministry of Communications & Transport P.O. Box 50346 Fairley Road Lusaka ZAMBIA Telephone: +260 1 250 716/251 444/251 022 Fax: +260 1 253 165/251 795 Email: dmiw@zamtel.zm
Associate Member HONG KONG, CHINA	The Director of Marine Marine Department GPO Box 4155 HONG KONG, CHINA Telephone: +852 2852 3085 Fax: +852 2815 8596 Telex: 64553 MARHQ HX

Notes

Notes